Space Science and Technologies

Series Editor

Peijian Ye, China Academy of Space Technology, Beijing, China

Space Science and Technologies publishes a host of recent advances and achievements in the field – quickly and informally. It covers a wide range of disciplines and specialties, with a focus on three main aspects: key theories, basic implementation methods, and practical engineering applications. It includes, but is not limited to, theoretical and applied overall system design, subsystem design, major space-vehicle supporting technologies, and the management of related engineering implementations.

Within the scopes of the series are monographs, professional books or graduate textbooks, edited volumes, and reference works purposely devoted to support education in related areas at the graduate and post-graduate levels.

More information about this series at http://www.springer.com/series/16385

Zezhou Sun

Technologies for Deep Space Exploration

BEIJING INSTITUTE OF TECHNOLOGY PRESS

版权专有　侵权必究

图书在版编目（CIP）数据

深空探测技术 = Technologies for Deep Space Exploration：英文 / 孙泽洲等编著. —北京：北京理工大学出版社，2020.9
　ISBN 978-7-5682-9029-6

　Ⅰ.①深…　Ⅱ.①孙…　Ⅲ.①空间探测–英文　Ⅳ.①V1

中国版本图书馆 CIP 数据核字（2020）第 173476 号

Sales in Mainland China Only
本书仅限在中国大陆地区发行销售

出版发行 / 北京理工大学出版社有限责任公司
社　　　址 / 北京市海淀区中关村南大街 5 号
邮　　　编 / 100081
电　　　话 / （010）68914775（总编室）
　　　　　　（010）82562903（教材售后服务热线）
　　　　　　（010）68948351（其他图书服务热线）
网　　　址 / http：//www.bitpress.com.cn
经　　　销 / 全国各地新华书店
印　　　刷 / 三河市华骏印务包装有限公司　　　丛书策划 / 李炳泉
开　　　本 / 710 毫米 × 1000 毫米　1/16　　　策划编辑 / 李炳泉
印　　　张 / 40　　　　　　　　　　　　　　　张海丽
字　　　数 / 758 千字　　　　　　　　　　　　责任编辑 / 张海丽
版　　　次 / 2020 年 9 月第 1 版　2020 年 9 月第 1 次印刷　责任校对 / 周瑞红
定　　　价 / 218.00 元　　　　　　　　　　　　责任印制 / 李志强

图书出现印装质量问题，请拨打售后服务热线，本社负责调换

Series Editor's Preface

China's space technology and science research have earned a place in the world, but have not been compiled into a series of systematic publications yet. In 2018, the series *Space Science and Technology* edited mainly by me and coauthored by the leading figures in China's space industry was published in China, when China Academy of Space Technology was celebrating the 50th anniversary of its founding. This collection contains 23 volumes in Chinese, only 10 of which have been selected, recreated and translated into English. In addition, each English volume has been recreated at the suggestion of the Springer, by deleting the contents similar to Springer's existing publications and adding the contents that are internationally advanced and even leading, and bear both Chinese characteristics and worldwide universality. This series fully reflects the knowledge and engineering experience recently accumulated by Chinese scientists and engineers in space technology and science research.

As the Editor-in-Chief of this series, I always insist that this collection must be of high quality, either in Chinese version or English version. First, the contents of this series must be condensed and sublimated based on the combination of theory and practice, so as to provide both a theoretical value and engineering guidance. Second, the relationships between past knowledge and state of the art and between other people's work and our own new findings should be properly balanced in the book contents to ensure the knowledge systematicness and continuity and to highlight new achievements and insights. Each volume intends to introduce the readers something new. Third, the English version should be customized for international exposure and play a solid supporting role for China to contribute to the world's space field.

This collection consists of 10 volumes, including *Spacecraft Thermal Control Technologies, Spacecraft Power System Technologies, Spacecraft Electromagnetic Compatibility Technologies, Technologies for Spacecraft Antennas Engineering Design, Satellite Navigation Systems and Technologies, Satellite Remote Sensing Technologies, Spacecraft Autonomous Navigation Technologies Based on Multi-source Information Fusion, Technologies for Deep Space Exploration, Space Robotics, Manned Spacecraft Technologies*.

To meet the special technical requirements of deep space exploration missions, *Technologies For Deep Space Exploration* systematically expounds the common technologies involved in the development of deep space probes from three aspects, namely system design technologies, key technologies, test and verification technologies on ground. This volume has the following features:

1. Based on the theory of system engineering, this volume elaborates on the connotation, elements and methods of system design technologies of deep space probes from the aspects such as mission analysis, specification decomposition, system operation procedure design, key technical approaches and system design validation in combination with four types of deep space mission examples including orbiting exploration, landing exploration, roving exploration and sample return exploration.
2. This volume summarizes and refines the theories and engineering design methods of common key technologies involved in deep space exploration, such as orbit design, TT&C and communication, navigation and control, thermal control, propulsion, atmospheric braking, payload, power supply, mechanism, tele-operation and autonomous management. Moreover, the future development trend of relevant specialized technologies is also discussed in the book.

The publication of this series adds a new member to the international family of space technology and science publications and intends to play an important role in promoting academic exchanges and space business cooperation. It provides comprehensive, authentic and rich information for international space scientists and engineers, enterprises and institutions as well as government sectors to have a deeper understanding of China's space industry. Of course, I believe that this series will also be of great reference value to the researchers, engineers, graduate students and university students in the related fields.

Peijian Ye
Academician
Chinese Academy of Sciences
Beijing, China

Preface

Deep space exploration is very challenging, innovative and leading in the field of the highest new science and technology in the world and is of great and far-reaching significance to the country's politics, economy, science and the sustainable development of human society. Today, man-made deep space probes have travelled to all the planets in the solar system, including many moons, comets and asteroids. These deep space exploration activities enable mankind to understand the Earth, the solar system and the universe and then explore and settle other planets in the solar system.

Chinese deep space exploration activities started from lunar exploration. At present, it has successfully achieved the goals of "orbiting" and "landing," and has mastered key technologies such as orbiting exploration around the moon and lunar surface exploration, lunar surface soft landing and reentry and return from the Moon. Moreover, it has made great progress in developing the space infrastructure and corresponding capabilities such as launch, TT&C, communication and recovery. The lunar exploration project phase "return" will be completed in 2020, in which the lunar sample return mission will be realized. The first Mars mission has also been carried out, and Chinese self-developed Mars probe will go to Mars in 2020. These deep space exploration activities have the characteristics of new target environment, wide detection range, long flight distance, long operation time and so on. Compared with near-Earth spacecraft, the required technology is more novel and complex. Among them, the development of the probe system is the key link of the engineering task.

In this book, the editors focuses on the general design, orbit design, TT&C and communication, navigation and control, thermal control, propulsion, atmospheric deceleration, payload and other common and difficult technologies involved in the development of the deep space detector system, taking into consideration of the technical requirements of deep space missions combined with rich engineering experience. The purpose of compiling this book is to provide a more systematic and comprehensive reference book for scientific and technological personnel engaged in the design, implementation and development of deep space probe systems, which can be used as a reference for aerospace scientific and technological workers. This book can also be used as a textbook and reference book for undergraduates and graduate students majoring in aerospace in colleges and universities and scientific research institutions.

The book is divided into three parts, a total of 14 chapters; the contents are as follows: The first part is the deep space probe system design technology, a total of 4 chapters, including an Introduction, Characteristics of Deep Space Environment and Corresponding Impacts, System Design Technology, Technology of Orbit Design; the second part is the Key System and Core Technology of Deep Space Probe, which consists of 9 chapters, including Payload Technology, Guidance, Navigation and Control Technology, Atmospheric Braking Technology, TT&C and Control Communication Technology, Thermal Control Technology, Propulsion Technology, Power Supply Technology, Autonomous Management and Tele-operation Technology, Landing Gear System; the third part is the Ground Test Verification Technology, a total of 1 chapter.

This book is edited and written by Zezhou Sun. In terms of specific chapters, Chap. 1 is written by Zezhou Sun; Chap. 2 is written by Shaojie Qu and Zhenbo Cai; Chap. 3 is written by Linzhi Meng, Baiyi Zhang, Yang Jia and Jing Peng; Chap. 4 is written by Wenyan Zhou, Baiyi Tian, Zhongsheng Wang, Shan Gao and Lei Zhang; Chap. 5 is written by Shuwu Dai; Chap. 6 is written by Xiaolei Wang, Jianxin Chen and Jie Dong; Chap. 7 is written by Wei Rao, Qi Li, Jian Li, Jie Dong, Xiangyu Huang, Minwen Guo and Changhao Yang; Chap. 11 is written by Zhigang Liu; Chap. 12 is written by Yang Jia and Linzhi Meng; Chap. 13 is written by Baofeng Yuan and Jianzhong Yang; Chap. 14 is written by Chuang Wang.

Beijing, China   Zezhou Sun

Acknowledgements

This book was completed with the consent of the editorial board of the Deep Space Technology and Scientific Research Series. In the process of writing and publishing, this book has received strong support and help from leaders and experts at all levels of the China Academy of Space Technology and Beijing Institute of Spacecraft System Engineering; in the process of examination and approval of this book, it has been guided and supported by National Academician Peijian Ye, National Academician Mengfei Yang, researcher Tingxin Zhang, Research Fellow Jiawen Qiu, Research Fellow Jiangchuan Huang and other experts.

The author would like to express his very special appreciation and gratitude to Dr. Zhongsheng Wang, who gave great support and help in the process of review and proofreading. The author would like to thank the following individuals for their help in reviewing the manuscripts, which crystallized into the chapters in this volume: Chenghao Dai, Jun Huang, Xiaoyu Jia, Xiaowei Ju, Fei Li, Yanshuo Ni, He Tian, Tianyi Zhang, Yang Zhao.

About the Author

Zezhou Sun, Doctoral supervisor, is a professor in deep space exploration at the Beijing Institute of Spacecraft System Engineering of the China Academy of Space Technology, the chief designer of Chang'e 4 and the Mars probes. He is mainly engaged in deep space exploration project demonstration, spacecraft overall design and other research work. He once presided over the project of "Chang'e 4" and successfully completed the first lunar soft landing and roving exploration by a man-made probe. He is now in charge of the Mars exploration mission to be implemented in 2020 opportunity. He has won many awards, such as the special prize of the National Science and Technology Progress Award and the National Defense Science and Technology Award. More than 20 core journal papers have been published.

Contents

1	**Introduction**		1
	1.1	The Significance of Deep Space Exploration	1
	1.2	Overview of Deep Space Exploration Development	2
		1.2.1 Overview of International Deep Space Exploration Development	2
		1.2.2 Overview of Deep Space Exploration Development in China	4
		1.2.3 Development Trends of Deep Space Exploration	6
	1.3	Future Development Requirements for Deep Space Exploration Technologies	9
	1.4	Prospects	11
	References and Related Reading		12
2	**Characteristics of Deep Space Environment and Corresponding Impact**		13
	2.1	Introduction	13
	2.2	Geospace Environment	14
		2.2.1 Main Geospace Environment Characteristics for Deep Space Probes	14
		2.2.2 Impact of Geospace Environment on Deep Space Probes	19
	2.3	Lunar Space Environment	22
		2.3.1 General	22
		2.3.2 Lunar Radiation Environment and Impacts	23
		2.3.3 Lunar Atmosphere and the Impact	26
		2.3.4 Lunar Soil/Lunar Dust and Impacts	27
	2.4	Space Environment of Mars	33
		2.4.1 Overview	33
		2.4.2 Mars Radiation Environment and Its Impact	34
		2.4.3 Impact of Mars Atmospheric Environment	35

	2.4.4	Impact of Mars Dust Environment	35
	2.4.5	Landforms on the Surface of Mars	36
2.5	Space Environment of Jupiter		37
	2.5.1	Overview	37
	2.5.2	Jupiter's Strong Magnetic Field Environment	37
	2.5.3	Strong Radiation Environment of Jupiter	38
	2.5.4	Jupiter Plasma Environment	38
	2.5.5	Jupiter Atmosphere	39
2.6	Space Environment of Venus		39
	2.6.1	Overview	39
	2.6.2	Magnetic Field of Venus	40
	2.6.3	Venus Atmosphere	41
	2.6.4	Venus Surface Topography	41
2.7	Other Interplanetary Space Environments		42
	2.7.1	Interplanetary Environment	42
	2.7.2	Asteroid Environment	42
	2.7.3	Comet Environment	43
2.8	Outlook		44
References and Related Reading			44

3 System Design Technology — 45

3.1	Introduction		45
3.2	Overview of the System Design of Deep Space Probes		46
	3.2.1	Characteristics of Deep Space Probe Missions	46
	3.2.2	System Mission Analysis	48
	3.2.3	Overall System Design Process	50
3.3	System Design of Orbiting Exploration Missions		51
	3.3.1	Mission Analysis	51
	3.3.2	Decomposition of Technical Specifications	54
	3.3.3	Flight Procedure Design	54
	3.3.4	Analysis of Key Technologies	55
	3.3.5	Design Validation	56
3.4	System Design of Landing Exploration Missions		56
	3.4.1	Mission Analysis	56
	3.4.2	Decomposition of Technical Specifications	60
	3.4.3	Flight Procedure Design	61
	3.4.4	Analysis of Key Technologies	63
	3.4.5	Design Validation	64
3.5	System Design of Rover Exploration Missions		65
	3.5.1	Mission Analysis	65
	3.5.2	Decomposition of Technical Specifications	68
	3.5.3	Work Procedure Design	69
	3.5.4	Analysis of Key Technologies	70
	3.5.5	Design Validation	71

	3.6	System Design of Sample Return Exploration Mission	71
		3.6.1 Mission Analysis	71
		3.6.2 Definition of Specifications	78
		3.6.3 Flight Procedure Design	79
		3.6.4 Analysis of Key Technologies	81
		3.6.5 Design Validation	86
	3.7	Prospects	86
		References and Related Reading	86
4	**Technology of Orbit Design**		**89**
	4.1	Introduction	89
	4.2	Classical Types of Orbits	90
		4.2.1 Lunar Exploration Orbit	90
		4.2.2 Planetary Exploration Orbit	91
		4.2.3 Asteroid Exploration Orbit	92
		4.2.4 Libration Point Exploration Orbit	94
	4.3	Brief Introduction to Orbit Design Procedures	95
	4.4	Design of Transfer Trajectories	97
		4.4.1 Direct Transfer	97
		4.4.2 Deep Space Maneuver	106
		4.4.3 Gravity Assist	109
		4.4.4 Low Thrust Transfer	115
	4.5	Design of Mission Orbits	130
		4.5.1 Planet Orbiting Missions	130
		4.5.2 Missions to Lagrange Libration Points	138
		4.5.3 Rendezvous and Docking	143
	4.6	Design of Orbital Maneuver Strategy	149
	4.7	Future Prospects	151
		References and Related Reading	152
5	**Payload Technology**		**155**
	5.1	Introduction	155
	5.2	Major Scientific Issues in Deep Space Exploration Research	156
		5.2.1 Scientific Issues of Deep Space Exploration from a Systematic Perspective	156
		5.2.2 Scientific Objectives and Payload Configuration for Lunar and Mars Exploration in China	159
	5.3	Topography Acquisition Technology	168
		5.3.1 Introduction	168
		5.3.2 Stereo Image Acquisition Technology	168
		5.3.3 Color CMOS Devices	169
		5.3.4 Camera System Design	169
		5.3.5 Automatic Exposure Technology	173
		5.3.6 Calibration and Ground Verification Test	174

5.4	Elemental Component Identification Technology		174
	5.4.1	Introduction	174
	5.4.2	Principles of Elemental Composition Identification	175
	5.4.3	Selection Strategy of Excitation Source	178
	5.4.4	Sensor Selection and Design Techniques	178
	5.4.5	System Design	180
	5.4.6	Calibration and Ground Verification Test	182
5.5	Lunar-Based Astronomical Observation Technology		182
	5.5.1	Introduction	182
	5.5.2	Selection of Spectral Segments and Observation Sky Regions	184
	5.5.3	Telescope Design	184
	5.5.4	Stray Light Suppression	185
	5.5.5	Calibration and Ground Verification Test	185
5.6	Prospects		185
References and Related Reading			186
6	**Guidance, Navigation and Control Technology**		**189**
6.1	Introduction		189
6.2	Orbital Control Technology		190
	6.2.1	Features of Deep Space Orbital Control	190
	6.2.2	Large-Impulse Orbital Control Strategy	191
	6.2.3	Precision Orbit Control	197
	6.2.4	Design of Orbit Control System	203
6.3	Entry and Landing GNC Technology		207
	6.3.1	Characteristics of Entry and Landing GNC Technology	207
	6.3.2	Atmospheric Entry Control	208
	6.3.3	Powered Descending Control	210
	6.3.4	Obstacle Identification and Avoidance	211
	6.3.5	Design of Entry and Landing GNC System	212
6.4	GNC Technology in Celestial Body Surface Roving		219
	6.4.1	Features of Rover GNC	219
	6.4.2	Environmental Perception	220
	6.4.3	Position/Attitude Determination and Estimation	221
	6.4.4	Path Planning	223
	6.4.5	Motion Control	224
	6.4.6	Design of GNC System for Celestial Body Surface Roving	227
6.5	Outlook		232
References and Related Reading			233

7	**Atmospheric Braking Technology**		235
	7.1	Introduction	235
	7.2	Aerodynamics and Aerodynamic Analysis	237
		7.2.1 Basic Concepts of Aerodynamics	237
		7.2.2 A Study on Aerodynamic Problems in Atmospheric Entry	243
		7.2.3 Atmospheric Entry Aerodynamic Analysis and Prediction	245
	7.3	Aerodynamic Thermal Protection Design	256
		7.3.1 Basic Theory of Thermal Protection Technology	256
		7.3.2 Aerodynamic Thermal Protection Technology	263
		7.3.3 Aerodynamic Thermal Protection Design	266
	7.4	Atmospheric Entry Guidance and Control Design	275
		7.4.1 Atmospheric Entry Guidance and Control Technology	275
		7.4.2 Atmospheric Entry Trajectory Design	277
		7.4.3 Atmospheric Entry Guidance and Control Design	283
	7.5	Parachute Deceleration System Design	285
		7.5.1 Overview of Parachute Deceleration Technology	285
		7.5.2 Atmospheric Entry Parachute Technology	288
		7.5.3 Deep Space Probe-Parachute Design	291
		7.5.4 Simulation Analysis of Parachute Design	302
	7.6	Prospects	304
	References and Related Reading		304
8	**TT&C and Communication Technology**		307
	8.1	Introduction	307
	8.2	Deep Space Radio Measurement Technology	308
		8.2.1 Deep Space Ranging	308
		8.2.2 Deep Space Velocity Measurement	311
		8.2.3 Deep Space Angle Measurement	315
	8.3	Deep Space RF System Technology	320
		8.3.1 Radio Frequency Modulation	320
		8.3.2 High-Sensitivity Reception	321
		8.3.3 High EIRP Emission	322
		8.3.4 Laser Communication	323
	8.4	Deep Space Telemetry and Telecommand and Data Communication Technology	325
		8.4.1 Data Format	325
		8.4.2 Channel Encoding	329
	8.5	Design of Deep Space TT&C and Communication System	330
		8.5.1 Mission Analysis	331
		8.5.2 System Scheme	339
		8.5.3 Simulation and Verification	343

	8.6	Prospects	346
		References and Related Reading	347
9	**Thermal Control Technology**		349
	9.1	Introduction	349
	9.2	Characteristics of Thermal Environment in Deep Space	349
		9.2.1 Mercury's Thermal Environment	350
		9.2.2 Venus's Thermal Environment	351
		9.2.3 Lunar Thermal Environment	352
		9.2.4 Mars Thermal Environment	354
		9.2.5 Thermal Environment of Exoplanets	355
	9.3	Key Technologies of Thermal Control S	355
		9.3.1 Gravity-Assisted Two-Phase Fluid Loop Technology	356
		9.3.2 Water Sublimator Technology	358
		9.3.3 Variable Conductivity Heat Pipe Technology	361
		9.3.4 Aerogel Technology	363
	9.4	Thermal Control System Design for Deep Space Probe	366
		9.4.1 Introduction of Typical Thermal Control Systems for Deep Space Probes	366
		9.4.2 Basic Principles for Thermal Design	371
		9.4.3 Thermal Design	372
		9.4.4 Thermal Analysis	375
		9.4.5 Ground Simulation Test	380
	9.5	Prospect	387
		References and Related Reading	388
10	**Propulsion Technology**		389
	10.1	Introduction	389
	10.2	Propulsion System Classification	389
		10.2.1 Cold Gas Propulsion	390
		10.2.2 Chemical Propulsion	391
		10.2.3 Electric Propulsion	399
		10.2.4 New Concept Propulsions	404
	10.3	Design and Verification of Deep Space Exploration Propulsion System	407
		10.3.1 Mission Analysis	407
		10.3.2 Propulsion System Selection	410
		10.3.3 Scheme Design	414
	10.4	Outlook	426
		References and Related Reading	427

11	**Power Supply Technology**		429
	11.1	Introduction	429
	11.2	Solar Cell Technology	429
		11.2.1 Spectral Matching	430
		11.2.2 Dustproof Techniques	432
	11.3	MPPT Technology	435
		11.3.1 Basic Principles of MPPT	435
		11.3.2 MPPT Implementation	435
		11.3.3 MPPT Topology	436
	11.4	Lithium-Ion Battery Technology	440
		11.4.1 Overview of Lithium-Ion Batteries	440
		11.4.2 Low-Temperature Resistance Technology for Lithium-Ion Cells	441
	11.5	Space Nuclear Power	446
		11.5.1 Overview of Space Nuclear Power	446
		11.5.2 RTG Technology	449
		11.5.3 Nuclear Reactor Power Supply	451
	11.6	Deep Space Power System Design	454
		11.6.1 Mission Analysis	454
		11.6.2 Solar Array Design	459
		11.6.3 Battery Pack Design	462
		11.6.4 Power Controller Design	464
		11.6.5 Example of Power System Design	465
	11.7	Prospects	470
	References and Related Reading		471
12	**Autonomous Management and Tele-operation Technology**		475
	12.1	Introduction	475
	12.2	Autonomous Management Technology for Deep Space Probes	476
		12.2.1 Development of Autonomy Capabilities	478
		12.2.2 Mobile Intelligent Agent	479
		12.2.3 Mars Mobile Intelligent Agent	480
		12.2.4 Autonomous Management Implementation Framework of Mars Rovers	488
		12.2.5 Mars Rover Autonomous Mission Planning	490
	12.3	Tele-operation Technology of Rover	492
		12.3.1 Tele-operation in Space Environment	493
		12.3.2 Planetary Surface Roving Tele-operation	493
		12.3.3 Key Technology for Rover Tele-operation System	495
		12.3.4 Tele-operation System	503
	12.4	Prospects for Technology Development	507
	References and Related Reading		508

13 Mechanism Technology ... 509
- 13.1 Introduction ... 509
- 13.2 Landing Gear System ... 510
  - 13.2.1 Functions and Composition Characteristics of Landing Gear System ... 511
  - 13.2.2 Design and Verification of Landing Gear System ... 513
- 13.3 Rover Transfer and Release System ... 525
  - 13.3.1 Functions and Composition Characteristics of Rover Transfer and Release System ... 527
  - 13.3.2 Design and Verification of Rover Transfer and Release System ... 528
- 13.4 Rover Mobility System Mobility System ... 536
  - 13.4.1 Functions and Composition Characteristics of Rover Mobility System ... 536
  - 13.4.2 Design and Verification of Rover Mobility System ... 538
- 13.5 Sampling Mechanism ... 552
  - 13.5.1 Functions and Composition Characteristics of Sampling Mechanism ... 553
  - 13.5.2 Design and Verification of Sampling Mechanism ... 556
- 13.6 Outlook ... 566
- References and Related Reading ... 567

14 Ground Test Verification Technology ... 569
- 14.1 Introduction ... 569
- 14.2 Technological Development Status ... 570
  - 14.2.1 Aerodynamic Deceleration Test Technology ... 570
  - 14.2.2 Test Technology for Dynamic Deceleration ... 572
  - 14.2.3 Verification Technology of Soft-Landing Process ... 573
  - 14.2.4 Validation Technology for Takeoff Process ... 575
- 14.3 Demand Analysis ... 575
  - 14.3.1 Principles of Test Planning ... 575
  - 14.3.2 Test Requirement Verification ... 576
- 14.4 Test Verification Technology ... 580
  - 14.4.1 Aerodynamic Deceleration Test Technology ... 580
  - 14.4.2 Verification Technology for Powered Deceleration, Soft Landing and Takeoff ... 592
- 14.5 Outlook ... 616
- References and Related Reading ... 616

Chapter 1
Introduction

1.1 The Significance of Deep Space Exploration

Deep space exploration is the first step to understand the Earth, the solar system and the universe and then to survey, explore and settle on other celestial bodies in the solar system. Deep space exploration will be an important approach for human beings to develop and utilize space resources and make space-based scientific and technological innovations in the twenty-first century. At present, deep space exploration goes forward in six key directions: lunar exploration, Mars exploration, exploration of asteroids and comets, solar exploration, exploration of Mercury and Venus, exploration of giant planets and their satellites.

Nowadays, deep space exploration activities are challenging, innovative and motivative in hi-tech fields around the world and are of great and profound significance for the sustainable development of politics, economics, science and human society for a nation.

Scientifically, deep space exploration activities directly address major frontier scientific issues such as the origins and evolutions of the universe and life, and the comparative study of planets and will promote the development of basic disciplines such as Earth science, space science and astronomy. These activities will also lead to a number of emerging disciplines such as comparative planetary science and solar system chemistry. In addition, the deep space exploration will also greatly boost the development of space science and make distinctive contributions to the cognitive world of the humans.

Technologically, deep space exploration is a multidisciplinary system engineering featuring highly technical integration. It will promote the development of a large number of key generic technologies such as advanced space propulsion, intelligent control and innovative power system and promote the development of basic technologies such as computers, MEMS, materials and precision manufacturing. Deep space exploration enables humans to move into unknown space, overcome the challenges of vast distances and extreme environmental conditions and achieve a new leap in space technology.

Economically, deep space exploration can widely drive the advancement of engineering technology, promote the application and transfer of a series of new technologies, new processes and new materials to the national economies. It will also generate long-term benefits and is especially beneficial to the demands of green growth featuring low energy consumption, low pollution and high added values. In addition, deep space exploration can enhance the abilities of humans to explore the space and utilize space resources. Emerging concepts such as space mining have brought potential economic growth possibilities for future human society. Once implemented and applied in a developed manner, the dilemma of increasingly scarce Earth resources may be greatly alleviated.

In addition, the scientific literacy of the public is the cornerstone of social progress. The major scientific problems solved by deep space exploration are of great social significance, public awareness and science literacy. The science ideologies contained thereof are easy to spread, the scientific methods are easy to publicize, and the scientific spirits are easy to advocate. Deep space exploration has fresh vitality and is conductive to popularizing knowledge and education for the public and improving the scientific literacy of the people.

1.2 Overview of Deep Space Exploration Development

To date, countries and organizations that have independently or cooperatively conducted deep space exploration activities mainly include the USA, the Soviet Union/Russia, the European Space Agency (ESA), Japan, China and India. Such countries and organizations have launched about 241 deep space probes. Presently, celestial bodies such as the Moon, Mars, Venus, Mercury, Jupiter, Saturn, Uranus, Neptune, asteroids and comets in the solar system have been explored, and successful landings have been made on the Moon, Mars, Venus, Titan, asteroids and comets. In addition, samples have been collected from the Moon, asteroids and comets. Some of the probes are even flying out of the solar system into extrasolar systems.

1.2.1 Overview of International Deep Space Exploration Development

The USA took the lead in deep space exploration in 1960 and implemented the most deep space exploration missions. After successful manned landings on the Moon in the 1960s and 1970s, the USA suspended lunar exploration activities. It was not until in 1994 that the USA restarted the lunar exploration activities and successively launched Clementine (1994), Lunar Prospector (1998), Lunar Reconnaissance Orbiter (LRO) (2009), Lunar Crater Observation and Sensing Satellite (LCROSS) (2009), Gravity Recovery and Interior Laboratory (GRAIL) (2011) and

Lunar Atmosphere and Dust Environment Explorer (LADEE) (2013). In addition to the lunar exploration missions, the USA also participated in 63 deep space exploration missions (57 independent missions and 6 international cooperation missions). With the most comprehensive targets, the USA is the only country that has explored the Sun and other seven major planets of the solar system in the world [1–4].

Russia has conducted a great number of launches on explorations of the Moon, Venus and Mars in diversified forms and achieved remarkable results. Presently, Russia has planned Luna-Resource and Luna-Glob missions for its unmanned lunar exploration, specifically including probes such as Luna-Glob lander (Luna-25), Luna-Glob orbiter (Luna-26), Luna-Resurs lander (Luna-27), etc. In terms of Mars exploration, Russia has participated in or independently implemented missions such as ExoMars and Forbes-Grunt-2 Phobos sample return mission (to be accomplished after 2020).

Starting from the flyover of the Halley comet in 1985, ESA has independently carried out five missions with great success and high levels. It has successfully flied around the Venus and Mars and the attachment of comets. In addition, as the members are globally leading in the R&D of a number of scientific tasks, which have been carried in exploration missions by the USA, Russia, Japan, etc., and fruitful research outcomes have been obtained. Presently, the first phase of Mars Biology has been accomplished. In the second phase, drilling tools and instruments carried by the Mars rover will drill the Martian soil for further analysis. The probe is scheduled to be launched in 2020.

Japan also started deep space exploration activities in 1985 and has carried out eight missions, including four missions on small celestial bodies. And it is also the earliest country to achieve sample return from the asteroids. However, it has repeatedly suffered setbacks in planetary exploration and failed in the only Mars and Venus exploration. In general, Japan is focusing on sample return missions for multiple types of asteroids and hopes to maintain its dominance in the field of asteroid exploration by continuously developing new technologies.

In 2008, the Chandrayaan-1 launched by India achieved successful flight around the Moon, marking the start of deep space exploration of India. In November 2003, India launched Mangalyaan Mars probe. In September 2014, the probe successfully achieved braking at periareon and eventually entered the orbit of Mars. Later on, India gradually clarified and broadened its thoughts on solar system exploration and shifted its focuses to the Venus and the Sun. In addition, it plans to make in-depth exploration of the Moon and Mars in the future and implement lunar soft-landing and rovering exploration of the lunar surface.

1.2.2 Overview of Deep Space Exploration Development in China

The Lunar Exploration Project of China (also known as the Chang'e Project) consists of three phases (Phase I, Phase II and Phase III), with the primary goals being global survey via orbiting the Moon, soft landing and rover survey on the lunar surface for in-situ exploration, collecting soil sample on lunar surface for laboratory study back onto Earth, respectively. Chang'e 1, the first lunar probe of China, achieved Moon-orbiting exploration successfully in October 2007. It obtained a large amount of lunar scientific exploration data and plotted the clearest and the most complete lunar image map published then in the world (see Fig. 1.1). In October 2010, the Chang'e 2 probe, as the forerunner of the second phase of the lunar exploration project, completed a number of technical verification tasks. During the extended mission stage, it flew by and detected the Tutatis asteroid and obtained its clear image (see Fig. 1.2). Multi-target explorations were achieved over the Moon, the Sun–Earth L2 point and Tutatis asteroid just in one mission. On December 14, 2013, the Chang'e 3 probe successfully landed in Sinus Iridum of the Moon (Figs. 1.3 and 1.4 show the images taken by the lander and rover on lunar surface for each other). The probe conducted explorations such as "lunar surveying, sky scanning and Earth observation" on lunar surface and obtained a great number of scientific data. This marked the full realization of the second strategic goal of the lunar exploration project. On November 1, 2014, the reentry capsule of the Chang'e 5 high-speed reentry test mission landed smoothly (see Fig. 1.6). On January 3, 2019, the Chang'e 4 lander safely landed in the preset area in the Von Karman Crater on the back of the Moon, marking the first soft landing and rover survey on the far side of the Moon by human beings [5–10] (Fig. 1.5).

China has started its first Mars exploration mission. The launch is scheduled in 2020, and the probe will orbit and land on Mars in 2021.

Full Moon image by the first lunar exploration project of China

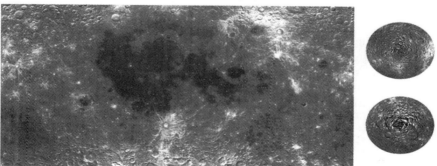

Fig. 1.1 Full moon image plotted by Chang'e 1

1.2 Overview of Deep Space Exploration Development

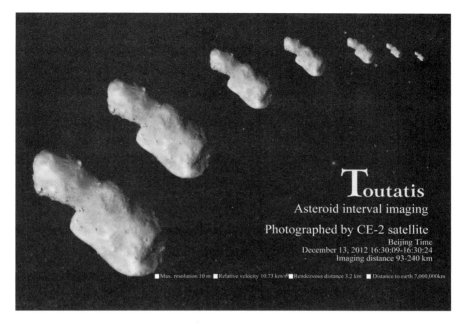

Fig. 1.2 Tutatis photographed by Chang'e 2

Fig. 1.3 Lander image captured by the rover on lunar surface

Fig. 1.4 Rover image captured by the lander on lunar surface

1.2.3 Development Trends of Deep Space Exploration

Throughout the history of deep space exploration in the world, the following major characteristics and trends may be concluded.

1. **Diversified exploration targets**

The exploration of the solar system by the humans started from the Moon, the satellite of the Earth, and then was extended to nearby planets (Mars and Venus) and then to the Sun. In the 1970 s, the exploration of other planets in the solar system (Jupiter, Saturn, Mercury, Uranus and Neptune) and space observations of the entire solar system were gradually carried out. Since the 1980 s, various types of small objects in the solar system have been explored, for example, the comets, Phobos, asteroids and Titan. At the beginning of twenty-first century, it has become increasingly eager for human beings to explore the origin and evolution of the universe, to find extraterrestrial life and to expand the space of activities. Major space powers have developed ambitious deep space exploration plans. Pluto exploration was implemented for the first time. In addition, more in-depth exploration was conducted for celestial bodies and interplanetary space at all levels of the solar system. Human beings have entered a new era of comprehensive and fine exploration of the solar system.

2. **Diversified exploration means**

Deep space was explored in diversified means such as flyby exploration, hard landing exploration, orbiting exploration, soft-landing and in-situ exploration, soft-landing and rover survey exploration, unmanned sample return and manned landing. Among

1.2 Overview of Deep Space Exploration Development

Fig. 1.5 Landing of reentry capsule of Chang'e 5

them, all the means mentioned above have been utilized in lunar exploration. Flyby, orbiting, soft-landing and in-situ exploration, soft landing and rover survey have been applied in Mars exploration. Fixed point and multiple-point observations and high orbit detection have been realized for solar exploration. Flyby, orbiting, soft-landing and in-situ explorations have been conducted for the Venus and Titan. Flyby and orbiting explorations have been achieved for Jupiter and its Galileo satellites. Presently, attachment and sample return explorations are being carried out for asteroids and comets.

Fig. 1.6 Rover Yutu rolls onto lunar surface

3. **Increasingly expanded exploration approaches**

For deep space probes, the major technologies have been upgraded from visible light remote sensing at the beginning gradually to the entire electromagnetic spectrum of γ-ray, X-ray, ultraviolet, infrared, microwave and radio waves. Whether it is image or composition exploration, the space-time resolution is getting increasingly high. Analytical tools also range from remote-sensing analysis, in-situ analysis to laboratory analysis of returned samples. With these advances, human beings have more profound understanding of the solar system and on this basis have developed the solar-terrestrial physics (STP), planetary science and comparative planetology. This has further deepened the understandings of the commonalities and characteristics of the planets and the Earth as well as the influence of the Sun on other planets, including the Earth.

4. **Constantly enriched exploration contents**

The contents of deep space exploration have gradually evolved from the exploration of celestial environment, topography and material composition at the earlier stage to the exploration of internal structure of celestial bodies, from single target exploration to multi-target and multi-task exploration. With increasingly enriched scientific results of deep space exploration, more new scientific problems have been spawned and new scientific goals of exploration activities have been formed, resulting in spiral and continuous improvement of the understandings of human beings on the solar system.

5. Further extended international cooperation

The deep space exploration activities have developed from the early contending for hegemony between the USA and Soviet Union to the current multi-polar competition, and its international cooperation feature has become increasingly distinctive. For example, the largest and most complex Cassini–Huygens Saturn Probe launched in 1997 was jointly carried out by the USA and European countries. Deep space exploration is a common goal of human beings, and generally, the tasks are complicated. Therefore, international cooperation has become an indispensable element thereof.

1.3 Future Development Requirements for Deep Space Exploration Technologies

The deep space exploration missions have typical characteristics such as wide target range, long flight distance and time. Compared with traditional near-Earth spacecraft, there are some differences in the probes and subsystem technologies; that is, higher requirements are proposed. These are all the technological demands for the development of deep space spacecraft technologies.

1. **Orbit design**

The design of the deep space spacecraft orbit requires the consideration of the restrictive three-body problem or the orbit design method under the action of multiple celestial bodies as the case may be. In recent years, new design concepts and methods have emerged such as gravity assist exploiting the celestial gravity field and aerobraking for celestial bodies with atmosphere. Combined with other constraints in mission planning, higher requirements have been imposed on the orbital design. The orbit design involved in the future deep space exploration missions includes the design of the swing-by orbit, the low-thrust transfer orbit, the high-speed entry technology of planetary atmosphere, trajectory design for soft landing on airless celestial body, libration point application technology, etc.

2. **Autonomous technology**

The deep space probe has a long flight distance, making it difficult for ground monitoring and control, and the communication delay is significant. It is also difficult to provide the high-precision orbit information required for deep space exploration missions in real time with ground monitoring and control. Especially in critical flight phases such as landing of the probe and approaching the target celestial body, it is impossible to rely on the ground for measurement and control like traditional spacecraft. Therefore, the probe is required to determine its own position and attitude autonomously, including probes of special forms such as rover working on the surface of the planet, which urgently requires navigation and positioning, especially autonomous navigation.

Deep space probes generally have limited communication opportunities with the ground and have relatively long communication intervals. In some missions, they can only communicate with the Earth once a week or even several weeks. This means that deep space probes have to be operated and work on scientific exploration for long periods of time without monitoring. Therefore, in addition to requirements of determining the flight attitude and orbit autonomously, the deep space probes should also conduct independent monitoring of the working state of the probes and decide its own tasks to some extent. They should also be able to autonomously locate, diagnose and repair or reconstruct some possible faults.

3. **TT&C technology**

The long-term goal of TT&C technology is to achieve continuous communication and precise navigation at any time in the solar system and to conduct TT&C communication with the deep space probes hundreds of millions of kilometers away. The requirements thereof are much more demanding than those applied on an Earth spacecraft. Therefore, most advanced technologies should be applied as possible and to continuously improve the communication link and TT&C accuracy so as to satisfy the needs of the growing and evolving deep space exploration missions. In response to the future demands, major breakthroughs are expected in interplanetary deep space communication in terms of system architecture and core technology. Importance should be attached to technologies such as long distance TT&C communication network architecture of 10 billion kilometers, light and small integrated X/Ka band measurement and control transponder and optical communication.

4. **Propulsion technology**

Propulsion technology has always been one of the decisive factors for human beings to enter the space. In general, deep space exploration missions have a high demand for transmitting energy, and it is difficult to meet all mission requirements by relying solely on improving the launch capability. Therefore, the probe is required to have strong maneuverability with velocity increments of several kilometers per second or more. However, the specific impulse (ISP) of traditional chemical propulsion is hard to exceed the limit of 500 s, making it difficult to meet the demands of deep space exploration missions. Therefore, it is urgent to develop new advanced propulsion technology featuring higher specific impulse, longer service life and superior performance. According to different energy sources, new advanced propulsion technology can mainly be divided into electric, solar and nuclear propulsions.

5. **Entry, descent, landing and ascent technology**

The means of deep space exploration in the future will be greatly enriched, and the explored breadth and depth will be further improved, by mastering the technologies of soft landing on celestial bodies without atmosphere such as the Moon as well as the entry, landing and ascent on Mars (thin atmosphere), Venus (dense atmosphere), small celestial bodies (weak gravitation). The major research directions include safe landing technology on complicated lunar terrain, integrated buffer and mobile mechanism technology, extraterrestrial celestial atmospheric deceleration technology, extraterrestrial celestial soft-landing technology, extraterrestrial celestial surface rover technology, extraterrestrial celestial surface takeoff technology, asteroid rendezvous/attachment/fly-around technology, interplanetary return and reentry to Earth technology.

6. **Payload technology**

To complete specific scientific target exploration, it is necessary to develop various scientific exploration instruments and equipment. Deep space explorations have various targets and may encounter various environmental conditions, posing diversified requirements on scientific payloads. Scientific exploration instruments featuring high precision, high performance and multi-functional are important reflections of the scientific and technological level for a nation. With the progress of basic technologies such as micromachines and microelectronics as well as the emergence of some new detection theories, new types of exploration payloads and technologies will continue to emerge. This will help to promote the innovation of scientific instrumentation and equipment as well as the technical level of engineering industrialization.

7. **Power source technology**

The intensity of sunlight is inversely proportional to the square of the distance from the Sun. The outer planets are far from the Sun, where the intensity of the sunlight is quite weak. Therefore, for future deep space exploration missions, efforts should be made to improve technologies such as ultra-high efficiency solar array technology, high-performance battery technology for deep space environment, isotope power generation technology and space nuclear reactor power supply.

8. **Thermal control technology**

Deep space probes are subject to multi-mission phases and huge changes in the external thermal environment, which has placed new requirements on thermal control technology. In the future, lightweight and efficient insulation technology, high-efficiency heat dissipation technology, high-reliability heat transfer technology, high-temperature thermal protection technology will be well applied in deep space exploration missions.

1.4 Prospects

In the new century, various space powers and organizations have developed their own deep space exploration plans. The USA still takes the extension of the scope of human activities in space as its long-term goal. Through continuous exploration of all major celestial bodies in the solar system, it aims to fully grasp the deep space exploration technology to ensure and strengthen its leading position in the aerospace field and lays its recent exploration targets on Mars and asteroids. ESA intends to achieve manned Mars flight and cooperates with Russia sharing the same goal to achieve a Mars landing and rover survey by 2020. Russia has proposed plans to explore other major celestial bodies in the solar system. The focus of Japan in the development of deep space exploration is to implement sample return missions for multiple types of asteroids and to continuously develop new technologies so as to maintain its dominant position in the field of asteroid exploration. India regards the improvement of space capabilities as a shortcut to realize the dream to be a powerful nation. Upon the success of Mars exploration in 2013, it has further accelerated the pace of deep space exploration. It has planned to implement lunar landing rover and the second Mars exploration by 2020. China has already taken the lead in the world in the new round of lunar exploration. Future plans for Mars exploration, asteroid exploration, Jupiter and planetary flyby exploration are also in steady progress.

The deep space exploration focusing on the solar system and the vast space of the universe is full of possibilities and challenges. Many unknown worlds wait there for human beings to explore, discover and understand. In terms of scientific exploration, deep space exploration provides an important way and advanced means for space science research, which will bring abundant scientific discoveries. In terms of space capabilities, deep space exploration is a comprehensive manifestation of space enter, explore and utilization capability. The major space powers in the world will continue to carry out deep space exploration and take it as an important way to enhance the comprehensive strength of the country. Under the impetus of deep space exploration missions, deep space exploration technology is expected to continue to develop and progress.

References and Related Reading

1. Surveyor Project Staff (1969) Surveyor VII mission report. Part 1: Mission Description and Performance. NASA
2. HUGHES (1964) Surveyor spacecraft A-21A model description. NASA
3. Milton Beilock (1964) Surveyor lander mission capability. NASA
4. Van Dorn JB (1967) Detail specification surveyor system functional requirements. NASA
5. Peijian Y, Jing P (2006) Deep space exploration and its prospect in China. Eng Sci 8(10):13–18 (Chinese vision)
6. Peijian YE, Jiangchuan H, Zezhou S et al (2014) The process and experience in the development of Chinese Lunar Probe. Science China: Technological Sciences 44(6):543–558 (Chinese vision)
7. Weiren WU, Dengyun YU (2014) Development of deep space exploration and its future key technologies. J Deep Space Explor 1(1):5–17. (Chinese vision)
8. YU Dengyun, WU Xueying, WU Weiren (2016) Review of technology development for chinese lunar exploration program. J Deep Space Explor 3(4):307–314 (Chinese vision)
9. Zhaoyu P, Qiong W, Yaosi T (2015) Technology roadmap for chang'e program. J Deep Space Explor 2(2):99–110 (Chinese vision)
10. Mengfei Y, Gao Z, Wu Z et al (2015) Technique design and realization of the circumlunar return and reentry spacecraft of 3rd phase of chinese lunar exploration program. Sci China: Technol Sci 45(2):111–123

Chapter 2
Characteristics of Deep Space Environment and Corresponding Impact

2.1 Introduction

Different from Earth-orbiting spacecraft, during missions, deep space probes are subject to the influences of geospace environment, interplanetary space environment, and the surrounding or surface environment of celestial bodies. Therefore, in deep space exploration missions, the adaptability of deep space probes to various types of space environments encountered in mission profiles must be considered [1].

Generally, the space environment experienced during a deep space mission may be divided into three stages:

(1) Launch and fly away from the Earth: At this stage, the deep space probe still flies within the magnetosphere of the Earth (the height of the magnetopause is about 10 Re, of which Re is the radius of the Earth) and the flight time is generally 5–6 h. And the probe is mainly in near-Earth space environment, which consists of radiation belt of the Earth, including the galactic cosmic ray, plasma, neutral atmosphere, atomic oxygen, ultraviolet radiation, thermal radiation, terrestrial infrared radiation, etc.

(2) Transfer from the Earth to celestial body: This stage starts from the breakthrough of the magnetopause to the arrival of the surrounding environment of the target celestial body. For the lunar exploration, this period would last for about several days; for Mars, several months; and for other planets, possibly several years. At this stage, the probe is in interplanetary environment, which mainly consists of solar wind, solar cosmic ray, galactic cosmic ray, thermal radiation, ultraviolet radiation, etc.

(3) Operation around or on the celestial body: In this stage, the probe fulfills its missions and involves activities such as orbiting around the celestial body, landing and roving exploration on the surface. The mission may last for several days to several years. In addition to the normal environment from the Sun, at this stage, the probe will also face the special space environment that is unique to the celestial body, for example, atmosphere and dust of Mars, intense magnetic field and radiation of the Jupiter, dense atmosphere and sulfuric acid cloud of the Venus, topography of specific celestial surface, etc.

© Beijing Institute of Technology Press and Springer Nature Singapore Pte Ltd. 2021
Z. Sun, *Technologies for Deep Space Exploration*, Space Science and Technologies

Table 2.1 demonstrates the main space environment elements bearing impacts on deep space probes. These space environmental elements may interact with deep space probes to induce various space environmental effects, affecting the normal operation or safety of deep space probes to different extents (mostly adverse).

For example, solar electromagnetic radiation may disturb the wireless communication of deep space probes, and long-term irradiation of solar ultraviolet light may degrade the performance of organic materials on the outer surface of probes, the atmospheres of Mars and the Venus may have impacts such as drag or aerodynamic heat on the orbiting and landing of probes, the intense radiation of the Jupiter may pose a major challenge to the anti-radiation design of probes, the dust of Mars and Lunar may have impacts such as pollution on the optical systems of the landing rovers, the complicated thermal environment of the planet surface may result in extreme environment for probes, and the surface topography of the planet may affect the abilities of the rover such as moving and avoiding obstacle on the celestial surface.

Therefore, to ensure the space environment adaptability of deep space probes, it is a necessary prerequisite to fully consider the environmental adaptability design in the process of engineering design, development and production of probes to improve their ability to survive and complete scheduled missions in deep space [2].

Therefore, understanding the deep space environment is an indispensable and important work in designing deep space probes. This chapter discusses the deep space environments that may adversely affect the design of deep space probes and their corresponding impacts from the perspective of engineering design of deep space probes.

2.2 Geospace Environment

2.2.1 Main Geospace Environment Characteristics for Deep Space Probes

The Earth is the "departure station" for deep space probes. During the launch from the Earth, the probe is within the scope of Earth magnetosphere and is mainly subject to the impact of the same geospace environment as the Earth-orbiting spacecraft. Due to the short duration, the extent of geospace environment influence on the probe is different from Earth-orbiting satellites. In combination with the requirements of discussion in subsequent chapters, this section only focuses on the charged particle radiation environment in geospace, mainly including Earth radiation belt, solar cosmic ray, galaxy cosmic rays, plasma, etc. [3]. Other environment elements are not discussed in a detailed manner herein.

2.2 Geospace Environment

Table 2.1 Major space environment elements affecting deep space probes

Space environment elements	Circum-terrestrial space (within the magnetosphere)	Interplanetary space (between planets)	Mercury surrounding	Venus surrounding	Lunar surrounding	Mars surrounding	Jupiter surrounding	Saturn surrounding	Uranus surrounding	Neptune surrounding
Global magnetic field	✓	–	✓	✓	–	–	✓	✓	✓	✓
Atmospheric environment	✓	–	✓	✓	–	✓	✓	✓	✓	✓
Atomic oxygen	✓	–	–	–	–	–	–	–	–	–
Solar ultraviolet radiation	✓	✓	✓	✓	✓	✓	✓	✓	✓	✓
Charged particle capture zone	✓	–	–	–	–	–	✓	✓	✓	✓
Galactic cosmic ray	✓	✓	✓	✓	✓	✓	✓	✓	✓	✓
Solar cosmic ray	✓	✓	✓	✓	✓	✓	✓	✓	✓	✓
Solar wind	–	✓	✓	✓	✓	–	–	–	–	–
Hot plasma	✓	–	–	–	–	–	–	–	–	–
High-energy electron	✓	–	–	–	–	–	✓	–	–	–
Planetary dust	–	–	✓	–	✓	✓	–	–	–	–

(continued)

Table 2.1 (continued)

Space environment elements	Circum-terrestrial space (within the magnetosphere)	Interplanetary space (between planets)	Mercury surrounding	Venus surrounding	Lunar surrounding	Mars surrounding	Jupiter surrounding	Saturn surrounding	Uranus surrounding	Neptune surrounding
Landform	–	–	✓	–	✓	✓	–	–	–	–

1. Earth radiation belt

The Earth radiation belt, also known as the "Van Allen Belt," is a concentrated area of charged particles formed around the Earth due to the binding and trapping of charged particles by the magnetic field of the Earth. The physical mechanism of the formation of the Earth radiation belt is so far regarded as the combined effect of spiral motion around magnetic lines of flux, bouncing motion between the south and north mirror points along the magnetic lines of flux, and the drifting motion along the east–west direction. Visually, the Earth radiation belts are equivalent to two charged particle belts surrounding the Earth over the equator, one near the Earth is the inner radiation belt and the other is the outer radiation belt (as shown in Fig. 2.1). The basic characteristics of the inner and outer radiation belts are shown in Table 2.2. Note that the lower boundary distortion caused by geomagnetic anomaly zone over the South Atlantic has little to do with deep space exploration.

2. Solar cosmic ray

The solar cosmic ray is a high-energy, high-flux charged particle stream that erupts in a short time from the surface of the Sun during the solar explosive activities. As most of these charged particles are composed of protons, such activities are often referred to as solar proton events. The duration of the solar cosmic rays observed

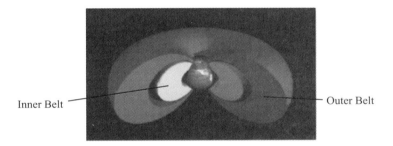

Fig. 2.1 Schematic layout of shape of Earth radiation belt

Table 2.2 Basic characteristics of the inner and outer radiation belts

Characteristics	Inner belt	Outer belt
Height range on the equatorial plane/km	600–10,000	10,000–60,000
Geomagnetic latitude range/(°)	40°N–40°S	From (55°–70°)N to (55°–70°)S
Particle composition	Electrons and protons	Mainly electrons, with a small amount of low-energy protons
Energy range/MeV	Electrons: 0.04–7 Protons: 0.1–400	Electrons: 0.04–4 Protons: flux drops sharply for energy greater than 1 MeV
Height of center position (height over the equator)/km	3000–5000	20,000–25,000

near the Earth's orbit is about several hours to several days. The solar cosmic rays entering the Earth magnetosphere are shielded by the magnetic field of the Earth. The solar cosmic rays can be ignored when the surface of the Sun is peaceful.

For the solar cosmic ray particles, mainly made up of protons, their storage energy ranges from 10 MeV to tens of GeV. In addition, there are also helions, which account for 3%–15% and heavy nucleus with atomic number $Z > 2$, of which the heavy nuclei of $Z = 6, 7, 8$ account for 0.05% of the total particle flux. Such particles have two sources: One is the direct eruption from the surface of the Sun, and the other is the acceleration of the shock wave propagating in the interplanetary space driven by the coronal mass ejection.

The occurrence of solar proton events is highly random and casual. The propagation of energy particles from the Sun to the Earth is modulated by solar winds and interplanetary magnetic fields. Therefore, it is unevenly distributed in space and sudden in time. According to statistics, in the high years of solar activity, there are relatively frequent solar proton events (over 10 times per year), while in the low years of solar activity, there are relatively infrequent solar proton events (3–4 times per year, or even less).

3. **Galactic cosmic ray**

The galactic cosmic rays are charged particles with low flux but high energy from the galaxy outside the solar system, and the particle energy range is generally 102 MeV–109 GeV. Most of the particle energies are concentrated at 103–107 MeV. The flux in free space is generally only 0.2–0.4 $(cm^2 \cdot sr \cdot s) -1$. The galactic cosmic rays contain almost all elements in the periodic table. However, the main component is protons, accounting for 84.3% of the total, followed by α particles, accounting for 14.4%. The other heavy nuclei components account for approximately 1.3%.

The intensity of the galactic cosmic rays is considered to be uniform and constant if not affected by the solar wind before entering the heliosphere. Specifically, the intensity does not change with time and space. However, after entering the heliosphere, due to the repulsive interaction of the interplanetary magnetic fields moving outward with the solar wind, the intensity of the galactic cosmic rays at the edge of the heliosphere is the strongest, which gradually decreases inward and presents an intensity gradient. In the high years of solar activities, due to the enhanced repulsion of the interplanetary magnetic fields, the intensity of the galactic cosmic rays in the inner heliosphere (including the Earth orbit of an astronomical unit from the Sun) is weaker than that in the low years of solar activities, resulting in a negative correlation with solar activities. Near the Earth, the isotropic flux of the galactic cosmic rays is 1.3×10^8 $(cm^2 \cdot year) -1$ (low year of solar activity) and 7×10^7 $(cm^2 \cdot year) -1$ (high year of solar activity).

The intensity variation of galactic cosmic rays may be periodic or non-periodic. Periodic variations include half-day variation, solar day variation, 27-day variation, 11-year variation, 22-year variation, etc., of which the 11-year variation is the most significant with the amplitude of variation less than 50%. Non-periodic variations include cosmic ray storm, etc.

4. Plasma

When the Earth magnetic field is inactive, the geostationary orbit (GEO) is filled with cold plasma (with energy less than 10 eV). During the sub-storm of the magnetosphere, a large amount of hot plasma (with energy of about 20 keV) is injected from the magnetic tail (see Fig. 2.2). When such hot plasmas meet the GEO spacecraft, the surface of the spacecraft will be charged. When the charging potential is high enough, discharge may occur, adversely affecting the electronic systems of the spacecraft [4].

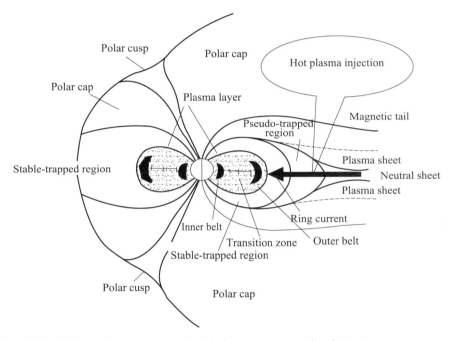

Fig. 2.2 Hot plasma injection at magnetic tail during a geomagnetic sub-storm

2.2.2 Impact of Geospace Environment on Deep Space Probes

Considering the characteristics of the orbits of the deep space probes, we can see that deep space probes have a noteworthy feature in geospace environment—probes fly only 5–6 h in the magnetosphere of the Earth and cross the inner and outer belts only once. Based on this characteristic, the impacts of the geospace environment on deep space probes are mainly shown as follows:

1. Impact of ionizing radiation dose

The radiation dose of the radiation belt of the Earth to deep space probes is much lower than that on ordinary Earth-orbiting spacecraft. Such radiations are generated and accumulated only as probes traverse the inner and outer belts, while in subsequent long-term deep space flight, probes are no longer affected by the radiation belts of the Earth.

Figure 2.3 gives the comparison of the radiation dose experienced by a deep space probe crossing the inner and outer belts and that by an Earth-orbiting spacecraft. As can be seen from the figure, the radiation dose absorbed by deep space probes is much lower than that by Earth-orbiting spacecrafts. For components inside the device (approx. 3 mm Al shield), the absorption dose is only about 0.1 krad (Si) in one radiation belt crossing. Even the deep space probe possibly experiences a radiation dose similar to that generated at the October 1989 extraordinary solar proton event during its mission [5], the radiation dose on the probe with the 3 mm Al shield is only 2–3 krad (Si). Such low radiation dose level is not likely to pose a threat to safety of deep space probes. Therefore, the deep space probes do not need to take too many protective measures, such as increasing radiation shielding for the near-Earth radiation environment.

2. Single particle effect

The charged particles inducing single-event effects in geospace environment are mainly from the Earth radiation belts, galactic cosmic rays and solar cosmic rays. The single-event effect of the deep space probe does not differ much from that of the Earth-orbiting spacecraft, in terms of the probability of the occurrence of the effects. It is required to identify any elements of the electronic systems of the deep space probe that are particularly sensitive to single-event effects, and take proper protection measure against the effects when necessary.

3. Impacts of surface charge/discharge effects

Deep space probes are supposed to fly through the height range of 20,000–60,000 km within 5–6 h after launching from the ground. Therefore, they may be affected by hot plasma. However, due to the short duration, the occurrence of hot plasma injection is random and accidental. Therefore, the impact of surface charge on deep space probes caused by hot plasma is generally relatively small and not required to be considered as an important risk.

4. Impacts of internal charging effects

The generation of internal charging effects caused by high-energy electrons requires the spacecraft to be retained in high-flux, high-energy electronic environment for a long time. This kind of electronic environment is usually the high-energy electron storm at the center of the outer radiation belt generated in the late stage of geomagnetic storm. The deep space probe will leave the Earth magnetosphere and enter the interplanetary space after a short while. Therefore, it is not likely to stay in the high-energy electron storm region in the late stage of geomagnetic storm for a long period. Therefore, it is not necessary to consider the effects of internal charging effects for

2.2 Geospace Environment

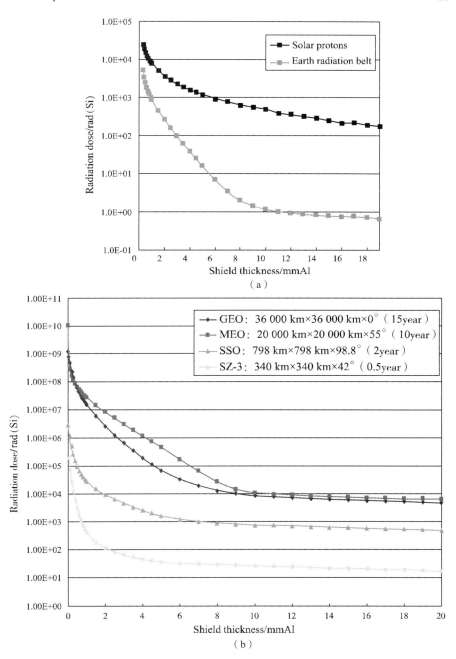

Fig. 2.3 Comparison of total ionizing doses. **a** The radiation dose experienced by the deep space probe in one radiation belt crossing; **b** The radiation dose experienced by the Earth orbit spacecraft during the mission

the flight phase when the probe is launched and flying way from the Earth. Note that due to the strong electron radiation environment around the Jupiter, it is necessary to fully consider the influence of internal charging caused by high-energy electrons.

5. **Impacts of other effects**

For the trajectory of deep space probes within the Earth magnetosphere, other near-Earth environments, such as the Earth's neutral atmosphere, vacuum, solar electromagnetic radiation and geomagnetic fields, are not much different from the environments faced by Earth-orbiting spacecraft. Therefore, the influences of these environments on deep space probes are similar to that on Earth-orbiting spacecraft.

2.3 Lunar Space Environment

2.3.1 General

As the natural satellite of the Earth, the lunar has some unique characteristics in terms of space environment compared with that of the Earth. Probes landing and conducting rovering exploration on the lunar are also subject to environmental impacts such as the lunar landform and lunar dust. The basic physical characteristics of the lunar are shown in Table 2.3.

Table 2.3 Basic physical characteristics of the lunar

Parameters	Value
Average distance from the Earth to the lunar/km	384,402
Average density/(g·cm^{-3})	3.34
Average radius/km	1737.5
Mass/kg	7.352 × 1022
Period of rotation/Earth day	27.32
Period of revolution/Earth day	27.32
Axial inclination/(°)	1.54
Orbital eccentricity	0.05
Average speed orbiting the Earth/(km·s^{-1})	1.02
Gravitational acceleration on the equator/(m·s^{-2})	1.6

2.3.2 Lunar Radiation Environment and Impacts

1. **Composition of lunar radiation environment**

The magnetic field of the lunar is extremely weak and cannot form a stable charged particle trapping zone around it. Therefore, the lunar has no charged particle radiation environment similar to the Earth radiation belt. The charged particle radiations of the lunar orbit and the lunar surface mainly come from the galactic cosmic rays, the solar cosmic rays and the solar wind.

As the distance between the lunar and the Sun is approximately the same as the distance between the Sun and the Earth, the galactic cosmic ray and solar cosmic ray environment of the lunar orbit and the lunar surface are basically the same as those in geospace environment of the Earth (especially the GEO space environment) and are not discussed in a detailed manner in this section.

The lunar has no global magnetic field or magnetosphere structure similar to the Earth. The solar wind particles are not shielded by the magnetic field and can reach the lunar orbit and surface of the lunar, directly irradiating the probe surface. The solar wind is a thin hot plasma that originates from high-temperature solar corona. When the solar corona temperature exceeds the constraint imposed by the solar gravitational force, the solar wind is emitted toward all directions. The solar wind is mainly up by electrons and protons, which account for over 95%. Heavy ions in the solar wind are mainly helium ions, accounting for about 4.8%. Other components such as oxygen ions and iron ions are quite low.

The main physical characteristics of the solar wind are shown in Table 2.4. The solar wind has uneven flow rate in terms of both spatial distribution and velocity magnitude, with the maximum value reaching 900 km/s and the minimum value as low as 200 km/s. The density of the solar wind particles fluctuates greatly, with an average density of $106/m^3$. The average value of the solar wind magnetic field is about 7 nT.

2. **Impacts of lunar radiation environment**

    1) Impact of ionizing radiation dose

As there is no continuous and stable charged particle radiation belt near the lunar, probes will not be subject to the long-term continuous strong radiation environment similar to the Earth radiation belt in lunar orbiting or lunar surface rovering. During

Table 2.4 Characteristics of solar wind (at 1 astronomical unit)

Electron density/cm^{-3}	7.1
Proton density/cm^{-3}	6.6
Flow rate/(km·s^{-1})	450
Electron temperature/K	1.4×10^5
Proton temperature/K	1.2×10^5
Magnetic field/nT	7

lunar orbiting and lunar surface operations, the probe is exposed to a continuous space radiation dose of 0 rad (Si) regardless of the time length of operation.

In the absence of a solar proton event, the charged particle radiation environment for the probe is only low-energy solar wind and high-energy low-flux galactic cosmic rays. The contribution of both radiation sources to the total radiation dose is small, and the total ionizing dose (TID) is generated only when the range of the probe surface thickness is 1–2 μm. Therefore, for most materials and electronic devices of probes, the influence of the total ionizing dose may be ignored during lunar orbiting and lunar surface rovering exploration.

In case of a great solar proton event during the mission, the radiation dose during the lunar orbiting and lunar surface operation will all come from the high-energy solar protons. Considering great solar flare eruptions similar to that in October 1989 might happen during lunar orbiting and lunar surface operation of the lunar probes, a solid ball shielding model is used in Table 2.5 to analyze the dose–depth relationship of the lunar probe during such solar flare eruption. Taking the general device shell thickness of 3 mm Al, as shown in Table 2.5, the radiation dose for the electronics inside the devices will increase by 2–3 krad during a solar flare event. For some radiation-sensitive, very large-scale and high integration density microelectronic devices (such as FPGAs, CCDs, large-capacity RAMs), this dose may cause performance degradation of the devices. Therefore, importance should be attached to the anti-TID protection design of such devices. A radiation dose of 2–3 krad (Si) generally does not result in significant performance degradation for other devices or materials, so the impact of the total ionizing dose is relatively small.

2) Impact of single-event effect

In the absence of solar disturbances, the charged particles that have a single-event effect (SEE) on the lunar probes are mainly high-energy heavy ions of the galactic cosmic rays. During the solar flare eruption, high-energy, high-flux and high-energy solar protons are injected into the lunar space and the lunar surface, which will be the main source of particles triggering the single-event effect of the lunar probe.

Figure 2.4 gives the linear energy transfer (LET) spectrum (integral spectrum) of all particle components with atomic number $Z = 1$-92 after 1 mm Al shielding in lunar orbit or on lunar surface without solar proton events (with the galactic cosmic ray (GCR) only) and with solar proton events (with both GCR and the solar cosmic

Table 2.5 Dose–depth relationship of lunar probe during the solar proton event

Equivalent aluminum thickness/mm Al	Shielding surface density/(g·cm^{-2})	High-energy solar proton dose/rad (Si)
0.10	0.027	5.96E + 04
1.00	0.270	7.52E + 03
3.00	0.810	2.27E + 03
5.00	1.350	1.13E + 03
10.00	2.700	4.52E + 02

2.3 Lunar Space Environment

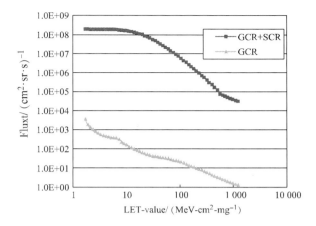

Fig. 2.4 LET integral spectrum for solar proton events in lunar orbit and on lunar surface with/without solar proton event (1mmAl shielding)

ray (SCR)). It is obvious from Fig. 2.4 that, during the solar proton event, the high-energy heavy ions may result in a sharp increase of single-event effect, with the flux achieving an increase of over four orders of magnitude. Therefore, importance must be attached to the single-event effect protection issues of lunar probes during the eruption of solar proton events, especially, for microelectronic devices in large scales and small sizes such as FPGAs, DSPs, CPUs, SRAMs and bus interface chips used in probes. If such devices are not manufactured with special anti-SEE-strengthening process, they may become highly sensible to single-event effect.

3) Impacts of solar cell radiation damage

The lunar probes mainly depend on solar cells for power generation. Therefore, the effect of space charged particle radiation on solar cells performance should be put into consideration. With high-energy particle radiation and under the action of both physical mechanisms of ionization and displacement, the open-circuit voltage V_{oc}, short-circuit current I_{sc}, maximum output power P_{max}, etc., of the solar cells may all be degraded.

Generally, according to the radiation test data of solar cells, the damage degree of space charged particles with different energies to solar cells is equivalent to the of 1 MeV electrons of certain flux to solar cell. The equivalent 1 MeV electron damage flux is used to describe the space radiation damage to solar cells.

For lunar probes in lunar orbit or on lunar surface, in the absence of a solar proton event, the charged particle radiation that the solar cells suffer is from low-energy solar wind and low-flux galactic cosmic ray. The performance degradation of solar cells from such particles is almost imperceptible and may be ignored in engineering design.

In case of great solar proton events, as a large number of high-energy, high-flux solar protons are injected into the range of lunar space and lunar surface; such high-energy protons will result in relatively greater damage to the performance of solar cells of the lunar probes.

Considering that lunar probes may encounter great solar proton events similar to that of October 1989 during lunar orbit flight and lunar surface operations, Table 2.6 quantitatively analyzes the performance damage to solar cells (Si) of lunar probes during the eruption of such events and gives the equivalent 1 MeV electron damage flux of the solar cell (Si).

Table 2.6 Equivalent 1 MeV electron damage flux of lunar probe solar cell (Si) (data dimension in the table is cm^{-2})

Shielding surface density of cover glass/(g·cm^{-2})	Without solar proton event		With solar proton event	
	Open-circuit voltage V_{oc}	Short-circuit current I_{sc}	Open-circuit voltage V_{oc}	Short-circuit current I_{sc}
5.59E − 03	Ignored	Ignored	8.40E + 14	3.10E + 14
1.68E − 02	Ignored	Ignored	3.07E + 14	1.35E + 14
3.35E − 02	Ignored	Ignored	1.15E + 14	7.64E + 13
6.71E − 02	Ignored	Ignored	7.40E + 13	4.14E + 13
1.12E − 01	Ignored	Ignored	3.92E + 13	2.49E + 13
1.68E − 01	Ignored	Ignored	2.52E + 13	1.76E + 13
3.35E − 01	Ignored	Ignored	1.20E + 13	9.23E + 12

4) Displacement damage effect of optical devices

In solar proton events, high-energy and high-flux solar protons may damage the optical components of the lunar probes (such as CCD cameras and optical sensors) to some extent. This effect is generally caused by the displacement effect of energetic protons on optical components.

Since the optical sensors such as CCD and CMOS-APS used in probes have relatively low anti-displacement damage thresholds, after the eruption of serious solar proton event, the displacement damage of the energetic solar protons to the sensors will significantly increase the dark current of the optical sensors (minority carrier devices) as well as the image background noises. Such damages generally refer to permanent physical damages to the devices.

Therefore, when optical systems are used on lunar probes, the impact of solar high-energy protons on the performance of optical sensors must be fully considered in design to avoid adverse impact on system performance and whole-probe performance under such state of image quality degradation.

2.3.3 Lunar Atmosphere and the Impact

The atmospheric pressure on the lunar surface ranges from 10^{-6} Pa to 10^{-10} Pa. As for atmospheric density, the typical data of the number density of lunar atmosphere is shown in Table 2.7. The lunar atmosphere mainly contains inert gases and H, CO_2,

2.3 Lunar Space Environment

Table 2.7 Ranges of gas density and height on lunar surface

Gas	Number density/(1/cm^3)		Height range/km	
	Day	Night	Day	Night
20Ne	4×10^3–10^4	105	100	25
He	8×10^2–4.7×10^3	4–7×10^4	511	128
H$_2$	2.5×10^3–9.9×10^3	10^4–1.5×10^5	1022	256
Ar	2×10^3	$<10^2$	55	–
CH$_4$	1.2×10^3	–	–	–
CO$_2$	10^3	–	–	–
NH$_3$	4×10^2	–	–	–
OH + H$_2$O	0.5	–	–	–

CH$_4$, NH$_3$, etc., and does not contain atomic oxygen which is common at low Earth orbit (LEO) and bears relatively great impact on spacecrafts.

To understand the lunar atmosphere from the perspective of aerospace engineering, the state of the Earth atmosphere at Earth orbit of 100–2500 km altitude is analyzed, and the results are shown in Table 2.8.

Comparing Table 2.7 with Table 2.8, we can see that the atmospheric pressure of the lunar (vacuum) is equivalent to the atmospheric pressure at the Earth at the height of 500–2500 km. However, the orbital height of Earth-orbiting spacecraft is usually above 500 km, or even up to 36,000 km. Therefore, the engineering design of the material outgassing, thermal design, vacuum cold welding and dry friction of the lunar probes may be the same as that of the Earth-orbiting spacecraft.

In addition, the number density of the lunar atmosphere is equivalent to that of the Earth orbit with an altitude above 1500 km in terms of the order of magnitude. Therefore, in the design of the lunar orbit, the lunar atmospheric drag may be treated the same as in the design of the Earth-orbiting spacecraft with an orbit height of over 1500 km.

2.3.4 Lunar Soil/Lunar Dust and Impacts

1. **Lunar soil impact analysis**

Lunar soil refers to a loose structure mixture covered on the lunar surface consisting of rock debris, powder, breccia and impact melting glass material. Lunar soil has the characteristics of looseness, non-consolidation, fine grains, etc., and is generally appears in the color of light brown and dark gray. Lunar soil is formed by the combined effects on lunar surface of meteorite and micrometeorite striking, continuous irradiation of cosmic rays and solar winds, and thermal expansion, contraction and breaking of lunar rocks due to distinct diurnal temperature variations.

Table 2.8 Earth atmosphere analysis results for orbit altitude of 100–2500 km

Height/km	Density/(g·cm^{-3})	Pressure/Pa	Number density of molecules/(cm^{-3})					
			O$_2$	N$_2$	O	Ar	He	H
100	5.99E − 10	3.47E − 02	2.25E + 12	9.72E + 12	7.15E + 11	1.16E + 11	1.61E + 08	0.00E + 00
200	3.52E − 13	1.23E − 04	4.18E + 08	4.27E + 09	4.95E + 09	0.00E + 00	7.68E + 06	0.00E + 00
300	2.91E − 14	1.36E − 05	1.11E + 07	1.76E + 08	7.65E + 08	0.00E + 00	4.63E + 06	0.00E + 00
500	8.55E − 16	4.82E − 07	1.80E + 04	6.34E + 05	3.05E + 07	0.00E + 00	2.05E + 06	1.95E + 04
800	1.41E − 17	1.53E − 08	2.54E + 00	2.70E + 02	3.62E + 05	0.00E + 00	6.77E + 05	1.47E + 04
1000	2.89E − 18	5.45E − 09	1.03E − 02	2.17E + 00	2.31E + 04	0.00E + 00	3.40E + 05	1.24E + 04
1500	4.86E − 19	1.15E − 09	3.72E − 08	3.73E − 05	4.38E + 01	0.00E + 00	7.09E + 04	8.34E + 03
2500	4.21E − 20	1.39E − 10	3.36E − 17	4.51E − 13	1.32E − 03	0.00E + 00	5.24E + 03	4.33E + 03

2.3 Lunar Space Environment

During the soft landing of the lander, the physicomechanical parameters of the lunar soil bear impacts on the landing impact and stability. Such parameters include particle composition, dry density, void ratio, cohesive forces, internal friction angle, bearing capacity, etc.

1) Particle composition

Generally, the sorting of lunar soil is relatively poor with wide particle size distribution. Diameters of most particles are less than 1 mm. Most of the particles have diameters ranging from 30 μm to 1 mm. The median diameter is 40 to 130 μm, and the average diameter is 70 μm. The lunar soil particle gradations are shown in Fig. 2.5.

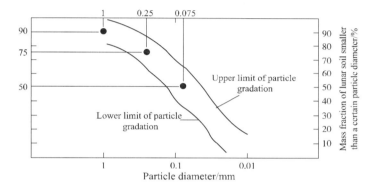

Fig. 2.5 Lunar soil particle gradations

Lunar soil particles with diameter less than 20 μm account for 10%–20%. Such particles are prone to floating and attaching themselves to mechanical equipment. The lunar soil particles have a variety of shapes, including shapes from spherical to extremely angular. However, strip-shaped, sub-angular and angular particle are relatively common.

2) Density

Lunar soil density refers to the mass of lunar soil per volume unit when it has not been disturbed. According to existing lunar soil sample analysis, the relationship between lunar soil density ρ (in g/cm^3) and depth z (in cm) may be expressed in two simplified functions, namely hyperbolic function (Eq. 2.1) and exponential function (Eq. 2.2).

$$\rho = 1.92 \frac{z + 12.2}{z + 18} \tag{2.1}$$

$$\rho = 1.39 z^{0.056} \tag{2.2}$$

The results of the lunar soil density analysis are shown in Fig. 2.6. Within the

Fig. 2.6 Lunar soil density analysis results

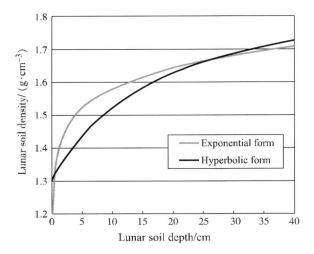

depth range from 0 to 40 cm, the density of lunar soil basically varies from 1.3 to 1.7 g/cm³. The best estimates of the average density of lunar soil at different depths in the impact craters are shown in Table 2.9.

Table 2.9 Best estimates of average density of lunar soil at different depths in the impact craters

Range of depth/cm	Average density/(g·cm⁻³)
0–15	1.45–1.55
0–30	1.53–1.63
0–60	1.61–1.71
30–60	1.69–1.79

3) Porosity properties

Porosity properties include void ratio and porosity factor. The void ratio e of lunar soil refers to the ratio of pore volume to particle volume in lunar soil. The porosity ratio of lunar soil in the natural state may be used to assess the degree of compaction of the lunar soil. Generally, if $e < 0.6$, the lunar soil is compact with low compressibility, and if $e > 1.0$, the lunar soil is loose with high compressibility. Porosity factor n is the ratio of the volume occupied by the pores in the lunar soil to the total volume, which is expressed as a percentage. The best estimates of the average porosity factor and void ratio of the lunar soil at different depths are shown in Table 2.10.

4) Shear resistance (cohesive force, internal friction angle)

Relative displacements between the particles of lunar soil occur subjected to external force result in the sliding of one part of the lunar soil against another. The property of lunar soil particles against such sliding is called the shear resistance of lunar soil, which is determined by two indicators, namely cohesive force c and internal friction angle φ. The shear characteristic of lunar soil is expressed as Eq. (2.3):

2.3 Lunar Space Environment

Table 2.10 Best estimates of average porosity factors and void ratios of lunar soil at different depths

Range of depth/cm	Average void ratio e	Average porosity factor n/%
0–15	1.00–1.14	50–54
0–30	0.89–1.03	47–51
0–60	0.80–0.94	44–48
30–60	0.71–0.85	42–46

$$\tau = c + \sigma \tan \phi, \tag{2.3}$$

where σ is normal stress.

The reference values for the cohesion and internal friction angle of lunar soil at different depths are shown in Table 2.11.

Table 2.11 Reference values for cohesive force and internal friction angle of lunar soil at different depths

Range of depth/cm	Cohesive force c/kPa		Internal friction angle φ/(°)	
	Range of variation	Average value	Range of variation	Average value
0–15	0.44–0.62	0.52	41–43	42
0–30	0.74–1.1	0.90	44–47	46
0–60	1.3–1.9	1.6	48–51	49
30–60	2.4–3.8	3.0	52–55	54

5) Bearing capacity

Bearing capacity refers to the ability of the lunar soil to carry loads. The relationship between static bearing capacity and depth of lunar soil in different regions is shown in Fig. 2.7.

The mechanical properties of lunar soil also vary across the lunar surface. In design and ground test verification, most unfavorable lunar soil parameters for landing impact and stability should be selected for simulation analysis and verification according to a more conservative design principle. For landing impact, the physicomechanical properties of lunar soil mainly affect the energy absorption of the landing buffering mechanism. The purpose of stricter assessment can be achieved by selecting to simulate lunar soil properties with greater bearing strength. As for stability analysis, the slippage and subsidence characteristics of the lander on the lunar surface are the major concerns. Lunar soil properties with smaller bearing strength should be selected in the simulation for stricter assessment.

2. **Lunar dust impact analysis**

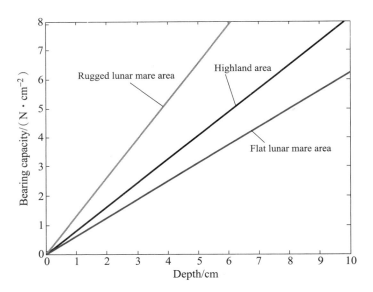

Fig. 2.7 Relationship between static bearing capacity and lunar soil depth

Lunar dust is a kind of lunar soil with relatively small particles. It is relatively loose and may produce flying dust under certain external force. Therefore, its basic physical and chemical properties are the same as those of lunar soil.

The excitation mechanism of lunar dust consists of two types: natural and artificial. The natural excitation mechanism includes secondary eruption and dust electrostatic floating due to the impacts of meteors and micrometeoroids. The artificial excitation mechanism includes four types, which can be listed in the order of increasing effects: walking of the astronauts, rotation of rover wheels, engine jet effect and soft-landing impacts.

Possible damages caused by lunar dust to probes include: Lunar dust may result in an increase of the measurement error or failure of the lunar distance and velocity measurement sensors and optical sensors. It may cause contamination of the surfaces of solar sensors, camera lenses, solar arrays, thermal control coatings (OSR sheets, etc.) and result in reduced light transmittance, reduced power supply capability, reduced thermal control performance, etc. In addition, fine lunar dust grains may also erode bearings, gears and other mechanical devices that are not completely enclosed, resulting in wear or seizure of the mechanical parts. Therefore, the characteristics of the lunar dust and its impacts on the probe thereof are important issues that must be considered in probe design.

2.4 Space Environment of Mars

2.4.1 Overview

Mars is the closest planet to the Earth in terms of the natural environment. The two planets have many similar features. Table 2.12 shows the basic physical properties of Mars [6].

(1) The revolution eccentricity of Mars is much greater than that of the Earth. The average distance of Mars from the Sun is about 1.52 AU (astronomical unit, 1 AU = 1.495 978 930 × 108 km) with the farthest distance of 1.67 AU and nearest distance of 1.38 AU.

(2) The angle between the equatorial plane and the orbital plane of Mars is 25.2°, which is quite similar to that of the Earth. Therefore, Mars also has a seasonal change similar to the Earth.

(3) A Martian year is equivalent to about 687 Earth days (669 Mars days), and the duration of each season on Mars is nearly twice that on the Earth.

(4) The direct sunlight received by Mars is only 43% of that of the Earth. Due to relatively large orbital eccentricity of Mars, the sunlight received by Mars changes greatly compared with the Earth in the course of a year.

Table 2.12 Basic physical characteristics of Mars

Parameters		Value
Distance to the Sun/km	Average (semi-major axis)	2.278×10^8
	Perihelion	2.065×10^8
	Aphelion	2.491×10^8
Distance to the Earth/km	Min.	5.581×10^7
	Max.	3.989×10^8
	Average	7.835×10^7
Period of rotation (h:m:s)		24:37:22.6689
Period of revolution/Earth day		686.9804
Angle between the ecliptic plane and the equatorial plane/(°)		25.2
Surface gravity acceleration/(m·s^{-2})		3.71
Mass/g		6.418×1026
Density/(g·cm^{-3})		3.933
Rotation angular velocity of Mars/(rad·s^{-1})		$7.088217\,66 \times 10^{-5}$

2.4.2 Mars Radiation Environment and Its Impact

The maximum solar radiation intensity of Mars orbit is 715 W/m^2, the minimum value is 491 W/m^2, and the average is about 591 W/m^2. It is about 43% of the solar radiation intensity near the Earth. The intensity of solar radiation near Mars varies by about ± 19% over a Martian year, while the variation of solar radiation intensity near the Earth is only ± 3.5% over an Earth year. The change in solar radiation intensity at the top of the Martian atmosphere over a Martian year is shown in Fig. 2.8.

The solar radiation environment has an impact on the thermal design, the optical sensors design and the solar cell circuit design of Mars probes.

For the absence of a global magnetic field, Mars has no radiation trapping zone. Therefore, charged particles in the solar wind can reach the surface of Mars directly. For a Mars probe launched from the Earth, the space charged particle environment that it faces during its flight includes electrons and protons trapped by the Earth radiation belt, galactic cosmic ray, solar cosmic ray (only available during the solar outbreak) and solar wind. During the mission, due to the changing spatial position of Mars probes, the space environment of charged particle radiation it encounters varies to some extent.

(1) Electrons and protons trapped by the Earth radiation belt: The Mars probe will traverse the inner and outer radiation belts of the Earth before crossing the top of the magnetosphere of the Earth (about 65,000 km in height) for about 5 h. During this period, it will encounter the electrons and protons trapped by the Earth radiation belt.

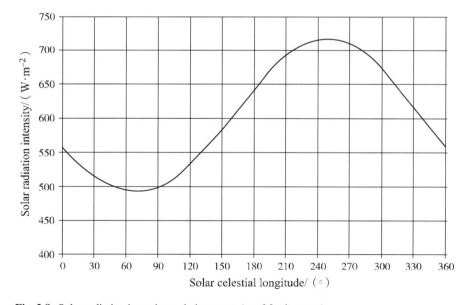

Fig. 2.8 Solar radiation intensity variation curve (one Martian year)

2.4 Space Environment of Mars

(2) Galactic cosmic ray: The spatial distribution is uneven and increases with the increase of heliocentric distance. However, the increase within the range of 1–1.5 AU does not exceed 3%. In addition, it may also change over time.

(3) Solar cosmic ray: It depends mainly on the solar activity and should be specifically analyzed when selecting the launch windows. The intensity of charged particles in the solar cosmic rays weakens with the increase of heliocentric distance.

(4) Solar wind and solar proton events mainly depend on the periodicity of solar activities. The extent of the impact of such events decreases with the increase of heliocentric distance.

2.4.3 Impact of Mars Atmospheric Environment

The Mars atmosphere is made of specific components in specific densities. It is regularly distributed along the horizontal and vertical directions. The atmosphere of Mars is thinner than that of the Earth. The surface atmospheric pressure of Mars is 5.6 mbar,[1] which is equivalent to the pressure at a height of 37 km in the atmosphere of the Earth. The surface acoustic velocity of Mars is 270 m/s, 20% lower than that of the Earth. The main component of Mars atmosphere is CO_2, and there are also other trace components such as N_2. The average relative molecular mass is 43.34. The atmospheric composition of Mars is shown in Table 2.13.

Table 2.13 Atmospheric composition of Mars

Composition	Content (percentage in volume)	Composition	Content (percentage in volume)
CO_2	0.9532	H_2O	0.0003
N_2	0.027	NO	0.00013
Ar	0.016	Ne	2.5 ppm
O_2	0.0013	Kr	0.3 ppm
CO	0.0007	Others (Xe, O_3, CH_4, etc.)	0.0013672

The main impacts of Mars atmosphere environment on Mars probes include low-pressure discharge, parachute system deceleration in the process of Mars lander entering the atmosphere, windward ignition of the braking engine, etc.

2.4.4 Impact of Mars Dust Environment

The surface of Mars is covered with thick Mars dust, and there is also dense dust in the atmosphere of Mars. Such dust is the product of the weathering of Martian

---

[1] 1 bar = 10^5 Pa.

rocks. Such feature affects landing safety and the normal movement of the rover on the surface of Mars.

In areas with thick dust on the surface, the mechanical equipment of Mars rover may be torn. The working performance of the optical instrument and the output power of the solar array may also be affected. Based on the data obtained from the US Mars Pathfinder mission, Mars dust deposition will affect the power generation of solar arrays. The decay rate per day for the first 30 days is 0.3%, and the decay rate thereafter is approximately 0.1% per day.

The thin, dry atmosphere and the wind on Mars may lift dust from the surface and produce dust storms. Mars dust storms include both global dust storms and local dust storms. According to researches and analyses, the dust storms originate from the southern hemisphere of Mars in summer and then extend to the entire surface.

The dust of Mars is charged, which absorbs the green light. The accumulation of the dust may affect the design of the solar array cell, mechanism and electrostatic protection of the rover. For Mars landers and rovers, in the design of the power distribution system, the influence of Mars dust should be considered to provide dust removal abilities or certain margins. For the charging effect of Mars dust, the electrostatic protection design should be considered in the design process. Moreover, dustproof design should be taken into account for related mechanism of Mars probes.

2.4.5 Landforms on the Surface of Mars

The surface landform of Mars is an asymmetrical structure which is low in the north and high in the south. The surface is divided by a great circle that is tilted at 30° to the equator, and the difference between the surface structure of the southern hemisphere and northern hemisphere is significant. As for the geologic history of Mars, the southern hemisphere is quite ancient, and the surface is rugged and craters are densely spread over the surface; the northern hemisphere is featured with large volcanic lava plains, dotted some dead volcanoes.

The topography and geological structure of Mars mainly include four categories, namely ancient unit, volcano unit, transformed unit (canyon and riverbed) and polar region unit. The scale of Martian plateaus and shield volcanoes is larger than that of the Earth, and the maximum elevation difference of the Martian surface is also greater than that of the Earth surface.

The ancient unit mainly refers to the heavily cratered regions in the highland of the southern hemisphere, including densely spread meteorite craters, impact basin and related mountain ring, which account for half of Martian surface and possibly represent the most ancient Martian crust. There are many large craters in the highland of the southern hemisphere of Mars, which indicates that the geological structure there is ancient and possibly formed 3800 million years ago.

The landform of the polar region is different from other regions. The top layer of the polar region surface is the polar cap that is composed of water ice, carbon ice and dust, whose size increases or decreases with the change of Martian seasons.

2.4 Space Environment of Mars

The landform of Mars is more complex than that of the lunar and Mercury, and is also quite different from the landform of the Earth. The impacts thereof should be considered in Mars lander and rover design in terms of landing site selection, landing buffering, mobile systems and communication systems.

2.5 Space Environment of Jupiter

2.5.1 Overview

Jupiter is located in the orbit that is 5.2 AU (AU: astronomical unit, the average distance between the Sun and the Earth) to the Sun. It is the largest among the planets in the solar system. But the average density is the lowest. The mass of Jupiter is 318 times that of the Earth, and the average density is only 1.33 g/cm^3. The volume of Jupiter is 1316 times the Earth and is therefore called the "giant planet." It has a fast rotation speed. The period of rotation of the equatorial part is 9 h 50 m 30 s, and the period of rotation of the two polar regions is slightly longer.

To compare Jupiter with the Earth, the basics of Jupiter and Earth are listed in Table 2.14.

Table 2.14 Comparison of basic information between Jupiter and Earth

Celestial body	Radius/km	Average heliocentric distance/AU	Period of rotation/h	Surface gravitational acceleration/(m·s^2)	Average density/ (g·cm^{-3})	Magnetic field dipole axis inclination/(°)
Earth	6712	1.0	24	9.8	5.5	11.3
Jupiter	71,400	5.2	10	25.6	1.3	9.6

2.5.2 Jupiter's Strong Magnetic Field Environment

As the magnetic field of Jupiter is much stronger than that of the Earth, and the dynamic pressure of the solar wind in Jupiter orbit is about 30 times smaller than that near the Earth orbit, the magnetosphere of the Jupiter is about 100 times larger than that of the Earth. The top position of the magnetospheric surface of Jupiter is about 50 R_j ($R_j = 71,400$ km, the radius of Jupiter), and its magnetic tail can reach 200 R_j. In addition to the influence of the interplanetary magnetic field, the Jupiter magnetosphere is also affected by the rapid rotation of Jupiter. In addition, the scale of the Jupiter magnetosphere far exceeds the orbits of Io, Europa, Ganymede and Callisto. These satellites also have significant impact on the Jupiter magnetosphere. In particular, Io, whose orbit is about 5.9 R_j from Jupiter, loses its atmospheric matter (mainly sulfur dioxide) at a rate of about 1 t/s. The sulfur dioxide is ionized to form sulfur and oxygen ions and then trapped by the magnetic field of Jupiter. Due to Jupiter's powerful magnetic field and rapid rotation, the plasma released from Io is co-rotating with Jupiter at a rapid rate (up to 54 km/s), forming a cold plasma torus around the Io orbit.

The magnetosphere of Jupiter is the largest and strongest among the solar system planets and is the second largest continuous structure in the solar system after the Sun. Compared with the magnetosphere of the Earth, the magnetosphere of Jupiter is wider and flatter, and is stronger by several orders of magnitude. Its magnetic moment is 18,000 times that of the Earth, and its surface magnetic field strength is about 14 times that of the surface of the Earth. The powerful current in the magnetosphere creates permanent aurora and intensely variable radio radiation near the polar axis of Jupiter, which means that Jupiter may be regarded as a weak radio pulsar. The aurora of Jupiter covers almost all of the electromagnetic spectrum, including infrared, visible, ultraviolet and soft X-rays.

2.5.3 *Strong Radiation Environment of Jupiter*

Due to its strong magnetic fields, Jupiter has a Van Allen radiation belt similar to that of the Earth. The flux of high-energy protons (>100 MeV) in Jupiter radiation belt is relatively small, but the low-energy proton flux is about ten times that of the Earth radiation belt. The low-energy electron flux of Jupiter radiation belt is basically equivalent to the Earth's radiation belt, but the high-energy electron flux is 2–3 orders of magnitude higher than that of the Earth; especially, there are plenty of relativistic electrons with energies exceeding 10 MeV in the Jupiter radiation belt.

The interaction between the high-energy particles of the Jupiter radiation belt and the surface of the largest satellite of Jupiter, Galilean, has a significant effect on the physical and chemical properties of probes. Similarly, these particles also affect the thin particle motion within the Jupiter ring. The strong radiation belt of Jupiter has an important influence on the probe, including total ionizing dose, single-event effect, displacement damage, surface charge and discharge, and internal charging. Therefore, special radiation-proof designs must be considered for Jupiter probes.

2.5.4 *Jupiter Plasma Environment*

Jupiter rotates very fast. Since there is an 10° angle between the spin axis and the dipole axis of the Jupiter, as a result, the ionized neutral atoms released from Io at 5.9 R_j are restrained by the magnetic field of Jupiter and co-rotates with Jupiter, forming a gigantic cold plasma torus. Due to the dip angle of the Jupiter dipole axis and the rapid rotation of Jupiter, the plasma ring around the Io orbit oscillates up and down, and the plasma parameters at the same spatial position fluctuate with the period of rotation of Jupiter (about 10 h).

The plasma environment of the Jupiter can be divided into three parts: cold plasma near the Io ring (with energy of $0 < E < 1$ keV), medium energy plasma (1 keV $< E < 60$ keV) and radiation belt particles ($E > 60$ keV). The cold plasma has a relatively high density of up to 2000 cm^{-3}, and the main components are hydrogen, oxygen, sulfur and sodium ions. The density of the electrons (energy of about 1 keV) and protons (energy of about 30 keV) in the medium energy plasma decreases exponentially with increase of the distance from Jupiter, which is about 5 cm^{-3} at less than 10 R_j, and further falls to 10^{-3} cm^{-3} at over 40 R_j. The co-rotation speed of the plasma particles with Jupiter is about 45 km/s at 4 R_j and reaches 200 km/s at 20 R_j.

2.5.5 Jupiter Atmosphere

The atmospheric structure of Jupiter is relatively simple. The troposphere is in convective equilibrium. There is a constant vertical temperature gradient, the tropopause temperature drops to a minimum of about 100 K, and then rises to 160 K above the tropopause. The temperature remains constant in the region with the greatest re-entry aerodynamic deceleration and heat flux density. At a height of about 300 km above the constant temperature zone, the atmospheric temperature starts to rise again.

As indicated by the measured data of the Galileo probe, the atmospheric composition of Jupiter is H2 (86%), He (13.6%), CH4 (0.18%) and NH3 (0.07%). The atmospheric density component is applicable for the troposphere and most of the stratosphere of the Jupiter atmosphere.

2.6 Space Environment of Venus

2.6.1 Overview

Venus is the closest planet to the Earth in the solar system. It has a radius of about 6052 km, a volume of about 88% of the Earth and a mass of four-fifths of the Earth. Like the Earth, the surface age of the Venus is also quite young. So, Venus is often referred to as the "sister star" of the Earth. The rotation of Venus is clockwise when viewed from its north pole, which means that the Sun is rising in the west and setting in the east. Another feature of Venus is that its period of rotation is longer than its period of revolution around the Sun.

Venus has quite a harsh space environment. The temperature is extremely high, the atmosphere is dense, and there are cloud layers consisting of SO_2 and liquid sulfuric acid in the atmosphere. For probes orbiting Venus, in addition to considering the radiation environment such as the galactic cosmic rays and the solar cosmic rays, the influences of the dense atmosphere and the acidic cloud layer on the probe itself and the observation performance of the payload should be considered in design. Basic physical characteristics of Venus are shown in Table 2.15 [7].

Table 2.15 Basic physical characteristics of the Venus

Parameters	Value
Revolution radius/km	1.08×10^8
Period of rotation/d	243.02
Period of revolution/d	224.701
Spin axis inclination/(°)	1.54
Surface gravitational acceleration/(m s^{-2})	8.78
Equatorial diameter/km	12,103.6
Escape speed/(km s^{-1})	10.4
Surface temperature/K	Minimum 738, maximum 758, average 748
Atmospheric pressure/kPa	9231.9
Atmospheric CO_2 content/%	96.5

2.6.2 Magnetic Field of Venus

The intrinsic magnetic field of Venus is quite weak with the dipole magnetic moment only about 10^{-4} of that of the Earth. There is no high-energy particle trapping area around Venus (i.e., the radiation belt). Therefore, the magnetic field of the Venus is not sufficient to block the solar wind at a far distance. The solar wind, ultraviolet rays and X-rays may penetrate deep into the atmosphere, ionizing part of the atmosphere and forming a very thin Venus ionosphere within the height range of 100–420 km.

The ionosphere of the Venus is very close to the surface of Venus, and the electron density of the ionosphere is small at night. At a height of 100–160 km, the electron density of the Venus ionosphere is 10^5–10^6 cm^{-3}, and the electron density above the height of 160 km is on the order of 10^5 cm^{-3} (equivalent to the F layer of the Earth ionosphere). When the height is increased to above 450 km, the Venus ionosphere basically disappeared. Venus also has a bow shock, which is caused by the blocking of solar wind by the Venus ionosphere. The electron density at the 300 km orbit altitude of a Venus probe is about 10^5 cm^{-3}, which is equivalent to the electron density at the low orbit of 600 km altitude of the Earth.

2.6.3 Venus Atmosphere

The atmospheric density near Venus surface is quite high with an average value of 63.4 kg/m^3, resulting in a surface pressure of over 9×10^6 Pa (about 90 times the atmospheric pressure of the Earth surface). The main components of the surface atmosphere of the Venus are carbon dioxide (with a volume mixing ratio of 96.5%) and nitrogen (3.5%). The average surface temperature of the Venus is up to 464 °C, but the surface temperature of Venus has little variation. The temperature difference between midday and midnight, between the near-equator zone and the polar zone, is only 5–15 °C.

With increasing height, the number densities of atomic oxygen, carbon monoxide, He and atomic H of the atmospheric composition of Venus increase gradually. At the height exceeding 300 km, the atmospheric composition of Venus is mainly He and atomic H as well as a small amount of nitrogen.

Compared to the Earth atmosphere, the significant difference of the Venus atmosphere is that there is a thick layer of sulfuric acid (H_2SO_4) droplets within the range of 50–70 km. According to current observation results, at least the upper portion of the cloud layer is composed of droplets of sulfuric acid on the order of magnitude of micrometers (over 75%). In addition, there is a significant violent movement in the Venus atmosphere. Within the height range of 0–70 km, the Venus atmosphere is mainly in belt-shaped rotation against the direction of rotation (in the direction of latitude lines). At a height of about 70 km at the top of the cloud layer, the speed can reach 100 m/s; and within a higher altitude range (100–180 km), the atmospheric movement of Venus is gradually replaced by the atmospheric circulation from noon to midnight.

With the increase of height, the atmospheric density of Venus declines much faster than that of the Earth atmosphere. At about 140 km, the atmospheric density of Venus is comparable to the Earth atmospheric density at the same height, and if the altitude increases to 300 km, the atmospheric density of Venus falls to the order of 10^{-14}–10^{-15} kg/m^3, which is equivalent to the atmospheric density of the Earth at a height of 700–800 km. The drags to probes are almost negligible.

There are significant atmospheric circulations at different heights in the Venus atmosphere, which mainly are the east–west winds along the rotation direction of the Venus and the south–north winds from the equator to the higher latitudes. The east–west winds have relatively high wind speed (up to the order of 100 m/s at 80 km above the Venus equator) and vary significantly with latitude and height.

2.6.4 Venus Surface Topography

As early as the early 1960s, people started to detect the surface of Venus from the Earth with radar. Topographical–geological (radar) images of Venus were drawn accordingly. The surface of the Venus has been detected by the radars on probes, and the resolution has gradually increased to 120–300 m since the late 1970s.

Radar images of Venus surface show various features such as plateaus, ridges, troughs, volcanoes and craters. Except for a few plateaus, the elevation difference of the Venus surface is quite small. The elevation difference of 60% of its surface is less than 500 m, and only 2% is over 2 km. The topography difference of the south and northern hemispheres of Venus are significant. The northern hemisphere is mainly plateau with many mountains, and the southern hemisphere is mainly relatively flat plain.

2.7 Other Interplanetary Space Environments

2.7.1 Interplanetary Environment

Interplanetary space refers to the space around the Sun and planets in the solar system, which is dominated by interplanetary media and extends outward to the edge of the solar system. The interplanetary space is almost pure vacuum, but not completely. The space is filled with sparse cosmic rays, solar wind plasma, dust, micrometeoroids, etc.

Cosmic ray particles in interplanetary space include the galactic cosmic rays and the solar cosmic rays, and their characteristics are similar to those in near-Earth space. For probes in the interplanetary environment, the space environment elements that should be concerned include the high-energy particle radiation of the galactic cosmic rays and the solar cosmic rays, the low-temperature environment, and the vacuum environment. The corresponding effects include single-event effect of electronic components, material outgassing and mass loss as well as cold welding and dry friction effects of moving parts. In addition, for probes working at a distance from the Sun that is no nearer than the Earth, the solar radiation declines because of the long distance from the Sun. The influence of lighting conditions should be put into consideration in designing the solar cells. The solar cells should be designed bearing in mind.

2.7.2 Asteroid Environment

Asteroids refer to a type of celestial bodies in the solar system with similar movement of the planets around the Sun. However, asteroids are much smaller in volume and mass. Most of the asteroids in the solar system fly between Mars and Jupiter, forming the Asteroid Belt. There are also asteroids distributed farther from the Sun than Neptune's orbit. This zone is called the Kuiper Belt.

As proved by the data obtained from spectral analysis, the surface compositions of asteroids are quite different. According to characteristics of their spectrum, asteroids can be divided into several categories:

C-type asteroids: These asteroids account for 75% of all asteroids and therefore have the greatest number among the asteroids. The surface of C-type asteroids contains carbon, and the albedo is quite low (only about 0.05). The composition of C-type asteroids is generally considered to be the same as that of carbonaceous chondrite (a type of aerolite). Generally, C-type asteroids are mostly distributed in the outer layer of the Asteroid Belt.

S-type asteroids: These asteroids account for 17% of all asteroids and have the second largest number among the asteroids. S-type asteroids are generally distributed in the inner layer of the Asteroid Belt. The albedo of S-type asteroids is relatively high, ranging from 0.15 to 0.25. Their composition is similar to that of ordinary chondrites. Such meteorites are generally composed of silicide.

M-type asteroids: Most of the remaining asteroids fall into this category. These asteroids may formerly be the metal cores of relatively large asteroids. Their albedo is similar to that of S-Asteroids. Their composition may be similar to nickel–iron meteorite.

In the past, people believed that an asteroid was a single piece of rock. However, the density of asteroids is lower than that of rock, and the huge craters on their surface indicate that the structure of relatively large asteroids is quite loose. They are more like giant rubble piles combined by gravity. Such a loose object will not break under a large impact and can absorb the energy of the impact. But an integral single object may break into pieces by a shock wave at a large impact. In addition, the rotation of large asteroids is very slow. If they rotate at a high speed, they may be disintegrated by centrifugal force. Astronomers generally believe that asteroids larger than 200 m are mainly composed of such rubble piles, and some of those relatively small fragments become satellites of some asteroids.

2.7.3 Comet Environment

Comets refer to a type of celestial bodies that entered the solar system from the galaxy, whose brightness and shape vary with change of the distance to the Sun. Comets are mainly composed of water, ammonia, methane, cyanide, nitrogen and carbon dioxide. Generally, a comet consists of two parts: the cometary head and tail. The head further consists of the cometary nucleus and the cometary coma. When the material of the comet evaporates, it forms a vague coma around the ice core and a tail of a stream of thin material flow.

The nucleus is the most central, essential and most important part of a comet. It is generally considered to be solid and consists of stones, iron, dust, ammonia, methane and ice. The diameter of the nucleus is quite small, ranging from a few kilometers to a dozen of kilometers. The diameter of the smallest nucleus is only a few hundred meters.

The cometary coma is a star-shaped mist made of gas and dust around the nucleus. The radius of the coma is up to several hundred thousand kilometers, and the average density is much smaller than that of the Earth atmosphere.

The cometary tail begins to appear as the comet approaches the Sun to a distance of about 300 million kilometers, and it gradually grows larger and longer. When comets are flying away from the Sun after the perihelion passage, the tail gradually becomes smaller until it disappears. With the impact of solar wind, the tail always points away from the Sun. Comets have no fixed volume. It is small when it is far away from the Sun. When it is close to the Sun, the coma becomes increasingly bigger and the tail becomes increasingly longer, and the volume becomes huge.

2.8 Outlook

Deep space probes face complicated and volatile space environments in missions. Space environment affects deep space exploration activities and bears significant impacts on the engineering design and on-orbit reliability of deep space probes. Space environment protection technology is crucial for deep space probes to complete exploration tasks in the presence of uncertainties in target space environment. The development of deep space exploration missions will drive the development of space environment protection technology. Meanwhile, the science objectives of deep space exploration activities themselves also include exploration of the space environment. With the increase of the distance travelled by probes, more and more celestial bodies explored, and the content of detection becoming richer in deep space exploration, our understandings on deep space environment will also be more comprehensive and more accurate.

References and Related Reading

1. Tan WZ, Hu JG (2009) Spacecraft system engineering. Science and Technology of China Press, Beijing (Chinese vision)
2. Chu GB, Ma SJ (2002) Introduction to aerospace technology. China Astronautic Publishing House, Beijing (Chinese vision)
3. Peng CR (2011) System design of spacecraft. Science and Technology of China Press, Beijing (Chinese vision)
4. Kivelson MG et al (2001) Introduction to space physics. Translated by Cao JB et al. Science Press, Beijing (Chinese vision)
5. Marvin DC, Corney DJ (1991) Solar proton events of 1989: effects on spacecraft solar arrays. J Spacecraft Rockets 28(6):11–12
6. Hou JW et al (2016) Deep space exploration—exploration of Mars. National Defense Industry Press, Beijing (Chinese vision)
7. Hou JW et al (2015) Deep space exploration—exploration of Venus. National Defense Industry Press, Beijing (Chinese vision)

Chapter 3
System Design Technology

3.1 Introduction

Deep space probe is a complex system consisting of multiple subsystems (modules) with different functions and performances. Generally, these subsystems (modules) are completed under the cooperation of different teams of professionals. To ensure that the subsystems are coordinated and unified and the probe design satisfies requirements of explorations, the system design of the probe is required; namely, the design should be broken down level by level, top to bottom, and then iterated bottom to top for multiple times using systems engineering approaches. In addition, specific verifications should be conducted accordingly to gradually form a coordinated, well-matched design scheme that is optimal at system level and meets customers' demands [1, 2].

System design refers to top-level design and technical analysis for the probe system. It is the top-level, systematic, and comprehensive design of the probe through all stages of its development. It can be seen from Chap. 2 that due to the unknown environment of exploration targets and diversified exploration approaches, etc., spacecraft systems in deep space exploration missions are more complicated than Earth-orbiting spacecraft, demanding more emerging technologies, and involving more difficulties. Thus, the overall system design featuring systems matching, reliability and robustness should be highlighted [3].

The system design involves multiple disciplines and specialties. For example, the orbital design to solve the problem of how to reach the target celestial body is a top-level design for deep space probes, which is the premise or basis for the system scheme design of the probe and will be explained in detail in Chap. 4. The ground special test verification technology for deep space probes is also within the scope of system design and will be specifically explained in Chap. 14. This chapter will focus on the highlighted elements, mission analysis, etc., in the system design of deep space probes with the demands and examples of different exploration forms.

3.2 Overview of the System Design of Deep Space Probes

3.2.1 Characteristics of Deep Space Probe Missions

Compared with spacecraft orbiting the Earth, deep space probes have distinct technical features, which mainly involve the characteristics described as follows.

3.2.1.1 Diversified Mission Forms

Extraterrestrial celestial bodies include terrestrial planets with solid surfaces, Jovian planets with gaseous surfaces, celestial bodies with a dense atmosphere such as Venus, and those with almost no atmosphere such as the Moon. Therefore, to explore such celestial bodies with different characteristics, deep space probes must be designed in accordance with a variety of mission forms. Typical mission forms include orbiting, landing, returning, rovering, attachment and flyby.

Among them, "orbiting" means that the probe flies around a certain celestial body and conducts scientific remote-sensing exploration at a certain orbital height; besides, "orbiting" may also mean traveling around a certain point in space, such as the Lagrangian point. The gravitational balance point lies between two celestial bodies. There are two types of "landing," namely hard landing and soft landing. In hard landing, the probe generally impacts the surface of the target celestial body at a relatively fast speed to obtain close-range measurement data within a short time; while in soft landing, the probe lands on the surface of the target celestial body with stable attitude and less impact, then scientific payload instruments will be turned on for in-situ scientific explorations. "Returning" refers to the process that, after the probe has collected a certain amount of samples from the surface of the celestial body, it carries the samples back to the Earth for in-depth scientific researches. "Rovering" refers to the mobile rovering exploration of the probe on the surface of the celestial body; generally, it is realized by the landing of the lander to the surface of the celestial body and releasing robots. In terms of "attachment," for small celestial bodies with weak gravitation (asteroids or comets), the probe can just be attached to the surface of the celestial body to carry out exploration by certain technical means. "Flyby" means that the probe flies through the space near a certain celestial body and acquires the exploration data in such process. For different exploration mission forms, different mission constraints and requirements should be considered in probe system design of probe systems.

3.2.1.2 Great Differences in Targets

From the perspective of celestial bodies in the solar system only, the distances of various targets such as planets, planetary satellites, comets, and asteroids from the Sun range from about 0.5 AU to tens of AU. As a result, lighting conditions vary drastically. The rotation and revolution characteristics, gravitational field, magnetic fields, atmospheric environment, surface morphology and material composition of the planets are different. Such differences pose huge challenges to the design of deep space probes.

3.2.1.3 Large Environmental Uncertainty

Although humans have completed over 200 deep space exploration missions, our knowledge of the vast universe is still very limited. Over 100 exploration missions have been carried out on the Moon, the nearest celestial body to the Earth, but only the knowledge of part of the Moon has been achieved. In addition, little is known about Mars, Jupiter, and small celestial bodies. Specifically, at the vicinity of a celestial body to be faced by a probe, there are insufficient supports with measured data, especially for the surface environment of the celestial body such as softness of the surface soil, degree of fluctuation of terrains and distribution of the magnetic field. Therefore, in the system design of the probe, uncertainty of the environment must be taken into account.

3.2.1.4 Long Mission Duration

Deep space probes are supposed to fly to distant target objects, which often require relatively long transfer time. In the long interplanetary flight, factors such as energy supply, temperature maintenance, autonomous management and space radiation environment should be considered, which pose new challenges to achieving long-life and high reliability of the probes.

In the system design of deep space probes, it is necessary to well consider the characteristics of the mission, identify its constraints and boundary conditions, work hard on mission analysis and top-level design, and strive to optimize system design.

The disciplines involved in the system design of deep space probes include systems engineering theories and methods, astrodynamics, engineering cybernetics, propulsion technology, heat transfer, structural mechanics, computers, mechanical engineering, electrical engineering, microwave communications, environmental engineering, etc. The implementation of deep space exploration missions involves not only the spacecraft itself but also other large systems such as launch vehicles, ground TT&C, launching sites and ground application systems.

The designing of a successful deep space probe system should consider the following basic principles:

(1) Meet the requirements of mission objectives. Generally, requirements for deep space exploration missions include two aspects: scientific exploration and driving technical development. Demands of scientific exploration should be satisfied to the best on the basis of a comprehensive assessment of technological feasibility.

(2) Highlight system design optimization. Deep space probe consists of several branch systems, subsystems, and hundreds of stand-alone equipment. The final design outcome should be a coordinated system with optimal comprehensive performance. The overall capabilities of the system should not be compromised just for optimization of parts instead of the whole.
(3) Adapt to special environments. Due to the environmental complexity and uncertainty of deep space exploration targets, in the design of deep space probes, sometimes specific environmental constraints are not well-defined and the probe should be designed with stronger environmental adaptability to reduce risks of missions.
(4) Ensure coordination of large systems. Deep space probes rely on launch sites and launch vehicles for launch and orbit insertion. It is necessary to rely on the ground TT&C system to enable on-orbit track and control. And the ground application system is required to receive and process scientific exploration data. These large systems are linked to the probes via the interfaces of mechanical, electrical, radio frequency, etc. Coordination of such interfaces must be guaranteed to complete the mission successfully.
(5) Ensure reliability and safety. Deep space probes have long flight time, complex environments, and often high mission costs, posing higher demands on reliability and safety of the probes.
(6) Focus on features such as lightweight, small size, high integration, and high degree of autonomy of the design. For every gram mass of deep space probes, hundreds or even thousands of times of costs are required for the launch vehicle and the probe itself to ensure that it can be carried to the mission orbit. Therefore, high integration, lightweight, small size are always the key objectives of the system design of deep space probes. In the long interplanetary flight process, the tracking and measurement accuracy and chronergy of the ground TT&C system often fail to meet the demands of the mission. The probe itself must be equipped with strong capabilities in autonomous navigation, autonomous control, autonomous management, and autonomous fault diagnosis and recovery abilities.

3.2.2 System Mission Analysis

In system design of deep space probes, importance should be attached first to mission analysis. Mission analysis aims to obtain the basic requirements for the function and performance of the probe design from the mission requirements under various constraints and through the analysis and demonstration of key technical approaches.

In the mission analysis of deep space probe systems, the common top-level elements mainly include the following:

1. **Demand analysis of science objective**

For any deep space exploration mission, scientific demand is one of its most important goals. The scientific goal of human beings to carry out deep space exploration activities is to explore the universe, study the origin and evolution of the solar system and the universe, seek extraterrestrial resources and extraterrestrial life and thus better serve the sustainable development of human civilization.

The selection of the form of exploration mission is basically determined by the requirements of the science objective of the deep space probe. For example, to study the relatively large-scale topography, surface spectral characteristics and surrounding environment of the celestial body, the form of orbiting exploration is generally adopted; to study the fine local characteristics and surface physicochemical properties, landing or rovering exploration is generally required; to study the internal structures and characteristics of the celestial body, a probe with penetrator to penetrate into the celestial body may be required.

A wide range of scientific explorations may be conducted under deep space exploration activities, which involve multiple disciplines such as space astronomy, astrophysics, astrochemistry and space biology. In the planning of scientific goals of a mission, efforts should be made to focus on the originality and feasibility of the scientific plan and to obtain more original scientific results.

The scientific goals, once defined and clarified, may be broken down into specific requirements on payload. Based on the demands of the payload, the requirements for the system design of the probe may be concluded, which includes the weight of the payload, power consumption, orientation requirements of the probe, field-of-view requirements, data storage and transmission requirements, special temperature environmental requirements, installation or release requirements, etc. After the scientific demands are basically defined and clarified, the system design of the probe may be continuously optimized along with the evolving demands.

2. **Analysis of large system design constraints**

In designing deep space probes, considerations should be taken on the constraints of many aspects such as large system interface, and efforts should be made mainly on identification, analysis and coordination of the constraints of engineering realization. The system scheme design of the probe should be carried out under such constraints, and it should be guaranteed that the probe developed could meet the requirements of the customers on the developers and is adapted to and coordinated with such constraints. Specifically, it mainly includes:

(1) Scale and orbit selection of the probe is directly determined by the launch capability and orbit insertion accuracy of the launch vehicles for different orbits.
(2) Design of the orbit of the probe, etc., is also restrained by geographical latitude of the launch site, launch azimuth and the environment of the launch area.
(3) Orbit determination and orbit prediction accuracy of ground TT&C system bear a direct impact on the flight orbit planning, orbit control, and autonomous management strategy of probes.

(4) Transmitting and receiving capabilities of the ground TT&C system and the ground application station constrain the design of the TT&C and communication channel to the Earth.
(5) Location of the recovery site of the recovery system determines the orbit and control design of the Earth return mission.

The design of probes requires continuous iteration with large system constraints. In addition, the requirements of the engineering system administration in large for the launch time and cost also constitute the constraints that must be considered in the system design of the probe. It bears a direct impact on the design of the flight mission, the feasibility of selected new technologies, etc.

3. **Analysis of characteristics of target celestial body**

The targets and objects of deep space exploration missions vary greatly, which may be the Sun, major planets and their satellites, or small celestial bodies in various shapes. For these celestial bodies, multiple aspects such as the gravitational field, atmosphere, and electromagnetic radiation characteristics near the planets are quite different from each other. In addition, deep space exploration is highly scientific and exploratory. With payloads of various detection mechanisms and in different modes, the feature of diversity in deep space exploration missions is greatly significant.

Therefore, in the system design of the probe system, key analysis should be made first to understand the characteristics of the target celestial body, such as the rotation/revolution period, gravity, shape and size, and basic structure. Analysis should also be made on the environmental characteristics of the target celestial body such as the presence/absence of atmosphere, atmospheric composition, radiation environment, temperature variation, and illumination variation. Such characteristic parameters and models are the basis for mission design and implementation. For the environmental characteristics and influencing factors of different celestial exploration objects, please refer to Chap. 2 for details.

4. **Selection of key technology approaches**

The deep space probe is an organic whole consisting of multiple subsystems with different functions and performances. And the selection of key technical approaches plays a decisive role in the system design of the probe. In the system design phase, the key technical approaches should be demonstrated and analyzed to define and clarify the technical routes. Different key technologies may be available for different forms of missions. Generally, flight mission design, propulsion system selection, navigation and control methods, satellite–Earth communication link planning, payload operation modes, etc., are all important parts to be considered.

3.2.3 Overall System Design Process

The system design process of deep space probes is a process of continuous trade-off and iteration. Generally, the design can be divided into phases of conceptual design, feasible scheme demonstration, prototype design, and flight model design. Each phase is the optimization and refinement of the design of the previous phase.

The objective of the feasibility demonstration phase is to determine the capabilities and technical specifications of the probe platform and payload through analysis on scientific objectives and engineering objectives, carry out comparisons of multiple alternative schemes and finally establish a preliminary feasible overall scheme that can accomplish the mission to support the comprehensive project authorization demonstration.

The overall scheme design phase aims to complete the system scheme design of the probe and provide input for the prototype development through the decomposition of the overall technical specifications, the development of related specifications, orbit design, the structure and layout development, the design of system assembly and the design of subsystem scheme. This phase requires the personnel from the system design and subsystems design divisions to work collaboratively and cooperatively, consider the system design and analysis elements comprehensively and form a complete and optimized system scheme through iterative work approaches.

The work objectives of the prototype phase are to carry out the detailed design of the overall system, subsystems, devices, as well as the design of data interfaces between the subsystems and devices to establish the product baselines of the probe system; complete the detailed design of the prototype of the probe and determine the prototype spacecraft model and product status in the prototype phase by system engineers, which is the basis for prototype product production and integration test.

The work objectives of the flight model phase are to complete the comprehensive flight mode design, detailed design of the flight models of each subsystem and development of the flight models of devices. Finally, system-level assembly, tests and trials shall be completed.

In the following, the main aspects, elements, and key technologies of the system design of probes are highlighted for exploration missions in different forms, with reference to successful missions and probe developments in China.

3.3 System Design of Orbiting Exploration Missions

3.3.1 Mission Analysis

In the system design of the orbiting exploration missions, it is necessary to focus on how to reach the intended orbit of target celestial body and to meet scientific exploration demands. Our analysis includes the following elements:

1. **Selection of flight orbit schemes**

The selection of the orbiting exploration flight orbit scheme is an important part of the system design of the deep space exploration orbiting mission. The selection of flight orbit scheme should meet related requirements of the mission of the probe, and the reasonable orbital type and orbital parameters should be determined according to the mission. The design process is as follows.

(1) Summarize and organize the requirements of the probe mission, analyze the mission characteristics and initially determine the preliminary overall flight profile;
(2) For specified flight phases, conduct comparison studies of multiple schemes for the orbit types of each phase, evaluate the schemes from various aspects such as mission satisfaction, implementation difficulty, and usage in past missions and finally determine the orbit type of each flight phase;
(3) Select mission orbit parameters suitable for the orbiting mission of the probe according to the characteristics and capabilities of the orbiting exploration payloads;
(4) Select transfer orbit parameters suitable for the mission of the probe according to the characteristics of the transfer orbit;
(5) Select appropriate capture orbit parameters and carry out preliminary patching of the capture orbit, mission orbit, transfer orbit, and capture orbit provided that the constraints of TT&C, lighting, etc., are satisfied;
(6) Analyze the launch capability and carry out the preliminary patching of the launch trajectory and the transfer orbit;
(7) Trajectory patching work in each flight phase should be carried out driven by orbit design of orbiting missions so that the orbits of different flight phases can form an integrated orbit satisfying the mission requirements.

2. **TT&C and data transmission technology approaches**

In the design and implementation of deep space exploration missions, it is of great importance to guarantee necessary TT&C and communication capabilities. After the launch of the deep space probe, the TT&C and communication system is the only link between the Earth and the probe. The main characteristics of deep space communication include long communication distance, great time delay, large path loss, huge loss of transmission power, etc.

In view of such characteristics, in the system design process of deep space probes, it is necessary to emphasize the analysis of the requirements from the top level of the mission for the TT&C and data transmission of probes (e.g., communication between probes and the Earth, between the probes, etc.), the matching performance with the interfaces of the ground TT&C system and ground application system (e.g., communication framework, angle measurement system, distance measurement system, ground station G/T value, EIRP indicators, etc.), as well as the matching performance of the communication interfaces between probes. The main analysis processes are as follows: (1) propose preliminary function and performance requirements for TT&C and data transmission according to the orbiting exploration requirements (data volume, bit rate, etc.); (2) analyze and determine the technical approaches, including working frequency band, modulation mode, onboard implementation of the ranging system and angle measurement system, information rate

3.3 System Design of Orbiting Exploration Missions 53

and coding mode, etc.; (3) work on link planning and link budget; (4) conduct analysis on frequency compatibility; (5) propose antenna layout requirements and make antenna directivity requirements analysis.

3. **Control technology approaches**

Define and clarify the major tasks of the control system according to the preliminary flight orbit and flight profile; analyze and determine the flight attitude according to the orbital maneuver, TT&C, power, scientific detection, etc. Considering the characteristics of the exploration mission, detailed analysis on the requirements of the control tasks should be made, namely define and clarify the main operation mode and flight attitude of the control system, make preliminary selection on the control system sensors, controller and actuators, and make preliminary estimate on the major parameters such as the propellant mass and the maximum loading of the fuel tank.

4. **Propulsion technology approaches**

It is necessary to define and clarify the task requirements of the propulsion system according to the preliminary analysis of the exploration mission and the flight orbit, determine the implementation approaches of the propulsion subsystems with reference to domestic and international propulsion system analysis as well as R&D progress of the propulsion products, and determine the preliminary configuration of the thrusters according to the preliminary scheme of the control system and the control accuracy requirements, preliminary analysis on mass property, etc.

5. **Power supply and distribution technology approaches**

It is required to select the power supply and distribution system and configuration scheme based on system mission analysis and power statistics analysis according to the design constraints such as the power demands of the orbiting exploration mission (peak power, average power, initial and end-of-life power requirements), orbital illumination conditions, probe attitude, lifetime and space radiation damage, etc.; determine the layout of the solar cell array, the selection of solar cells, the composition of the solar array and its size and weight accordingly; determine the battery pack selection and configuration, power regulation control scheme as well as the design of power converters, distributors, initiating explosive device managers and cable networks. As such, establish the subsystem scheme in line with the probe mission as well as reliability and safety requirements.

Elements above are the technical route analysis and multi-scheme comparison analysis highlighted in the final system scheme design. For more detailed design approaches, design elements, etc., refer to the corresponding subsequent chapters.

3.3.2 Decomposition of Technical Specifications

The system specifications allocation process is a process of gradual deepening, refining, and repeated iteration closely related to the development phase and the cognitive process of design schemes and products.

A system may include many technical specifications. On one hand, it refers to the system resource specifications that the system design attaches importance to, which includes system weight, system rigidity, power consumption, command telemetry resources, TT&C channel resources, and data volume resources. Taking weight specifications as an example, when decomposing and determining such specifications, efforts should be made to focus on aspects such as the launch capability, payload weight, weight of newly developed or inherited platforms, propellant budget for conventional attitude and orbit control, propellant budget for special orbital control events (celestial body capture, etc.) as well as the extent of using new processes, new materials, new equipment use and the design margin of the overall spacecraft weight. Major design considerations are as follows.

(1) Estimate the weight of the platform structure and the basic equipment of the platform according to the project objectives;
(2) Estimate the allocated weights of the payloads of the entire probe according to the scientific objectives;
(3) Make propellant estimation iteration for the conventional orbital control events and special orbital control events of the probe, obtain the preliminary estimation results of the mass of the entire probe and judge the rationality of the mass allocation preliminarily;
(4) Take full consideration of the balance mass demands in various configurations;
(5) Make propellant budget iteration of the conventional and special attitude and orbit control tasks according to the estimated dry mass of the probe, and then the estimated mass of the whole spacecraft is obtained;
(6) Consider the mass margin and finally work out the mass estimation and allocation of the entire probe according to the R&D experience and large system constraints.

On the other hand, key system specifications of GNC, propulsion, TT&C and power supply should be decomposed from top to bottom and determined, considering the mission objectives and the characteristics of the target object.

3.3.3 Flight Procedure Design

According to the requirements of the orbiting exploration mission, system engineers need to determine the various operation modes of the probe from the time of orbit insertion to the normal operation in the mission orbit, define and clarify the working state, function and performance requirements of the probe under different operation

3.3 System Design of Orbiting Exploration Missions

modes, which is the basis for flight program design, power balance calculation of the probe, etc. Tasks of flight procedure design are as follows.

(1) Determine the operation modes, working hours, transmission requirements, etc., of the various types of payloads for each flight phase;
(2) Determine the operation mode, switching timing and related constraints of each subsystem of the platform for each flight phase;
(3) Propose the typical operating modes of the orbiter as the basis for the thermal design, thermal verification, energy balance budget and flight procedure design of the entire probe in combination with the mission characteristics, orbit design and flight attitude, etc., of the orbiter;
(4) Analyze the operation mode of the probe under special astronomical phenomena (such as lunar eclipse), etc., based on the actual conditions of the mission.

The flight procedure design provides an important basis for the detailed design of the overall system and each subsystem of the probe in the scheme design phase. Based on the mission of the probe, it schedules all the on-orbit events to be experienced by the probe from pre-launch to the completion of the mission, which mainly include defining and clarifying the TT&C arcs and lighting conditions experienced by the probe, specifying the status of each subsystem before launch, establishing the working state after launch and orbit insertion and scheduling the execution timing, execution conditions, specific execution actions, success criterion, etc., of each on-orbit event. The design elements include: the state setting requirements and setting timing of each equipment before launch, the initial orbit insertion state setting requirements of the probe, the working sequence of establishing the probe state, the use conditions and requirements of the new product or new function of the probe platform, the lighting conditions of the probe, the TT&C arcs experienced by the probe (and whether the major on-orbit event can be completed in the arcs), the ground test coverage of the probe on-orbit tasks, the orbit control constraints and orbit prediction precision, etc.

3.3.4 Analysis of Key Technologies

Compared with conventional near-Earth spacecraft, orbiting probes in deep space missions are subject to technical challenges in terms of flight orbit design, guidance, navigation and control (GNC) technology, etc. The main challenges are as follows.

1. **Orbit design**

Orbit transfer and braking near the celestial body is a prerequisite for orbiting explorations. For example, in the orbital design of the Earth–Moon transfer, consideration must be given to the influence of the gravity of the Earth and the Moon, and it is necessary to solve the three-body problem of the Earth, the Moon and the satellite. After lunar orbit insertion (LOI), due to the complexity of the lunar gravitational field, the orbit evolutions are greatly different from those for Earth-orbiting spacecraft and intensive analysis is required. In addition, the orbit design is also restrained by a series of conditions such as the relative position of the Earth and the Moon, TT&C requirements, rocket launch conditions, propellant carrying capacity, lunar shadow distribution and lunar eclipse timing, etc. It is necessary to comprehensively consider various constraints and make selection and optimization accordingly [4].

2. **Guidance, navigation and control technology**

Deep space probes are expected to undergo multiple complicated orbit and attitude maneuvers throughout the entire flight, and the control tasks are quite complicated. Especially for the braking and capturing control near the target celestial body, the reliability requirements are demanding with strong real-time nature. During the orbiting exploration of the target celestial body, there may be many requirements and constraints on the attitude control of the probe; for example, the scientific detection task requires the probe to be oriented toward the target celestial body, the energy acquisition requires the solar cell array to be Sun-oriented, and the TT&C and communication requires the directional antenna to be Earth-oriented.

3.3.5 Design Validation

For the orbiting exploration missions, based on the decomposition of the system technical specifications and mission design scheme, test coverage analysis, etc., efforts should be made to carry out the necessary special test verification in a comprehensive and effective manner. Taking Chang'e 2 as an example, in response to the requirements of the mission and the new environment, some special tests were planned and completed during the R&D process, such as high-temperature verification of the solar wing, high-temperature verification of TT&C omnidirectional antenna, monitoring camera imaging of 490 N engine hot-firing test, ground verification of velocity–height ratio compensation in CCD stereo camera imaging [5].

3.4 System Design of Landing Exploration Missions

3.4.1 Mission Analysis

In view of the characteristics of landing exploration missions for extraterrestrial celestial bodies, the mission analysis elements highlighted for realizing "reliable landing" are as follows.

3.4 System Design of Landing Exploration Missions   57

1. **Landing area selection**

To complete the design of the safe landing on an extraterrestrial celestial body, the landing area must be clearly defined and clarified. Different landing areas have substantial impacts on the design of the orbit, communication link and configuration layout of the lander.

Basic principles of landing area selection: Under the premise of taking safe landing as the main objective, the selection of soft-landing area should not only provide the feasibility of engineering implementation but also satisfy the demands of scientific exploration.

First of all, engineering feasibility and reliable landing should be guaranteed. When considering the engineering feasibility of the landing area in the soft-landing mission, the main factors under consideration include the topographical and geomorphic conditions of the landing area, the orbit reachability, the TT&C conditions of the powered descent, the solar illumination conditions, the thermal control conditions and the lunar surface communication conditions. The impacts thereof on lander design are mainly in the following aspects.

(1) Landform: The lunar surface is covered with large and small craters and rocks, the lunar mare is relatively flat, and the highland is relatively rugged. The local topography of the landing site, including the slope, crater and rock morphology, bears an important impact on whether the lander can land safely on the lunar surface while maintaining a stable attitude.
(2) Orbital reachability: The lander can only land at a landing site whose lunar latitude is less than the lunar orbital inclination. Theoretically, a polar orbit may be used for landing at any location on the Moon provided that comprehensive consideration have been made on regression period, TT&C conditions, etc., for a specific landing site. Therefore, the orbital reachability should be considered in landing site determination.
(3) TT&C and communication conditions: As the rotation period of the Moon is basically the same as the period of revolution of the Moon around the Earth, the Moon always has one side facing the Earth and the other invisible from the Earth. If landing is made on the near side of the Moon, the communication with the ground stations can be guaranteed. In contrast, if landing is made on the far side of the Moon, a relay satellite may be needed to maintain the communication between the probe and the ground stations.

In addition to considering lunar surface communication, the communication conditions of the landing process should also be considered. During the landing process, which is also known as the powered descent, the lander will experience large variation in altitude and speed, and there will be many control operations of sensors and engines, so the powered descent is the most risky stage of the entire landing mission. Before the actual "powered" descent, a large amount of data and instructions should be uploaded to the lander, which are supposed be completed within the TT&C arc that is visible to the Earth. For the powered descent, from the perspective of safety, it is also desired that this process will be carried out within the TT&C arc, so that it is possible for the personnel on the Earth to take some operations in some cases to cope with possible on-orbit failures.

(4) Solar illumination conditions: There is sufficient sunlight during the daytime on the Moon, so solar cells may be used to satisfy the energy demands during the daytime of the Moon. The maximum angle between the equator of the Moon and the ecliptic plane is only 1.5°, so the sunlight almost directly illuminates the equator of the Moon. The higher the latitude is, the smaller the solar elevation angle is. In the design of the size and orientation of the solar panel, the variation of the solar elevation and azimuth of the landing area should be considered. Generally, it is desired to select the landing time to be the local "daytime" of the landing site so that sufficient energy can be achieved. When selecting the landing area, it is also necessary to consider the influence of the local solar elevation angle on the imaging of the optical sensors. If the elevation angle is too high or too low, it may be difficult to ensure good imaging of the optical sensors.

(5) Temperature environment: The Moon has no atmosphere. Therefore, the temperature of the Moon surface is mainly determined by solar radiation. The maximum temperature of the Moon surface during the daytime of the Moon varies greatly with respect to the latitudes. The higher the latitude is, the lower the maximum temperature is. The closer the area is to the equator, the higher the temperature is. However, the temperature difference on the Moon surface at night is not significant. In the thermal design of the lander, the thermal flow conditions on the lunar surface must be considered.

For the constraints of the landing area selection, it is necessary to comprehensively evaluate the impacts on the lander design, assign different weights and reasonably select the landing area with the optimal comprehensive conditions.

To carry out scientific exploration activities in the landing area, the selection principle of the landing area may be summarized into the following two aspects:

(1) Landing area selected should be rich in geological phenomena, such as contact belts of the lunar mare and the highland, large mountain ranges and typical crater structure areas, which can satisfy the demands of lunar geological research.
(2) Landing areas selected should be in mature lunar soil areas rich in mineral resources to meet the demands of lunar resource utilization and research.

2. **Landing/attachment options**

For different target celestial bodies, namely large celestial bodies such as the Moon and Mars, or small celestial bodies such as asteroids and comets, there are two different approaches fort the probe to contact the surface of the target celestial body for sampling purposes: soft landing or attachment. For large celestial bodies, the soft-landing methods of atmospheric celestial bodies (Mars, Venus, Titan, etc.) and airless objects (such as the Moon and Mercury) are different.

For celestial bodies with atmosphere, it is theoretically possible for the probe to realize a soft landing by itself through braking to achieve decelerated descent, or by a

3.4 System Design of Landing Exploration Missions

combination of atmospheric deceleration from the atmosphere near the target celestial body, parachute deceleration and decelerated descent from the braking of the probe. However, usually because the gravitational field of the large celestial body is quite strong, before the soft landing, whether the probe enters the orbit around the celestial body or directly enters the landing trajectory at the end of the transfer orbit from the Earth to the target celestial body, the required velocity increment is of the magnitude of several km/s. Taking the Mars landing missions as an example, the velocity increment needed to achieve the descent from the orbit around to the surface is greater than 3.5 km/s, and the propellant required for the deceleration of the probe system itself is too great. Therefore, a combination method of atmospheric deceleration from the atmosphere near the target celestial body, parachute deceleration and decelerated descent from the braking of the probe is applied for soft landing.

For the exploration mission on celestial bodies with no atmosphere, it can only rely on itself to realize a soft landing through braking to achieve decelerated descent.

The final landing method includes landing buffer mechanism, airbag cushioning, etc. Considering the need for the subsequent takeoff of the sample return mission, there is a certain requirement for the landing attitude, and the sample return mission generally adopts the landing buffer mechanism.

If the object to be detected is a small celestial body with weak gravitational force, such as an asteroid and a comet, when the probe system completes the transfer flight, a close-range rendezvous with the target celestial body can be achieved, and the probe is attached to the target celestial surface by using an anchoring mechanism.

3. **Landing control technology approaches**

For the lander, the braking deceleration process of soft landing is the core to ensure a safe landing, which poses a new set of requirements that are different from previous spacecraft maneuver processes.

1) Navigation and control demands

The duration of the lander braking process is short with large speed variation, and the relationship between the relative altitude and speed with respect to the target celestial surface is difficult to predict, so it is impossible to achieve the guidance, navigation, and control by commands from the Earth. It is necessary to equip the probe with navigation sensors for speed and distance measurement to realize autonomous guidance, navigation and control. In addition, the landing surface has complicated terrains such as slopes, rocks and craters, making it unpredictable. Therefore, the lander should be provided with the capabilities of autonomous terrain recognition and autonomous obstacle avoidance so that it will not fall over after landing.

2) Propulsion capability demands

In the process of landing, to achieve a smooth and controllable braking process and adapt to the constant changes in the mass of the lander, the thrust should be adjustable within a certain range. For the landing process, the greater the thrust-to-weight ratio is, the shorter the landing process is, and the smaller the gravity loss is. However, the overload of the entire probe will be intensified, and the dynamic environment of the landing process will be aggregated. Therefore, it is necessary to make comprehensive assessment and determine the capabilities to be provided by the propulsion system such as the total impulse, thrust magnitude and its adjustable range.

3) Landing buffer demands

In the final stage of the landing, the landing buffer should be able to absorb the landing impact load so as to ensure that the equipment on the lander is not damaged, the lander does not tip over or sink into ground and provide strong support for the lander's subsequent work.

3.4.2 Decomposition of Technical Specifications

The technical specifications of landing probes generally include aspects such as the landing capability, navigation, and control capability, TT&C capability, power source capability, scientific exploration capability, and data management capability. The technical specifications with the most important lander features are the quantitative requirements of landing buffering capability, navigation and control capability, and propulsion capability. The landing buffering capacity mainly includes the adaptability to the landing surface for safe landing; the navigation and control capability mainly includes the control errors of the landing speed, the triaxial angular velocity, and attitude angles; the propulsion capability mainly includes the velocity increment capability, the main performance of the orbit control engine, and the attitude control thruster, etc. Technical specifications above are used to determine the main capabilities of the lander in the mission execution process, which need to be comprehensively considered together with the mission requirements and demands and are demonstrated and determined by means such as simulations and tests.

Taking the propulsion capacity as an example, the main specifications are the specific impulse performance of the orbit maneuver engine, which directly affects the propellant carrying capacity and the total mass of the lander. After the engine propellant is determined, the specific impulse performance of the engine is improved mainly by improving the combustion efficiency of the engine and increasing the area ratio of the nozzle. However, the increase of combustion efficiency often results in the excessive high engine operating temperature, further affecting the life of the high-temperature resistant oxidation coating. And the increase of the nozzle area ratio will increase the engine size and structure mass, reducing the fundamental frequency of engine structure and affecting the mechanical resistance of the engine. Therefore, to determine the engine specific impulse performance reasonably, it is necessary to combine the requirements of multiple aspects such as overall propellant demand, structural layout, propulsion system mass (wet weight, dry weight), coating life, and mechanical and thermal environmental conditions.

3.4 System Design of Landing Exploration Missions

3.4.3 Flight Procedure Design

The landing process is a mission process unique to the lander. The main tasks of this process can be summarized as the following aspects (illustrated by the lunar soft landing as an example):

(1) Reduce the speed of the lander relative to the Moon to an acceptable range with its own propulsion system;
(2) Measuring the altitude and speed of the lander relative to the Moon surface by means of the onboard lunar distance and speed sensors and an inertial measurement unit, and independently calculate the corresponding control strategy by the onboard computer;
(3) Identify the terrain of the landing area with the optical and laser sensors on the lander, independently search for a safe landing area, and formulate corresponding obstacle avoidance strategies;
(4) Autonomously complete the main engine shutdown at a certain altitude before the lander touchdown on the Moon, and land on the lunar surface at a given speed;
(5) Rely on the landing buffer mechanism to absorb the landing impact load and guarantee a stable attitude after landing so as to ensure that the subsequent scientific detection tasks can be carried out effectively.

Different landing process control strategies are developed according to different navigation and control strategies, navigation sensors and propulsion system configurations. Taking the landing mission of the Chang'e 3 probe as an example, the landing process is divided into eight phases: ignition preparation phase, main braking phase, rapid adjustment phase, approaching phase, hovering phase, obstacle avoidance phase, slow descending phase and landing buffer phase [6].

1) Ignition preparation phase

The main task of the ignition preparation phase is to correct the ignition timing and the corresponding orbit uploaded from the ground before the ignition of the powered descent, calculate the desired ignition attitude and adjust the attitude in place. The attitude is determined by the combination of the gyro and the star sensor, and the attitude control is executed by a 10 N engine.

2) Main braking phase

The main braking phase is from about 15 km to about 3 km in altitude from the lunar surface. The main task of this phase is soft-landing braking, in which the speed of the lander (about 1.7 km/s) is reduced to a preset value, and the altitude is lowered to about 3 km. According to the requirements of this task, the trajectory should be a gentle descending trajectory naturally formed by the main engine braking, with an altitude drop of about 12 km and a flight range of about 430 km.

3) Quick adjustment phase

For quick adjustment phase, the height from the actual lunar surface is from about 3 km to about 2.7 km. The main task of this phase is to quickly connect the main braking phase and the approaching phase. The fast attitude maneuver is made to acquire the entry attitude of the approaching phase, and the engine thrust is synchronously reduced to a low-thrust level. According to the requirements of this task, the trajectory should be a gentle descending trajectory formed by the main engine deceleration, the height is reduced by about 300 m, and the range is about 1 km.

4) Approaching phase

For approaching phase, the height from the actual lunar surface is dropped from about 2.7 km to about 100 m. The main task of this trajectory is to roughly avoid obstacles. According to the requirements of the rough obstacle avoidance, to ensure that the field of view of the optical imaging sensor is aligned with the landing area, the approaching trajectory to the target landing area is designed to meet the requirements of the specific attitude and the descending trajectory. The lander is approaching the landing area in an approximate straight line with a pitch angle of 45°, and the optical imaging sensor is used to detect large obstacles (stones or pits larger than 1 m in diameter) in order to determine the safe landing area and avoid obstacles; finally, the lander reaches a height of about 100 m over the landing area. At this point, the velocity of the lander relative to the lunar surface is close to zero, and the distance travelled along-track is about 3 km.

5) Hovering phase

For hovering phase, the height of the lander from the lunar surface is about 100 m. The main task of this phase is to conduct fine obstacle detection in the landing area. The variable-thrust engine offsets the lander gravity and maintains the lander in a hovering state. The three-dimensional imaging sensor is used to observe the landing area and select a safe landing point.

6) Obstacle avoidance phase

The obstacle avoidance phase is from about 100 m to about 30 m in altitude over the lunar surface. The main task of this phase is to avoid obstacles and descend. According to the relative position information of the safe landing point given by the hovering phase, the lander descends to an altitude of 30 m above the landing point, the velocity of the lander relative to the lunar surface follows a preset value with the horizontal velocity approaching zero, and the trajectory falls obliquely to the landing point.

7) Slow descending phase

The vertical distance from the lunar surface is dropped from about 30 m to the height at which the signal of the gamma shutdown sensor becomes effective. The main task of this phase is to ensure that the lander is smoothly and slowly descending to the lunar surface, and the velocity and attitude control accuracy of landing to the lunar surface satisfy the related requirements. If the gamma shutdown sensor signal does not take effect, the touchdown sensor will provide a shutdown signal instead.

8) Landing buffer phase

When the lander touches down on the Moon, the footpad touches the lunar surface first, and the ball joint can rotate freely to adapt to the lunar terrain. Then the foot pad pushes the inner cylinder of the main pillar to slide relative to the outer cylinder, and the relative motion between the two cylinders results in the depression and deformation of the aluminum honeycomb energy absorbing component, thereby absorbing the impact energy and achieving the purpose of buffering.

3.4.4 Analysis of Key Technologies

With the consideration of all aspects of the landing mission, the key technologies requiring breakthrough are as follows.

1. **Soft-landing technology**

The soft-landing process is related to the exploration target and landing mode. It can be generally divided into two types of technology, namely the powered descent and landing on airless celestial bodies and atmospheric entry, braking descent and landing on celestial bodies with atmosphere.

For landing missions on celestial bodies without atmosphere, taking the Moon as an example, it generally involves key technologies in guidance, navigation and control, propulsion and landing buffer for powered descent. It is necessary to determine the corresponding technical approaches and technical specifications according to the specific analysis of soft-landing accuracy and landing mass.

For the landing mission of the Chang'e 3, the landing process was decomposed into seven mission phases in the control scheme. The guidance law design of the seven mission phases was completed by using the guidance techniques of the constant-thrust fuel-suboptimal explicit guidance, the fourth-order polynomial guidance, etc.; according to the main control object of the current guidance task, the landing guidance is achieved with focuses under the time and fuel constraints, also achieved are landing hovering, obstacle avoidance and slow descending to the lunar surface. Besides, multi-information fusion technology was adopted, and the navigation scheme of inertial navigation assisted by ranging and speed measurement correction was adopted. The obstacle recognition algorithm based on optical image and 3D elevation data was developed, and relay obstacle avoidance strategy using optical imaging rough obstacle avoidance plus laser three-dimensional imaging precision obstacle avoidance was applied. The "PID + PWM" attitude control algorithm using the joint control of the "10 N + 150 N" thrusters was adopted to realize the attitude control of the landing process.

To address the problem of landing deceleration, a variable thrust engine with high specific impulse and high control precision was developed. The engine applied a pintle-type flow regulating device to precisely control the flow of oxygen, fuel and cooling paths so as to achieve continuous and variable thrusts. The thrust variation range was 1500–7500 N, and the thrust control accuracy was 7.5 N.

For the final buffer phase, comprehensive comparisons were made on the advantages and disadvantages of different configurations of landing buffer mechanisms. The landing buffer mechanism adopts a "cantilever" configuration design to reduce system mass, improve landing stability and facilitate the release of the rover. The compression release and deployment locking problems were solved by the integrated design of the compression release deployment locking device and the auxiliary buffer. The problem of tensile energy absorption buffer was solved by developing a new type of ordinary temperature superplastic material with an elongation of exceeding 70%.

For landing missions on celestial bodies with atmosphere, taking Mars as an example, it is necessary to solve the technical difficulties such as aerodynamic deceleration, parachute deceleration, powered deceleration, obstacle avoidance and landing buffer with the available technologies of Earth atmospheric deceleration and lunar powered descent. The Mars atmospheric entry, descent and landing mission process is more complicated, and the requirements for system robustness and matching attribute of the specifications of different phases are more demanding. For the solution approaches of the related key technologies, please refer to Chap. 7 for details.

2. **Attachment technology**

For exploration missions of small celestial bodies, in surface landing, the objective of collecting samples can generally be achieved through the two different approaches of non-contact and contact. As the gravitation on the surface of a small celestial body is quite weak, this process is often called "attachment."

To enable surface attachment on small celestial bodies, it is necessary to carry out fine mapping of the surface morphology in advance, select suitable attachment points, and then stably connect the probe to the target small celestial body with specific attachment fixtures such as claw thorn, ultrasonic adhesion, solid glue attachment, laser perforation attachment and harpoon anchor attachment. Efforts should be made to focus on addressing the integrated attachment control and mechanism design technology and meet the mission requirements of different attachment mechanisms such as adapting media, time required for attachment, adhesion and repeatability.

3.4.5 Design Validation

For the landing exploration missions, intensive special test verifications are required in accordance with the mission objectives and environmental characteristics in the development process. Taking the Chang'e 3 and the Mars mission as examples,

3.4 System Design of Landing Exploration Missions 65

in the engineering development, the pneumatic and ablation test, the parachute system test and key separation surface separation test were carried out for the aerodynamic deceleration phase, the comprehensive verification test of hovering, obstacle avoidance and slow speed descent were carried out for the powered deceleration phase, and the landing impact test, landing stability test, airbag cushion test, plume impingement test, etc., were carried out for soft-landing phase, referring to Chap. 14 for details [6].

3.5 System Design of Rover Exploration Missions

3.5.1 Mission Analysis

Depending on the difference of target celestial bodies on which the mission is performed, the rover is often referred to as a lunar rover or a Mars rover. The exploration rover is required to be able to adapt to the surface environment of the target celestial body, carry out exploration activities in the roving area and transmit the exploration data back. Based on the actual use of the rover and whether it is manned or not, the rover is simply classified as shown in Table 3.1.

According to whether the rovers are manned or not, they can be divided into manned and unmanned rovers; according to their purpose for use, they can be divided into rovers for research, transportation, construction, mobile relay station and mobile shelter. The Lunar Roving Vehicle (LRV) used in the Apollo Program of the USA was operated by astronauts and therefore was a manned rover for transportation purposes; the lunar rover Lunokhod of the Soviet Union, the Mars rovers Sojourner,

Table 3.1 Classification of exploration rovers

Whether manned or unmanned		Actual usage				
		Research	Transportation	Construction	Mobile relay station	Mobile shelter
Unmanned	Remote or automatic control	√	√	√	√	–
	External console	–	√	√	–	–
Manned	Open cockpit	–	√	√	–	–
	Sealed cockpit	√	√	–	–	√

Note The item marked with "√" in the table indicates that the scheme has been applied or is considered to be the scheme to be most likely adopted

MER, Curiosity, etc., of the USA and the Yutu lunar rover of China are all unmanned rovers for research purposes. Lunar surface rovers of different countries are shown in Fig. 3.1.

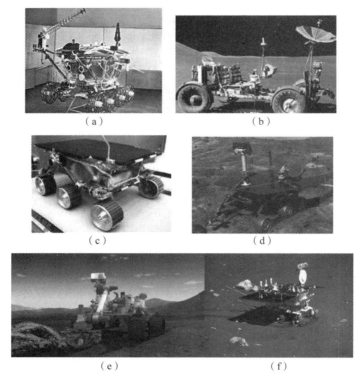

Fig. 3.1 Outline drawings of rovers of different countries. **a** Lunokhod 1; **b** LRV; **c** Sojourner; **d** MER; **e** Curiosity; **f** Yutu

In view of the demands of rover exploration missions on extraterrestrial celestial bodies, taking Chang'e 3 mission of China as an example, the elements to be highlighted in the mission analysis for realizing "surface roving" are discussed in the following.

1. **Technical approaches for the survival on surfaces of extraterrestrial celestial bodies**

The period of rotation of the Moon is about 27.3 days. Therefore, in the landing area a lunar day is equivalent to about 14 Earth days and a lunar night is also equivalent to about 14 Earth days. There is no sunlight during the lunar night, and the extreme temperature can reach $-180\ °C$.

Compared with cases where other spacecraft entering the shadow of the Earth or the Moon, there are relatively large differences for rovers entering the lunar night, which include the following aspects.

3.5 System Design of Rover Exploration Missions

(1) Long duration: The duration of a lunar night on the lunar surface is about 14 Earth days. There is no sunlight during such period, which is much longer than the orbital shadow time of spacecraft orbiting the Earth or the Moon.
(2) Lack of space infrared radiation heat flow: Low Earth orbit spacecraft may obtain certain infrared radiation heat of the Earth while they are in the Earth shadow. The temperature on the surface of the Moon may reach $-180\,°C$ during the lunar night, and there is little infrared radiant heat flow to the rover.

In response to the particular conditions of the lunar night environment on the lunar surface, the energy required to survive the lunar night must be guaranteed. If a storage cell-type energy storage device is used to provide the power and heat required in lunar night, a large-capacity cell should be provided; if the power-off sleep mode is adopted, the storage temperature environment of the devices may only be required to be maintained during the lunar night, thereby greatly reducing the demand of energy. However, the autonomous wake-up issue should be addressed as a result. In the case of power-off of the rover, to autonomously wake up the rover at the "dawn" on the lunar surface when both the lighting condition and the lunar surface temperature condition meet the needs of the rover, the criteria for waking up the rover may be considered mainly from two aspects, namely solar illumination intensity and temperature.

In case of lunar night environment, with no temperature control measures taken, the devices and equipment in the rover would be damaged or lose effectiveness. Therefore, it is necessary to adopt an energy source that is highly adaptable to the environment and that does not depend on conditions such as sunlight or temperature to ensure the safety of the rover during the lunar night. The isotope heat source is a type of nuclear energy source featuring long service life, strong environmental adaptability, high reliability, no external energy requirement and comprehensive energy supply of heat and power. Therefore, isotope thermoelectric cells and isotope heat sources are able to provide electrical and thermal energy to the rover in extreme environments, thus ensuring the survival of the rover during the lunar nights.

2. Technical approaches for light weight and miniaturization design

Lightweight, small volume, and low power consumption have always been important topics in the design of deep space probes. As limited by the rocket carriers and landers, importance should be attached to such issues in the design of lunar exploration rovers. In the process of equipment development, focus should be made on features of lightweight and low power consumption to realize the goal of comprehensive integration and satisfy the demands of lunar rovers.

During development of the Yutu lunar rover, the directive functions required for the directional antenna, panoramic camera and navigation camera were all integrated on the mast at a system level. The solar wing simultaneously assumes the functions of power output, thermal insulation during lunar nights, sun shading under extreme temperature conditions. These functions are activated using a time-sharing scheme to reduce the resource consumption. The measures taken at the design level of stand-alone equipment include the integration of electronic equipment, downsizing of the mechanisms, the unification of power supply, etc.

3.5.2 Decomposition of Technical Specifications

The technical specifications of the rover generally include aspects such as mobile capabilities, navigation and control capabilities, TT&C capabilities, energy capabilities, scientific detection capabilities and data management capabilities. The technical specifications that are of rover characteristics most are the quantitative requirements of mobile capabilities, navigation and control capabilities. The mobile capabilities mainly include the moving speed, obstacle crossing ability, turning radius, endurance mileage, climbing ability, static stability angle, etc. Navigation and control capabilities include static attitude determination ability, dynamic attitude determination ability, emergency obstacle avoidance time, emergency obstacle avoidance distance, autonomous driving distance, etc. Such technical specifications determine the main capabilities of the rover in the mission execution process and should be considered comprehensively together with the mission requirements, technical basis, etc., and demonstrated and confirmed by means of simulations, tests, etc.

Take the determination of the obstacle crossing ability of the Mars rover as an example to illustrate the process to determine related technical specifications. The topography of the surface of the celestial body is not only the important aspect for scientific exploration but also a factor to be closely concerned in the engineering design of the probe. The selection of the landing area, the safe landing of the lander and the safe driving of the rover are all closely related to the topography. To carry out the roving exploration mission on the surface of Mars, it is necessary to attach importance to the terrain of the landing area and reasonably configure the traffic ability of the Mars rover. For this reason, the mean free path of the Mars rover should be studied and the obstacle crossing ability of the rover should be reasonably determined.

On the surface of Mars, within a unit area the area ratio $F_k(D)$ of rocks with diameter larger than a certain value D generally satisfies:

$$F_k(D) = k e^{-qD} \tag{3.1}$$

where k is the rock area factor; q is a function of k and satisfies:

$$q(k) = 1.79 + \frac{0.152}{k} \tag{3.2}$$

The rock height is generally 0.5 times the diameter. Let the number of rocks with diameter D be $n(D)$, and the corresponding area satisfies:

3.5 System Design of Rover Exploration Missions

$$qke^{-qD} = \frac{\pi D^2 n(D)}{4} \quad (3.3)$$

Then the distribution density of the number of rocks is:

$$n(D) = \frac{4qke^{-qD}}{\pi D^2} \quad (3.4)$$

The mean free path analysis of the rover can be further carried out by applying the topographical distribution model of the Mars represented by Eq. (3.4).

The mean free path refers to the average distance covered by the Mars rover when it encounters an impassable obstacle while traveling in a straight line. The mean free path is usually used to describe the ability of the rover to move in a particular environment. When the mean free path is non-dimensionalized, the mean free path is divided by the turning radius of the rover, which is called "mean free path coefficient." If the mean free path coefficient is much greater than 1, it can be interpreted that the terrain is not dangerous for the rover, the rover can travel freely in most cases, and therefore, a simple obstacle avoidance algorithm can be used to improve the work efficiency of the rover. When the mean free path coefficient approaches 1, the terrain is difficult to pass for a given Mars rover, and the rover may encounter obstacles at any time, so it is necessary to consider a targeted obstacle avoidance algorithm in navigation and control to ensure that the rover can cross the area smoothly.

For the design of the rover, traffic ability of the rover should be increased as possible, so that a larger mean free path can be obtained in a given area, thereby improving the efficiency of the exploration mission.

3.5.3 Work Procedure Design

Take the Chang'e 3 lunar rover of China as an example to outline the work procedures and operation mode design.

After landing, the rover performs system checks with the assistance of the lander (or landing platform), deploys solar cell arrays and directional antennas and establishes communication links. After confirming that the working conditions are normal, the connection with the lander (or landing platform) is cut off, and the rover moves to the lunar surface through the transfer mechanism.

After rolling onto the lunar surface, the rover mainly works in three modes: mobile mode, exploration mode and communication mode. The three modes are independent from each other and are performed serially. After the next stop point and path are selected, the rover is controlled to the scheduled location; the rover stops and then starts the scientific instruments to carry out the detections (certain individual scientific instruments work during the movement of the rover); finally, the exploration data information is transmitted to the Earth or other probes such as the orbiter and lander.

The typical working process of the Yutu lunar rover is as follows. First, the panoramic camera and the navigation sensor are used to establish environmental information, which is transmitted to the Earth, and the ground control station completes the path planning according to the environmental information and the state of the lunar rover, determines the task commands that are then uploaded to the rover. Second, the rover works according to commands, after reaching the target point, various scientific instruments start to work, and the exploration data is stored, then the scientific instruments stop working. Third, the directional antennas are aligned with the Earth in a controlled manner and transmit the exploration data and navigation information back to the Earth. Then the above steps are repeated.

In addition, when the lunar night falls, the lighting and temperature environment of the lunar surface cannot satisfy the working demands of the Yutu lunar rover, and the rover enters the sleep mode. On the next lunar day when the lighting and temperature conditions meet the requirements, the rover wakes up autonomously, and a new round of work is resumed.

3.5.4 Analysis of Key Technologies

For the roving exploration on the surface of celestial bodies, generally speaking, the special key technologies requiring breakthroughs are as follows.

1. **Rover mobility technology**

To complete scientific investigations, the rovers should be able to move forward, backward, turn, climb and overcome obstacles with limited energy. They should also have the stability of motion under complicated celestial surface conditions to ensure safe and smooth motion of themselves. The mobility performance of the rover directly affects the realization of scientific goals. Mobility technology is of great importance to achieve safe and reliable movement of the rover. The specific contents of mobility technology include wheel form, gear shape and layout scheme, suspension-type research and parameter optimization, mobile performance test technology, mobile performance simulation technology, evaluation system, etc.

2. **Rover navigation and control technology**

It is a major goal of the mission for the rover navigation and control system that the rover normally travels in a natural terrain environment and safely reaches a predefined working point. Therefore, the rover must be able to identify hazards and obstacles in the environment, determine its own attitude and position, identify the target location, plan the route to the target location, move along the planned path, detect and avoid obstacles, etc. Technical highlights should include densely restored binocular stereo vision algorithms, obstacle feature extraction, slip estimation and compensation, global and local path planning algorithms, motion coordination control, etc. Please refer to Chap. 6 for details.

3. **Rover thermal control technology**

As affected by the rotation and revolution of the celestial body, the surface temperature and radiation heat flux change drastically; the driving directions of the rover are not fixed; the ground slope and obstacles will affect the attitude of the rover. These factors increase the difficulty in the heat control task of the rover. Some special thermal design measures and special thermal control hardware should be applied to solve such problems. For example, to safely endure the lunar night of 14 days, it is necessary to apply isotope energy technology. To achieve thermal insulation in the low atmospheric pressure environment of the martian surface, new thermal control materials such as aerogel should be adopted.

3.5.5 Design Validation

For the rover exploration mission, due to the difference in gravity environment between the Earth ground and the celestial surface, an equivalent simulation of the low-gravity environment of the rover is required. The terrain and geomorphic features of the celestial body surface are required to be simulated on the ground to achieve effective verification of the rover movement and navigation performance. In addition, the soil and rock environment of the celestial surface should also be simulated with engineering simulation soil. A large-area light array should be constructed, and the light source and its position should be optimized to simulate the direct and scattered light illumination environment on the celestial surface. Taking the Chang'e 3 rover as an example, the special tests carried out at system level mainly include three types of tests, namely infield test, field test, and combined surface test. Please refer to Chap. 14 for details.

3.6 System Design of Sample Return Exploration Mission

3.6.1 Mission Analysis

According to the requirements of sample return missions from extraterrestrial celestial bodies, the key elements of mission analysis are described in the following.

1. **Selection of mission scenario**

The target celestial body has a great influence on the probe design, especially the sample return mission on large bodies such as the Moon, Mars and Venus, the gas giant planets and their satellites. As the target to be sampled is far away from the Earth, and the probe must pass flight phases such as arrival at the target celestial surface, lift-off and escape from the gravitational field of target bodies, returning to the Earth, Earth atmosphere reentry, landing and recovery, the probes must be able to fly under complicated and various environmental conditions. Therefore, the size of the probe is relatively large. If the target celestial body is a small celestial body, such as an asteroid, a comet or even cosmic dust, the probe can make contact with the target celestial surface and sample by short-distance rendezvous or accompanying flight. If comet dust is to be collected as samples, sampling may be realized just by rendezvous without direct contract, and the size of the probe for such missions may be greatly reduced.

For example, for the lunar sample return mission, if there are several available launch vehicles with different launch capabilities, the optional mission scenarios may include:

(1) Option 1: one launch without rendezvous and docking.
(2) Option 2: one launch with rendezvous and docking.
(3) Option 3: two or more launches with rendezvous and docking.

In Option 1, the probe typically consists of the lander, ascent module and reentry capsule. Wherein, the lander carries the ascent module and the reentry capsule. After the lander is landed on the lunar surface, the lunar sample will be collected and transferred to the reentry capsule. Then the ascent module carries the reentry capsule to lift off and ascend into circumlunar orbit or directly into Earth transfer orbit. The reentry capsule is separated from the ascent module when it reaches the Earth, and the reentry capsule with the sample reenters the Earth atmosphere and is landed on ground for recovery.

In Option 2, the probe typically consists of the orbiter, lander, ascent module and reentry capsule. The orbiter carries the lander, ascent module and reentry capsule into circumlunar orbit and then is separated from the lander and the ascent module. The orbiter with the reentry capsule orbits the Moon and waits for rendezvous and docking with the ascent module. The lander with the ascent module is landed on the lunar surface. After lunar sample is collected, the sample will be transferred to the ascent module. Then the ascent module with the sample will life off and enter circumlunar orbit. After the ascent module is docked with the orbiter and reentry capsule, the sample will be transferred into the reentry capsule. Then the ascent module is separated from the orbiter, and the orbiter with the reentry capsule and the sample is inserted into Earth transfer orbit reentry capsule. Later, the reentry capsule is separated from the orbiter after it reaches the Earth. The reentry capsule with the sample will reenter the Earth atmosphere and be landed on ground for recovery.

In Option 3, the probe consists of a probe (orbiter and reentry capsule) and a second probe (lander and ascent module), which are launched separately. Among them, the orbiter with the reentry capsule is launched first and inserted into circumlunar orbit after approaching the Moon. The lander with ascent module is subsequently launched and landed on lunar surface after approaching the Moon. The lunar sample is collected and stored in the ascent module. The ascent module lifts off and ascends. After entering circumlunar orbit, it will be docked with the orbiter and reentry capsule. After the lunar sample is transferred into the reentry capsule, the orbiter will be inserted into Earth transfer orbit. The reentry capsule is separated after the orbiter

3.6 System Design of Sample Return Exploration Mission

approaches the Earth, then the reentry capsule with the lunar sample will reenter the Earth atmosphere and be landed on ground for recovery.

In the 1960s, the unmanned lunar sample return missions of the Soviet Union adopted Option 1 (Fig. 3.2a). The Phase III of the China Lunar Exploration Program (Chang'e 5) and the Apollo manned lunar missions of the USA adopted Option 2

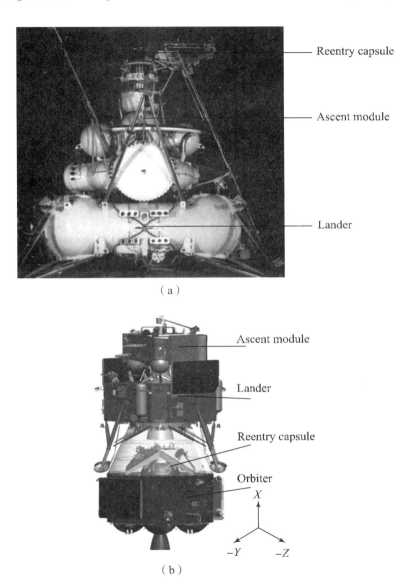

Fig. 3.2 Typical lunar sample return probes. **a** Lunar 24; **b** Chang'e 5

(Fig. 3.2b). Option 3 is more common in the future scenario of Mars sampling return missions, which have not been implemented yet [7].

The above three mission scenarios can all accomplish the sample return missions. When a sample return mission is proposed, it is necessary to conduct comprehensive comparison and trade-off according to the engineering objectives and actual mission constraints to determine the final option.

2. **Selection of sampling approaches for extraterrestrial celestial bodies**

The selection of sampling method is closely related to the mission objectives, and it is also related to the environment and sample characteristics of the target celestial body. There are many options available to be analyzed and compared.

For example, between 1969 and 1976 there were total 9 lunar sample return missions accomplished, in which 382 kg lunar samples were collected with diversified sampling methods such as drilling, scooping, clamping and extraction. Generally, subsurface sample can be obtained by means of drilling, and multi-point surface samples can be obtained by means of scooping, clamping and harrowing.

In the Moon and Mars exploration missions, the objects to be sampled are typically rocks or soil particles. In future exploration missions, new challenges will occur for the sampling methods. For example, the mainstream view currently is that there is water–ice in the polar regions of the Moon, and there is even water under the ice layer of the satellites of the Jupiter. To obtain such samples, more different sampling tools and methods are required to be considered.

To ensure that the status of the samples is consistent with the original status, in the sample return missions, the samples are generally required to be sealed to ensure that the samples are not contaminated during return to the Earth, reentry, landing and recovery process. In addition to the traditional sealing rings, the reliability of the sealing may be guaranteed by one or the combination of the approaches such as metal extrusion sealing and metal fusion sealing. Different sealing methods determine the complexity of the sealing surface structures of the sample containers. The final sealing approaches should be determined according to the characteristics of the samples, leakage rate requirements related to the flight process and time.

3. **Analysis of Lift-off, Ascent and Escape**

When the target celestial bodies are different, different issues should be addressed. If the target celestial body has a gravitational field, such as the 1/6 gravity field of the Moon, rather than the microgravity of the small celestial body, an independent lift-off and ascent process must be designed for the probe to leave from the surface of the target celestial body and enter the orbit around the target body or directly enter the Earth transfer orbit.

The mission analysis of lift-off and ascent focuses on the basic design of lift-off and ascent process, guidance law, the initial reference for ascent, lift-off stability, and related subsystems such as guidance, navigation and control (GNC) and propulsion.

The above analysis is for lift-off from celestial bodies without atmosphere such as the Moon. For the celestial bodies with atmosphere, factors to be considered should

3.6 System Design of Sample Return Exploration Mission

include atmospheric drag during lift-off, weather phenomena on the surface of the celestial body (wind, sandstorm), etc., and their influences on the preparation before lift-off and the ascent process, for which corresponding analysis and design should be conducted.

For the sample return missions from small celestial bodies, the timing of leaving the small celestial body is relatively flexible; the detachment mode is related to the previous attachment mode and usually involves unlocking or unfolding of the attachment mechanism. In addition, the detachment method should ensure that the detachment process is safe and reliable and does not affect subsequent flight back to the Earth.

4. **Analysis of rendezvous, docking and sample transfer**

Generally, only for sample return missions on large celestial bodies, it should be considered to reduce the size of the mission through rendezvous and docking. However, the rendezvous, docking and sample transfer processes are complex, and they are difficult key points in the whole mission process. For rendezvous and docking operations, it is generally necessary to analyze the characteristics of the rendezvous and docking missions and focus on the analysis of active and passive spacecraft selection, rendezvous orbit selection, rendezvous strategy, rendezvous and docking GNC design, evacuation process, docking mechanism, sample transfer mechanism, etc.

(1) Active and passive spacecraft selection: To reduce the scale of the lander and ascent module, it is more advantageous to select the ascent module as the target spacecraft. Even in some particular cases, the ascent module finally releases a small target as a passive spacecraft to cooperate with the orbiter, which acts as the active spacecraft, to complete the rendezvous and docking process. Such design is common in general extraterrestrial celestial body sample return missions. Based on the above design principle, the heavier docking mechanism and the active part of the transfer mechanism are usually installed in the orbiter or orbiter–reentry capsule complex, and the active part of the relative measurement sensor required for the rendezvous and docking is also configured in the orbiter, which is advantageous for mass resource allocation.

(2) Selection of rendezvous and docking orbit: The target orbit for rendezvous and docking is the orbit where the orbiter and the ascent module finally complete rendezvous and docked. For rendezvous and docking in extraterrestrial celestial body orbit, the main factors to be considered for rendezvous and docking orbit include altitude, inclination, right ascension of ascending node, lift-off window and rendezvous orbit strategy.

(3) Selection of rendezvous and docking GNC: Typical rendezvous and docking GNC subsystem configuration requirements include the control computer, IMU, microwave relative measurement equipment, laser relative measurement equipment, optical imaging sensor, star sensor and solar sensor. After the configuration of the active and passive spacecraft is determined, the configuration and layout of the active/passive parts of the relative measurement sensors are

completed according to the selection of the GNC design, and the guarantee requirements of relative measurement on the environmental conditions such as microwave and optics of the probe and the configuration of the thruster are considered. After the flight hardware components of the GNC subsystem are determined, attitude determination method, relative navigation and attitude control method should be further determined. Then specific design of the GNC subsystem for the guidance and control of autonomous rendezvous and docking at each phase should be determined, and corresponding simulations shall be conducted to determine the technical specifications.

(4) Flight plan in evacuation process: To maximize the advantages of rendezvous and docking, after it is confirmed by ground control that the sample is transferred successfully, the orbiter is separated from the docked ascent module, leaving the ascent module on the orbit. The mechanisms used for evacuation, the timing of evacuation, and the orbit/attitude control schemes shall be determined.

(5) Docking mechanism and sample transfer mechanism: Since the first space rendezvous and docking in 1966, with the development of the docking mechanism for over 50 years, the docking mechanisms have evolved into various types in accordance with the differences in the installation position, transmission principle, application scope, etc., of the functional mechanism. By the typical classification method, the docking mechanisms are divided into impact type docking mechanisms and weak impact type docking mechanisms in accordance with the difference in the applicable precision of the docking initial conditions. With the development of space mechanisms and artificial intelligent technologies, more and more new docking mechanisms and even innovative technologies for the integration of docking and sample transfer have emerged, for example, capturing and transferring the target spacecraft or sample container with a space robot or flying net. Such approaches may maximize the advantages of the rendezvous and docking solutions. In the future sample return missions from extraterrestrial celestial bodies, efforts may be made to combine the development of such innovative technologies to conduct comprehensive analysis to select feasible technical approaches accordingly.

5. Analysis of high-speed reentry and return to Earth atmosphere

As the reentry speed of the reentry capsule to Earth atmosphere is ultra-high, the technical challenge is high and risky. According to the scale of the reentry capsule, the overload requirement, the recovery zone constraint, etc., the reentry mode, the shape of the reentry capsule, the thermal protection, GNC, reentry flight process, and TT&C shall be analyzed to determine the scheme of the reentry process.

(1) Size of reentry capsule: The size of the reentry capsule depends mainly on the mass, types, and packaging form of the samples. For unmanned sample return missions, the samples and sealed container are generally small, so the size of the reentry capsule is generally smaller than recovery satellites and manned spacecraft. To avoid the aerothermodynamic impact during the reentry process on the samples and the sealed containers, the sealed containers are generally

3.6 System Design of Sample Return Exploration Mission

located on the downwind side or deep inside the reentry capsule. Therefore, specific requirements must be met when designing the configuration of the reentry capsule.

(2) Reentry overload: Different reentry modes and speeds will result in different maximum overload during the reentry; generally, the maximum overload of the ballistic reentry process is greater than 10 g, while the maximum overload of the semi-ballistic skipping or lifting reentry processes are relatively small, generally less than 10 g. The maximum reentry overload is determined based on the sample characteristics and the design of the sealed containers, which is then used as a constraint for selecting reentry mode and detailed design.

(3) Recovery area: According to the different reentry methods, the optional recovery area includes land recovery landing site and marine recovery area.

(4) Reentry mode: The reentry modes of the return spacecraft include the ballistic type, semi-ballistic type (semi-ballistic reentry can be further divided into direct reentry and skip reentry types) and lifting type. For most of the deep space exploration probes that have been launched in the world, ballistic returning type is applied, of which the typical probes include Stardust (USA), Genesis (USA), Luna (Soviet Union) and Hayabusa (Japan). Internationally, probes applying semi-ballistic reentry mode for high-speed return includes the Apollo command modules (USA). When the reentry method is determined, it is necessary to comprehensively consider the constraints of the reentry capsule size, reentry overload, recovery area, flight range and TT&C conditions.

(5) Shape of Reentry capsule: In reentry capsule design, aerodynamic shape is very crucial. The selection of the aerodynamic shape will affect the thermal environment analysis, stability design and flight overload design of the reentry capsule. Typical aerodynamic shapes include spherical, small blunt-head and cone, large blunt-head and cone combination plus inverted cone, ball-end inverted cone, etc.

(6) Thermal protection: The thermal environment of the reentry capsule is determined by the aerodynamic shape of the reentry capsule and the reentry trajectory. The results of the thermal environment analysis determine the selection of the thermal protection material of the reentry capsule, the design of the thickness of thermal protection material as well as the thermal protection structure of the reentry capsule. The thermal environment of the reentry capsule is mainly described by the stagnation point peak heat flux and stagnation point maximum heating amount in the reentry process of the reentry capsule. In addition, the reentry capsule generally has to pass a complex flight process, and it is likely to simultaneously be faced with the harsh environment of extremely low temperature, extremely high temperature and alternation of high and low temperatures. Therefore, it is necessary to comprehensively consider the issue of sharp temperature alternation in the selection of thermal protection (ablative) materials and the design of thermal protection system.

(7) Reentry GNC: For ballistic return, generally before the separation from the orbiter carrying the reentry capsule, to facilitate the stable flight of the reentry capsule, the flight status of the reentry capsule before reentry is established by the control of separation attitude and rotating mechanism. Then, the inertial

flight process before reentry is completed by fully relying on the gravity of the Earth. For a semi-ballistic or lifting return, after the reentry capsule is separated from the orbiter, the major task of the return GNC system is to complete the attitude control required by the reentry capsule before the parachute is opened.

(8) Reentry flight process: The reentry flight process generally consists of the following steps: Before separation, the necessary initial status is established, the reentry capsule is separated from the orbiter; from the moment of separation, the reentry capsule enters the free flight phase and reenters the atmosphere at the predetermined altitude (generally determined as 120 km); when the axial overload reaches the preset value, the recovery controller is activated; the deceleration umbrella is ejected and opened; the recovery beacon starts to work; remove the deceleration parachute, and open the main parachute; the reentry capsule is landed, the recovery beacon antenna is deployed and assist searching and recovery.

(9) TT&C of reentry and return process: To realize the telemetry, telecommand and tracking after the reentry capsule is separated with the orbiter, and positioning during the landing process, the corresponding TT&C equipment shall be configured. For example, by transmitting a radio beacon on the reentry capsule, the ground search and recovery team can quickly locate the reentry capsule, and an international rescue beacon or a Beidou navigation short message transmitter may be configured to assist positioning of the reentry capsule after landing. After the reentry capsule is positioned, the recovery time is determined according to the mission requirements on sample storage, and the search and recovery team may be arranged to perform search and recovery of the reentry capsule by helicopter or ground vehicle. For unmanned sample and return missions, there are also methods of using ground optical observation equipment or airborne optical observation equipment to complete positioning and recovery of the reentry capsule.

3.6.2 Definition of Specifications

The technical specifications of sample return probes generally include sampling capability, navigation and control capability, TT&C capability, energy capability, scientific exploration capability and onboard data handling capability. As the configuration of the sample return probe is complex, it is necessary to consider its composition and functions and clearly define the technical specifications.

Typical functions of the probe in the sample return missions are as follows. With the supports from other engineering systems such as launch vehicle, launch site, ground stations and ground application, the probe sequentially passes the phases of launch, transferring to target celestial body, orbiting target celestial body, landing/attachment, lift-off/escape, return to the Earth, reentry and recovery, and the collected sample is safely sent back to the Earth for ground laboratory analysis and research.

3.6 System Design of Sample Return Exploration Mission

After the composition and functions of the probe are defined, at first it is necessary to complete the preliminary design of the flight orbit, determine the ΔV budget at each phase, and then make propellant budget according to the type of propulsion system and the specific impulse of engines. Before the propellant budget is determined, the approximate mass allocation shall be completed. Considering the functional requirements of the different probes/modules, the flight profile and using some mature spacecraft or probes as the baseline, the mass of the reentry capsule and ascent module shall be determined first. Then, according to the basic result of the propellant budget, the mass of other modules is allocated. Finally, the launch mass of the probe shall meet the constraint of the launch capability of the launch vehicle, while meeting the propellant requirements at each phase with certain margins. When the margin is allocated, the amount of propellant margin may be less for the modules with relatively certain orbit changes; vice versa, more propellant margins shall be reserved for the modules with more uncertain changes. Separate propellant consumption analysis shall be conducted for powered flight (e.g., powered descent and ascent), extraterrestrial celestial atmosphere entry, and high-speed reentry to the Earth atmosphere.

After the approximate mass allocation and propellant requirements of each module in the probe are determined, the existing equipment development level and existing product status may be considered, and the mass of each subsystem of the probe can be allocated. The technical specifications in different aspects such as navigation and control capability, TT&C capability, energy capability, scientific detection capability, and onboard data handling capability in different flight phases are defined according to the functions of each module.

It should be noted that, due to the complex composition of the sample return probe, the process of determining the system specifications may be repeated. Therefore, in the probe design process, design iterations are often required; after the preliminary system design of the probe is completed, the corresponding designs are conducted successively with the subsystems, and then the system specifications such as mass allocation are adjusted according to the design results of the subsystems and the equipment to determine the final system specifications.

3.6.3 Flight Procedure Design

Typical sample return missions include the flight phases and events as follows.

(1) Launch: This phase refers to the lift-off of the launch vehicle and separation of the probe from the launch vehicle.
(2) Transfer trajectory from the Earth to the target celestial body: This phase covers from the separation of the probe from the launch vehicle, to the time when the probe arrives at a position near the target celestial body. During the transfer flight, the probe system shall perform trajectory correction maneuvers if necessary.

(3) Orbit insertion to achieve the mission orbit around the target celestial body: After the probe approaches the target celestial body, it is inserted into an orbit around the target celestial body. The orbit insertion is performed by its own propulsion system or aerobraking to reduce the flight speed so that the probe can enter the orbit around the target celestial body.

(4) Orbital flight around the target celestial body: This phase covers from the orbit insertion of the probe into the orbit around the target celestial body to the starting point of the descent of the lander. Usually in orbital flight, the complex will be separated into two parts: the orbiter–reentry capsule complex and the lander–ascent module complex. The orbiter–reentry capsule complex continues the circling flight on orbit, and the lander–ascent module complex is preparing for descent.

(5) Descent and landing: The lander–ascent module complex is decelerated and landed softly to the surface of the target body. If the target celestial body has no atmosphere (e.g. the Moon), it is necessary for the probe to decelerate by its own propulsion, and finally achieve a soft landing through a landing buffer mechanism or airbag. If the target celestial body has atmosphere (e.g. Mars), the lander–ascent module complex is generally in an encapsuled module and a soft landing is achieved by atmospheric deceleration, parachute deceleration, powered descent and final buffer. The landing buffer may be landing legs, a skycrane or an airbag.

(6) Operation on the target body surface: From landing to the ignition and lift-off of the ascent module. Generally, several procedures such as sample collection, packaging and transfer to the ascent module are followed. In addition, in-situ exploration and lift-off preparation may also be implemented.

(7) Lift-off and ascent: From lift-off of the ascent module to the insertion of the ascent module to the orbit around the target celestial body. This phase helps make preparation for subsequent rendezvous and docking as well as sample transfer operations.

(8) Rendezvous, docking and sample transfer: This phase covers from the ascent module entering the target orbit to sample transfer completed. During this phase, the orbiter–reentry capsule complex and the ascent module will establish the conditions for the close-range rendezvous of the two spacecraft through ground control and autonomous navigation and rendezvous control. After that, the two complexes are docked through the docking mechanism, and then the sample is transferred to the reentry capsule; later the ascent module is separated from the orbiter–reentry capsule complex; or the sample container is directly released and will be captured and transferred by the orbiter–reentry capsule complex without docking so that the sample container can enter the reentry capsule.

(9) Cruise on orbit: Due to the constraints of the return window to the Earth from the target celestial body, generally after the sample transfer is completed, the orbiter–reentry capsule complex shall wait for a period of time on the orbit till such window is available, and then it is accelerated to escape from the gravitational field of the target celestial body.

3.6 System Design of Sample Return Exploration Mission

(10) Transfer to the Earth: The orbiter–reentry capsule complex escapes from the gravitational field of the target celestial body and enters the transfer orbit to the Earth until it reaches the reentry point of the Earth atmosphere. Trajectory correction maneuvers are performed during the transfer if necessary, and the end of the transfer orbit shall meet the reentry requirements. When approaching the Earth, the orbiter is separated from the reentry capsule, and the orbiter needs to perform an escape maneuver or reenter the Earth atmosphere in advance to ensure the safety of the reentry capsule.

(11) Reentry and recovery: This phase is from the arrival of the reentry capsule at the reentry point to the completion of the recovery of the reentry capsule. Generally, the reentry point altitude of the deep space exploration mission is 120 km, which is specified based on the fact that the reentry speed of the deep space sample return mission is relatively high, greater than the second cosmic velocity. After the reentry capsule reenters the atmosphere, it will be landed at the designated area. And search and recovery are completed with the cooperation of the ground force.

The above flight process is for the sample return missions on large celestial bodies. For the sample return missions of celestial bodies with weak gravitational field such as small celestial bodies, it is not necessary to include the phases of descent, landing, lift-off and ascent in the flight profile, and sampling is achievable just by attachment and then detachment. Generally, rendezvous and docking/sample transfer is unnecessary and the samples collected may be directly transferred to the attached reentry capsule.

In different flight phases, each module shall complete the corresponding tasks independently or in cooperation with other modules. There are various options for the corresponding flight procedures and operation modes, which may be designed for the specific mission scenarios. The difference with other types of missions is that due to the greater number of modules and the multiple cooperation between different modules, it is necessary to comprehensively design power, information and control modes from the aspects of mechanics, electricity, heat, etc., in order to complete the mission reliably.

3.6.4 Analysis of Key Technologies

To complete the sample return missions from the surface of the extraterrestrial celestial bodies, according to the technical development roadmap, it is usually necessary to develop the key technologies of the "orbiting" and "landing" exploration of the celestial bodies in advance. In addition to the key technologies mentioned in Sects. 3.3 and 3.4, the key technologies for sample return missions are as follows.

1. **Sample collection and packaging technology**

To achieve the collection and packaging of extraterrestrial celestial samples, at first the key processes shall be analyzed according to the mission objectives, functional requirements and working processes, such as mechanism held-on, collection of different types of samples, sample shaping and primary package and sealed package. The key processes shall be adapted to the mission profiles and objectives. After the key processes are defined, the technical approaches for the implementation can be determined for each process and the interface between different processes. For example, in the Phase III of China Lunar Exploration Program (Chang'e 5), a drilling mechanism is used to obtain lunar subsurface samples with a maximum depth of about 2 m. The robotic arm with samplers is applied for lunar surface sampling. In case of sampling on the surface of the asteroid, different forms such as projectile impact, auger drilling, gas excitation, ultrasonic drilling, tape adsorption, abrasion wheel grinding and mesh cover clamping may be considered. The appropriate sampling method should be selected according to the mission requirements and characteristics.

After the sampling is completed, the packaging form of the samples should also be considered. Generally, to ensure that the characteristics of the sample on the surface of the target celestial body are not affected by the subsequent flight processes, necessary sealing means should be taken to seal the sample, the preliminary packages or the reshaped sample package, thereby resulting in the requirement for sample transfer. Flight procedure of the probe, the system configuration, the characteristics of the sampling device and the requirement for leak rate of the sealing shall be comprehensively considered to design a safe and reliable transfer mechanism or path to ensure that the samples collected are eventually transferred to the reentry capsule or sample container and returned back to the Earth.

When the sample collection, packaging, and transfer solutions, especially the corresponding mechanisms are designed, in addition to the environmental conditions such as mechanics, temperature, vacuum and radiation experienced during the flight processes, the effects of environmental factors on mechanism design such as dust, gravity, illumination, terrain and geological characteristics of the surface of the celestial body shall also be considered to ensure that the relevant mechanisms can adapt to the uncertainties in the environment and processes accordingly.

2. **Lift-off and ascent technology**

There are both similarities and differences in terms of lift-off on the surface of celestial bodies with and without atmosphere. In the following, the technical features of lift-off and ascent on the surface of the Moon and Mars are analyzed.

The key points of the lunar lift-off technology include lift-off stability, engine plume deflection, lift-off guidance, navigation and control. The lift-off stability design shall ensure that in the absence of control, during the ignition and ascending of the ascent module, no overturn will occur. The design parameters mainly include the attitude angles and attitude angle rates at the separation of the ascent module from the lander. In addition, the stability boundary conditions of the initial status parameters before lift-off shall also be determined. In the design of lift-off engine plume deflection, considerations shall also be taken on factors such as configuration of the related components, plume deflection space, support bracket and separation mechanism to ensure that the plume does not affect the performance of the engine. Moreover, the impact on the stability of the lift-off shall be minimized. If a special

3.6 System Design of Sample Return Exploration Mission

flow guiding device is designed, it shall adapt to the high-temperature wash of the plume of the engine during lift-off as well as various environmental conditions before lift-off and during ascent flight. The major challenges in lift-off guidance, navigation and control include the initial positioning and alignment before lift-off, dynamics and attitude control methods for lift-off separation process as well as the guidance methods for the ascending process. The related parameter ranges, the control methods and guidance laws shall be determined by means of analysis, simulation and ground test to satisfy the initial conditions at the activation of the guidance, navigation and control system for lift-off and the functional and performance requirements of the subsequent powered ascent process.

For lift-off and ascending from the martian surface, in addition to design elements such as lift-off stability, engine plume deflection, guidance, navigation and control of lift-off and ascending. Similar to those for lift-off and ascending from the lunar surface, the aerodynamic shape and propulsion shall be considered as well. As the atmosphere of the Mars is relatively thin (only about 1% of the density of the Earth atmosphere), during the ascending process from the martian surface, the initial speed is relatively low, and the atmospheric density is quite high, while in the later phase, the speed is high but the atmospheric density is relatively low. Therefore, the dynamic pressure is relatively small, resulting in negligible impact of aerodynamic heat. However, the impact of aerodynamic force on the acceleration of the ascent process shall not be ignored, and the impact of aerodynamic force and torque on attitude control shall also be considered. As such, it is necessary to select the appropriate shape and ascending process to ensure that the aerodynamic characteristics (including lift–drag ratio, etc.) can adapt to the corresponding lift-off and ascending process. The gravitational field of Mars is greater than that of the Moon. Therefore, the gravity loss during ascent is greater. In the selection of propulsion system, to achieve greater efficiency, the thrust–weight ratio shall be greater, the propulsion system elements, especially the propellant and propulsion products shall also be can adapt to the extreme environments. For example, generally the temperature of the probe is relatively high when it is near the Earth for the Mars sample return mission, and it is not difficult to store the propellant products. However, on the surface of Mars, the temperature is generally low and the long-term storage temperature may be below -100 °C, a new challenge to propellant or solid fuel. In addition, ignition and operation of the engine under such low-temperature conditions shall also be considered. To solve this problem, it is necessary to improve the adaptability to low-temperature environment by applying innovative propulsion technologies such as changing the composition of propellant or solid fuel; in the probe design, a temperature environment that the propulsion system can adapt to shall also be provided by active thermal control measures to guarantee the temperature of the propulsion products, for example, electric heating, isotope heat source, fluid circuit, etc.

3. **Rendezvous, docking and sample transfer technology**

The key technologies of rendezvous, docking (RVD) and sample transfer include RVD guidance, navigation and control, docking and sample transfer mechanism.

After the functions and performance determined from the flight procedure design, the relevant design shall be conducted accordingly.

In the process of rendezvous and docking, generally the major tasks of guidance, navigation and control include the orbit and attitude control of two spacecraft modules or complexes to achieve phasing orbit control and short-range (homing, closing, final approach) control, including capture, tracking and measurement, short-range guidance until approaching the ascent module and satisfying the docking conditions. Among them, in conventional orbit and attitude control, reference can be made from the related technology of traditional spacecraft. Even for lunar sample return missions, may also be completed under the guidance from ground control. However, for other celestial missions, such as the Mars sample return mission, due to the long distance and time delay, it is impossible to completely rely on ground support on the Earth to complete the orbit and attitude control. Therefore, the requirements on autonomous guidance, navigation and control are even higher. Generally, short-range guidance and control is realized fully relying on the autonomous operation of the probe. Besides a higher requirement on the autonomous of the GNC computer, the control accuracy of the guidance control methods should meet the requirements of subsequent docking operation. In addition, the most important technical challenge is the relative measurement, i.e., measuring the motion parameters between two modules by onboard at present the main relative measurement technologies include microwave ranging and velocity measurement, laser ranging and velocity measurement, optical imaging ranging and velocity measurement, etc. The operating distance, measurement accuracy and adaptability to the environment of such methods may vary greatly. It is necessary to configure one or more sensors according to the mission requirements to jointly perform relative measurement at close range, thereby realizing short-range autonomous rendezvous control.

Once the relative relationship between the two modules meets the initial conditions of docking or sample transfer, the two modules shall be reliably connected together through the docking mechanism and the samples and containers can be transferred from one module to another through the sample transfer mechanism, or directly transferred without docking. Docking mechanisms available for consideration include cone-type docking mechanisms (impact type), which is widely applied in space industry, gripper or three-finger type. Another option is direct capture in non-docking conditions using various types of mechanisms, for example, rigid robotic arm capture, flexibility net capture, etc. Regardless of the specific form of the mechanisms, an integrated design of complicate mechanisms is often required to achieve docking and sample transfer.

4. **Ultra-high-speed Earth atmosphere reentry and return technology**

In the sample return missions of extraterrestrial celestial bodies, before the landing of the samples on the surface of the Earth, ultra-high-speed Earth atmosphere reentry is necessary to recover the samples to the ground for research. According to the results of mission analysis, there are usually two optional technical approaches: semi-ballistic reentry and ballistic reentry. On the basis of technologies of aerodynamic force, heat and thermal protection of reentry from low Earth orbit, the challenges faced by different reentry approaches are different for ultra-high-speed reentry.

In case of semi-ballistic reentry, main efforts shall be devoted to solve technical challenges such as high-precision aerodynamic/thermal environment prediction, lightweight thermal protection, semi-ballistic reentry GNC and miniaturized recovery technology. In terms of the high-precision aerodynamic/thermal environment prediction, several entry points such as six-degree-of-freedom aerodynamic deviation calculation and processing method for reentry capsule center of gravity offset and shape change, aerodynamic force and moment coefficient prediction algorithm for free molecular flow zone, transitional flow zone and slip flow zone, real gas effect aerodynamic and thermal high-precision numerical simulation methods based on chemical reaction kinetics model and ground wind tunnel test shall be taken to establish the aerodynamic database for the reentry capsule, which is used as the input of other designs. In terms of lightweight thermal protection technology, it is necessary to determine the heatproof material from the aspects of heat-resistant material selection, ablation model and mechanism research, ground ablation test research, etc., which are suitable for heat flux and large area, and then optimize the heatproof structure to determine thermal protection material configuration, adjacent material ablation rate matching and other design to ensure the accuracy of ablation back-off prediction and achieve lightweight thermal protection. In terms of semi-ballistic reentry GNC, solutions shall be established from aspects such as longitudinal/transverse guidance method, reentry corridor planning and evaluation method, and landing point prediction algorithm and high-reliability fusion navigation method under large dynamic range in order to satisfy the requirements of long range and high-precision navigation of semi-ballistic reentry. In terms of miniaturized recovery, research shall be made on aspects such as parachute-opening load design method, multistage parachute inter-stage connection method and cover ejection parachute release to determine the form, material, fabrication and packaging of the parachute and to adapt to the functional and performance requirements of the reentry capsule for final parachute deceleration.

In terms of ballistic reentry, technical challenges such as transition under ultra-high-speed and ultra-high Reynolds number turbulent flow, reentry structure heat-resistant design and high-strength supersonic parachute design shall be resolved. A method combining numerical simulation analysis considering the coupling of thermochemical non-equilibrium and turbulence model with sorghum transfer heat test shall be considered, and efforts should be made to determine the variation law of momentum Reynolds number during the transition process, set the transition altitude of the reentry process and the change of the heat flux distribution of the rear body before and after the transition, and solve the difficulties of aerodynamic analysis of ultra-high-speed and ultra-high Reynolds number. Efforts shall also be made to develop functionally graded materials that can withstand high heat flux peaks and high enthalpy. Based on the thermal variations of different parts, different thermal protection materials and structural forms shall be applied to optimize the configuration and the force transfer path; moreover, the thermal protection materials are configured in layers with variable densities to improve the thermal protection and insulation performance. The design of the high-strength supersonic parachute can be realized by technical means such as improving the drag coefficient of the parachute, applying lightweight and high-strength materials for the parachute, and increasing packing density.

3.6.5 Design Validation

For sample return missions such as Chang'e 5 lunar sample return mission, the validation tests in the engineering development which are different from those of the probes in other mission types include lift-off stability test, lift-off plume deflection test, comprehensive test of lift-off, RVD and sample transfer test, aerodynamic test of reentry capsule, ablation test, parachute airdrop test, etc. Among them, tests such as aerodynamic test of reentry capsule, ablation test and parachute airdrop test are similar to the tests of recovery satellites and manned space flights with only minor differences in terms of test conditions, test scenario setting, and referring to Chap. 14 for details.

3.7 Prospects

As the systematic, top layer and comprehensive design, the system design of deep space probes plays an important role in the overall design process of the probes. It involves multiple disciplines, interfaces, and systems and requires uniform coordination and treatment on the relationship between innovation and inheritance, as well as global and local problems. To improve the level of the system design, consummated norms and methods should be established for the system design of the probes, and technological reserve should be built up in advance for different targets, starting with conducting researches and building various types of models of the target celestial bodies, and advanced technologies such as computer, intelligent control, and information communication should be integrated to establish mission simulation analysis methods and professional analysis models, which would better satisfy the demands of future deep space explorations.

References and Related Reading

1. Ye P, Peng J (2006) Deep space exploration and its prospect in China. Eng Sci 8(10):13–18 (Chinese vision)
2. Ye P, Huang J, Sun Z et al (2014) The process and experience in the development of Chinese lunar probe. Sci China Technol Sci 44(6):543–558 (Chinese vision)
3. Wu W, Yu D (2014) Development of deep space exploration and its future key technologies. J Deep Space Explor 1(1):5–17 (Chinese vision)

4. Ye P, Sun Z, Rao W (2007) Research and development of Chang'e-1. Spacecraft Eng 16:9–10 (Chinese vision)
5. Ye P, Huang J, Zhang T et al (2013) Technical achievements of Chang'e-2 probe and prospect on China deep space exploration. Sci China Technol Sci 43:467–477 (Chinese vision)
6. Sun Z, Zhang T, Zhang H et al (2014) The technical design and achievements of Chang'e-3 probe. Sci China Technol Sci 44:331–343 (Chinese vision)
7. Yang M, Zhang G, Zhang W et al (2015) Technique design and realization of the circumlunar return and reentry spacecraft of 3rd phase of Chinese lunar exploration program. Sci China Technol Sci 45(2):111–123 (Chinese vision)
8. Peng C (2011) System design of spacecraft. Science and Technology of China Press, Beijing (Chinese vision)

Chapter 4
Technology of Orbit Design

4.1 Introduction

Orbit design is very crucial for spacecraft design. Orbit design for deep space exploration missions includes transfer trajectory design, mission orbit design, orbital control strategy design, etc. General procedures of orbit design are as follows. First, actual mission requirements and constraints of launch vehicle, TT&C and spacecraft systems are analyzed with a reasonable tool model during the conceptual design stage, based on which a proper orbit and a preliminary propellant budget that satisfy the mission requirements are proposed to lay a foundation for future designs. Second, in the preliminary design stage, by orbit property analysis and close cooperation with the engineers from other systems to further clarify the relevant constraints, the orbit designers optimize the initial orbit design to obtain an improved flight profile and preliminary orbit design report. Third, the orbit design model and algorithms are fine-tuned during the final design stage to be used for further optimizing the orbit design; in this stage, orbit maneuver strategies for the contingency are devised under the consideration of possible extended missions.

Improvements and developments are constant in today's orbit design technology. This chapter discusses the basic approaches in orbit design of deep space exploration missions, basing on the engineering design practices of Chinese deep space missions. Problems that need to be considered in orbit design are also addressed. The chapter concludes with a discussion of the future of orbit design technology [1–7].

4.2 Classical Types of Orbits

Before discussing details of some orbit design techniques, a few examples of classical types of orbits for deep space missions are presented in the following.

4.2.1 Lunar Exploration Orbit

The mission objective of a lunar probe could be lunar hard landing, lunar soft landing, lunar orbit flying, or lunar flyby. For different lunar missions, the flight orbit may be composed of all or part of the following segments: launch trajectory, Earth-to-Moon transfer orbit, lunar orbit, lunar landing trajectory and Moon-to-Earth transfer orbit. A brief description of typical flight orbit of a lunar mission is given in the following by using the Chang'e 3 mission as an example.

The flight profile of the Chang'e 3 mission consists of four phases, namely launch, Earth-to-Moon transfer, lunar orbit flying and powered descent to lunar surface, as shown in Fig. 4.1. The launch phase started at launch vehicle lift off and ended when Chang'e 3 was separated from the launch vehicle and achieved the required initial trans-lunar velocity, entering a large elliptical Earth-to-Moon transfer trajectory with about 200 km perigee altitude and about 380,000 km apogee altitude. The Earth-to-Moon transfer phase started at the Trans-Lunar Insertion (TLI) and ended right before the probe executed lunar braking, with a flight time of about 5 days, during which

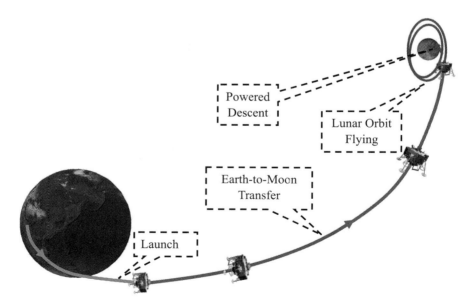

Fig. 4.1 Chang'e 3 flight orbit

three trajectory correction maneuvers were carried out. The lunar orbit flying phase started when Chang'e 3 performed Lunar Orbit Insertion (LOI) at perilune and ended right before the powered descent. When LOI was completed, Chang'e 3 entered a circular lunar orbit with 100 km orbit altitude. The probe flew in its lunar orbit for several days, waiting for the landing opportunity, during which several operations such as on-board device checkout and precision orbit determination were conducted. Later, at a proper orbital position, Chang'e 3 executed a descent maneuver and entered an elliptical lunar orbit, with 15 km perilune altitude and 100 km apolune altitude (denoted as 15 km × 100 km); then, the probe started the powered descent from the 15 km perilune and achieved a soft landing on the lunar surface minutes later.

4.2.2 Planetary Exploration Orbit

Flight orbit of a planetary mission are the same with lunar exploration missions, except for the frequent adoption of orbit control techniques such as planetary gravity assist (GA), deep space maneuver (DSM) and low thrust.

The first deep space probe that applied planetary GA is the Mariner 10 spacecraft. The flight orbit of Mariner 10 is illustrated in Fig. 4.2, which includes a Venus GA (1974-2-5) and a Mercury GA (1974-3-29). During each flyby, observation of the planet was conducted.

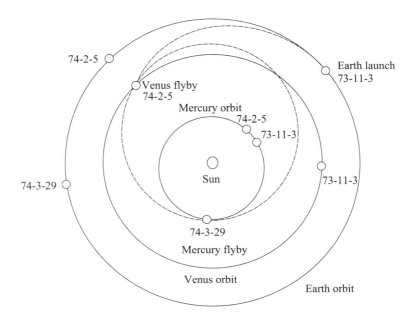

Fig. 4.2 Mariner 10 trajectory

The flight orbit of China's first Mars mission consists of the launch trajectory, the Earth-to-Mars transfer trajectory, Mars orbit insertion, the Mars parking orbit, the relay communication orbit, the remote-sensing mission orbit and the Martian atmosphere entry trajectory, as illustrated in Fig. 4.3. A deep space maneuver is arranged during the Earth-to-Mars transfer, which helps achieve a wider launch window and reduce the total delta-V budget of the whole mission.

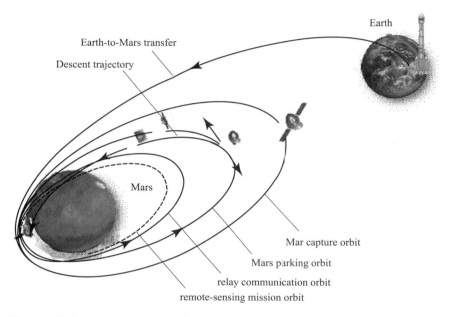

Fig. 4.3 Flight orbit of first mars mission in China

4.2.3 Asteroid Exploration Orbit

Different transfer orbit schemes are selected in asteroid exploration missions, depending on the type of the destination asteroid and their distance from the Earth. For near-Earth asteroids, a direct transfer trajectory is usually selected, in which deep space maneuvers (DSMs) can be included and several trajectory correction maneuvers (TCMs) can be arranged in the real flight. For the exploration of asteroids far away from the Earth, such as the main belt asteroids, the orbit energy of the asteroid is much higher than the Earth orbit about the Sun. Therefore, the velocity impulse (delta-V) budget is unrealistically large if a direct transfer trajectory is selected. Therefore, multiple planetary GAs and electric propulsion are usually applied in designing the transfer trajectory to reduce the total fuel consumption for the entire mission [8–10].

The Near Earth Asteroid Rendezvous (NEAR) mission of the USA is the first asteroid exploration mission in the world. The NEAR spacecraft was launched on February 17, 1996, and arrived at the S-type asteroid Eros on February 14, 2000; it became the first spacecraft to rendezvous with and achieve orbit around an asteroid; later, the probe conducted a comprehensive observation and measurement operation on the asteroid. During the transfer flight to Eros, one Earth GA and multiple DSMs are utilized, and the probe flew by asteroid Mathilde. The flight orbit scheme of the NEAR mission is illustrated in Fig. 4.4.

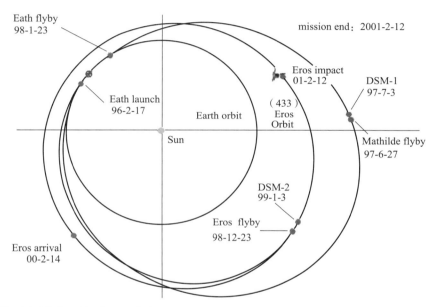

Fig. 4.4 Flight orbit of near mission

The HAYABUSA mission of Japan is the first asteroid sample return mission in the world, in which a full-scale exploration of asteroid Itokawa was realized and invaluable information was acquired. The HAYABUSA spacecraft was launched on May 9, 2003; one Earth GA was performed during the flight, and the probe arrived at Itokawa on September 14, 2005. Then, it started in situ investigations, landed on Itokawa and collected some tiny grains of asteroid material, which were successfully returned to Earth aboard the probe on June 13, 2010. The HAYABUSA spacecraft was equipped with Ion engines as its main propulsion, and its flight orbit design is a typical example of joint applications of planetary GA and low-thrust trajectory techniques.

In 2015, the Dawn, a US spacecraft flew to Vesta and Ceres successively and achieved a great success in exploring these two protoplanets, using the orbit design techniques of planetary GA and Low-Thrust (LT) trajectory. This represents the latest achievement of orbit design technology in multi-target and multi-task missions. Ion propulsion is adopted in the Dawn mission, which enables the spacecraft to enter and leave the orbits about multiple celestial bodies at ease because of the property of LT trajectory. In contrast, a spacecraft utilizing traditional chemical propulsion, generally speaking, can only perform planetary GA and flyby before it reaches the destination celestial body. The GA-LT trajectory of Dawn is illustrated in Fig. 4.5.

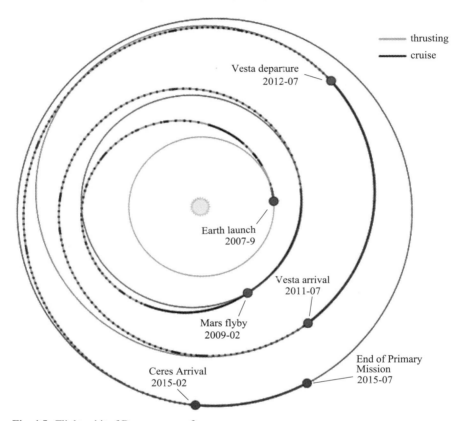

Fig. 4.5 Flight orbit of Dawn spacecraft

4.2.4 Libration Point Exploration Orbit

The main orbit types in missions to Lagrange libration points include Lissajous orbits and halo orbits, both of which have been successfully realized in real deep space missions. For example, China has accomplished three missions to Lagrange libration points. Chang'e 2 entered a Lissajous orbit about the Sun–Earth L2 libration point, and the flight test vehicle of Chang'e 5 entered a Lissajous orbit about the Earth–Moon L2 libration point.

Moreover, in China's Chang'e 4 mission, a relay communication satellite was launched in May 2018 and later deployed in a halo orbit about the Earth-Moon L2 libration point. The flight orbit of the Chang'e 4 relay communication satellite includes a launch trajectory, an Earth-to-Moon transfer trajectory, a Moon-to-L2 transfer trajectory, a capture (Lissajous) orbit and the primary mission orbit, which is a halo orbit about the L2 libration point with the out-of-plane amplitude A_z being 13,000 km. The flight orbit is illustrated in Fig. 4.6.

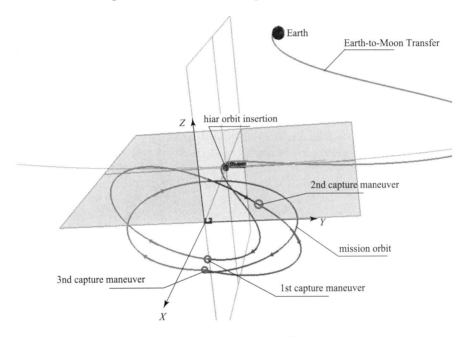

Fig. 4.6 Flight orbit of Chang'e 4 relay communication satellite

4.3 Brief Introduction to Orbit Design Procedures

The first step in orbit design is to define the mission objective. Different mission objectives may lead to quite different orbit design results. For deep space exploration missions, different destination celestial bodies, different detection methods or different propulsion systems may necessitate selecting different flight profiles, which has to be clarified in the very beginning of orbit design.

Then, an appropriate dynamic model should be selected according to the computation accuracy requirements for different orbit schemes. A suitable computation model is needed to numerically simulate the flight orbit in orbit propagation or optimization.

For different candidate orbit schemes, intensive numeric analyses are carried out to find out the properties of each orbit scheme, which are studied jointly with the science objective of the specific mission to determine the orbit scheme that best satisfy the mission requirements and constraints.

Specifically speaking, the numeric analyses may include a study of the implications of the launch window selection, the characteristics of orbit evolution, orbital error propagation, delta-V budget for different orbit parameters, tracking and lighting conditions for different launch windows, etc.

During the entire orbit design phase, sufficient considerations should be given to the requests from the systems engineering division, the capability of the TT&C system, the capability of the launch vehicle and the requirements from the spacecraft subsystems and payload. Constant analyses and iterations with the other systems have to be carried out until all design requirements and constraints are satisfied. Major engineering constraints are as follows.

1) Mission Objective

   There is no such thing as a general orbit design that meets the needs of different missions. Different mission objectives necessitate selecting different orbit types and orbit schemes. Take lunar exploration as an example. If the mission objective is to complete a global remote-sensing task, then a polar circular orbit about the Moon should be selected.

2) Requirements on Launch windows and Mission Duration from Systems Engineering

   Again, take lunar exploration as an example. If the mission objective is to perform remote-sensing tasks from a lunar orbit, then the selection of launch windows determines the initial BETA angle of the lunar orbit. For lunar landing missions, the lighting condition at the landing site is also related to launch windows; moreover, the mission duration may affect the initial flight direction of the lunar probe in its lunar orbit.

3) Constraints of Launch Site and Launch Vehicle Capability

   The location of the launch site and the limitation on the achievable argument of perigee of the initial transfer orbit are directly related to the search of feasible launch windows; therefore, these constraints have a great influence on orbit design results.

4) Constraints of TT&C System

   Based on the requirements of TT&C system on tracking period duration and the corresponding orbit determination accuracy, the spacecraft maneuver strategies can be devised, and the needed delta-V can be computed.

5) Constraints of Spacecraft Subsystems

   The constraints from spacecraft subsystems should be considered carefully in orbit design, including the inputs from the payload, thermal control system, power system, GNC and propulsion system.

   To sum up, orbit design tasks include the selection of orbit type, the determination of orbit elements, orbit evolution analysis, orbit error analysis, delta-V budget analysis, orbital maneuver strategy analysis, analysis of tracking and lighting conditions, etc. Orbit design engineers need to take into account all the design constraints, perform intensive analyses on the parameters related to orbit design and propose a flight orbit scheme that meets the mission requirements. The proposed orbit scheme should be optimized constantly to achieve (for example) a minimum total delta-V.

4.4 Design of Transfer Trajectories

Design of transfer trajectory is critical in orbit design for deep space exploration missions. In engineering design of transfer trajectory, different types of transfer trajectory and different orbit parameters are selected according to the mission objective, the requirements from launch system, the capability of propulsion system and the mission constraints on the lighting and tracking conditions.

Types of transfer trajectories include direct transfer trajectory, transfer trajectory with GA and DSM, and LT trajectory, applicable for different mission objectives. Direct transfer trajectory is usually used in lunar exploration missions and single planet missions. However, due to the constraints from launch vehicle, the launch window may be quite short. A transfer trajectory scheme that includes DSMs can compensate for a weak launch capability, and the launch window can be extended to include a longer period of days. Orbital design techniques of GA and LT trajectory are more widely applied in the exploration of asteroids and multi-target multi-tasks deep space exploration missions. Detailed discussions on these different types of transfer trajectory are conducted in the following.

4.4.1 Direct Transfer

1. **Hohmann transfer**

Taking Earth-to-Moon transfer trajectory as an example, the most energy-efficient Earth-to-Moon transfer for a lunar probe is to fly a Hohmann transfer trajectory, in which the elliptical transfer orbit is tangential to the parking orbit of the probe about the Earth at perigee and is also tangential to the orbit of the Moon about the Earth at apogee, as illustrated in Fig. 4.7.

The orbital parameters of Hohmann transfer trajectory can be computed as follows.

The eccentricity is given by

$$e = \frac{R_a - R_p}{R_a + R_p} = \frac{D - R_p}{D + R_p} = \frac{D - R_e - H}{D + R_e + H}, \quad (4.1)$$

where e is the eccentricity of the transfer trajectory; R_p is the radius of the parking orbit; R_a is the radius of the orbit of the Moon about the Earth; D is the distance

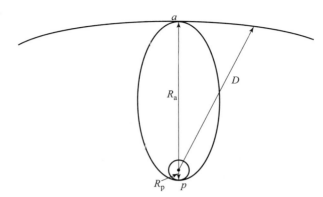

Fig. 4.7 Hohmann transfer trajectory

between the Earth and the Moon; R_e is the radius of the Earth; H is the altitude of the parking orbit.

The semi-major axis is given by

$$a = \frac{R_a + R_p}{2} = \frac{D + R_e + H}{2} \tag{4.2}$$

The perigee speed is

$$v_p = \sqrt{\frac{2\mu}{R_p} \frac{R_a}{R_a + R_p}} = \sqrt{\frac{2\mu}{R_e + H} \frac{D}{D + R_e + H}}, \tag{4.3}$$

where μ is the gravitational constant of the Earth. With $R_p = 6600$ km, then

$$v_p = 10.90 \text{ km/s} \tag{4.4}$$

The time of flight from perigee to perilune is

$$T = \pi \sqrt{a^3/\mu} = 5 \text{ days} \tag{4.5}$$

If a shorter transfer time is desired for, the semi-major axis can be increased such that the velocity of the probe (with respect to a geocentric inertial frame) is not parallel to the velocity of the Moon when the probe arrives at the Moon. In this way, the time of flight on the transfer trajectory can be significantly reduced, with only limited increase in braking delta-V at perilune. As can be seen from Eq. (4.3), when R_a is increased to infinity, the perigee velocity would be $v_p = \sqrt{2\mu/R_p}$.

Homann transfer trajectory can be utilized in the design of transfer trajectory between planets. Considering the Hohmann transfer trajectories between the Earth and other planets, the interplanetary transfer trajectory parameters such as the escape hyperbolic excess velocity, time of flight, launch window data can be computed using the two-body model, as illustrated in Table 4.1.

4.4 Design of Transfer Trajectories

Table 4.1 Hohmann transfer between Earth and planets

	Escape hyperbolic excess velocity/(km/s)	Time of transfer month	Frequency of launch opportunity/month
Mercury	7.5	3.5 months	4
Venus	2.5	4.9 months	19
Mars	2.9	8.6 months	26
Jupiter	8.8	2.7 years	13
Saturn	10.3	6 years	13
Uranus	11.3	16 years	12
Neptune	11.7	30.7 years	12

2. **Lambert transfer**

The Lambert's problem can be stated as: Given two position vectors r_1, r_2 and time of flight Δt, determine the transfer trajectory, as illustrated in Fig. 4.8.

Fig. 4.8 Lambert's problem

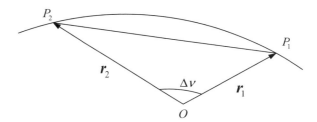

In the following, the universal variable approach is employed to solve the Lambert's problem.

First, the universal variable approach is applied to solve the problem of orbit propagation in two-body model, namely given the position vector r_0 and the velocity vector v_0 at instant t_0, determine the position vector r and velocity vector v at instant t.

A universal variable x is defined as follows.

$$x = \sqrt{a}\Delta E, \tag{4.6}$$

where a is the semi-major axis, E is the eccentric anomaly. The time derivative of the universal variable is

$$\dot{x} = \frac{\sqrt{\mu}}{r}, \tag{4.7}$$

where μ is the gravitational constant of the central celestial body, r is the distance between the probe and the center of the central celestial body.

Also, a variable z is defined as

$$z = \frac{x^2}{a} \tag{4.8}$$

then

$$z = \Delta E^2 \tag{4.9}$$

The following equations can be derived:

$$\sqrt{\mu}(t - t_0) = x^3 S + \frac{r_0 \cdot v_0}{\sqrt{\mu}} x^2 C + r_0 x (1 - zS) \tag{4.10}$$

$$r = x^2 C + \frac{r_0 \cdot v_0}{\sqrt{\mu}} x (1 - zS) + r_0 (1 - zC), \tag{4.11}$$

where r_0 and v_0 are the position and velocity vectors of the spacecraft at the instant t_0, respectively; r is the distance between the spacecraft and the center of the primary celestial body at the instant t; S and C are the two series with respect to the variable z, as illustrated in Eqs. (4.12) and (4.13).

$$C(z) = \sum_{k=0}^{\infty} \frac{(-z)^k}{(2k+2)!} = \begin{cases} \dfrac{1 - \cos \sqrt{z}}{z} & z > 0 \\ \dfrac{1 - \cosh \sqrt{-z}}{z} & z < 0 \\ \dfrac{1}{2} & z = 0 \end{cases} \tag{4.12}$$

$$S(z) = \sum_{k=0}^{\infty} \frac{(-z)^k}{(2k+3)!} = \begin{cases} \dfrac{\sqrt{z} - \sin \sqrt{z}}{\sqrt{z^3}} & z > 0 \\ \dfrac{\sinh \sqrt{-z} - \sqrt{-z}}{\sqrt{(-z)^3}} & z < 0 \\ \dfrac{1}{6} & z = 0 \end{cases} \tag{4.13}$$

The, n the position vector r and velocity vector v of the spacecraft at the instant t can be written as

$$r = f r_0 + g v_0 \tag{4.14}$$

$$v = \dot{f} r_0 + \dot{g} v_0, \tag{4.15}$$

where $f, g, \dot{f}, \dot{g}$ are the scalars related to time, and the calculation formulas are as follows.

4.4 Design of Transfer Trajectories

$$f = 1 - \frac{x^2}{r_0}C \tag{4.16}$$

$$g = t - \frac{x^3}{\sqrt{\mu}}S \tag{4.17}$$

$$\dot{f} = \frac{\sqrt{\mu}x(zS-1)}{r_0 r} \tag{4.18}$$

$$\dot{g} = 1 - \frac{x^2}{r}C \tag{4.19}$$

Next, two variables A and y are introduced:

$$A = \frac{\sqrt{r_1 r_2}\sin\Delta\nu}{\sqrt{1-\cos\Delta\nu}} \tag{4.20}$$

$$y = \frac{r_1 r_2(1-\cos\Delta\nu)}{a(1-e^2)}, \tag{4.21}$$

where $\Delta\nu$ is the angle between the two position vectors, namely $\cos\Delta\nu = r_1 \cdot r_2$; e is the eccentricity of the orbit. It can be shown that

$$y = x^2 C \tag{4.22}$$

and Eq. (4.21) is simplified to

$$y = r_1 + r_2 - A\frac{(1-zS)}{\sqrt{C}} \tag{4.23}$$

Applying the definitions of the above variables, Eq. (4.10) is simplified to

$$\sqrt{\mu}\Delta t = x^3 S + A\sqrt{y} \tag{4.24}$$

Similarly, the variables f, g and $\dot{g}$ can be written in terms of A and y as

$$f = 1 - \frac{y}{r_1} \tag{4.25}$$

$$g = A\sqrt{\frac{y}{\mu}} \tag{4.26}$$

$$\dot{g} = 1 - \frac{y}{r_2} \tag{4.27}$$

Then, the velocity vectors at point P_1 and P_2 (Fig. 4.8), namely v_1, v_2 can be calculated using the following two equations:

$$v_1 = \frac{r_2 - f r_1}{g} \tag{4.28}$$

$$v_2 = \frac{\dot{g} r_2 - r_1}{g} \tag{4.29}$$

The first step in the orbit design for planetary exploration missions is to design the transfer trajectory, for which Lambert's method can be employed to compute the departure velocity and arrival velocity, given the departure date and arrival date. Repeat the computation for different combinations of departure date and arrival date from the eligible departure date range and arrival date range and plot the energy contour diagram, from which the energy-optimal departure date and arrival date can be identified. Lambert's method results in a heliocentric trajectory that is less than one revolution about the Sun, and this type of transfer is termed single-revolution Lambert transfer. For more complicated deep space exploration missions, multi-revolution Lambert transfer finds broad applications, especially for sample return missions, in which multi-revolution transfer can be utilized to adjust the waiting time between two successive launch opportunities, or optimize the departure parameters to satisfy launch capability.

Once the departure velocity $\bar{v}_1$ is determined using Lambert's method, the right ascension of ascending node (RAAN) and argument of perigee can be computed for a specified value of departure orbit inclination, which in turn can be used as the initial values for the precise orbit computation. Taking Earth departure orbit as an example (Fig. 4.9), the departure orbital parameters can be computed as follows.

$$\bar{v}_\infty^+ = \bar{v}_1 - \bar{v}_e \tag{4.30}$$

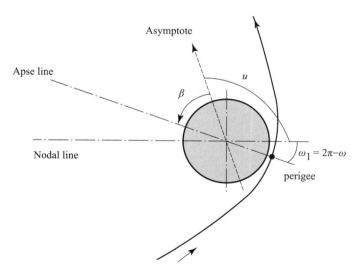

Fig. 4.9 Departure hyperbolic trajectory

4.4 Design of Transfer Trajectories

$$a = -\frac{\mu}{v_\infty^2} \quad (4.31)$$

$$e = 1 - \frac{r_p}{a} \quad (4.32)$$

$$\cos \beta = \frac{1}{e} \quad (4.33)$$

$$\vec{v}_\infty = A_z(-\Omega) A_x(-i) A_z(-u) \begin{pmatrix} v_\infty \\ 0 \\ 0 \end{pmatrix} = v_\infty \begin{pmatrix} \cos\Omega \cos u - \sin\Omega \cos i \sin u \\ \sin\Omega \cos u + \cos\Omega \cos i \sin u \\ \sin i \sin u \end{pmatrix}$$
(4.34)

$$\omega = u + \beta + \pi \quad (4.35)$$

3. **Numerical computation**

From the discussions above, the initial values of departure orbital parameters can be obtained based on two-body model analysis of the transfer trajectory. These initial values are to be corrected using orbital integration and state transition matrix in order to calculate the precise transfer trajectory. Taking Earth-to-Moon transfer trajectory as an example, the initial state of the transfer trajectory is given by the osculating orbital inclination, perigee altitude, perigee speed, argument of perigee and right ascension of ascending node (RAAN); the desired end state on the transfer trajectory is given by the specified values of perilune altitude, lunar orbit inclination, lunar orbit true anomaly or the angle between perilune velocity and position vector.

The problem statement is: given the transfer time, the initial orbital inclination and perigee altitude, the other initial orbital parameters (perigee speed, argument of perigee, RAAN) are to be found so that the desired values of the orbital parameters at the end of the transfer trajectory are achieved and the transfer trajectory is completely determined. There is no analytical solution to this problem and a numeric approach can be employed to find the solution. The key to this numeric solution is to determine the relationship between the end state error and the initial state error, which is implicated by the state transition matrix relating the initial state to the end state. As shown in the following discussion, it is essential to carry out an iterative calculation process utilizing this state transition matrix in order to find the precise solution of the transfer trajectory [1, 11].

Assume that $(\vec{r}, \vec{v})$ are the geocentric position vector and velocity of the lunar probe with respect to the geocentric inertial frame, and $(\vec{r}_0, \vec{v}_0)$ are the geocentric position vector and velocity of the probe at perigee; also assume that $(\vec{r}_1, \vec{v}_1)$ are the position vector and velocity of the lunar probe with respect to the Moon-centered inertial frame, with their Cartesian frame components expressed in the geocentric equatorial inertial frame. Moreover, let $\vec{p}, \vec{q}$ be the state vectors at perigee and perilune, respectively, as illustrated in Eqs. (4.36) and (4.37):

$$\vec{p} = \begin{bmatrix} v_p \\ \Omega \\ \omega \end{bmatrix}, \quad \vec{q} = \begin{bmatrix} q_1 \\ q_2 \\ q_3 \end{bmatrix} \quad (4.36)$$

$$q_1 = H_m, \quad q_2 = \frac{\vec{r}_1 \cdot \vec{v}_1}{r_1 v_1}, \quad q_3 = \vec{k} \cdot \frac{\vec{r}_1 \times \vec{v}_1}{|\vec{r}_1 \times \vec{v}_1|}, \quad (4.37)$$

where v_p is perigee speed, Ω is RAAN, ω is argument of perigee, H_m is perilune altitude, $\bar{k}$ is the unit vector of z-axis of the Moon-centered equatorial inertial frame, with its Cartesian frame components expressed in the geocentric equatorial inertial frame. As can be seen from Eq. (4.37), q_2 is the cosine of the angle between the velocity $\vec{v}_1$ and the position vector $\vec{r}_1$ of the probe at perilune, q_3 is the cosine of the osculating orbital inclination of the probe at perilune with respect to the Moon-centered equatorial inertial frame.

The end state $\vec{q}$ is related to the initial state $\vec{p}$ as follows.

$$\vec{q} = \vec{q}(\vec{p}) \quad (4.38)$$

The errors of the two states are related in linearization:

$$\Delta \vec{q} = \frac{\partial \vec{q}}{\partial \vec{p}} \Delta \vec{p} \quad (4.39)$$

Applying this relationship, the precise numeric solution can be found after several iterations, based on appropriate initial values. This approach is frequently employed in solving functional equations. The partial derivative in Eq. (4.39) is the error propagation matrix, the numeric solution of which can be obtained from numerical integrations as shown in the following.

Applying the chain rule of differentiation gives

$$\Delta \vec{q} = \frac{\partial \vec{q}}{\partial \begin{pmatrix} \vec{r} \\ \vec{v} \end{pmatrix}} \cdot \frac{\partial \begin{pmatrix} \vec{r} \\ \vec{v} \end{pmatrix}}{\partial \begin{pmatrix} \vec{r}_0 \\ \vec{v}_0 \end{pmatrix}} \cdot \frac{\partial \begin{pmatrix} \vec{r}_0 \\ \vec{v}_0 \end{pmatrix}}{\partial \vec{p}} \cdot \Delta \vec{p} = T_s \cdot \Delta \vec{p} \quad (4.40)$$

Obviously, the error propagation matrix T_s is the multiplication of three matrixes S, R, T:

$$T_s = S \cdot R \cdot T, \quad (4.41)$$

where

4.4 Design of Transfer Trajectories

$$S = \frac{\partial \vec{q}}{\partial \begin{pmatrix} \vec{r} \\ \vec{v} \end{pmatrix}} = \begin{pmatrix} \frac{\partial q_1}{\partial x} & \frac{\partial q_1}{\partial y} & \frac{\partial q_1}{\partial z} & \frac{\partial q_1}{\partial v_x} & \frac{\partial q_1}{\partial v_y} & \frac{\partial q_1}{\partial v_z} \\ \frac{\partial q_2}{\partial x} & \frac{\partial q_2}{\partial y} & \frac{\partial q_2}{\partial z} & \frac{\partial q_2}{\partial v_x} & \frac{\partial q_2}{\partial v_y} & \frac{\partial q_2}{\partial v_z} \\ \frac{\partial q_3}{\partial x} & \frac{\partial q_3}{\partial y} & \frac{\partial q_3}{\partial z} & \frac{\partial q_3}{\partial v_x} & \frac{\partial q_3}{\partial v_y} & \frac{\partial q_3}{\partial v_z} \end{pmatrix} \quad (4.42)$$

The state transition matrix is

$$R = \frac{\partial \begin{pmatrix} \vec{r} \\ \vec{v} \end{pmatrix}}{\partial \begin{pmatrix} \vec{r}_0 \\ \vec{v}_0 \end{pmatrix}} = \begin{pmatrix} \frac{\partial \vec{r}}{\partial \vec{r}_0} & \frac{\partial \vec{r}}{\partial \vec{v}_0} \\ \frac{\partial \vec{v}}{\partial \vec{r}_0} & \frac{\partial \vec{v}}{\partial \vec{v}_0} \end{pmatrix} = \begin{pmatrix} R_{11} & R_{12} \\ R_{21} & R_{22} \end{pmatrix} \quad (4.43)$$

$$\dot{R} = \begin{pmatrix} \frac{\partial \vec{v}}{\partial \vec{r}_0} & \frac{\partial \vec{v}}{\partial \vec{v}_0} \\ \frac{\partial \vec{g}}{\partial \vec{r}_0} & \frac{\partial \vec{g}}{\partial \vec{v}_0} \end{pmatrix} = \begin{pmatrix} \frac{\partial \vec{v}}{\partial \vec{r}_0} & \frac{\partial \vec{v}}{\partial \vec{v}_0} \\ \frac{\partial \vec{g}}{\partial \vec{r}} \cdot \frac{\partial \vec{r}}{\partial \vec{r}_0} & \frac{\partial \vec{g}}{\partial \vec{r}} \cdot \frac{\partial \vec{r}}{\partial \vec{v}_0} \end{pmatrix} = \begin{pmatrix} R_{21} & R_{22} \\ \frac{\partial \vec{g}}{\partial \vec{r}} R_{11} & \frac{\partial \vec{g}}{\partial \vec{r}} R_{12} \end{pmatrix} \quad (4.44)$$

Similar to the derivation of S expression, the expression of T can be found as follows.

$$T = \frac{\partial \begin{pmatrix} \vec{r}_0 \\ \vec{v}_0 \end{pmatrix}}{\partial \vec{p}} = \begin{pmatrix} \frac{\partial \vec{r}_0}{\partial v_p} & \frac{\partial \vec{r}_0}{\partial \Omega} & \frac{\partial \vec{r}_0}{\partial \omega} \\ \frac{\partial \vec{v}_0}{\partial v_p} & \frac{\partial \vec{v}_0}{\partial \Omega} & \frac{\partial \vec{v}_0}{\partial \omega} \end{pmatrix} \quad (4.45)$$

In the following, the orbital dynamic equation is given for a lunar probe:

$$\begin{cases} \dot{\vec{r}} = \vec{v} \\ \dot{\vec{v}} = \vec{g} = -\frac{\mu_e}{r^3}\vec{r} - \mu_m \left(\frac{\vec{r}_{md}}{r_{md}^3} + \frac{\vec{r}_m}{r_m^3}\right) - \mu_s \left(\frac{\vec{r}_{sd}}{r_{sd}^3} + \frac{\vec{r}_s}{r_s^3}\right) \\ \quad - \left(\frac{3\mu_e J_2 R_e^2}{2r^5} - \frac{15\mu_e J_2 R_e^2 z^2}{2r^7}\right)\vec{r} - \frac{3\mu_e J_2 R_e^2}{r^5}\begin{pmatrix} 0 \\ 0 \\ z \end{pmatrix} \end{cases} \quad (4.46)$$

where μ_e, μ_m and μ_s are the gravitational constants of the Earth, the Moon and the Sun, respectively; J_2 is the zonal harmonic coefficient of the Earth; R_e is the mean equatorial radius of the Earth; $\vec{r}_m$, $\vec{r}_s$ are the position vectors from the Earth to the Moon and the Sun, respectively; $\vec{r}_{md}$ and $\vec{r}_{sd}$ are the position vectors of the probe from the Moon and the Sun, respectively.

To compute the needed correction of the initial state, firstly, Eqs. (4.46) and (4.44) are integrated simultaneously for a given initial condition to obtain the position, velocity of the probe and the state transition matrix at perilune; then, matrix T is obtained by substituting the given initial condition into Eqs. (4.45), and (4.46) is integrated to obtain the state at perilune, which is substituted into Eq. (4.42) to get matrix S; the multiplication of matrix S, R, T gives the error propagation matrix T_s; finally, the state error at perilune $\Delta \bar{q}$ is found by subtracting the desired perilune state from the perilune state obtained by integration, and the needed correction of the initial state is given by:

$$\Delta \bar{p} = T_s^{-1} \Delta \bar{q}, \tag{4.47}$$

where T_s^{-1} is the inverse matrix of the error propagation matrix. The initial state is to be updated by applying the correction in Eq. (4.47), and then, the preceding integration procedure will be repeated; by this way, the precise solution of the transfer trajectory will be found after a few iterations that achieves the desired values of perilune altitude and inclination (within convergence tolerance) for the specified perigee altitude, time of flight.

The iterative method of initial value correction with state transition matrix calculation has been widely applied in the solution of transfer trajectory, trajectory correction maneuver and orbit design for libration point exploration missions.

4.4.2 Deep Space Maneuver

A DSM can be arranged in the transfer trajectory of a deep space probe in order to solve some engineering problems, such as the narrow launch window for a direct transfer orbit scheme due to the constraints from the launch vehicle and the probe mass budget, the poor tracking and poor lighting conditions at the Mars arrival. For a deep space maneuver, a velocity impulse is inserted somewhere in the transfer trajectory to cause a change in orbit plane orientation and orbit parameters. The purpose of a DSM is to achieve the maximum possible payload mass for the probe so that the poor launch conditions (e.g., the constraints on launch azimuth, duration of un-powered flight) could be compensated to some extent.

The research on the application of deep space maneuver in transfer trajectory design was conducted in NASA during the mission design of the Mars orbiting exploration mission in 1970s. Research findings show that there is a significant improvement to the overall mission design, and more flexibility is available to the mission design after DSMs are introduced to the orbit scheme.

Deep space maneuvers were applied in the orbit design of the Galileo spacecraft, in which a delta-V of about 230 m/s was applied during the transfer to Jupiter, and the Earth departure C_3 (square of departure V_∞) was reduced from 118 to 80 km^2/s^2 [12, 13].

Some Chinese scholars studied the optimization method employed in the computation of deep space maneuvers, the improvements of orbital conditions at departure and arrival induced by DSMs, and a comparison of the delta-V budgets was made between the scheme of direct transfer and the scheme of interplanetary transfer with DSMs.

4.4 Design of Transfer Trajectories

The orbit design with DSMs should take into account the specific mission requirements. For different design constraints, the design results of orbit scheme may be quite different. Take the Mars exploration mission as an example. The grid search algorithm was studied first, taking into account the requirements and constraints from the launch vehicle and spacecraft system. In the study, intensive computations were carried out for grid search to find minimum-fuel solution of the transfer trajectory for varying positions of the target and DSM. Moreover, some optimization algorithms are applied to achieve the optimal transfer trajectory with DSM. It is found from the comparison study that the result from grid search is quite stable but the cost of computation time is significant; in contrast, applying the optimization algorithms in the transfer trajectory design leads to fast convergence; however, the result might be a locally optimal solution only.

A transfer trajectory with DSM is illustrated in Fig. 4.10, in which a velocity impulse is inserted at an appropriate position along the transfer trajectory that splits the transfer trajectory into two segments, namely the transfer trajectories before and post-DSM. Two velocity impulses are needed for the probe to enter the orbit about Mars, namely the velocity impulse for DSM and the velocity impulse for Mars Orbit Insertion (MOI). The objective of optimal orbit design is to achieve a maximum remaining mass of the probe after it has performed the two maneuvers, taking into account the constraints mentioned previously.

In the following discussions, it is assumed that the position of the probe coincides with the center of the planet at either Earth departure or Mars arrival. The design variables for trajectory optimization are as follows:

(1) Launch time of the probe;
(2) Time of flight for Earth-to-Mars transfer TOF;

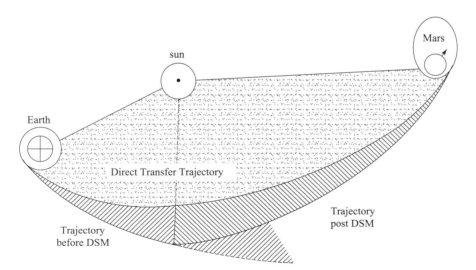

Fig. 4.10 Transfer trajectory with DSM

(3) Time of DSM;
(4) Velocity impulse vector of DSM;
(5) Hyperbolic excess velocity vector at Mars arrival;
(6) Inclination at launch.

The trajectory optimization model is constructed as follows. First, the heliocentric position and velocity of the Earth at departure (R_e, V_e) can be retrieved from a JPL ephemeris such as JPL/DE405 according to the given launch time t_L; the time of Mars arrival t_A can be calculated from t_L and the given Earth-to-Mars transfer time T_{OF}, and then, the heliocentric position and velocity of Mars at arrival (R_m, V_m) can be retrieved from the same JPL ephemeris.

Second, given the hyperbolic excess velocity of the probe at Mars arrival $V_{\infty A}$, the heliocentric position and velocity of the probe at Mars arrival (R_f, V_f) can be determined; then, a backwards integration of the dynamic equation from t_A to T_{DSM} can be carried out, using (R_f, V_f) as the initial condition, and the position and velocity of the probe right after DSM (R_{DSMt}, V_{DSMt}) can be obtained.

Third, given the velocity impulse vector of DSM ΔV_{DSM}, the position and velocity of the probe right before DSM (R_{DSM0}, V_{DSM0}) can be obtained; then, continue to carry out the backwards integration from T_{DSM} to t_L, and the position and velocity of the probe at Earth departure (R_0, V_0) can be found.

At this point, the total velocity impulse ΔV, the hyperbolic excess velocity $V_{\infty D}$ and the corresponding C_3 energy C_{3L} of the probe at Earth departure can be calculated; then, the initial mass of the probe m_0 can be figured from the launch vehicle manual based on C_{3L} just found, and the declination of $V_{\infty D}$ (denoted as i_{DECL}) can also be calculated.

At last, the total propellant expenditure m_p can be calculated, and the initial argument of perigee ω_0 at launch can be found for the given inclination at launch i_L and $V_{\infty D}$ just found.

Based on the above optimization model, the orbit design problem becomes a multi-dimensional nonlinear planning problem as stated in the following:

Design variables:

$$X = [t_L, T_{OF}, T_{DSM}, \Delta V_{DSM}, V_{\infty A}, i_L] \tag{4.48}$$

Performance measure:

$$J = m_p \rightarrow \min, \text{ or } J = m_0 - m_p \rightarrow \max, \text{ or } J = \Delta V \rightarrow \min \tag{4.49}$$

Design constraints:

$$R_0 = R_e$$
$$C_{3L} \leqslant C_{3L\max}$$
$$\max(i_{L\min}, i_{DECL}) \leqslant i_L \leqslant i_{L\max}$$
$$\omega_{\min} \leqslant \omega_0 \leqslant \omega_{\max}$$

4.4 Design of Transfer Trajectories

$$\Delta V_{\text{DSMmin}} \leq \|\Delta V_{\text{DSM}}\| \leq \Delta V_{\text{DSMmax}} \quad (4.50)$$

This nonlinear planning problem can be solved using the sequential quadratic programming (SQP) method.

Fast convergence and better design results can be achieved when applying the optimization algorithms in the transfer trajectory design with DSM. However, some results might be locally optimal solutions only, which can be verified according to the continuity of the numerical results.

4.4.3 Gravity Assist

The researches of Leverrier, Tisserand and other scholars in the nineteenth century on orbital perturbation theory of planets and comets had laid the foundation for the orbital design technique of gravity assist (GA). In the 1950s, the famous American scholar Battin first proposed the method of exploiting planetary GA to return a probe back to the Earth without propellant expenditure; later on, many scientists conducted relevant studies on GA technique for interplanetary orbit design and had applied this technique in the orbit design of several deep space missions, which were accomplished with great success. The basic principle of GA and relevant methods will be presented in the following discussions [14–16].

1. **Basic principle**

The trajectory of a probe flying by a planet is hyperbolic with respect to the inertial frame centered at the planet. The velocity of the probe relative to the planet at an infinite distance from the planet is termed as hyperbolic excess velocity and denoted as $\bar{v}_\infty$. Since the gravitational field of the planet is conserved, the incoming hyperbolic excess velocity $\bar{v}_\infty^-$ of the probe arriving at the sphere of influence of the planet has the same magnitude as the outgoing hyperbolic excess velocity $\bar{v}_\infty^+$ of the probe departing from the sphere of influence of the planet; on the other hand, there is a turn angle δ between $\bar{v}_\infty^-$ and $\bar{v}_\infty^+$ due to the influence of planet gravity.

Based on the patched conic method, the heliocentric orbit of a probe can be determined by its position and heliocentric velocity at the sphere of influence of a planet. It is assumed that the probe has the same heliocentric position vector as the planet, and its heliocentric velocity equals to the vector addition of the heliocentric velocity $\bar{v}_P$ of the planet and hyperbolic excess velocity $\bar{v}_\infty$. Therefore, the heliocentric velocity of the probe before planet flyby is $\bar{v}^- = \bar{v}_\infty^- + \bar{v}_P$, and its heliocentric velocity after planet flyby becomes $\bar{v}^+ = \bar{v}_\infty^+ + \bar{v}_P$. Consequently, the heliocentric trajectory after planet flyby is different from the trajectory before planet flyby, and a new orbit period P and heliocentric periapsis radius R_P are achieved.

The technique of GA is widely applied in interplanetary transfer trajectory design to achieve low propellant consumption due to the fact that GA can help realize the change of orbital energy and change of orbit plane orientation, with no extra propellant expenditure during the flyby.

The geometry of a hyperbolic flyby trajectory is illustrated in Fig. 4.11. The definitions of the parameters shown in the figure are given as follows:

r_p Periapsis radius of the hyperbolic flyby trajectory;
δ Turn angle between $\bar{v}_\infty^-$ and $\bar{v}_\infty^+$;
θ_∞ True anomaly of the asymptote of the hyperbolic flyby trajectory;
V_∞^- Fly-in hyperbolic excess velocity;
V_∞^+ Fly-out hyperbolic excess velocity;
Δ Distance from the planet center to the asymptote of the hyperbolic flyby trajectory;
a Semi-major axis of the hyperbolic flyby trajectory;
p Semi-latus rectum of the hyperbolic flyby trajectory;
r Distance between the probe and the center of the planet;
θ True anomaly of the probe.

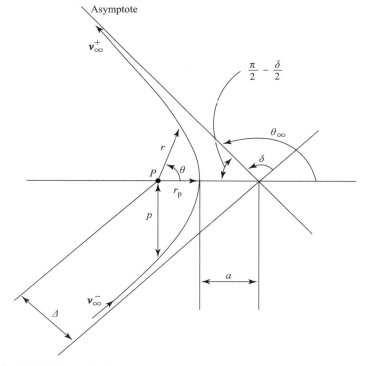

Fig. 4.11 Hyperbolic flyby trajectory

According to the basic principle of GA, Eq. (4.51) is valid:

$$|V_\infty^-| = |V_\infty^+|. \tag{4.51}$$

4.4 Design of Transfer Trajectories

The orbital equation for a hyperbolic flyby orbit is

$$r = \frac{p}{1+e\cos\theta} = \frac{a(e^2-1)}{1+e\cos\theta}, \quad (4.52)$$

where e is the eccentricity of the hyperbolic flyby trajectory.
Let $r \to \infty$ and a formula for evaluating θ_∞ can be derived as follows.

$$\cos\theta_\infty = \lim_{r\to\infty}\left\{\frac{1}{e}\left[\frac{a(e^2-1)}{r}-1\right]\right\} = -\frac{1}{e} \Rightarrow \theta_\infty = \cos^{-1}\left(-\frac{1}{e}\right). \quad (4.53)$$

From Fig. 4.11, it can be seen that

$$\frac{\pi}{2} - \frac{\delta}{2} = \pi - \theta_\infty \quad (4.54)$$

Namely

$$\theta_\infty = \frac{\pi}{2} + \frac{\delta}{2} \quad (4.55)$$

Substituting Eq. (4.55) into Eq. (4.53) gives

$$\frac{1}{e} = \sin\frac{\delta}{2} \quad (4.56)$$

From the vis-viva equation, the semi-major axis of the hyperbolic flyby trajectory is given by

$$a = \frac{\mu_P}{V_\infty^2}, \quad (4.57)$$

where μ_P is the gravitational constant of the flyby planet. It can be shown that the eccentricity can be found from

$$e = 1 + \frac{r_P V_\infty^2}{\mu_P}. \quad (4.58)$$

Assumed the hyperbolic excess velocities $\bar{v}_\infty^-$ and $\bar{v}_\infty^+$ are known, then the turn angle δ can be evaluated using dot product calculation, and the flyby periapsis radius r_P can be found from Eqs. (4.56) and (4.58). On the contrary, if r_P is specified as a mission parameter, then the eccentricity of the hyperbolic flyby trajectory can be calculated from Eq. (4.58), and the turn angle δ can be obtained from (4.56) thereafter.

The heliocentric trajectory of the probe post-planet flyby is determined by its heliocentric velocity $\overline{V}^+$. Referring to Fig. 4.12, the heliocentric velocity of the

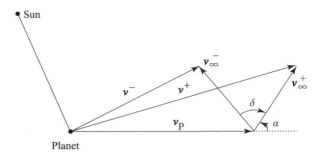

Fig. 4.12 Change of heliocentric velocity due to planet flyby

planet is $\overline{V}_P$, and $\overline{V}^-$, $\overline{V}^+$ are the heliocentric velocities of the probe before and post-planet flyby, respectively.

When a probe flies through the gravitational field of a planet, the direction of its hyperbolic excess velocity $\bar{v}_\infty$ is deflected and the turn angle is δ, which leads to a change in the heliocentric velocity of the probe consequently. As illustrated in Fig. 4.12, assume α is the angle between $\bar{v}_\infty^+$ and $\overline{V}_P$, if the magnitude of $\bar{v}_\infty^+$ is kept constant but its direction is varied continuously, namely let angle α vary from $0°$ to $180°$, then different heliocentric velocity $\overline{V}^+$ can be achieved, with different magnitudes and directions.

Therefore, the energy of the heliocentric orbit of a probe can be increased or reduced after performing a planet flyby. If $\alpha = 0°$, then $\bar{v}_\infty^+$ and $\overline{V}_P$ have the same direction, and the heliocentric orbit with the highest energy is achieved for the given v_∞; moreover, the heliocentric speed of the probe reaches the maximum value, so the probe is at the periapsis of the new heliocentric orbit right after the GA. On the other hand, if $\alpha = 180°$, then the direction of $\bar{v}_\infty^+$ is opposite to the direction of $\overline{V}_P$, and the heliocentric orbit with the lowest energy is achieved for the given v_∞; moreover, the heliocentric speed of the probe reaches the minimum value, so the probe is at the apoapsis of the new heliocentric orbit right after the GA.

2. **$P - R_p$ Diagram**

A critical problem in orbit design with GA is the selection of the sequence of celestial bodies for GA. To solve this problem, an energy analysis-based approach can be taken and a so-called $P - R_p$ diagram can be introduced to search for possible sequences of celestial bodies for GA.

As mentioned above, for the same hyperbolic excess speed v_∞, different heliocentric velocities $\overline{V}^+$ can be obtained when the angle α varies, and then, different heliocentric orbits can be achieved with different orbit period P and periapsis radius R_P. A v_∞ contour diagram of orbit period P versus periapsis radius R_P can be constructed for different planets and is termed $P - R_p$ diagram. The following discussion shows how to construct a $P - R_p$ diagram.

If the flyby planet is selected and the hyperbolic excess speed v_∞ is given, then the heliocentric velocities $\overline{V}^+$ of the probe post-GA is

4.4 Design of Transfer Trajectories

$$V = \sqrt{(V_P + V_\infty \cos \alpha)^2 + (V_\infty \sin \alpha)^2}. \quad (4.59)$$

The mechanical energy E of the heliocentric orbit can be evaluated from

$$E = \frac{V^2}{2} - \frac{\mu_S}{r}, \quad (4.60)$$

where μ_S is the gravitational constant of the Sun, r is the distance between the probe and the Sun that is considered to be equal to the radius of planet orbit about the Sun.

The mechanical energy E is directly related to the semi-major axis a of the heliocentric elliptical orbit of the probe post-planetary GA as follows.

$$a = -\frac{\mu_S}{2E}. \quad (4.61)$$

The angular momentum of the probe is

$$\boldsymbol{h} = \boldsymbol{r} \times \boldsymbol{V}, \quad (4.62)$$

with the magnitude given by

$$h = r \cdot (V_P + V_\infty \cos \alpha), \quad (4.63)$$

The semi-latus rectum p of the orbit is

$$p = \frac{h^2}{\mu_S}. \quad (4.64)$$

The eccentricity e of the orbit is found once the semi-latus rectum p and the semi-major axis a are known:

$$e = \sqrt{1 - \frac{p}{a}}. \quad (4.65)$$

The orbital period of the heliocentric elliptical orbit of the probe is

$$P = 2\pi \sqrt{\frac{a^3}{\mu_S}}, \quad (4.66)$$

The periapsis radius R_P of the heliocentric elliptical orbit of the probe is

$$R_P = a(1 - e) \quad (4.67)$$

Let α assume different numeric values in $[0, \pi]$, and different combinations of P and R_P can be computed. These combinations can be represented by the points in a coordination system, with the horizontal axis corresponding to R_P values and the vertical axis corresponding to P values. The collection of these points forms a curve, on which all the points correspond to the same V_∞ value, and the curve is called V_∞ contour curve. One or multiple V_∞ contour curves can be displayed in a single diagram, namely $P - R_p$ diagram. As illustrated in the $P - R_p$ diagram in Fig. 4.13, several families of V_∞ contour curves are displayed, corresponding to Venus, Earth, Mars and Jupiter, respectively. Note that the numbers labeled by the curves are the hyperbolic excess velocity of the probe flying by the relevant planet.

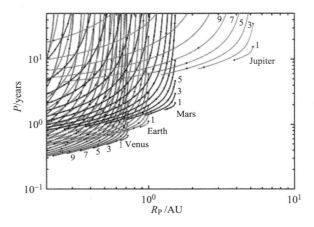

Fig. 4.13 $P - R_p$ diagram

Several observations can be made from a $P - R_p$ diagram:

1) The existence of a possible transfer trajectory with GAs between two planets

Each point on the V_∞ contour curve in a $P - R_p$ diagram represents a heliocentric elliptical orbit after the flyby of some planet. Therefore, if there is an intersection point between the V_∞ contour curves for two planets, then there is a possible transfer trajectory with GA between these two planets that satisfy the energy requirement. The feasibility of applying planetary GAs to design a transfer trajectory between two planets can be determined by studying the relationship between the V_∞ contour curves for these planets in a $P - R_p$ diagram.

2) Maximum orbit change from a GA

The gravity of a planet determines the possible change to the heliocentric orbit of the probe from the planet flyby, without the probe crashing into the planet obviously. If the lowest safety altitude of a planet flyby is given (e.g., 200 km for Earth-like planets and 5 times radius for Jupiter), then the maximum turn angle of the hyperbolic excess velocity δ_{max} can be determined for the same magnitude of V_∞, which corresponds to the black dot on the V_∞ contour curve in the $P - R_p$ diagram; and the maximum change to the heliocentric orbit of the probe after the planet flyby is indicated by the two adjacent dots. If the heliocentric orbit of the probe corresponds to one of

4.4 Design of Transfer Trajectories

the black dots on the V_∞ contour curve, then the possible change to the heliocentric orbit only occurs between the black dot and the two adjacent black dots, with the flyby altitude greater or equal to the minimum safety altitude.

The number of GAs needed to achieve the desired heliocentric orbit can be determined from the spacing between the dots on the V_∞ contour curves, the gap between the initial orbit and the target orbit along the V_∞ contour curves.

3) Good reference for selecting the optimal planet sequence of GAs

Due to the flexibility of selecting a planet sequence in the orbit design with GAs, if there were no restrictions and the planet combination in a sequence is randomly chosen, the computation time needed for orbit design could be intolerably long, and a major portion of the computation work is wasted on meaningless sequences; therefore, a pre-selection of the sequences of GAs is necessary to identify those reasonable candidate sequences of GAs for further analysis [17].

On the other hand, a $P - R_p$ diagram provides useful information on the relationship between the flyby trajectories for different planets from an energy perspective, and can be used to evaluate the effect on orbit changes from the GAs. Therefore, some reasonable candidate sequences of GAs can be identified from the $P - R_p$ diagram, which provides a valuable reference in designing the optimal planet sequence of GAs and helps improve the efficiency of orbit design with GAs.

4.4.4 Low Thrust Transfer

The research on transfer trajectory design for deep space missions has deepened with the development of propulsion technology. At the beginning of interplanetary exploration, rapid developments were achieved in the orbit design technology that is based on chemical rocket engines. However, with the progress made in deep space exploration missions, it becomes evident that the chemical rocket engines cannot meet the requirements of many future deep space missions in planning, due to its inherent characteristics of low specific impulse and low thrusting efficiency.

The rapid development in low-thrust (LT) propulsion technologies, including electric propulsion, ion propulsion and solar sail propulsion, has drawn broad attentions from space exploration community. Compared with chemical rocket engines, LT propulsion can help increase the payload mass significantly, and achieve high control accuracy. Therefore, LT engines possess superior characteristics and are well suited for future deep space missions [18].

1. **Introduction to LT Trajectory Design**

For orbital dynamic model, both two-body Kepler's vector equation and Gaussian planetary perturbation equations can be used. For of transfer orbit design, an appropriate coordinate system is selected. The Cartesian coordinate system is frequently adopted in trajectory optimization. Different coordinate systems selected in the study of LT trajectory optimization may lead to different convergence properties of the optimization process.

Kluever adopted a spherical frame and a two-dimensional polar frame in his study of Earth-to-Moon transfer. Broucke, Cefola and Battin selected equinoctial orbital elements in their researches, respectively, and Walker, Ireland and Owens proposed the modified equinoctial orbital elements, which can be applied for all types of conic curve trajectory. Later, Betts found the solutions of interplanetary and Earth-to-Moon transfer trajectories with the modified equinoctial orbital elements, and Kechichian also adopted the modified equinoctial orbital elements in his study of LT transfer orbit about the Earth.

Both the classical orbit elements and the modified equinoctial orbital elements have the corresponding differential forms. The variations of these orbital elements are relatively small so that a large integration step size can be used. There is no singularity when the eccentricity or inclination becomes zero for the modified equinoctial orbital elements, which is the difference between the modified equinoctial orbital elements and classical orbit elements [19].

Orbit optimization can start after the orbital dynamic model is selected. The methods of orbit optimization can be classified into three types, namely indirect method, direct method and hybrid method according to the way that the dynamic model is applied.

1) Indirect methods

Applying the calculus of variations and Pontryagin's Maximum Principle, the problem of LT transfer trajectory optimization becomes a two-point boundary value problem (TPBVP) for which the solution satisfies the first-order necessary conditions for optimality, and then the optimal control solution can be found by solving the two-point boundary value problem. This approach is termed indirect method.

Intensive researches have been conducted on the design theory of indirect method. Kechichian applied nonlinear programming (NLP) in solving the TPBVP of LT transfer trajectory about the Earth that satisfies certain boundary constraints. Chuang, Goodson and Hanson studied the TPBVP in which switching curves are involved [20]. Brown, Harrold and Johnson proposed an indirect method named as OPGUID/SWITCH. McAdoo, Jezewski and Dawkins proposed an indirect method that is termed OPBURN. Topputo applied the indirect method in solving several optimal control problems in astrodynamics. Theoretically, the solutions of indirect method rigorously satisfy the optimality conditions. However, sometimes convergence of the practical computation of the optimal solutions is difficult to achieve due to the sensitivity of the co-states. Therefore, applications of indirect method are restricted in some cases [21].

2) Direct methods

For direct methods, there are no co-states and co-state equations. Instead, the dynamic equations and control variables are directly discretized and the parameter optimization method is applied to obtain the optimal solution of these discrete parameters. As a result, the optimal trajectory is obtained.

4.4 Design of Transfer Trajectories

Among the direct methods, direct collocations nonlinear programming (DCNLP) is frequently applied for LT trajectory optimization, in which the techniques of fitting polynomial integration and estimated node control are utilized. Dickmanns and Wells conducted many studies on the DCNLP method based on Lagrange multipliers and states estimated from Hermite polynomials. Enright and Conway studied the planar transfer problem between two circular orbits in an idealized gravitational field applying the DCNLP and DTNLP methods, respectively. They found that this type of method is very effective in solving the optimal transfer problem in which the trajectory consists of two or three thrust arcs [22]. Kluever applied the DCNLP method in solving the problem of interplanetary transfer trajectory optimization, in which the control angles are parameterized and the problem of LT transfer trajectory optimization becomes a parameter optimization problem. Betts applied the DCNLP method that is based on sparse nonlinear programming in the study of several LT transfer trajectory optimization problems, including Earth-to-Mars transfer and Earth-to-Moon transfer. Tang also studied the design of interplanetary transfer trajectory using a collocation method. Ilgen applied the nonlinear programming method and the optimality condition of TPBVP in his study of LT transfer trajectory design. Herman and Conway found the solution of Earth-to-Moon transfer trajectory using a collocation method, and then, they improved the collocation method using a high-order Gauss–Lobbato method; Herman and Spencer found the solution of optimal LT transfer trajectory about the Earth, applying a high-order collocation method [23].

Differential inclusion is another type of direct method that is often used in LT trajectory design. Coverstone-Carroll and Williams suggested that differential inclusion could be utilized while optimizing the LT interplanetary transfer trajectories that include a single thrust arc or two thrust arcs with direct methods. Kuma and Seywald also introduced the concept of differential inclusion in their study on LT trajectory design. Coverstone studied the design of optimal interplanetary transfer trajectory, applying differential inclusion method. Later, a comparison study of the collocation method and differential inclusion method was done by Conway and Larson.

Other types of direct methods include parallel shooting method, pseudo-spectral method, etc. For example, Sheel and Conway applied the parallel shooting method in solving the design problem of two-dimensional LT transfer trajectory about the Earth. Betts did a survey of the algorithms that can be utilized in trajectory optimization. Hull systematically summarized the methods to convert optimal control problems to parameter optimization problems [24].

In summary, the direct methods are well suited for designing LT trajectories using a precise dynamic model. However, the computation cost could be significant to achieve high-precision trajectories.

3) Hybrid Methods

In the hybrid methods, co-states are introduced and the control law is determined from the co-state equations; the TPBVP is not solved directly, the optimal control is found via parameter optimization method instead.

Applying the hybrid method, Zondervan, Wood and Caughey studied the optimal transfer problem in which the trajectory includes three non-coplanar thrust arcs of low thrust or moderate thrust, using an idealized gravitational field model. Ilgen adopted a hybrid strategy termed HYTOP to compute the LT transfer trajectory, in which the transfer trajectory is optimized using the principal vector function derived from the principle of optimal control. Kluever studied several Earth-to-Moon transfer problems using hybrid methods that include the elements from both direct methods and indirect methods. In these hybrid methods, the co-state equations are derived, and the performance measure is optimized using nonlinear programming. In recent years, the study of the relationships among the co-states derived in different coordinate systems has also attracted the attention of some scholars [25].

In summary, applying the hybrid methods in LT trajectory optimization can help obtain high-precision trajectory, and it is easier to achieve convergence than applying indirect methods.

4) Other Types of Optimization Methods

Besides the methods in which the control variables and states are directly parameterized, some other methods can be applied in trajectory optimization.

Edelbaum, Sackett and Malchow proposed to use the mean equatorial orbital elements as the state variables in solving minimum-time transfer problems in which the transfer orbits include one thrust arc only. The perturbation effect of solar radiation pressure was taken into account in the study. The method is mainly based on the technique of averaging and is named SECKSPOT method. Later, the SECKSPOT method was improved by Horsewood, Suskin and Pines so that it can be applied in solving the rendezvous problem between two non-coplanar circular orbits, in which the transfer trajectory consists of multiple LT arcs.

Ilgen designed a Lyaponov controller to guide a spacecraft in orbital transfer between different Earth orbits. Chang proposed to use a Lyaponov controller based on Laplace vector to realize the orbital transfer between different elliptical orbits. Schaub and other scholars designed Lyaponov controllers for spacecraft formation flying using LT propulsion. Petropoulos proposed the so-called Q-law method for LT orbit design, which is based on Lyapunov feedback control. Lee employed the genetic algorithm to optimize the LT trajectory achieved by Q-law [26].

A hybrid optimal control law was constructed by Kluever and Oleson for LT transfer orbit design, utilizing the extremum variation rate of the classical orbit elements. A similar strategy was adopted by Spencer and Culp to design the LT transfer trajectory from a low Earth orbit to the geosynchronous orbit. Kechichian also used this type of controller in LT transfer orbit design.

Fahroo and Ross proposed an orbit optimization method based on the pseudo-spectral method by Chebyshev. Jacobson proposed a second-order algorithm based on differential dynamic programming to solve optimal control problems. In recent years, genetic algorithms and neural network methods are also applied in orbit optimization.

Some orbit optimization methods from the above discussions have found important applications in the mission and trajectory designs of many practical space missions, including the designs of optimal transfer trajectories between Earth orbits,

4.4 Design of Transfer Trajectories

Earth-to-Moon transfer trajectories, interplanetary transfer trajectories, Earth departure and planet capture orbits and transfer trajectories in asteroids and comets exploration missions. Moreover, LT propulsion systems can be applied in some new types of space missions such as interception of small celestial bodies, LT and GA transfer, etc.

2. Preliminary Orbit Design Based on Fitting Polynomials

1) Basic theory

The dynamic model of a spacecraft flying in a gravitational field is given in a cylindrical coordinate system as follows, as illustrated in Fig. 4.14.

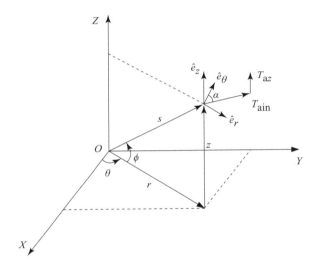

Fig. 4.14 Cylindrical coordinate system

$$\ddot{r} - r\dot{\theta}^2 = -\frac{\mu}{s^3}r + T_{\text{ain}} \sin \alpha$$
$$\frac{1}{r}\frac{d}{dt}(r^2\dot{\theta}) = T_{\text{ain}} \cos \alpha \ \cos \alpha, \quad (4.68)$$
$$\ddot{z} = -\frac{\mu}{s^3}z + T_{az}$$

where $s = \sqrt{r^2 + z^2}$ is the distance between the probe and the Sun, T_{az} is the Z component of the thrust acceleration. It is difficult to find a converged solution by directly integrating this dynamic model. To solve this problem, Pertropoulos and Bradley proposed to use sine exponential function and reciprocal polynomial to mimic the flight trajectory of the probe, respectively, so that approximate solutions to the dynamic model can be obtained. Their orbit designs were successful [27].

According to the fitting theory of reciprocal six-order polynomials, each thrust arc of the probe can be approximately described by a reciprocal polynomial and a polynomial:

$$r(\theta) = \frac{1}{a + b\theta + c\theta^2 + d\theta^3 + e\theta^4 + f\theta^5 + g\theta^6}$$
$$z(\theta) = a_z + b_z\theta + c_z\theta^{q-1} + d_z\theta^q \tag{4.69}$$

It is assumed that the flight direction of the probe is counterclockwise in its transfer trajectory. Given the departure time t_L and arrival time t_A, the position vectors of the probe at the start and end of the transfer trajectory $(\bar{s}_1, \bar{s}_2)$ can be obtained from some JPL ephemeris such as DE405. Then, the projections of $(\bar{s}_1, \bar{s}_2)$ onto $O - XY$ plane $(\bar{r}_1, \bar{r}_2)$ can be calculated, and the phase angle $\tilde{\theta}$ travelled by the probe in $O - XY$ plane can be obtained. Assume the initial phase angle of the probe is $\theta_1 = 0°$, the end phase angle is

$$\theta_f = \tilde{\theta} + 2\pi \cdot N_{\text{rev}}, \tag{4.70}$$

where Nrev is the number of revolutions of the transfer trajectory about the Sun. Note that Nrev is an integer equal or greater than zero, and the subscripts "1" and "f" refer to the initial and end parameters, respectively.

Consider the motion of the probe on $O - XY$ plane. The flight path angle γ of the probe can be derived from the reciprocal six-order polynomial as follows:

$$\tan \gamma = \frac{\dot{r}}{r\dot{\theta}} = -r \cdot (b + 2c\theta + 3d\theta^2 + 4e\theta^3 + 5f\theta^4 + 6g\theta^5) \tag{4.71}$$

The coefficients in Eq. (4.71) can be found from the given boundary conditions as follows:

$$a = \frac{1}{r_1}, \quad b = -\frac{\tan \gamma_1}{r_1}, \quad c = \frac{1}{2r_1}\left(\frac{\mu}{r_1^3\dot{\theta}_1^2} - 1\right)$$

$$\begin{bmatrix} e \\ f \\ g \end{bmatrix} = \frac{1}{2\theta_f^6} \begin{bmatrix} 30\theta_f^2 & -10\theta_f^3 & \theta_f^4 \\ -48\theta_f & 18\theta_f^2 & -2\theta_f^3 \\ 20 & -8\theta_f & \theta_f^2 \end{bmatrix} \begin{bmatrix} \frac{1}{r_2} - \left(a + b\theta_f + c\theta_f^2 + d\theta_f^3\right) \\ -\frac{\tan \gamma_2}{r_2} - \left(b + 2c\theta_f + 3d\theta_f^2\right) \\ \frac{\mu}{r_2^4\dot{\theta}_2^2} - \left(\frac{1}{r_2} + 2c + 6d\theta_f\right) \end{bmatrix}$$

$$a_z = z_1, \quad b_z = \frac{\dot{z}_1}{\dot{\theta}_1}$$

$$\begin{bmatrix} c_z \\ d_z \end{bmatrix} = \frac{1}{\theta_f^q}\begin{bmatrix} q\theta_f & -\theta_f^2 \\ -(q-1) & \theta_f \end{bmatrix} \begin{bmatrix} z_2 - a_z - b_z\theta_f \\ (\dot{z}_2/\dot{\theta}_2) - b_z \end{bmatrix} \tag{4.72}$$

The unknown parameter d can be found by solving the following equation with the given time of flight TOF on the transfer trajectory:

4.4 Design of Transfer Trajectories

$$T_{OF} = \int_0^{\theta_f} \sqrt{\frac{r(\theta)^4}{\mu}\left[\frac{1}{r(\theta)} + 2c + 6d\theta + 12e\theta^2 + 20f\theta^3 + 30g\theta^4\right]} d\theta \quad (4.73)$$

The magnitude of the thrust acceleration of the probe can be computed as follows.

$$\begin{aligned}
T_{\text{ain}} &= \frac{-\mu}{2r^3 \cos\gamma} \cdot \frac{6d + 24e\theta + 60f\theta^2 + 120g\theta^3 - (\tan\gamma)/r}{\left(1/r + 2c + 6d\theta + 12e\theta^2 + 20f\theta^3 + 30g\theta^4\right)^2} T \\
T_{\text{az}} &= \frac{\mu}{r^3}z - \left[c_z(q-1)(q-2)\theta^{q-3} + d_z q(q-1)\theta^{q-2}\right]\dot\theta^2 \\
&\quad + \left[b_z + c_z(q-1)\theta^{q-2} + d_z q\theta^{q-1}\right]\ddot\theta \\
T_{\text{a}} &= \sqrt{T_{\text{ain}}^2 + T_{\text{az}}^2}
\end{aligned} \quad (4.74)$$

where $\dot\theta, \ddot\theta$ are given by Eqs. (4.75):

$$\begin{aligned}
\dot\theta^2 &= \left(\frac{\mu}{r^4}\right)\frac{1}{1/r + 2c + 6d\theta + 12e\theta^2 + 20f\theta^3 + 30g\theta^4} \\
\ddot\theta &= \left(-\frac{\mu}{2r^4}\right)\left[\frac{4\tan\gamma}{1/r + 2c + 6d\theta + 12e\theta^2 + 20f\theta^3 + 30g\theta^4}\right. \\
&\quad + \left.\frac{6d + 24e\theta + 60f\theta^2 + 120g\theta^3 - \tan\gamma/r}{\left(1/r + 2c + 6d\theta + 12e\theta^2 + 20f\theta^3 + 30g\theta^4\right)^2}\right]
\end{aligned} \quad (4.75)$$

The preceding discussion is a brief review of the theory of approximating the spacecraft trajectory in three-dimensional space with a reciprocal six-order polynomial. Note that z is a relatively small quantity when compared to r, and the difference between s and r is less than 5% when the orbit inclination of the target planet is less than 15°. Therefore, z can be ignored for general application scenarios and the orbit design can be carried out in a simpler two-dimensional space.

The phase angle $\tilde\theta$ travelled by the spacecraft on the transfer trajectory can be determined using an analogy of the rotation of the hour hand on a clock: assume X-axis points to 0 h and Y-axis points to 3 h, and it takes 12 h for the hour hand to complete one revolution counterclockwise; suppose r_1, r_2 point to the hour numbers on a clock time_r_1 and time_r_2, respectively, then $\tilde\theta$ can be computed using Eq. (4.76):

$$\begin{cases} if\ \text{time\_}r_2 - \text{time\_}r_1 > 6\ or\ 0 \leqslant \text{time\_}r_1 - \text{time\_}r_2 \leqslant 6 \\ \tilde\theta = 2\pi - \arccos\left[\frac{r_1 \cdot r_2}{|r_1||r_2|}\right] \\ if\ 0 \leqslant \text{time\_}r2 - \text{time\_}r1 \leqslant 6\ or\ \text{time\_}r_1 - \text{time\_}r_2 > 6 \\ \tilde\theta = \arccos\left[\frac{r_1 \cdot r_2}{|r_1||r_2|}\right] \end{cases} \quad (4.76)$$

2) Performance measure

A minimum-fuel performance measure can be specified according to the fuel consumption model of the engine.

Assuming the specific impulse of the engine of a probe using electric propulsion is a constant, the thrust acceleration is

$$T_a = \frac{T}{m} = \frac{-\dot{m} \cdot I_{sp} g_0}{m}, \tag{4.77}$$

where T is the thrust magnitude, I_{sp} is the specific impulse of the engine, g_0 is the gravitational acceleration at sea level on Earth.

From Eq. (4.77),

$$\frac{dm}{m} = -\frac{T_a \cdot dt}{I_{sp} g_0} = -\frac{T_a}{I_{sp} g_0 \dot{\theta}} d\theta \tag{4.78}$$

Integrating Eq. (4.78) gives

$$\int_{m_0}^{m_f} \frac{dm}{m} = -\int_0^{\theta_f} \frac{T_a}{I_{sp} g_0 \dot{\theta}} d\theta \tag{4.79}$$

$$\frac{m_f}{m_0} = \exp\left(-\frac{1}{I_{sp} g_0} \int_0^{\theta_f} \frac{T_a}{\dot{\theta}} d\theta\right) \tag{4.80}$$

Therefore, minimum fuel is achieved when the following equation reaches its minimum value:

$$\delta m = 1 - \exp\left(-\frac{1}{I_{sp} g_0} \int_0^{\theta_f} \frac{|T_a|}{\dot{\theta}} d\theta\right) \tag{4.81}$$

As a result, the minimum-fuel performance measure used in the flight orbit design of a probe with electric propulsion is given as follows.

$$J = 1 - \exp\left[\sum_{i=1}^{k}\left(-\frac{1}{I_{sp} g_0} \int_{\theta_i}^{\theta_{fi}} \frac{|T_{ai}|}{\dot{\theta}_i} d\theta\right)\right], \tag{4.82}$$

where k is the number of thrust arcs of the probe; $|T_{ai}|$ is the magnitude of the thrust acceleration in the ith thrust arc.

4.4 Design of Transfer Trajectories

An ceiling bound is set for the thrust acceleration of each thrust arc during the optimization of a LT trajectory as the acceleration of an electrically propelled probe cannot be infinitely large, as illustrated in Eq. (4.83).

$$\max(|T_{ai}|) \leqslant T_{ai\,\max}, \qquad (4.83)$$

where $T_{ai\,\max}$ is the maximum thrust acceleration in the ith thrust arc.

3) Optimization Model for Trajectory Design

It is assumed that the electric propulsion system works in the entire transfer flight. Similar to the classical Lambert's problem, the basic calculation procedure for the LT transfer trajectory design is as follows. Given the departure time and arrival time, the departure and arrival positions of the probe can be obtained from some JPL ephemeris such as DE405; the departure and arrival hyperbolic excess velocity vectors are also given, and the departure and arrival states of the probe can be determined; after that, the coefficients of the reciprocal six-order polynomial are calculated, and the transfer trajectory equations can be determined, as illustrated in Eq. (4.69).

The optimization model for the transfer trajectory design of an electrically propelled probe is as follows.

(1) The selected design parameters are:

$$Z = [t_0, T_{OF}, N_{rev}, v_\infty^0, v_\infty^f, w_{in}^0, w_{in}^f, w_z^0, w_z^f, q], \qquad (4.84)$$

where t_0 is the departure time, T_{OF} is the time of flight on the transfer orbit, N_{rev} is the number of revolutions of the transfer trajectory, v_∞^0 is the departure hyperbolic excess speed, v_∞^f is the arrival hyperbolic excess speed, w_{in}^0 is the angle between the X-axis and the projection of the departure hyperbolic excess velocity onto O-XY plane, w_{in}^f is the angle between the X-axis and the projection of the arrival hyperbolic excess velocity onto O-XY plane, w_z^0 is the angle between the Z-axis and the departure hyperbolic excess velocity, w_z^f is the angle between the Z-axis and the arrival hyperbolic excess velocity, q is the exponent of the polynomial in Eq. (4.69).

(2) The minimum-fuel performance measure is given by Eq. (4.82).
(3) A global optimization algorithm is selected to perform the optimization computation of the transfer trajectory to find the optimal transfer trajectory. For each sample of the design parameters, the preceding basic calculation procedure is carried out and the performance measure is evaluated; the minimum value of the performance measure gives the optimal solution.
(4) Numeric example

A numeric example of orbit design from Mars exploration missions is given in the following. It is assumed that the probe mass is 4000 kg, the thrust magnitude is 400 mN, the specific impulse is 5000 s, and the possible launch dates is in 2028–2031.

The preliminary trajectory design is conducted applying the above method of fitting the reciprocal six-order polynomial. The preliminary design results are given in Table 4.2, also displayed is the results from further optimization (discussed in next section) based on the preliminary design results.

Table 4.2 Preliminary orbit design results for Mars mission

Orbit parameters	Preliminary design results	Results from further optimization
Launch date/(y/m/d)	2028/07/11	2028/06/11
Arrival date/(y/m/d)	2030/01/14	2030/03/14
Time of flight/(d)	551	641
Delta-V/(km/s)	5.586	5.876
Propellant consumption/(kg)	431	452
Remaining spacecraft mass/(kg)	3569	3548

3. **Trajectory Optimization for Electrically Propelled Spacecraft Based on Trajectory Patching**

1) Orbital dynamic model

Consider the motion equation of a spacecraft in the heliocentric ecliptic coordinate system. The heliocentric ecliptic coordinate system is defined as follows: The OX-axis is lying in the ecliptic plane and pointing to the vernal equinox direction, the OZ-axis is perpendicular to ecliptic plane and parallel to the angular velocity vector of the Earth's motion about the Sun, the OY-axis completes the right-hand set. Neglecting perturbation accelerations, the motion equation of the probe in the heliocentric ecliptic frame is given as

$$\dot{r} = v$$
$$\dot{v} = -\frac{\mu}{r^3}r + \frac{T}{m}\alpha, \quad (4.85)$$
$$\dot{m} = -\frac{T}{g_0 I_{sp}}$$

where r, v are the position and velocity vectors, respectively; m is the mass of the probe, including the propellant mass; μ is the gravitational coefficient of the central celestial body; T is the thrust magnitude; α is the unit vector of thrust direction; g_0 is the gravitational acceleration at sea level; I_{sp} is the specific impulse of the engine.

2) Problem statement of optimal control

4.4 Design of Transfer Trajectories

The objective of transfer trajectory design is to find a fuel-minimum transfer trajectory on which the probe can fly from a specified initial position to a target position during a given time of flight. For a spacecraft that is not subject to thrust control and perturbations, the flight orbit is a type of conic sections, and the design of this type of orbits can be achieved by solving a Lambert's problem. In contrast, there is no analytical solution to the transfer trajectory design of electrically propelled spacecrafts, and algorithms of numeric analysis are employed to find the solutions instead. The optimization of transfer trajectory for electrically propelled spacecraft is an optimal control problem, namely to find a control vector $u(t)$ such that the performance measure achieves the minimum value, and the design constraints have to be satisfied at the same time. The performance measure is given as follows:

$$J = \phi(x(t_f)) + \int_{t_0}^{t_f} L(x, u, t) dt \tag{4.86}$$

The constraints to be satisfied are listed in the following:
Dynamic model:

$$\dot{x} = f(x, u, t) \quad t \in [t_0, t_f] \tag{4.87}$$

Path constraints:

$$\begin{cases} g(x, u, t) = 0 \\ h(x, u, t) \leq u \end{cases} t \in [t_0, t_f] \tag{4.88}$$

Initial condition of the trajectory:

$$\psi_0(x(t_0)) = 0 \tag{4.89}$$

End condition of the trajectory:

$$\psi_f(x(t_f), t_f) = 0, \tag{4.90}$$

where $x(t)$ is the state vector of the probe at instant t.

3) General trajectory optimization model

For most trajectory optimization problems, it is required that certain end state conditions be achieved. Therefore, the performance measure is usually related to the end state and end time:

$$J = \phi[r(t_f), v(t_f), t_f] \tag{4.91}$$

Applying Pontryagin minimum value principle, the Hamilton function is written as

$$H = \lambda_r^T v + \lambda_v^T \left(-\frac{\mu}{r^3} r + \frac{T}{m} \alpha + f_p\right) - \lambda_m \frac{T}{g_0 I_{sp}}, \tag{4.92}$$

where $\lambda_r^T, \lambda_v^T, \lambda_m$ are the co-states corresponding to position, velocity and mass. Since α is the unit vector in thrust direction, it is required that $\alpha^T \alpha = 1$, and the Hamilton function can be rewritten as

$$\widetilde{H} = \lambda_r^T v + \lambda_v^T \left(-\frac{\mu}{r^3} r + \frac{T}{m} \alpha + f_p\right) - \lambda_m \frac{T}{g_0 I_{sp}} + \gamma \left(1 - \alpha^T \alpha\right) \tag{4.93}$$

The optimal thrust direction is obtained from the control equation:

$$\frac{\partial \widetilde{H}}{\partial \alpha} = 0 \rightarrow \alpha^* = \frac{\lambda_v}{\|\lambda_v\|} \tag{4.94}$$

As can be seen from Eq. (4.94), the thrust direction is related to co-state λ_v only, and λ_v is termed the principal vector in many references.

Deriving the partial derivatives of the Hamilton function with respect to the state variables gives the co-state equations:

$$\begin{cases} \dot{\lambda}_r = -\frac{\partial H}{\partial r} = \left(\lambda_v \frac{\mu}{r^3} - \frac{3\lambda_v^T r}{r^5} r\right) - \frac{\partial f_p}{\partial r} \lambda_v \\ \dot{\lambda}_v = -\frac{\partial H}{\partial v} = -\lambda_r - \frac{\partial f_p}{\partial v} \lambda_v \\ \dot{\lambda}_m = -\frac{\partial H}{\partial m} = \|\lambda_v\| \frac{T}{m^2} \end{cases} \tag{4.95}$$

in which $-\frac{\partial f_p}{\partial r} \lambda_v$ and $-\frac{\partial f_p}{\partial v} \lambda_v$ are related to perturbations, and the other terms can be found from the two-body orbit model.

The equality constraint for the end states:

$$\psi[r(t_f), v(t_f)] = 0 \tag{4.96}$$

The transversality condition is:

$$\begin{aligned} \lambda_r(t_f) &= \left.\frac{\partial \phi}{\partial r}\right|_{t=t_f} + \gamma_r^T \left.\frac{\partial \psi}{\partial r}\right|_{t=t_f} \\ \lambda_v(t_f) &= \left.\frac{\partial \phi}{\partial v}\right|_{t=t_f} + \gamma_v^T \left.\frac{\partial \psi}{\partial v}\right|_{t=t_f} \\ \lambda_m(t_f) &= \left.\frac{\partial \phi}{\partial m}\right|_{t=t_f} + \gamma_m \left.\frac{\partial \psi}{\partial m}\right|_{t=t_f} \end{aligned} \tag{4.97}$$

If the transfer time is not fixed, the following constraint applies:

4.4 Design of Transfer Trajectories

$$\left[\frac{\partial \phi}{\partial t} + \left(\frac{\partial \psi}{\partial t} \right)^T \gamma + \tilde{H} \right]_{t=t_f} = 0, \quad (4.98)$$

where $\gamma = [\gamma_r, \gamma_v, \gamma_m]^T$ is a constant multiplier vector.

The dynamic equation (Eq. 4.85), co-state equation (Eq. 4.95) and boundary conditions (Eqs. 4.96–4.98) constitute a two-point boundary value problem (TPBVP). Solving the TPBVP gives the optimal flight trajectory of the electrically propelled spacecraft in the heliocentric ecliptic frame, and the variations of the co-states can be obtained at the same time. The time history of the optimal control (thrust) can be determined from the control equation (Eq. 4.94).

4) Trajectory optimization of electrically propelled spacecraft using nonlinear programming

Based on the above analysis, a numeric method for trajectory optimization is introduced in the following. Using the Earth-to-Mars transfer trajectory design as an example, a trajectory-patching model can be adopted for the trajectory optimization, as illustrated in Fig. 4.15.

The design variables are:

(a) Earth departure time of the probe t_L;
(b) Transfer time T_{OF};
(c) Time ratio of the first thrust arc δ_1;
(d) Time ratio of the second thrust arc δ_2;
(e) Mass of the probe at Mars arrival m_f;
(f) Co-states λ_L at Earth departure (7 × 1 matrix);
(g) Co-states λ_A at Mars arrival (7 × 1 matrix).

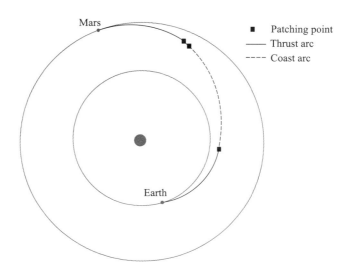

Fig. 4.15 Trajectory-patching model

For a given sample of the design variables, the position and velocity vectors at Earth departure (R_L, V_L) and Mars arrival (R_f, V_f) can be obtained from some JPL ephemeris, and the corresponding states and co-states are:

(a) States and co-states at Earth departure:

$$X_L = [R_L, V_L, m_0, \lambda_L]^T \tag{4.99}$$

(b) States and co-states at Mars arrival:

$$X_f = [R_f, V_f, m_f, \lambda_f]^T \tag{4.100}$$

Given the initial state and co-state of the probe (point 1 in Fig. 4.15), the dynamic equation (Eq. 4.85) and co-state equation (Eq. 4.95) can be integrated, and the state and co-state at point 2 in Fig. 4.15 can be obtained; after that, the integration continued and the state at point 3 $[R_3, V_3, m_3]^T$ are obtained. On the other hand, given the end state and co-state of the probe (point 5 in Fig. 4.15), integrating the state equation and co-state equation backwards in time gives the state and co-state of the probe at point 4 in Fig. 4.15. Since the state of the probe at the patching points (point 3 and point 4) has to be continuous, the following state constraint equation is valid.

$$X_4(1:7) - [R_3, V_3, m_3]^T = 0 \tag{4.101}$$

If there are no coast arcs, both the state and co-state have to be continuous, namely

$$X_4 - X_2 = 0 \tag{4.102}$$

In summary, if the transversality condition is neglected, the trajectory optimization problem for an electrically propelled probe becomes a multi-parameter optimization problem as stated in the following.

(a) Design variables: $Z = [t_L, T_{OF}, \delta_1, \delta_2, m_f, \lambda_L, \lambda_f]^T$
(b) Performance measure: $J = -m_f \to \min$ or $J = T_{OF} \to \min$
(c) Inequality constraint: $\begin{cases} 0 \leqslant \delta_1, \delta_2, \delta_1 + \delta_2 \leqslant 1 \\ \delta_1, \delta_2 \leqslant \frac{T_{on\ max}}{T_{OF}} \end{cases}$

where $T_{on\ max}$ is the allowable maximum thrust duration.

(d) Equality constraint: Eqs. (4.101) or (4.102)

Sequential quadratic programming (SQP) algorithm can be employed to solve this nonlinear programming problem. A relatively accurate initial value should be provided to guarantee the convergence of the computation process. The initial values of the states of the probe can be obtained from the energy contour graph or the shape algorithm; however, an independent guess of the initial values of the co-states is necessary.

4.4 Design of Transfer Trajectories

5) Numerical simulation

(a) Interplanetary Transfer Trajectory Design of Mars Rendezvous Mission

The solar panel will provide the power to the electric propulsion system during the interplanetary transfer to Mars. Assume the engine works continuously during the transfer, and the input power is proportional to the reciprocal of the square of the distance between the probe and the Sun, namely $P = P_0/r^2$; also assume that the initial input power P_0 is 6500 w, for which the distance between the probe and the Sun is 1AU); the thrust efficiency is 65%, the constant specific impulse is 3100 s, and the initial probe mass is 1200 kg.

For possible launch dates from January 1, 2009, to January 1, 2011, the energy contour graph of C_3 for the Mars rendezvous mission is given in Fig. 4.16.

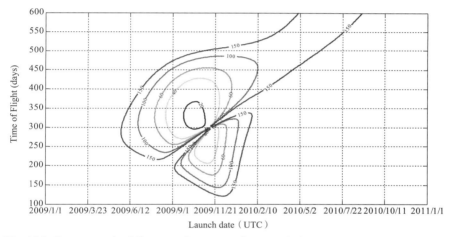

Fig. 4.16 Contour graph of C_3 energy for mars rendezvous mission

(b) Interplanetary Transfer Trajectory Design of Apophis Rendezvous Mission

Consider a rendezvous mission to asteroid Apophis. Input for the interplanetary transfer trajectory design is: The initial mass of the probe is 1500 kg, 6 electric thrusters are available with 100 mN thrust magnitude for each thruster, the specific impulse is 3500 s, 3 thrusters may work at the same time with the combined thrust totaling 300 mN, and the candidate launch dates are between 2017 and 2019.

Applying the preceding trajectory optimization method for electric propulsion systems (HTOM-POT), the results of trajectory optimization are displayed in Table 4.3.

As can be seen from the trajectory optimization results of the Apophis rendezvous mission, for the launch window in 2017–2019, the optimal transfer durations and propellant consumptions needed for the two thrusting modes are 472 days, 356 kg and 806 days, 186 kg, respectively.

Table 4.3 Trajectory optimization results for Apophis rendezvous mission

Design parameters and results	Thrusting mode	
	Continuous thrusting	Thrusting-coast-thrusting
Thrust magnitude	300 mN	300 mN
Specific impulse	3500 s	3500 s
Launch date (y/m/d)	2018/04/26	2018/03/31
Departure C_3 (km^2/s^2)	0	0
Arrival date (y/m/d)	2019/08/11	2020/06/14
Arrival C_3 (km^2/s^2)	0	0
Thrust duration of first thrust arc (d)	–	116.8534
Coast duration (d)	–	560.0585
Thrust duration of second thrust arc (d)	–	129.4068
Total transfer time (d)	471.7815	806.3186
Engine operation time (d)	471.7815	246.2601
Propellant mass ratio	0.2377	0.1241
Propellant consumption (kg)	356.5182	186.0951
Total delta-V (km/s)	9.3086	4.5434

Compared with the continuous thrusting mode, the "thrusting-coast-thrusting" mode can significantly reduce the propellant consumption of the probe and the total delta-V; however, the transfer time is much longer.

The interplanetary transfer trajectory of the Apophis rendezvous mission is illustrated in Fig. 4.17.

4.5 Design of Mission Orbits

For different deep space missions, the types of the mission orbits may be different, which include closed orbit about a planet, descent trajectory onto the surface of a celestial body, asteroid companion orbit, flyby trajectory, orbit about a Lagrange libration point, etc.

4.5.1 Planet Orbiting Missions

1. **Orbital dynamic model**

For the orbit design of planet orbiting missions, an appropriate orbital dynamic model should be selected first, according to the characteristics of the orbit and the requirements of computational precision. Moreover, the dynamic model should take

4.5 Design of Mission Orbits

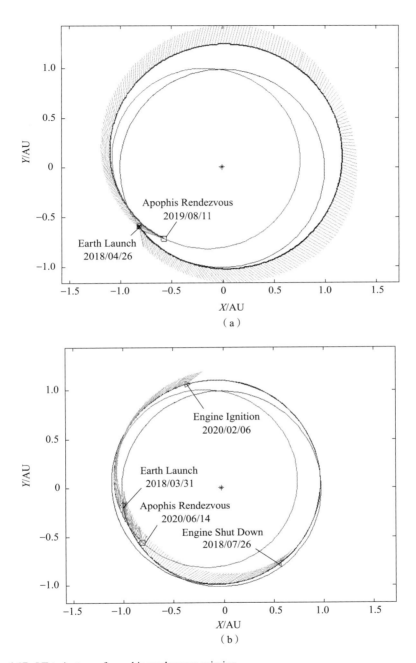

Fig. 4.17 LT trajectory of apophis rendezvous mission

into account the major perturbations, including the non-spherical gravitational perturbation of the central celestial body, the gravitational perturbation from the natural satellites of the central celestial body, the gravitational perturbations from the Sun and other major planets, the perturbation from solar radiation pressure and attitude control's influence on orbit. The dynamic model should be tailored to meet the specific needs.

The general form of the orbital dynamic equation of a spacecraft can be written as

$$\ddot{r}_p = -\mu \frac{r_p}{r_p^3} + A_N + A_{NSC} + A_{NSS} + A_D + A_R + A_P, \tag{4.103}$$

where μ is the gravitational constant of the central celestial body, r_p is the position vector of the probe, A_N is the N-body gravitational perturbation acceleration, A_{NSC} is the non-spherical gravitational perturbation acceleration of the central celestial body, A_{NSS} is the non-spherical gravitational perturbation from the natural satellites of the central celestial body, A_D is the atmospheric drag perturbation acceleration, A_R is the solar radiation pressure perturbation acceleration, A_P is the thrust acceleration.

1) N-body Gravitational Perturbation Acceleration A_N

The N-body gravitational perturbations come from the Earth, the Sun and the other planets. The combined perturbation acceleration can be written as

$$A_N = -\sum_{j=1}^{N} \mu_j \left(\frac{r_j}{r_j^3} + \frac{r_j - r_p}{\|r_j - r_p\|^3} \right), \tag{4.104}$$

where N is the number of the celestial bodies that cause the N-body gravitational perturbations, μ_j is the gravitational constant of the jth celestial body, r_j is the position vector of the jth celestial body with respect to the central celestial body, which can be found by approximate calculation or from some JPL ephemeris.

For a lunar probe flying in the cislunar space, the N-body gravitational perturbations usually include the perturbations from the Sun, the Earth and the Moon only because perturbations from the other celestial bodies are negligible.

2) Non-spherical gravitational acceleration of the central celestial body A_{NSC}

The gravitational potential function of the central celestial body is a solution of the Laplace equation, and its non-spherical part is

$$U_{NSE} = \frac{\mu_e}{R} \sum_{n=2}^{N} \sum_{m=0}^{n} \left(\frac{R}{R_p} \right)^n \overline{P}_n^m (\sin \varphi) (\overline{C}_{nm} \cos m\lambda + \overline{S}_{nm} \sin m\lambda) \tag{4.105}$$

where N is the degree of the model of the non-spherical gravitational potential of the central celestial body, R is the radius of the central celestial body, and R_p is the

4.5 Design of Mission Orbits

distance between the probe and the center of the central celestial body, (λ, φ) are the longitude and latitude of the probe in the reference frame fixed to the central celestial body, $(\overline{C}_{nm}, \overline{S}_{nm})$ are the normalized coefficients of the gravitational potential from the gravitational model of the central celestial body, $\overline{P}_n^m(\sin \varphi)$ is the normalized Legendre polynomial.

Taking the Earth as an example, its non-spherical gravitational perturbation acceleration can be written as

$$A_{\text{NSC}} = (T_{\text{eei}})^T \nabla U_{\text{NSC}} \quad (4.106)$$

where T_{eei} is the transformation matrix from the Earth-centered equatorial inertial frame to the Earth-centered equatorial fixed frame, ∇U_{NSC} is the gradient of the Earth non-spherical gravitational potential function with respect to the Cartesian coordinates of the Earth-centered equatorial fixed frame.

3) Non-spherical Gravitational Acceleration of the Natural Satellite of the Central Celestial Body A_{NSS} [28–31]

For the natural satellite of the central celestial body, the non-spherical part of the gravitational potential function is

$$U_{\text{NSS}} = \frac{\mu_S}{r} \sum_{n=2}^{N} \sum_{m=0}^{n} \left(\frac{R_S}{r_p}\right)^n \overline{P}_n^m (\sin \varphi') \left(\overline{C}'_{nm} \cos m\lambda' + \overline{S}'_{nm} \sin m\lambda'\right), \quad (4.107)$$

where N is the degree of the model of the non-spherical gravitational potential of the natural satellite, R_S is the radius of the natural satellite, and r_p is the distance between the probe and the center of the natural satellite, (λ', φ') are the longitude and latitude of the probe in the reference frame fixed to the natural satellite, $(\overline{C}'_{nm}, \overline{S}'_{nm})$ are the normalized coefficients of the gravitational potential from the gravitational model of the natural satellite, $\overline{P}_n^m(\sin \varphi')$ is the normalized Legendre polynomial.

Taking the Moon as an example, as the natural satellite of the Earth, its non-spherical gravitational perturbation acceleration can be written as

$$A_{\text{NSS}} = (T_{\text{mmei}})^T \nabla U_{\text{NSS}} \quad (4.108)$$

where T_{mmei} is the transformation matrix from the Moon-centered inertial frame to the Moon-centered equatorial fixed frame, ∇U_{NSS} is the gradient of the Moon non-spherical gravitational potential function with respect to the Cartesian coordinates of the Moon-centered equatorial fixed frame.

4) Solar Radiation Pressure Perturbation Acceleration A_R

$$A_R = \kappa \rho_\oplus r_s^2 C_R \left(\frac{S_R}{m}\right) \frac{\boldsymbol{r}_p - \boldsymbol{r}_s}{\|\boldsymbol{r}_p - \boldsymbol{r}_s\|^3}, \quad (4.109)$$

where $\rho_\oplus$ is the intensity of solar radiation pressure (4.560×10^{-6} N/m² in near-Earth space), C_R is the reflection coefficient of the probe surface (typical value 1.33), S_R is the area of the cross section of the probe perpendicular to the sunlight direction, m is the mass of the probe, κ is the shadow factor. Note that a cylindrical shadow model can be adopted in computing the shadow factor since the magnitude of the solar radiation pressure perturbation acceleration is relatively small.

5) Atmospheric drag perturbation acceleration A_D

$$A_R = -\frac{1}{2} C_D \left(\frac{S_D}{m}\right) \rho V_r V_r \tag{4.110}$$

where C_D is the drag coefficient of the probe (typical value 2.2), S_D is the area of the cross section of the probe perpendicular to the atmospheric velocity, ρ is the atmospheric density, V_r is the velocity of the probe relative to the atmosphere.

6) Thrust acceleration A_P

$$A_P = \frac{P}{m} \Delta \overline{V}, \tag{4.111}$$

where P is the thrust magnitude, $\Delta \overline{V}$ is the direction of velocity impulse.

2. **Closed orbit about Mars**

The possible orbit types for a planet orbiting mission include planet synchronous orbit, Sun-synchronous orbit, pseudo-Sun-synchronous orbit (large elliptical orbit with shifting periapsis), critical inclination orbit, frozen orbit. A few types of closed orbit about Mars are discussed in the following as examples.

1) Mars Synchronous Orbit

A nominal design of Mars synchronous orbit is as follows:
Angular velocity (mean motion) of the orbit:

$$n = \dot{\Omega}_T = 350.89189°/\text{d} \tag{4.112}$$

Semi-major axis:

$$a_0^3 = \mu/\dot{\Omega}_T^2 = 8.52426 \times 10^{21} \tag{4.113}$$

$$a_0 = 20428 \text{ km} \tag{4.114}$$

Orbit altitude:

$$h_0 = 17031 \text{ km} \tag{4.115}$$

4.5 Design of Mission Orbits

If the influence of J_2 term on orbit period is included in the nominal design, then the design value of semi-major axis would be slightly greater than a_0 obtained above.

For Mars stationary orbit (MSO), which is a special type of Mars synchronous orbit ($i = 0°$), the influence of the J_2 perturbation is greater than the gravitational perturbation from the Sun, which is different from the characteristics of geostationary orbit. For a geostationary satellite, the gravitational perturbation from the Sun and Moon is greater than the J_2 perturbation.

The nominal design of Mars stationary orbit is as follows:

$$a_{GS} = 20430.99 \text{ km} \tag{4.116}$$

$$h_0 = 17034.79 \text{ km} \tag{4.117}$$

$$\eta_{GS} = \frac{a_{GS}}{R} = 6.016, \tag{4.118}$$

where η_{GS} is the ratio of the semi-major axis of MSO to Mars radius, which is very close to the ratio of the semi-major axis of geostationary orbit to Earth radius. The reason behind this is that the spin rates and the mean densities of these two planets are close.

At present, no spacecraft has been launched to orbit Mars in MSO. A spacecraft flying in MSO can act as a relay communication satellite of Mars, the feasibility of which is being investigated by many space institutes.

2) Sun-synchronous orbit about Mars

The definition of a Sun-synchronous orbit (SSO) about Mars is defined as follows [32]:

$$\dot{\Omega} = \dot{\Omega}_S \tag{4.119}$$

$$\dot{\Omega} = -\frac{3nJ_2R^2}{2a^2(1-e^2)^2} \cos i, \tag{4.120}$$

where $\dot{\Omega}_S = 0.52404°/\text{d}$ is the angular rate of the apparent motion of the mean Sun on the Mars-centered celestial sphere. For a circular SSO about Mars, the relationship between the orbit altitude and inclination is illustrated in Fig. 4.18. For an elliptical SSO about Mars, the relationship between the apoapsis altitude and inclination is illustrated in Fig. 4.19, in which a periapsis altitude of 380 km is assumed as an example.

Note that the mission orbits of most launched Mars orbiting missions were Sun-synchronous orbits about Mars.

3) Large elliptical orbit with shifting periapsis

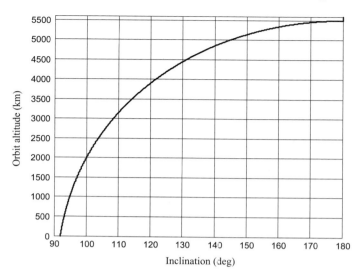

Fig. 4.18 Orbit altitude variation with respect to inclination for circular SSO

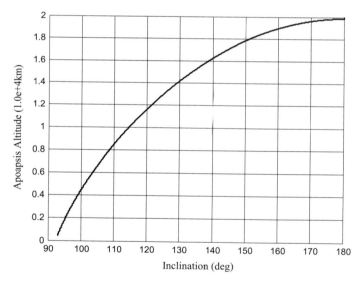

Fig. 4.19 Apoapsis altitude variation with respect to inclination for elliptical SSO

In orbit design of a Mars mission, the property of shifting periapsis of a large elliptical orbit about Mars can be exploited to design a polar orbit about Mars, with the periapsis shifting for one revolution in one Martian year and synchronized with the apparent movement of the Sun on the Mars-centered celestial sphere; as a result, the solar elevation angle at the subpoint of the periapsis would meet the specified requirement.

4.5 Design of Mission Orbits

Figure 4.20 shows the relationship between the inclination and semi-major axis for an elliptical orbit about Mars, with 265 km periapsis altitude and the periapsis shifting for one revolution in one Martian year.

As long as the periapsis shifts for one revolution in one Martian year precisely, global observation can be achieved for an inclination of 90° ± 5°, and the semi-major axis should be in the range of 9427–9615 km; also, global observation can be achieved for an inclination of 90° ± 10°, and the semi-major axis should be in the range of 8858–9615 km, as illustrated in Table 4.4. Provided that the solar elevation angle at the subpoint of the periapsis meets the design requirement, the shifting rate of the periapsis can be slightly adjusted.

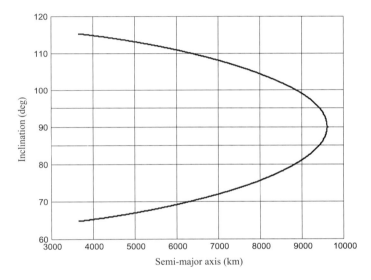

Fig. 4.20 Relationship between semi-major axis and inclination for orbit with shifting periapsis

Table 4.4 Design of semi-major axis versus inclination

Inclination/(°)	Semi-major axis/km
90 ± 10	8858–9615
90 ± 8	9133–9615
90 ± 6	9344–9615
90 ± 5	9427–9615
90 ± 4	9494–9615
90 ± 2	9584–9615
90 ± 0	9615

4.5.2 Missions to Lagrange Libration Points

1. **Introduction to the three-body problem**

The dynamic equation of the so-called circular restricted three-body problem in the rotating frame is:

$$\ddot{x} - 2\dot{y} = U_x \tag{4.121}$$

$$\ddot{y} + 2\dot{x} = U_y \tag{4.122}$$

$$\ddot{z} = U_z, \tag{4.123}$$

where the pseudo-potential function U is given by [34].

$$U = \frac{(1-\mu)}{r_1} + \frac{\mu}{r_2} + \frac{1}{2}(x^2 + y^2) \tag{4.124}$$

If the partial differential derivatives of the pseudo-potential function (U_x, U_y and U_z) are all zero, namely $\nabla U = 0$, there will be some equilibrium solutions for the above dynamic equation.

These solutions corresponds to the Lagrange libration points in the rotating frame where the gravitational forces and the centrifugal force (due to the rotation of the rotating frame) reaches an equilibrium, as a result, a particle placed at one of these points would remain stationary with respect to the rotating frame theoretically.

Among the five Lagrange libration points, there are three collinear libration points, whose location can be computed from the following equation

$$x - \frac{(1-\mu)(x+\mu)}{|x+\mu|^3} - \frac{\mu(x-(1-\mu))}{|x-(1-\mu)|^3} = 0 \tag{4.125}$$

The general form of motion near a collinear Lagrange point is a Lissajous trajectory, whose mathematical description is as follows [35].

$$\xi = A_1 \cos \lambda t + A_2 \sin \lambda t \tag{4.126}$$

$$\eta = -kA_1 \sin \lambda t + kA_2 \cos \lambda t \tag{4.127}$$

$$\varsigma = C_1 \sin \nu t + C_2 \cos \nu t, \tag{4.128}$$

where λ is the frequency of in-plane motion, ν is the frequency of out-of-plane motion. A first-order periodic orbit can be constructed by a proper selection of the

4.5 Design of Mission Orbits

initial orbit condition related to the Lissajous motion. Moreover, based on the first-order solution, the third-order periodic orbit about the collinear Lagrange point can be obtained by employing a computation procedure of consecutive approximation; based on the third-order solution, higher-order periodic solutions could be found by repeating the application of the Lindestedt–Poincaré method [36].

The differential correction of a halo orbit can exploit the symmetry of halo orbit about the *XOZ* plane, namely the trajectory crosses the XOZ plane perpendicularly. As a result, the following condition is valid at the point of crossing:

$$y = \dot{x} = \dot{z} = 0 \tag{4.129}$$

From this observation, Howell proposed the trajectory correction strategy for halo orbits in which the correction maneuver occurs when the probe is crossing the *XOZ* plane, namely $\bar{x}(t_0) = \begin{bmatrix} x_0 & 0 & z_0 & 0 & \dot{y}_0 & 0 \end{bmatrix}^T$, $\bar{x}(t_f)_{des} = \begin{bmatrix} x_f & 0 & z_f & 0 & \dot{y}_f & 0 \end{bmatrix}^T$. In the computation of the correction strategy, half revolution is taken as the unit of numeric integration, the condition for terminating the integration is $y_f = 0$, and the control objective is to make the terminal state $(\dot{x}_f, \dot{z}_f)$ approach zero.

2. **Numeric Solutions of Orbits about Lagrange Libration Points**

A two-layer differential correction approach can be employed to find the numeric solutions of Lissajous orbits, which was proposed by Howell and Pernicka in 1986, when they developed this method to compute the precise solutions of Lissajous orbits in circular restricted three-body problems (CR3BP). This method has been widely accepted and applied in the numeric computation of three-body problems.

The calculation of this two-layer differential correction approach is summarized as follows:

1) The first layer of correction is termed position correction, in which the position is made continuous between any two adjacent orbit segments and delta-Vs are inserted at the connecting points to match the velocities of these two segments. Moreover, approximate initial values of each orbit segment are obtained by analytical or numeric methods.
2) The second layer of correction is termed velocity correction, in which the delta-Vs inserted at the connecting points of adjacent segments in the first layer of correction are made minimum so that the natural solution of the Lissajous orbit can be found near the initial guess of the orbit. The delta-Vs are made minimum by adjusting the positions and times of the connecting points.

A detailed discussion of the numeric algorithm mentioned above is given in the following.

In the rotating reference frame in which the origin is located at a Lagrange libration point, assume the position is described by the Cartesian coordinates: $x = x(t)$, $y = y(t)$, $z = z(t)$.

The initial position is denoted as: $x_0 = x(t_0)$, $y_0 = y(t_0)$, $z_0 = z(t_0)$. The state is denoted as $X = \begin{bmatrix} x & y & z & \dot{x} & \dot{y} & \dot{z} \end{bmatrix}^T$.

The state transition matrix is:

$$\Phi(t, t_0) = \frac{\partial X(t)}{\partial X(t_0)} = \frac{\partial [x\ y\ z\ \dot{x}\ \dot{y}\ \dot{z}]^T}{\partial [x_0\ y_0\ z_0\ \dot{x}_0\ \dot{y}_0\ \dot{z}_0]^T} = \begin{bmatrix} \frac{\partial x}{\partial x_0} & \frac{\partial x}{\partial y_0} & \frac{\partial x}{\partial z_0} & \frac{\partial x}{\partial \dot{x}_0} & \frac{\partial x}{\partial \dot{y}_0} & \frac{\partial x}{\partial \dot{z}_0} \\ \frac{\partial y}{\partial x_0} & \frac{\partial y}{\partial y_0} & \frac{\partial y}{\partial z_0} & \frac{\partial y}{\partial \dot{x}_0} & \frac{\partial y}{\partial \dot{y}_0} & \frac{\partial y}{\partial \dot{z}_0} \\ \frac{\partial z}{\partial x_0} & \frac{\partial z}{\partial y_0} & \frac{\partial z}{\partial z_0} & \frac{\partial z}{\partial \dot{x}_0} & \frac{\partial z}{\partial \dot{y}_0} & \frac{\partial z}{\partial \dot{z}_0} \\ \frac{\partial \dot{x}}{\partial x_0} & \frac{\partial \dot{x}}{\partial y_0} & \frac{\partial \dot{x}}{\partial z_0} & \frac{\partial \dot{x}}{\partial \dot{x}_0} & \frac{\partial \dot{x}}{\partial \dot{y}_0} & \frac{\partial \dot{x}}{\partial \dot{z}_0} \\ \frac{\partial \dot{y}}{\partial x_0} & \frac{\partial \dot{y}}{\partial y_0} & \frac{\partial \dot{y}}{\partial z_0} & \frac{\partial \dot{y}}{\partial \dot{x}_0} & \frac{\partial \dot{y}}{\partial \dot{y}_0} & \frac{\partial \dot{y}}{\partial \dot{z}_0} \\ \frac{\partial \dot{z}}{\partial x_0} & \frac{\partial \dot{z}}{\partial y_0} & \frac{\partial \dot{z}}{\partial z_0} & \frac{\partial \dot{z}}{\partial \dot{x}_0} & \frac{\partial \dot{z}}{\partial \dot{y}_0} & \frac{\partial \dot{z}}{\partial \dot{z}_0} \end{bmatrix}$$

(4.130)

Define the pseudo-potential in three-body problem as:

$$U = \frac{x^2 + y^2}{2} + \frac{1-\mu}{d_1} + \frac{\mu}{d_2}, \tag{4.131}$$

where $\mu = \frac{m_2}{m_1 + m_2}$ is the mass ratio, $d_1 = \sqrt{(x+\mu)^2 + y^2 + z^2}$, $d_2 = \sqrt{(x-1+\mu)^2 + y^2 + z^2}$.

Denote the symmetric matrix of the partial differentiations of U with respect to x, y, z as U_{XX}, namely

$$U_{XX} = \begin{bmatrix} \frac{\partial^2 U}{\partial x^2} & \frac{\partial^2 U}{\partial x \partial y} & \frac{\partial^2 U}{\partial x \partial z} \\ \frac{\partial^2 U}{\partial y \partial x} & \frac{\partial^2 U}{\partial y^2} & \frac{\partial^2 U}{\partial y \partial z} \\ \frac{\partial^2 U}{\partial z \partial x} & \frac{\partial^2 U}{\partial z \partial y} & \frac{\partial^2 U}{\partial z^2} \end{bmatrix} \tag{4.132}$$

Moreover, U_{XX} can be simplified to a diagonal matrix since there are no coupled terms of x, y, z in the expressions of U:

$$U_{XX} = \begin{bmatrix} \frac{\partial^2 U}{\partial x^2} & 0 & 0 \\ 0 & \frac{\partial^2 U}{\partial y^2} & 0 \\ 0 & 0 & \frac{\partial^2 U}{\partial z^2} \end{bmatrix} = \mathrm{diag}(U_{xx}, U_{yy}, U_{zz}) \tag{4.133}$$

Denote as:

$$A(t) = \begin{bmatrix} 0 & I \\ U_{XX} & 2\Omega \end{bmatrix} \tag{4.134}$$

$$\Omega = \begin{bmatrix} 0 & 1 & 0 \\ -1 & 0 & 0 \\ 0 & 0 & 0 \end{bmatrix} \tag{4.135}$$

Then

4.5 Design of Mission Orbits

$$A(t)\Phi(t,t_0) = \begin{bmatrix} 0 & 0 & 0 & 1 & 0 & 0 \\ 0 & 0 & 0 & 0 & 1 & 0 \\ 0 & 0 & 0 & 0 & 0 & 1 \\ U_{xx} & 0 & 0 & 0 & 2 & 0 \\ 0 & U_{yy} & 0 & -2 & 0 & 0 \\ 0 & 0 & U_{zz} & 0 & 0 & 0 \end{bmatrix} \begin{bmatrix} \dfrac{\partial x}{\partial x_0} & \dfrac{\partial x}{\partial y_0} & \dfrac{\partial x}{\partial z_0} & \dfrac{\partial x}{\partial x_0} & \dfrac{\partial x}{\partial y_0} & \dfrac{\partial x}{\partial z_0} \\ \dfrac{\partial y}{\partial y} & \dfrac{\partial y}{\partial y} & \dfrac{\partial y}{\partial y} & \dfrac{\partial y}{\partial y} & \dfrac{\partial y}{\partial y} & \dfrac{\partial y}{\partial x} \\ \dfrac{\partial x_0}{\partial x_0} & \dfrac{\partial y_0}{\partial y_0} & \dfrac{\partial z_0}{\partial z_0} & \dfrac{\partial x_0}{\partial x_0} & \dfrac{\partial y_0}{\partial y_0} & \dfrac{\partial z_0}{\partial z_0} \\ \dfrac{\partial z}{\partial z} & \dfrac{\partial z}{\partial z} & \dfrac{\partial z}{\partial z} & \dfrac{\partial z}{\partial z} & \dfrac{\partial z}{\partial z} & \dfrac{\partial z}{\partial z} \\ \dfrac{\partial x_0}{\partial x_0} & \dfrac{\partial y_0}{\partial y_0} & \dfrac{\partial z_0}{\partial z_0} & \dfrac{\partial x_0}{\partial x_0} & \dfrac{\partial y_0}{\partial y_0} & \dfrac{\partial z_0}{\partial z_0} \\ \dfrac{\partial x}{\partial x_0} & \dfrac{\partial x}{\partial y_0} & \dfrac{\partial x}{\partial z_0} & \dfrac{\partial x}{\partial x_0} & \dfrac{\partial x}{\partial y_0} & \dfrac{\partial x}{\partial z_0} \\ \dfrac{\partial y}{\partial x_0} & \dfrac{\partial y}{\partial y_0} & \dfrac{\partial y}{\partial z_0} & \dfrac{\partial y}{\partial x_0} & \dfrac{\partial y}{\partial y_0} & \dfrac{\partial y}{\partial z_0} \\ \dfrac{\partial z}{\partial x_0} & \dfrac{\partial z}{\partial y_0} & \dfrac{\partial z}{\partial z_0} & \dfrac{\partial z}{\partial x_0} & \dfrac{\partial z}{\partial y_0} & \dfrac{\partial z}{\partial z_0} \end{bmatrix}$$

$$= \begin{bmatrix} \dfrac{\partial x}{\partial x_0} & \dfrac{\partial x}{\partial y_0} & \dfrac{\partial x}{\partial z_0} & \dfrac{\partial x}{\partial x_0} & \dfrac{\partial x}{\partial y_0} & \dfrac{\partial x}{\partial z_0} \\ \dfrac{\partial y}{\partial x_0} & \dfrac{\partial y}{\partial y_0} & \dfrac{\partial y}{\partial z_0} & \dfrac{\partial y}{\partial x_0} & \dfrac{\partial y}{\partial y_0} & \dfrac{\partial y}{\partial z_0} \\ \dfrac{\partial z}{\partial x_0} & \dfrac{\partial z}{\partial y_0} & \dfrac{\partial z}{\partial z_0} & \dfrac{\partial z}{\partial x_0} & \dfrac{\partial z}{\partial y_0} & \dfrac{\partial z}{\partial z_0} \\ \dfrac{\partial^2 U}{\partial x \partial x_0}+2\dfrac{\partial y}{\partial x_0} & \dfrac{\partial^2 U}{\partial x \partial y_0}+2\dfrac{\partial y}{\partial y_0} & \dfrac{\partial^2 U}{\partial x \partial z_0}+2\dfrac{\partial y}{\partial z_0} & \dfrac{\partial^2 U}{\partial x \partial x_0}+2\dfrac{\partial y}{\partial x_0} & \dfrac{\partial^2 U}{\partial x \partial y_0}+2\dfrac{\partial y}{\partial y_0} & \dfrac{\partial^2 U}{\partial x \partial z_0}+2\dfrac{\partial y}{\partial z_0} \\ \dfrac{\partial^2 U}{\partial y \partial x_0}-2\dfrac{\partial x}{\partial x_0} & \dfrac{\partial^2 U}{\partial y \partial y_0}-2\dfrac{\partial x}{\partial y_0} & \dfrac{\partial^2 U}{\partial y \partial z_0}-2\dfrac{\partial x}{\partial z_0} & \dfrac{\partial^2 U}{\partial y \partial x_0}-2\dfrac{\partial x}{\partial x_0} & \dfrac{\partial^2 U}{\partial y \partial y_0}-2\dfrac{\partial x}{\partial y_0} & \dfrac{\partial^2 U}{\partial y \partial z_0}-2\dfrac{\partial x}{\partial z_0} \\ \dfrac{\partial^2 U}{\partial z \partial x_0} & \dfrac{\partial^2 U}{\partial z \partial y_0} & \dfrac{\partial^2 U}{\partial z \partial z_0} & \dfrac{\partial^2 U}{\partial z \partial x_0} & \dfrac{\partial^2 U}{\partial z \partial y_0} & \dfrac{\partial^2 U}{\partial z \partial z_0} \end{bmatrix}$$

$$= \begin{bmatrix} \dfrac{\partial x}{\partial x_0} & \dfrac{\partial x}{\partial y_0} & \dfrac{\partial x}{\partial z_0} & \dfrac{\partial x}{\partial x_0} & \dfrac{\partial x}{\partial y_0} & \dfrac{\partial x}{\partial z_0} \\ \dfrac{\partial y}{\partial x_0} & \dfrac{\partial y}{\partial y_0} & \dfrac{\partial y}{\partial z_0} & \dfrac{\partial y}{\partial x_0} & \dfrac{\partial y}{\partial y_0} & \dfrac{\partial y}{\partial z_0} \\ \dfrac{\partial z}{\partial x_0} & \dfrac{\partial z}{\partial y_0} & \dfrac{\partial z}{\partial z_0} & \dfrac{\partial z}{\partial x_0} & \dfrac{\partial z}{\partial y_0} & \dfrac{\partial z}{\partial z_0} \\ \dfrac{\partial x_0}{\partial x} & \dfrac{\partial y_0}{\partial x} & \dfrac{\partial z_0}{\partial x} & \dfrac{\partial x_0}{\partial x} & \dfrac{\partial y_0}{\partial x} & \dfrac{\partial z_0}{\partial x} \\ \dfrac{\partial x_0}{\partial y} & \dfrac{\partial y_0}{\partial y} & \dfrac{\partial z_0}{\partial y} & \dfrac{\partial x_0}{\partial y} & \dfrac{\partial y_0}{\partial y} & \dfrac{\partial z_0}{\partial y} \\ \dfrac{\partial x_0}{\partial z} & \dfrac{\partial y_0}{\partial z} & \dfrac{\partial z_0}{\partial z} & \dfrac{\partial x_0}{\partial z} & \dfrac{\partial y_0}{\partial z} & \dfrac{\partial z_0}{\partial z} \end{bmatrix}$$

$$= \Phi(t, t_0) \tag{4.136}$$

Namely,

$$\frac{\mathrm{d}}{\mathrm{d}t}\Phi(t, t_0) = A(t)\Phi(t, t_0) \tag{4.137}$$

Assume the differentiation correction starts at the instant t_0 and ends at the instant t_f, with the subscripts "0" and "f" denoting the start instant and end instant, respectively, then

$$\delta X(t_f) = \frac{\partial X(t_f)}{\partial X(t_0)}\delta X(t_0) + \frac{\partial X(t_f)}{\partial t_f}\delta(t_f - t_0) = \Phi(t_f, t_0)\delta X(t_0)$$
$$+ \frac{\partial X(t_f)}{\partial t_f}\delta(t_f - t_0) \tag{4.138}$$

Denote that

$$L = \begin{bmatrix} \frac{\partial x_f}{\partial \dot{x}_0} & \frac{\partial x_f}{\partial \dot{y}_0} & \frac{\partial x_f}{\partial \dot{z}_0} & \dot{x}_f \\ \frac{\partial y_f}{\partial \dot{x}_0} & \frac{\partial y_f}{\partial \dot{y}_0} & \frac{\partial x_f}{\partial \dot{z}_0} & \dot{y}_f \\ \frac{\partial z_f}{\partial \dot{x}_0} & \frac{\partial z_f}{\partial \dot{y}_0} & \frac{\partial z_f}{\partial \dot{z}_0} & \dot{z}_f \end{bmatrix} \tag{4.139}$$

$$u = \begin{bmatrix} \delta \dot{x}_0 \\ \delta \dot{y}_0 \\ \delta \dot{z}_0 \\ \delta(t_f - t_0) \end{bmatrix} \tag{4.140}$$

$$b = \begin{bmatrix} \delta x_f \\ \delta y_f \\ \delta z_f \end{bmatrix} \tag{4.141}$$

From Eq. (4.138), the following linearized equation can be derived:

$$\begin{aligned} Lu &= \begin{bmatrix} \frac{\partial x_f}{\partial \dot{x}_0} & \frac{\partial x_f}{\partial \dot{y}_0} & \frac{\partial x_f}{\partial \dot{z}_0} & \dot{x}_f \\ \frac{\partial y_f}{\partial \dot{x}_0} & \frac{\partial y_f}{\partial \dot{y}_0} & \frac{\partial x_f}{\partial \dot{z}_0} & \dot{y}_f \\ \frac{\partial z_f}{\partial \dot{x}_0} & \frac{\partial z_f}{\partial \dot{y}_0} & \frac{\partial z_f}{\partial \dot{z}_0} & \dot{z}_f \end{bmatrix} \begin{bmatrix} \delta \dot{x}_0 \\ \delta \dot{y}_0 \\ \delta \dot{z}_0 \\ \delta(t_f - t_0) \end{bmatrix} \\ &= \begin{bmatrix} \frac{\partial x_f}{\partial \dot{x}_0}\delta \dot{x}_0 + \frac{\partial x_f}{\partial \dot{y}_0}\delta \dot{y}_0 + \frac{\partial x_f}{\partial \dot{z}_0}\delta \dot{z}_0 + \dot{x}_f \delta(t_f - t_0) \\ \frac{\partial y_f}{\partial \dot{x}_0}\delta \dot{x}_0 + \frac{\partial y_f}{\partial \dot{y}_0}\delta \dot{y}_0 + \frac{\partial x_f}{\partial \dot{z}_0}\delta \dot{z}_0 + \dot{y}_f \delta(t_f - t_0) \\ \frac{\partial z_f}{\partial \dot{x}_0}\delta \dot{x}_0 + \frac{\partial z_f}{\partial \dot{y}_0}\delta \dot{y}_0 + \frac{\partial z_f}{\partial \dot{z}_0}\delta \dot{z}_0 + \dot{z}_f \delta(t_f - t_0) \end{bmatrix} \\ &= \begin{bmatrix} \delta x_f \\ \delta y_f \\ \delta z_f \end{bmatrix} \\ &= b \end{aligned} \tag{4.142}$$

The minimum Euclidean norm solution of the above equation is

$$u = L^{\mathrm{T}}(LL^{\mathrm{T}})^{-1}b, \tag{4.143}$$

which gives the optimal correction for the position error.

4.5 Design of Mission Orbits

4.5.3 Rendezvous and Docking

At present, the USA, Russia and China have succeeded in accomplishing many rendezvous and docking (RVD) missions in near-Earth orbits. The USA also accomplished manned rendezvous and docking missions in lunar orbits. In the following discussions, a brief review of the typical phasing strategies in RVD missions is given first, and then the computation algorithms of the maneuver strategies are introduced [37–39].

1. **Typical Phasing Strategies in RVD Missions**

1) US RVD Missions in Near-Earth Orbits

During the orbital phase of phasing in the RVD mission of the US spaceship Gemini, several separate adjustments were made to the in-plane orbital parameters and the out-of-plane orbital parameters. Each maneuver had a specific objective, and the adjustments of in-plane orbital parameters occurred near the perigee or apogee, which is termed special point maneuvers.

The scheme of special point maneuvers was also adopted during the orbit phase of phasing in the RVD operations of American space shuttles and the International Space Station. After the space shuttle entered its Earth orbit, it would execute a transverse maneuver (OMS2) at the first apoapsis passage, whose purpose is to raise the perigee altitude so that the atmospheric drag will not cause the orbit altitude to decrease too fast; later, several transverse maneuvers were carried out at apoapsis (NC-1, NC-2, NC-3, NC-4), for which the purpose is to adjust the phase angle of the shuttle with respect to the space station, the execution of NC-2 and NC-3 could be canceled if the corresponding delta-Vs turned out to be too small; in the later RVD missions, a single maneuver in orbital normal direction (NPC) was scheduled between NC-2 and NC-3 in order to correct the orbit plane error, which was actually executed at two different orbital positions during earlier missions to adjust the orbital inclination and RAAN separately; also, before NC-4, a transverse maneuver (NH) at perigee was arranged in order to raise the apogee altitude to target orbit altitude. After NC-4, the initial aim point was achieved and relative navigation between the space shuttle and the space station became available.

The orbital phase of phasing ended at achieving the initial aim point, and the following orbital phases include far range rendezvous, close range rendezvous and docking, which are not the focus of this discussion.

2) Russian RVD Missions in Near-Earth Orbits

In the RVD missions of Russian spaceships Soyuz and Progress, the velocity impulses executed during the orbital phase of phasing included both in-plane component and out-of-plane component and were jointly applied to achieve the desired terminal orbital states at the initial aim point, as such each maneuver contributed to the correction of terminal orbital states, and the phasing scheme is termed comprehensive maneuvers or combination maneuvers.

Specifically speaking, the phasing maneuvers included five maneuvers (M1–M5), among which M1, M2, M4 and M4 were the main maneuvers. The main maneuvers were executed at apogee or perigee, and the velocity impulse of each main maneuver included both in-plane component and out-of-plane component. M3 is a minor maneuver, which was applied to correct the control errors from the preceding two maneuvers (M1 and M2).

3) Apollo Manned Lunar Missions

Lunar orbit RVD operations were required in US Apollo manned lunar missions, in which the ascent stage of the lunar landing module took off from the surface of the Moon and achieved the same position and velocity with the command module after several orbital maneuvers. The orbital maneuver scheme was developed based on the flight experiences of the RVD missions of Gemini spaceships.

The tracking condition for lunar orbit RVD is not as good as for near-Earth RVD, so that the orbital control strategy has to be simplified to avoid long flight time for the RVD operation and increased risks to the astronauts' safety. Therefore, the orbital phase of phasing and the following orbital phases were designed jointly in the RVD operations of the Apollo missions.

As illustrated in Fig. 4.21, the ascent stage of the lunar landing module took off from the surface of the Moon and entered an initial lunar orbit, at apoapsis it executed a maneuver (CSI) to enter the co-elliptical orbit, then the ascent stage performed an orbit plane adjustment maneuver (PC), and later, it executed an orbit altitude adjustment maneuver (CDH). These maneuvers were planned jointly to help achieve the desired initial condition of the terminal rendezvous phase.

Note that the above RVD scheme including a three-maneuver phasing phase was greatly simplified after the Apollo 14 mission due to matured tracking and control techniques: The ascent stage took off from the surface of the Moon, entered the initial lunar orbit, and then, the terminal rendezvous phase was immediately initiated.

2. **Orbital Maneuver Strategy**

Theoretical studies of phasing strategy include linear model-based studies and nonlinear two-body model-based studies. The results from the theoretical studies cannot be directly applied in practical phasing maneuvers since no orbital perturbations are taken into account. However, the basic methods and conclusions from theoretical studies are very importance references for practical orbit design, such as:

1) Theoretical studies found that the fuel-minimum solution of an optimal coplanar rendezvous problem is the two-impulse Hohmann maneuver strategy if both the chaser orbit and the target orbit are circular. In practical engineering designs, the in-plane phasing maneuvers are usually Hohmann maneuvers.
2) Theoretical studies found that the number of optimal maneuvers is 4 for coplanar multi-revolution rendezvous problems, and the number of optimal maneuvers is 5 or 6 for non-coplanar multi-revolution rendezvous problems, which is an important reference in planning the number of maneuvers in practical RVD missions.

4.5 Design of Mission Orbits

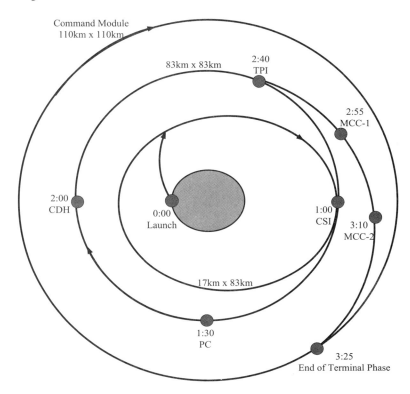

Fig. 4.21 Orbit scheme of lunar orbit RVD in Apollo missions

3) To compute the phasing maneuver strategy in a real RVD mission, the results obtained using the methods from theoretical studies can be used as the initial values for a full orbital dynamic model, in which perturbations are included, and the precise solution used in the actual executions can be obtained via an iteration process.

An introduction to the computation methods of the phasing maneuver strategies is given in the following. As mentioned in the preceding discussion, the phasing maneuver strategies applied in the real RVD missions include the strategy of special point maneuvers and the strategy of comprehensive maneuvers.

1) Special point maneuvers

For the phasing strategy of special point maneuvers, the maneuvers are executed at some special points of the orbit, such as perilune, apolune or ascending node. The principle of the phasing strategy of special point maneuvers is to exploit the property of orbital dynamics to separate the adjustments of in-plane orbital parameters and out-of-plane orbital parameters. As a result, the orbital controls of in-plane motion and out-of-plane motion are not coupled, and the computation of the orbital control parameters becomes much convenient and efficient.

The equations of variation of parameters (VOP) can be rewritten in terms of the velocity impulses as follows.

$$\Delta i = \frac{r \cos u}{na^2\sqrt{1-e^2}} \Delta v_n \tag{4.144}$$

$$\Delta \Omega = \frac{r \sin u}{na^2\sqrt{1-e^2}\sin i} \Delta v_n \tag{4.145}$$

$$\Delta \omega = \frac{\sqrt{1-e^2}}{nae}\left\{-\cos f \Delta v_r + \left(1 + \frac{1}{1+e\cos f}\right)\sin f \Delta v_t\right\}$$
$$-\cos i \frac{r \sin u}{na^2\sqrt{1-e^2}\sin i} \Delta v_n \tag{4.146}$$

$$\Delta a = \frac{2}{n\sqrt{1-e^2}}\{e \sin f \Delta v_r + (1+e\cos f)\Delta v_t\} \tag{4.147}$$

$$\Delta e = \frac{\sqrt{1-e^2}}{na}\left\{\sin f \Delta v_r + \left(\cos f + \frac{e+\cos f}{1+e\cos f}\right)\Delta v_t\right\} \tag{4.148}$$

As can be seen from the above equations, if a transverse velocity impulse Δv_t is applied at perilune or apolune, the semi-major axis and eccentricity (a, e) can be adjusted without affecting the orbit plane and argument of perilune (i, Ω, ω). Therefore, the orbital period can be adjusted to correct the phase angle of the chaser spacecraft relative to the target spacecraft. Similar to the two-impulse Hohmann maneuver scheme, the directions of the transverse velocity impulses Δv_t applied at perilune or apolune are parallel to the orbital velocity directions and energy-optimal solutions can be achieved.

Moreover, as illustrated in Eqs. 4.144 and 4.145, it is the most efficient to carry out inclination correction at the orbital position, at which the argument of latitude is 0° (ascending node) or 180° (descending node), and to carry out RAAN correction at the orbital position at which the argument of latitude is 90° or 270°. This arrangement can make the adjustment of inclination independent from the adjustment of RAAN, eliminating the coupled effect. On the other hand, a single velocity impulse in orbit normal direction can be applied at the orbital position at which the argument of latitude is $u = \tan^{-1}\left(\frac{\Delta \Omega}{\Delta i}\sin i\right)$ to adjust inclination and RAAN simultaneously.

The phasing strategy of special point maneuvers is simple and easy to understand, in which the purpose of each velocity impulse is justified and has a clear physical meaning. The main advantages of this strategy is that it is easy to find energy-optimal or suboptimal solutions, and it is convenient for scheduling flight events since the orbital position of the maneuvers are basically fixed. Typical real flight examples of applying the phasing strategy of special point maneuvers in near-Earth orbits include the manned RVD missions of Chinese spaceships Shenzhou and space

4.5 Design of Mission Orbits

stations Tiangong, US Gemini spaceships, US space shuttles and the international space station.

As can be seen from the applications of the phasing strategy of special point maneuvers in practical engineering, usually a transverse velocity impulse is applied at the apoapsis of the initial orbit to adjust the phase angle of the chaser spacecraft with respect to the target spacecraft, and at least two transverse velocity impulses should be applied later to adjust the in-plane orbital elements (a,e,ω). Moreover, the orbit plane correction (i,Ω) can be achieved using one or two velocity impulses applied in orbit normal direction.

Specifically, the first velocity impulse for phasing maneuver can be calculated as follows. Assume the initial phase angle difference between the chaser spacecraft and the target spacecraft is $\Delta\varphi$, the difference in orbit rates (mean motion) of the chaser orbit and target orbit is Δn, and the time of flight for phasing is T, then

$$\Delta n = \frac{\Delta\varphi}{T} \tag{4.149}$$

On the other hand, from the definition of mean motion:

$$n = \sqrt{\frac{\mu_m}{a^3}} \Rightarrow \Delta a = \frac{\Delta n}{\sqrt{\mu_m}(-1.5)a^{-2.5}} \tag{4.150}$$

From the above equation the semi-major axis a^* of the chaser orbit after phasing maneuver can be calculated. From the energy equation,

$$\frac{v^2}{2} - \frac{\mu_m}{r} = -\frac{\mu_m}{2a^*} \Rightarrow v = \sqrt{2\left(\frac{\mu_m}{r} - \frac{\mu_m}{2a^*}\right)} \tag{4.151}$$

Let the parameter r be the radius of the apoapsis of the initial orbit of the chaser spacecraft; Eq. (4.151) can be used to calculate the apoapsis speed right after the phasing maneuver, from which subtracting the apoapsis speed of the initial orbit of the chaser spacecraft gives the approximate value of the velocity impulse Δv_{t1} of the phasing maneuver.

The corrections of the in-plane orbital elements (a, e, ω) are achieved using two transverse velocity impulses Δv_{t2}, Δv_{t3}:

(a) If $\left(\frac{\Delta a}{a}\right)^2 > \Delta e_x^2 + \Delta e_y^2$

$$\Delta v_{t2} = \frac{V}{4} \cdot \frac{\left(\frac{\Delta a}{a}\right)^2 - \left(\Delta e_x^2 + \Delta e_y^2\right)}{\frac{\Delta a}{a} - \left(\Delta e_x \cos u_1 + \Delta e_y \cos u_1\right)} \tag{4.152}$$

$$\Delta v_{t3} = \frac{V \Delta a}{2a} - \Delta v_{t2} \tag{4.153}$$

(b) If $\left(\frac{\Delta a}{a}\right)^2 < \Delta e_x^2 + \Delta e_y^2$

$$\Delta v_{t2} = \frac{V}{4}\left(\frac{\Delta a}{a} + \sqrt{\Delta e_x^2 + \Delta e_y^2}\right) \qquad (4.154)$$

$$\Delta v_{t3} = \frac{V}{4}\left(\frac{\Delta a}{a} - \sqrt{\Delta e_x^2 + \Delta e_y^2}\right), \qquad (4.155)$$

where $e_x = e\cos\omega$, $e_y = e\sin\omega$, $V = na$. The velocity impulses calculated using the above equations can be taken as the initial values and substituted into the two-body model and full dynamic model (with perturbations included) successively for iteration computations, from which the precise solution can be obtained.

2) Comprehensive maneuvers

As mentioned earlier, the velocity impulses of phasing maneuvers executed in the RVD missions of Russian spaceships Soyuz and Progress included both in-plane component and out-of-plane component and were jointly applied to achieve the desired terminal orbital states. Each maneuver contributed to the correction of terminal orbital states, and the phasing scheme is termed comprehensive maneuvers. This phasing strategy of comprehensive maneuvers was devised by famous Russian scholar Baranov, based on the optimization method of near-circular small-deviation linear equations and geometric analysis [40].

Taking a typical application of Baranov's theory as an example, the computation method of the in-plane maneuver strategy is introduced in the following. The basic idea is to achieve the desired orbital parameters at the initial aim point by adjusting four transverse velocity impulses, and the maneuver positions can be optimized to minimize the total velocity impulse. The equations for calculating the velocity impulses of in-plane maneuvers are as follows:

$$3\alpha_1\Delta v_{t1} + 3\alpha_2\Delta v_{t2} + 3\alpha_3\Delta v_{t3} + 3\alpha_4\Delta v_{t4} = \Delta t \qquad (4.156)$$

$$2\Delta v_{t1} + 2\Delta v_{t2} + 2\Delta v_{t3} + 2\Delta v_{t4} = \Delta a \qquad (4.157)$$

$$2\cos\alpha_1\Delta v_{t1} + 2\cos\alpha_2\Delta v_{t2} + 2\cos\alpha_3\Delta v_{t3} + 2\cos\alpha_4\Delta v_{t4} = \Delta q \qquad (4.158)$$

$$2\sin\alpha_1\Delta v_{t1} + 2\sin\alpha_2\Delta v_{t2} + 2\sin\alpha_3\Delta v_{t3} + 2\sin\alpha_4\Delta v_{t4} = \Delta g \qquad (4.159)$$

Equation (4.156) is the phasing equation, in which α_i ($i = 1, 2, 3, 4$) are the phase angles from the velocity impulses to the end of phasing (initial aim point), and Δt gives the total phase angle travelled by the chaser spacecraft from its orbit insertion to the initial aim point. The principle behind this equation is that the phase angle of the chaser spacecraft is adjusted by the velocity impulses so that the chaser can reach the desired in-plane orbital position at the specified time of the initial aim point. Equation (4.157) describes the adjustment of the semi-major axis of the chaser orbit by applying the transverse velocity impulses to achieve the same semi-major axis of the target orbit. Equations (4.158) and (4.159) describe the adjustment of the eccentricity vector of the chaser orbit by applying the transverse velocity impulses to achieve the same eccentricity and argument of periapsis of the target orbit.

Note that the velocity impulses computed from the above four equations are approximate values based on a minor deviation linearization model. Similar to the special point maneuver method, the approximate velocity impulse values can be used as the initial values and substituted into the two-body model and perturbation model for iteration successively, which would give the precise values of the velocity impulses.

In summary, this strategy is rigorous in theory, and very flexible in applications. It can be used to replan a maneuver strategy conveniently and implement a comprehensive maneuver. In practical applications, some modified phasing strategy can be employed. For example, the three velocity impulses lying in orbit plane and the argument of latitude of the third velocity impulses can be selected as the design variables to achieve the desired orbital elements (a, e, ω, u) at the aim point; moreover, a velocity impulse in the normal direction of the orbit plane and its argument of latitude can be used as the design variables to achieve the desired orbital elements (i, Ω).

On the other hand, if the minor deviation linearization assumption are not valid (e.g., the eccentricity of the orbit is not close to zero), the approximate delta-V values obtained from the preceding four equations might not be reliable and may lead to convergence difficulty during the iterations, in which case an optimization algorithm can be employed to find the precise solution.

4.6 Design of Orbital Maneuver Strategy

In the design of orbital maneuver strategy, the control parameters related to a maneuver such as the engine ignition time, orbital location, thrust magnitude and direction, thrust duration are to be determined to help achieve the mission objective and the propagation of orbit determination error, control execution error and the constraints from the TT&C system and spacecraft system should be taken into consideration. The orbital maneuver strategy should be optimized to achieve minimum propellant consumption. The methods of covariance analysis and Monte Carlo shooting analysis are commonly used in the error propagation analysis.

For example, the orbital maneuver strategies of the Chang'e 3 mission include the strategy of trajectory correction maneuver (TCM), the strategy of Lunar Orbit Insertion (LOI), and the strategy of descent maneuver on lunar orbit. Taking the TCM strategy as an example, the specific process is as follows.

(1) Orbit determination should be carried out once the Chang'e 3 spacecraft entered the Earth-to-Moon transfer trajectory, basing on the orbit determination data from the TT&C system, the velocity impulse of the first trajectory correction maneuver (TCM-1), and the desired orbit parameters at perilune.

If the velocity impulse found is greater than some specified value Δv_1^*, the 7500 N main engine is selected to execute TCM-1.

After TCM-1 is completed, orbit determination and propagation should be conducted, and then, the velocity impulse for the third trajectory correction maneuver (TCM-3) is calculated. If the velocity impulse is greater than some specified value Δv_3^*, then the second trajectory correction maneuver (TCM-2) should be carried out. Otherwise, TCM-2 would be cancelled.

(2) If the velocity impulse found for TCM-1 is less than Δv_1^*, orbit propagation is continued without TCM-1 and the velocity impulse for TCM-2 is calculated.

If the velocity impulse found for TCM-2 is greater than Δv_1^* and less than some specified value Δv_2^*, TCM-1 would be cancelled, and the 7500 N main engine is selected to execute TCM-2 at the specified time.

(3) If the velocity impulse found for TCM-2 is greater than Δv_2^* or less than Δv_1^*, then TCM-1 and TCM-2 should be planned jointly. Based on orbit determination data, the velocity impulse component of TCM-1 in the direction of velocity (Δv_1^t), the azimuth and declination angles (α_2, δ_2) of the velocity impulse of TCM-2 ($\Delta \bar{v}_2$) are adjusted such that the desired orbit parameters at perilune can be achieved using differential corrections, with the magnitude of the velocity impulse of TCM-2 (Δv_2) fixed at Δv_1^*. Also, the azimuth and declination angles (α_1, δ_1) of the velocity impulse of TCM-1 is adjusted to minimize the magnitude of the velocity impulse of TCM-1 (Δv_1) and make the total velocity impulse ($\Delta v_1 + \Delta v_2$) of TCM-1 and TCM-2 less than Δv_2^*.

(4) After the preceding trajectory correction maneuver is completed, orbit determination and propagation should be conducted, and the velocity impulse for TCM-3 ($\Delta \bar{v}_3$) is calculated by differential corrections, the target being the desired orbit parameters at perilune. TCM-3 is executed using the 8 × 150 N engine combination.

Note that Δv_1^*, Δv_2^* and Δv_3^* are determined from the propellant mass and error propagation analysis. Practical engineering requirements should also be met. The flowchart of the TCM strategies in Earth-to-Moon transfer is given in Fig. 4.22.

Orbit maneuvers are necessary for trajectory corrections, orbit maintenance and transitions between the different flight phases of a space mission. As demonstrated in the above discussion, the objective of orbital maneuver strategy design is to help achieve the mission objectives by performing engineering design of the relevant control parameters, considering practical constraints and requirements.

4.7 Future Prospects

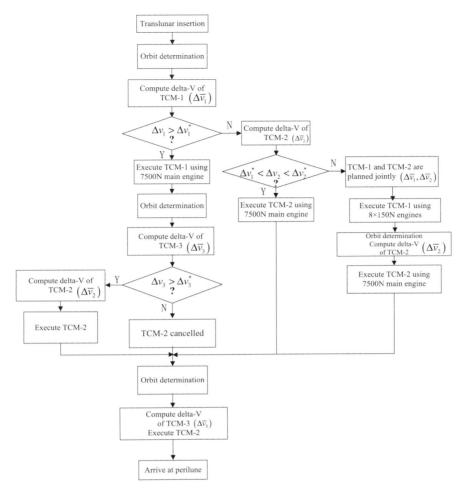

Fig. 4.22 TCM flowchart in Earth-to-Moon transfer

4.7 Future Prospects

As can be seen in the preceding discussions, progresses made in recent years in orbit designs of successful deep space missions revealed the possible trends in the development of orbit design technologies for deep space missions.

(1) The expansion of mission objectives leads to the fusion of different orbit design technologies.

On one hand, the application of the traditional near-Earth orbit design techniques is extended to deep space missions. For example, in the Chang'e 2 mission, the design method of repeating ground track orbit for Earth remote-sensing satellites was employed in the lunar orbit design for the remote-sensing task, and the flight test was a great success; in the trajectory design of the extended mission, in which the Chang'e 2 spacecraft was to flyby asteroid Toutatis, the flyby trajectory was offset to lower the risk of collision with the asteroid, applying the orbit design techniques of space debris collision prediction and collision probability analysis for Earth orbits.

On the other hand, the orbit design techniques for different types of orbits are applied in the same mission. For example, in the Chang'e 4 mission, the relay communication satellite is flying in a halo orbit about the Earth–Moon L2 libration point, and the lander flew a flight orbit similar to the Chang'e 3 mission to land on the far side of the Moon.

The fusion of different orbit design technologies and comprehensive application of different types of orbits help achieve optimal mission designs, which satisfy the requirements of multi-task and multi-target missions.

(2) With the advancement of space technologies, multi-task and multi-target missions become common, and deep space missions are becoming more complicated than ever before. The orbit design tools need constant improvements to suit the requirements of challenging mission design tasks. A typical deep space mission consists of several flight phases, and the total flight time can be several years or even more than ten years. Orbit computation cost is in a demanding situation. Convenience and efficiency are stressed in orbit computation tools and methods to perform rapid orbit computation, flight parameter analysis, orbit optimization and determine the orbit schemes in a short period of time.

References and Related Reading

1. Yang WL (1997) Study of the transfer trajectory for launching a lunar probe into polar lunar orbit. Spacecr Eng 12(3):19–33. (Chinese vision)
2. Yang JC (1995) Orbital dynamics and control of aerospace vehicle. China Astronautic Publishing House, Beijing. (Chinese vision)
3. Chu GB, Zhang H (2007) Lunar probe technologies. Science and Technology of China Press, Beijing. (Chinese vision)
4. Liu L (2015) Theory and applications of deep space exploration orbits. Electronics Industry Press, Beijing. (Chinese vision)
5. Li JF, Baoyin HX, Jiang FH (2014) Dynamics and control of interplanetary flight. Tsinghua University Press, Beijing. (Chinese vision)
6. Peng CR (2011) Spacecraft systems engineering. Science and Technology of China Press, Beijing. (Chinese vision)
7. Systems Engineering Division of Lunar and Deep Space Exploration (2014) Chinese Academy of Science. Lunar and deep space exploration. Guangdong Science and Technology Press, Guangdong. (Chinese vision)
8. Zhang Y (2005) Current status and development trend of the research and application of electric propulsion technology. Rocket Propuls 31(2):27–36. (Chinese vision)
9. Duan CH, Chen LY (2013) Study of full electrical propulsion technology in GEO satellite design. Spacecr Eng, 22(3):99–104. (Chinese vision)
10. Tian BY (2012) Research on design and optimization of low thrust transfer trajectory with GA. Harbin Institute of Technology, Harbin. (Chinese vision)

11. Zhou WY (2004) Design and analysis of transfer trajectory for a lunar probe. China Academy of Space Technology, Beijing. (Chinese vision)
12. Meltzer M (2007) Mission to Jupiter: a history of the Galileo Project, NASA SP-2007-4231. NASA, Washington DC
13. Miller LJ, Miller JK, Kirhofer WE (1983) Navigation of the Galileo mission. AIAA, Washington, DC
14. Dong J, Meng LZ, Zhao Y et al (2015) Study of mission planning of international Jupiter system missions. Spacecr Eng 24(3):85–92. (Chinese vision)
15. The JUICE Scicence Study Team (2011) JUICE exploring the emergence of habitable worlds around gas giants. ESA, Paris
16. Kowalkowski T, Johannesen J, Lam T (2008) Launch period development for the Juno Mission to Jupiter. AIAA, Washington, DC
17. Ceriotti M (2010) Global optimisation of multiple GA trajectories. University of Glasgow, Glasgow
18. Ren Y (2007) Research on design and optimization of low thrust transfer trajectory in interplanetary missions. Harbin Institute of Technology, Harbin. (Chinese vision)
19. Walker MJH, Ireland B, Owens J (1985) A set of modified equinoctial orbit elements. Celest Mech 36:409–419
20. Goodson T (1992) Fuel-optimal control and guidance for low-and medium-thrust orbit transfer. Ph.D. Dissertation of Georgia Institute of Techonlogy, 2–16
21. Roberto R, Topputo F (2006) A sitth-order accurate scheme for solving two-point boundary value problems in astrodynamics. Celest Mech Dyn Astron 96:289–309
22. Desai PN, Conway BA (2005) A two-timescale discretization scheme for collocation. Adv Astronaut Sci 119(2):2053–2063
23. Betts JT, Huffman WP (1993) Path constrained trajectory optimization using sparse sequential quadratic programming. J Guid Control Dyn 16(1):59–68
24. Betts JT (1998) Survey of numerical methods for trajectory optimization. J Guid Control Dyn 21(2):193–207
25. Russell RP (2007) Primer vector theory applied to global low-thrust trade studies. J Guid Control Dyn 30(2):460–472
26. Lee S, Petropoulos AE, Allmen PV (2006) Low-thrust orbit transfer optimization with refine Q-law and multi-objective genetic algorithm. Adv Astronaut Sci 123(3):2249–2264
27. Petropoulos AE, Longuski JM (2004) Shape-based algorithm for automated design of low-thrust, gravity-assist trajectories. J Spacecr Rockets 41(5):787–796
28. Hu WD (2009) Characteristics of orbital dynamics of spacecraft near an asteroid. Prog Astronomy 27(2):152–166. (Chinese vision)
29. Scheeres DJ, Williams BG, Miller JK (2000) Evaluation of the dynamic environment of an asteroid: applications to 433 Eros. J Guid Control Dyn 23(3):466–475
30. Scheeres DJ (1999) Stability of hovering orbits around small bodies. Spaceflight mechanics 1999: Advances in the Astronautical Sciences, Part II, 102:855–875
31. Werner RA, Scheeres DJ (1997) Exterior gravitation of a polyhedron derived and compared with harmonic and mascon gravitation representations of asteroid 4769 Castalia. Celest Mech Dyn Astron 65:313–344
32. Yang, W. L. ZY-1 Satellite ORBIT: theory and practice. Spacecr Eng 10(1):30–43. (Chinese vision)
33. Szebehely V (1967) Theory of orbits: the restricted problem of three bodies. Academic Press, New York
34. Farquhar RW (1998) The flight of ISEE-3/ICE: origins, mission history, and a legacy. AIAA, Boston
35. Farquhar RW, Kamel AA (1973) Quasi-periodic orbits about the translunar libration point. Celest Mech Dyn Astron 7(4):458–473
36. Richardson DL, Cary ND (1975) A uniformly valid solution for motion about the interior libration point of the perturbed elliptic-restricted problem. In: AAS/AIAA astrodynamics specialist conference, Bahamas

37. Zhang J (2008) Strategy and planning of space rendezvous phasing maneuvers. Changsha:National University of Defense Technology. (Chinese vision)
38. Gong SP, Baoyin HX, Li JF (2010) Rapid targeting method for multi-impulse near circular orbit rendezvous. Space Control Technol Appl 36(5):1–6. (Chinese vision)
39. Wang ZS, Meng ZF, Gao S (2014) Study of orbit maneuver strategy for lunar orbit rendezvous mission. Spacecr Eng 23(5):103–110. (Chinese vision)
40. Baranov AA (1986) Algorithm for calculating the parameters of four-impulse transitions between close almost-circular orbit. Cosm Res 24(3):324

Chapter 5
Payload Technology

5.1 Introduction

In deep space exploration, the roles of the payloads are mainly to carry out on-orbit exploration or surface exploration of extraterrestrial objects and obtain exploration data so as to provide basic materials for solving scientific problems under the conditions provided by the probe.

The types of payloads in deep space exploration are diversified and can be divided by multiple means. According to the objects of deep space exploration, payloads can be divided into lunar exploration, Mars exploration, Venus exploration, asteroid exploration, Jupiter exploration, and interplanetary exploration payload, etc. According to the exploration missions and the scientific problems to be solved, payloads can be divided into payloads for extraterrestrial celestial topography and surface structure exploration, celestial surface material composition exploration, celestial internal structure and geological structure, atmospheric composition structure and climate exploration, celestial physical characteristics and space environment exploration, as well as life information exploration that is emerging in recent years. According to the working principles and the spectrum segments used, payloads can be divided into payloads for γ- and X-ray exploration, UV exploration, visible spectrum segment exploration, infrared spectrum segment exploration, microwave and radio exploration, energy particle exploration, and various gravitational fields, magnetic fields and electric fields. According to the relative positions of the probes and the detected objects, payloads can be divided into two categories: in-situ exploration and remote-sensing exploration. The methods of remote sensing are mainly used in orbiting exploration; however, mass spectrometry can also be used for in-situ exploration of the atmospheric components in orbiting exploration. Moreover, landing exploration is mainly based on in-situ exploration. In addition, deep space explorations of human beings aim not only to recognize such extraterrestrial objects but also to conduct scientific exploration and resource development based on the special and convenient conditions for such celestial bodies, which depend on payloads for lunar-based astronomical exploration and Earth environmental exploration, as well as lunar and asteroid resource exploration.

This chapter first introduces the main scientific problems of deep space exploration research and then introduces the scientific missions and payload configurations of lunar exploration and Mars exploration of China. Finally, this chapter details the representative morphological feature acquisition, element component identification and lunar-based astronomical observation payloads.

5.2 Major Scientific Issues in Deep Space Exploration Research

In astronomy, in addition to the Sun itself, the celestial bodies in the solar system are further subdivided into eight major planets and their satellites, dwarf planets, asteroids, comets and interplanetary dust. However, in a broad sense, these celestial bodies all belong to the planetary system and are the main research objects of planetary science.

The main contents of planetary science research include the physical and chemical properties of the planetary system, the surface morphology and internal structure of the planet, the formation and evolution of the planet, and the origin of the solar system, etc. The research contents of planetary science are fundamental. For example, the origin and evolution of the solar system has always been a basic problem in human civilization research. In addition, the scientific issues to be researched are also dynamic. With the development of Earth-based and space-based exploration activities and the deepening of research, some of the existing problems have been solved. Meanwhile, with the deepening of understandings, new questions have been proposed. Therefore, the scientific problems of deep space exploration are developing dynamically.

5.2.1 Scientific Issues of Deep Space Exploration from a Systematic Perspective

The origin and evolution of the solar system, the origin and evolution of life and the influence of the solar system on human living environment and space exploration have always been the main objectives of deep space exploration. Although there are different emphases in lunar, Mars and asteroid exploration, however, these problems can be classified from the system perspective of human scientific research.

5.2.1.1 How Did the Planets and Small Bodies in the Solar System Originate?

The origin of the solar system is one of the key issues in deep space exploration. It is generally believed that the solar system is a "disk" derived from the collapse of the nebulas, the main part of which forms the Sun and the peripheral nebulas formed planets. Some of the fragments that were not merged or formed by collision events further constituted small bodies and interstellar dusts. To get to know the origin of the solar system, it is of vital importance to understand the traces remained from the early stages of planets and their satellites determine the chemical compositions and physical characteristics of various celestial bodies and research the original states of planet formation.

For this type of scientific problem, both macroscopically and microscopically the chemical composition of each celestial body (mainly the early components of the inner solar system and the outer solar system celestial bodies) as well as the special physical characteristics that affect the formation of celestial bodies should be detected by the payloads, including the detection of magnetic fields and gravitational fields, etc.

5.2.1.2 How Did the Celestial Bodies Within the Solar System Evolve into the Current State?

There are huge differences among major celestial bodies within the solar system. Although all belong to terrestrial planets, Mercury, Venus, Earth and Mars have great differences in terms of the atmosphere, internal structure and physical characteristics of the planet. How did such differences evolve? And how did gaseous giant planets and their satellite systems evolve? By studying such problems, we can understand the evolution process of the solar system. In addition, it is also of important significance to predict the future development of the Earth.

For this type of scientific problems, the payloads must be designed so as to support comparative planetary studies, detect the atmospheric composition and climatic characteristics, surface morphology and composition of each celestial body, and apply geological activities, natural or artificial impact to detect the internal structure of the celestial body.

5.2.1.3 How Did Life Originate and Evolve? Has Life Ever Existed in Other Parts of the Solar System?

The origin of life is one of the ultimate problems that human beings are exploring. According to the existing understanding of life on the Earth, water, heavy elements, organic matter and energy, etc., are the necessary conditions for maintaining life. The chemical and isotopic compositions of small bodies and comets must be studied to determine their contribution to the sources of water and that of organic molecules on the Earth. By studying the atmospheric chemistry and isotopic composition of Venus, the results can be used to determine if the existence of life was supported historically on Venus. By studying the physical and chemical characteristics of potential life zones such as Europe and Titan, the results can be used to determine if the existence of life is supported on these celestial bodies. Finally, life detection can be conducted on Mars on which it is most probable to have (or have had) existing lives.

For these scientific issues, the payloads should support researches of life origin, mainly related to the detection of chemical constituents and contents thereof, including water and water-bearing minerals, atmospheric constituents, especially organic matter, microorganisms and their metabolites, and fossils with life evidence, etc. To solve these scientific issues, the payloads should be able to detect the energy required for life in outer solar system (inferring the geological activity energy, chemical energy and tide energy by identifying the surface morphology of the celestial body).

5.2.1.4 What Factors in the Solar System Can Cause Harms to Human Beings? What Resources Can Be Utilized?

The exploitation and utilization of extraterrestrial celestial bodies based on relevant knowledge is of great significance to the sustainable development of human beings. In the process of survival and space exploration of human beings, in addition to the effects of solar explosion activities, small bodies also bear profound influence on the Earth and human beings. Therefore, it is necessary to understand the motion characteristics of small bodies and their potential impact on the Earth and spacecraft in a detailed manner. In addition, it is necessary to investigate what resources are available for human exploitation and utilization on the Moon, small bodies and Mars.

These scientific problems require the payloads to be able to identify and track small bodies and detect their compositions, to detect resources and their contents on the Moon, and to exploit and utilize resources based on the characteristics of the small bodies and the Moon.

It can be seen from the above problems that deep space exploration mainly focuses on the origin and evolution of the solar system, the origin and evolution of life, and the impact of the solar system on human beings. From the perspective of the payloads required for exploration, the main requirements include: (1) to obtain the morphological characteristics of the celestial body, including the macroscopic morphology and microscopic details, to judge the geological activities and impact processes experienced; (2) to identify the compositions and contents of celestial bodies, including elements, minerals, water, organic matters, isotopes, etc.; (3) to determine the physical characteristics of the celestial body, including gravity field, magnetic field, atmospheric and climatic conditions, internal structures, geological activities, etc.; and (4) to identify and utilize celestial resources and favorable conditions, so as to carry out scientific researches accordingly. As a matter of course, these scientific issues are not isolated from each other, and the required payloads are also interrelated.

5.2.2 Scientific Objectives and Payload Configuration for Lunar and Mars Exploration in China

The scientific questions in Sect. 5.2.1 are the basic problems faced by human beings in the process of deep space exploration. In addition, answering these questions is complex and arduous, requiring the joint efforts of the entire human race. The deep space explorations of China are also based on these basic problems faced by human civilization; taking the origin and evolution of the Moon, the material composition, and the resources available to human beings as the starting points, from near to far, efforts are being made step by step [1].

5.2.2.1 Chang'e 1 and Chang'e 2

Chang'e 1 marked the first activity of exploring the Moon in China. The scientific objectives are as follows [2, 3].

(1) To obtain a three-dimensional image of the lunar surface, partition the lunar topography and geomorphology units; count the size and density of the craters, calculate the age of the lunar surface, restore the early evolution history of the Moon; analyze the lunar surface structure, compile the lunar surface fault and annular structure images and the lunar tectonic zoning maps; study the evolution history of the lunar geological structure.
(2) To analyze the distribution of surface element content and material type of the Moon, plot the full Moon content and distribution of related elements, roughly classify rock types, and study the chemical evolution of the Moon.
(3) To detect the characteristics of lunar soil, measure the microwave radiation brightness and temperature of different bands of the whole Moon, infer the information of the thickness of the lunar soil and evaluate the resources and distribution of ^{3}He in the lunar soil.
(4) To detect the cislunar space environment, detect the composition, flux, energy spectrum and temporal and spatial variation characteristics of high-energy particles and solar wind ions in the cislunar space, and study the effects of solar activities and Earth magnetosphere on the lunar space environment.

Chang'e 1 is configured with eight scientific exploration payloads, in which the CCD stereo camera and the laser altimeter jointly complete the 3D image acquisition of the lunar surface, the interference imaging spectrometer and the γ-ray spectrometer, the X-ray spectrometer completes the detection of the content and distribution of elements and substance types, and the microwave detector completes the detection of the thickness of the lunar soil, and the high-energy particle detector and the solar wind ion detector complete the cislunar space environment exploration.

Chang'e 1 payloads have undergone a technological leap from scratch. The CCD stereo camera obtains three two-dimensional raw data of the same lunar surface target, namely sub-satellite point, front view and rear view by a large-area array

CCD camera. The image lunar surface resolution is 120 m (at the orbital altitude of 200 km), the Moon surface imaging swath is 60 km, the image data of the whole Moon including the 90° north and south latitudes is obtained, and the full Moon image map was spliced by the front view image. The laser altimeter has a ranging frequency of 1 Hz and a range resolution of better than 1 m. A full Moon digital elevation model with a spatial resolution of about 3 km is fabricated with laser altimetry data. The microwave probe applies the characteristics of different frequency bands of microwave having different penetration depths in lunar soil and infers the spatial distribution information of the thickness of the lunar soil by detecting the brightness and temperature of the microwave radiation in a specific frequency band of the lunar soil, including four frequencies at 3.0, 7.8, 19.35 and 37.0 GHz, which was the first time in the world to apply passive microwave remote-sensing technology to measure the information of the entire Moon microwave radiation, from which the microwave radiation brightness and temperature of the entire Moon under different lighting conditions were obtained after processing. The γ-ray spectrometer and the X-ray spectrometer detect the γ-rays and X-rays generated by the elements of the lunar surface spontaneously or induced by the cosmic rays. The energy resolution of the instrument is better than 9% @ 662 keV (γ-ray), 10% @ 59.5 keV (hard X-ray) and 600 eV @ 5.95 keV (soft X-ray), respectively. And the global content distributions of U, Th, K and other elements were completed. The high-energy particle detector and the solar wind ion detector obtained a large amount of meaningful data during the lunar orbiting, so the space environmental factors experienced by the Chang'e 1 were understood. On this basis, in-depth scientific research was carried out.

The Chang'e 2 mission includes four scientific goals as described in the following [4, 5].

(1) To obtain a three-dimensional image of the lunar surface. The working principle of the Chang'e 2 stereo camera is basically the same as that of the Chang'e 1 CCD stereo camera, to which self-pushing mode was applied, and two TDI CCDs were arranged in parallel on its focal plane, but the resolution was greatly improved to 7 m (at the orbital altitude of 100 km) and 1.05 m (at the orbital altitude of 15 km). A 7m-resolution image of the whole Moon and a 1m-resolution HD image of the pre-selected landing area of Sinus Iridium were drawn.

(2) To analyze the distribution of useful element content and material types on the lunar surface. Due to the lowering of the orbit, the improvement of the detection instrument (e.g., the γ-ray spectrometer uses the cesium bromide crystal to replace the cesium iodide crystal, the exploration energy resolution is 4% @ 662 keV) and the fact that Chang'e 2 happened to be operated in the lunar orbit in the peak year of solar activities, Chang'e 2 mission achieved finer detection results than Chang'e 1 and determined the aluminum and magnesium elements distribution on the entire Moon based on X-ray data for the first time, thus providing a basis for the study of lunar rock-type determination and lunar formation evolution.

(3) To detect the characteristics of lunar soil. Same as that in the Chang'e 1, the microwave detector was still used for the detection of lunar soil characteristics. Due to the lowering of the orbit, a higher-resolution detection result than the Chang'e 1 was obtained.

(4) To detect the lunar space environment. High-energy particle detectors and solar wind ion detectors were still used to detect the lunar space environment, and the lunar space environment data of the solar peak year was further obtained. It applied the solar wind proton data for the first time in the world and discovered the existence of micromagnetic layers caused by the remnant magnetization of the lunar surface at the antipodes of Mare Serenitatis, and further confirmed the existence of the lunar surface micromagnetic layer at the antipodes of large craters, which provides new references on the formation and evolution of lunar inner structures.

The payload configurations of Chang'e 1 and Chang'e 2 are shown in Table 5.1.

Table 5.1 Payloads configuration for Chang'e 1 and Chang'e 2

Chang'e 1 payload	Chang'e 2 payload	Detection tasks
CCD stereo camera	CCD stereo camera	Obtain the 3D image of the lunar surface.
Laser altimeter	Laser altimeter	
Interferometric imaging spectrometer	–	Analyze the distribution characteristics of useful elemental contents and material types on the lunar surface.
γ-ray spectrometer	Gamma ray spectrometer	
X-ray spectrometer	X-ray spectrometer	
Microwave detector	Microwave detector	Detect lunar soil characteristics.
High-energy particle detector	High-energy particle detector	Detect the cislunar space environment.
Solar wind ion detector	Solar wind ion detector	

5.2.2.2 Chang'e 3 and Chang'e 4 Probes [6–9]

The Chang'e 3 mission realized the first lunar soft landing and lunar rovering exploration of China, and there are three scientific exploration objectives as described in the following.

(1) Survey of lunar surface morphology and geological structure. The lunar surface morphological and geological structure of the landing zone and the inspection area were surveyed to obtain the data of the lunar surface morphology, geological structure, shell structure, crater and soil thickness of the Moon, thereby establishing the regional lunar morphology and geological evolution modes.

(2) Comprehensive in-situ analysis of mineral composition and chemical composition of the landing zone and the rovering zone. The most fundamental work of lunar science is to obtain the chemical composition, mineral composition, rock type and distribution of the Moon, which is the key to understanding the history of the evolution of the Moon.

(3) Exploration of the solar–Earth–lunar space environment and lunar-based astronomical observation. The space environment of the Sun, the Earth and the Moon is an important factor affecting the survival and development of human beings. Solar flares and coronal mass ejections release huge amounts of energy and matters, restricting the near-Earth space and the lunar surface environment. By studying stellar seismology,

we can directly understand the internal structure of the celestial body; Active galactic nucleuses are the oldest and far-reaching celestial body in the universe, which is of great significance to the study of the origin and evolution of the universe, and the Moon-based astronomical observation are of unique advantages.

The Chang'e 3 probe consists of a lander and a rover. The lander is equipped with four payloads: topographic camera, lunar-based optical telescope, extreme ultraviolet camera and landing camera. The rover is equipped with four payloads, namely panoramic camera, lunar-penetrating radar, infrared imaging spectrometer and particle-excited X-ray spectrometer. The payloads work together to complete the scientific exploration missions, and each has its own emphases. The specific tasks are shown in Table 5.2.

The Chang'e 3 mission is different from that of the Chang'e 1 or Chang'e 2 missions. It applied landing exploration and rovering exploration. The payloads are designed according to this feature. The infrared spectrometer, lunar penetrating radar, extreme ultraviolet cameras and lunar-based optical telescope all applied lighter and smaller new designs. For the first time, acousto-optic tunable filter and time domain

Table 5.2 Payloads configuration of Chang'e 3 mission

Payload	Detection tasks
Topographic camera	Obtain an optical image of the lunar surface topography of the area around the landing point, and observe the rover and its movement process on the lunar surface
Extreme ultraviolet camera	The working wavelength is 30.4 nm, and the detection target is the resonance scattering of the Earth plasma layer He + to 30.4 nm solar radiation, which can be continuously tracked and detected
Lunar-based optical telescope	Work in the near-ultraviolet spectrum, observe the changes in the brightness of various celestial bodies and enable continuous tracking and observation
Landing camera	During the landing of the lander, the lunar surface optical image of the landing area is continuously acquired at different heights
Panoramic camera	Acquire an optical image of the lunar surface topography around the landing zone and the rovering area, and image the lander
Lunar-penetrating radar	During the operation of the rover, the thickness and structure of the lunar soil and the shallow structure of the lunar crust are detected along the rovering path
Infrared imaging spectrometer	Obtain the infrared spectrum and image of the lunar surface around the inspection area for the mineral composition and distribution analysis of the lunar surface
Particle-excited X-ray spectrometer	Conduct field analysis of the content of major elements of the lunar surface material around the inspection area for the identification and verification of full rock composition of lunar rocks, full lunar rock composition and mineral composition of the lunar soil

5.2 Major Scientific Issues in Deep Space Exploration Research

detection radar are, respectively, used to spectral and lunar soil thickness detection; Earth plasma layer observation was achieved based on single sphere reflection and spherical photon counting imaging detectors, and lunar-based autonomous astronomy observation in the extreme ultraviolet range was also realized, etc.

In the Chang'e 3 mission, the landing camera, topographical camera and panoramic camera were used to obtain high-resolution landform data of the landing zone, and the detection results of panoramic cameras, infrared spectrometers, particle-excited X-ray spectrometers and lunar-penetrating radar are comprehensively utilized to reveal the volcanic evolution history of the Mare Imbrium of the Moon. The results indicated that there was still a large-scale volcanic eruption on the Moon about 2.5 billion years ago, which might be related to the rich radioactive elements in the region. This research result is of great significance for understanding the evolution of the Moon.

Another feature of the Chang'e 3 is to apply the favorable conditions of the Moon and use the Moon as a platform for Earth observation and astronomical observation. The lunar-based optical telescopes found a series of new astronomical phenomena in the study of variable stars. For example, a rare celestial body in the process of binary star rapid mass exchange evolution was discovered; a group of samples in the process of binary star slow material exchange evolution was discovered; a semi-phased close binary star in a six-star system was found; and the possibility that close binary stars are common in multi-star systems was discovered, etc. In addition, the lunar-based telescope detected a very low density of hydroxyl groups in the outer layer, again demonstrating that there are very few minerals containing structural water in the lunar rocks.

The probe Chang'e 4 consists of the relay satellite, lander and rover. It landed on the far side of the Moon in the Aitken Basin and conducted in-situ exploration and rovering exploration at the landing and rovering zone on the far side of the Moon with the communication support of the relay satellite. The scientific goals of Chang'e 4 include: lunar-based low-frequency radio astronomical observation research, shallow-structure detection in the rovering area on the far side of the Moon, and the morphology as well as mineral component detection in the rovering area on the far side of the Moon.

Chang'e 4 was equipped with eight payloads to accomplish these three scientific goals: four devices are deployed on the lander, which respectively are a landing camera, a topographic camera, a low-frequency radio spectrometer and a neutron and radiation dose detector; and there are four devices on the rover, which respectively are a panoramic camera, a lunar-penetrating radar, an infrared imaging spectrometer and a neutral atom probe. The payload configuration and tasks are shown in Table 5.3.

5.2.2.3 Mars exploration

The spacecraft system of the first Mars autonomous exploration of China consists of a landing rover and an orbiter. The scientific goals are to conduct a global and comprehensive exploration of Mars through orbiting exploration and conduct in-situ exploration of some particular area of Mars through rovering exploration, achieving

Table 5.3 Payloads configuration for Chang'e 4

Payload	Detection tasks
Low-frequency radio spectrometer	Simultaneously obtain radio sky-survey images with continuous frequencies in the very low-frequency band to study the large-scale structure of the universe and the frequency correlation of radio radiation
Landing camera	Make continuous imaging during the landing of the Chang'e 4 probe, positioning the landing zone and reconstructing the fine landing trajectory
Topographic camera	Acquire high-resolution topographic and image data of the landing zone and study the morphology and geological structure of the landing zone
Neutron and radiation dose detector	Measure the combined particle radiation dose and LET spectrum of the lunar surface; measure the fast neutron energy spectrum and thermal neutron flux of the lunar surface
Lunar penetrating radar	Obtain the thickness distribution of the lunar shallow structure on the walking path of the rover and provide scientific data for the study of lunar topography and geological structure
Panoramic camera	Make stereoscopic imaging of the surface of the lunar rovering area for terrain, structure, geological characteristics, volcanic pit investigation and monitor the lander status
Infrared imaging spectrometer	Analyze the mineral composition and distribution of the lunar surface and determine the rock types.
Neutral atom detector	Study the distribution function of energetic neutral atoms in the scattering process of lunar surface and its relationship with topography and local time

scientific exploration of surface morphology, soil properties, material composition, water ice, atmosphere, ionosphere, magnetic field, etc. as detailed in the following.

(1) Studying the morphological and geological structure characteristics of Mars;
(2) Studying the soil characteristics and water ice distribution on the surface of Mars;
(3) Studying the composition of the surface materials of Mars;
(4) Studying the atmospheric ionosphere and surface climate and environmental characteristics of Mars;
(5) Studying the physical fields and internal structures of Mars.

The payload configurations of the landing rover and orbiter are detailed in Tables 5.4 and 5.5. Compared with lunar exploration, the payload configuration of the Mars exploration mission is more well-rounded and more comprehensive. In terms of surface topography acquisition, the resolution level of 0.5 m will be achieved; for the in-situ measurement, the magnetic fields on the surface of Mars and in near-Mars space will be detected by the magnetometer, and the physical properties of Mars atmosphere will be detected with the meteorological measuring instrument.

5.2 Major Scientific Issues in Deep Space Exploration Research

Table 5.4 Payloads of the landing rover

Payload	Detection tasks
Navigation terrain camera	(1) Obtain topographic features in the rovering area through high-resolution stereo imaging close-range detection (2) Establish a three-dimensional topographic map near the rovering area on Mars (3) Obtain color image data of the rovering area
Multi-spectral camera	(1) Conduct the exploration of the surface geomorphology of Mars and the changes thereof, and conduct research on the geological structure and geomorphology of Mars (2) Perform high-precision imaging and detailed surveys in the inspection area, including high-resolution imaging of areas with possible water, sedimentary strata and water-flow geomorphology, volcanic landforms, crater geomorphology and wind erosion landforms on Mars
Mars subsurface-penetrating radar	(1) Detect the sub-surface structure of Mars (2) Detect underground water ice of Mars
Mars surface component detector	(1) Analyze the chemical element composition near the rovering area (2) Analyze mineral near the rovering area (3) Identify rock near the exploration area (4) Search for the environmental conditions necessary for the existence of liquid water in the history of Mars (5) Detect the dynamic characteristics of atmospheric dust of Mars
Mars surface magnetic field detector	Conduct magnetic field observation on the surface of the planet, detect the magnetic field characteristics of Mars and study the internal structure of Mars
Mars meteorological measuring instrument	(1) Monitor the surface meteorological elements on the Mars (2) Monitor the Martian surface meteorological elements (3) Acquire atmospheric profiles and accumulate data for Martian meteorology and atmospheric physics research

Table 5.5 Payloads of the orbiter

Payload	Detection tasks
Medium-resolution camera	(1) Obtain the color and full-color image of Mars with the resolution of better than 100 m, covering the whole Martian surface as possible (2) Explore Martian topography and its variations, including Martian surface imaging, Martian geological structures and topography study
High-resolution camera	(1) Obtain the optical images with a pixel resolution better than 0.5 m of some local areas on Mars; perform detections on Martian surface topography and its variations, including high-precision imaging and detailed survey in some key areas (2) Identify characteristic geomorphic units, such as fluvial landforms, volcanic landforms, crater landforms and wind erosion landforms (3) Make high-resolution imaging of areas with possible water, sedimentary topography and fluvial landforms
Orbiter subsurface exploration radar	Utilize high-frequency electromagnetic waves to detect surface and underground characteristics, mainly including: (1) Martian surface soil thickness and surface geological structure (2) Martian groundwater detection (3) Detect ultra-low-frequency radio spectrum in interplanetary space
Mars Mineral Spectrum Analyzer	Acquire on orbit the infrared spectrum information reflected and radiated from the minerals on the surface of Mars, and provide scientific basis for the analysis of Martian surface minerals and its distribution
Mars magnetometer	Make high-resolution measurement of the magnetic field environment near-Mars space, such as magnetic sheath, magnetic barrier layer, magnetic tail and residual magnetic field of Mars
Mars ion and neutral particle analyzer	Detect charged ions and neutral particles in the Martian space environment, on the basis of which the escape of the Mars atmosphere is studied

(continued)

5.2 Major Scientific Issues in Deep Space Exploration Research

Table 5.5 (continued)

Payload	Detection tasks
Mars Energy Particle Analyzer	Study the characteristics and variation of the energy spectrum, elemental composition and flux of the energy particles in near-Mars space environment and the Earth–Mars transfer orbit, mainly including: (1) Map the spatial distribution of the radiation of different types of energy particles in the Mars global or Earth–Mars transfer orbit (2) Detect joint with magnetometers, ions and neutral particle analyzers to study the relationship between the energy particle radiation near-Mars space and in the atmosphere, and the influence of the solar storm energy particle event on the escape of the Mars atmosphere and the law of interaction

5.3 Topography Acquisition Technology

5.3.1 Introduction

Morphological information, especially image information in the optical bands, is the most direct and effective means for human beings to perceive unknown celestial bodies and unknown environments. In addition to being able to obtain visual impressions, it is of great significance for judging the geological activities and impact processes experienced by celestial bodies and studying the evolution history of celestial bodies to obtain the morphology and topography of celestial bodies, including macroscopic features and microscopic details. In addition, the exploration of material composition should be combined the topography information to better study the origin of the celestial body. The topography information also greatly helps to determine whether there are volcanic activities necessary for sustaining life on outer solar system celestial bodies. The acquisition of topographical information requires as high an image resolution as possible (to be optimized according to the constraint on data volume), the capability to acquire stereoscopic information and the capability to possibly acquire color images or spectral information simultaneously.

The Chang'e 3 panoramic camera is a typical topography acquisition payload. This section takes the panoramic camera as an example to introduce the topography acquisition technology, including stereo image acquisition technology, sensor selection strategy, automatic exposure technology for extraterrestrial imaging, calibration technology as well as considerations on temperature control and weight reduction in the process of system design [10].

5.3.2 Stereo Image Acquisition Technology

Binocular vision captures two images of the same scene at different positions or angles simultaneously, then computes and extracts three-dimensional scene information from two two-dimensional images. Stereoscopic image inversion requires a certain distance between the two cameras or two images. However, the spacing cannot be too large due to actual resource constraints.

The Chang'e 3 rover was equipped with two panoramic cameras mounted on the mast head of the rover, and the optical axis was perpendicular to the mounting surface. The distance between the optical axes of the panoramic camera A and the panoramic camera B on the mast head was (270 ± 2) mm, the two optical axes were, respectively, tilted inwards by $1°$, the mounting accuracy was $\pm 10'$, and the measurement accuracy was $0.5'$.

While obtaining the two-dimensional planar image of the target, the distance of the target could also be obtained, and various functions such as panoramic imaging, stereo image pair acquisition and auxiliary navigation were provided. After acquiring the two-dimensional raw data images, the high-resolution three-dimensional lunar surface images of the landing zone and the rover zone were restructured by post-processing.

5.3.3 Color CMOS Devices

The CMOS-APS detector occupies much less resources than the CCD. Therefore, the CMOS-APS detector is preferred on the premise of satisfying the exploration mission requirements. Currently, CMOS-APS detectors available in the market include Dalsa, Micron, Cypress, Kodak, etc. in which Cypress is mainly used in the aerospace industry. Dalsa is also preparing to re-improve its CMOS-APS detector to satisfy aerospace application requirements. A comparison of Cypress's STAR1000, LUPA4000 and Dalsa's IA-G3 detectors in terms of main performance indicators is shown in Table 5.6.

5.3.4 Camera System Design

The camera system consists of optical lens, mechanical system, electronic system and thermal control system. The camera system design is introduced with the "Chang'e 3" panoramic camera as an example. The typical block diagram of the system is shown in Fig. 5.1, and the structural diagram is shown in Fig. 5.2.

5.3.4.1 Choice of Relative Aperture of Optical System

The Chang'e 3 panoramic camera is mounted on the mast of the rover. To minimize the temperature control system, the optical system itself should be provided with strong temperature adaptability. For the rovering exploration, the panoramic camera must image both close-range targets (3 m) and long-range targets (2–3 km).

Both requirements are satisfied, only when the F number of the camera and the depth of focus of the camera itself are increased. Since the focal depth of the optical system is proportional to the square of the F number, the increase of the F number can increase the depth of field of the optical system, and accordingly increase the temperature range the system can adapt to. Table 5.7 shows the relationship between the F number and the depth of focus of the camera. Table 5.8 shows the distance between the best image surface and the actual image surface at different distances (assuming that the best image surface is the image surface with a target distance of 6 m).

As can be seen from Table 5.8, because the image plane of the close target is too far away from the image plane of the distant target, it is quite difficult to obtain excellent image quality at different distances. With the increase of F number, the depth of field will be increase rapidly, and the increase of the F number is very

Table 5.6 Comparison of performance indicators of CMOS-APS detectors

Chip type	LUPA4000Cypress	STAR1000Cypress	IA-G3Dalsa
Number of pixels	2048 × 2048	1024 × 1024	2352 × 728
Pixel size/(μm × μm)	12 × 12	15 × 15	7.4 × 74
Pixel structure	6-transistor	3-transistor	5-transistor
Pixel rate (ADC)/MHz	66	12	2 × 160
Frame rate/fps	15	12	62
Spectral response range/nm	400–1000	400–1000	400–1000
QE × FF	QE × FF average ≥ 35%	QE × FF ≥ 20%	FF = 45% > 30% (400–750 nm)
Full well charge/ke	80	135	60
Saturation output	1 V	1.1 V	1023 DN
dynamic range/dB	66	72	57
FPN	<1.25%Rms	<0.56%PP	<2.4%Rms
PRNU	<2.5%Rms	Partial: < 0.67% Rms overall: < 3.93% Rms	<2.5%Rms
ADC/bit	10	10	10
Readout channel	2	1	2
Electronic shutter	Global electronic shutter	Yes (rolling electronic shutter)	Global synchronous shutter
Working voltage/V	+3.3	+5	+3.3
Power consumption/mW	250	350	2200 (62 fps)
Anti-radiation dose/krad	230	–	–
Existing application	Astronomical observation	Aerospace grade products including star sensor	–

Note QE is quantum efficiency, FF is the aperture factor, the effective photosensitive area is proportional to the area of the pixel, QE × FF is the index for measuring the photoelectric conversion efficiency; RMS is the root mean square; and PP is the peak-to-peak value

effective for expanding the imaging range. The disadvantage of increasing F number is that the limit MTF of the camera will be reduced. After taking into account the increased imaging range and system MTF, the F number of the optical system is finally selected to be F10, that is, the relative aperture of the optical system is 1:10.

5.3 Topography Acquisition Technology

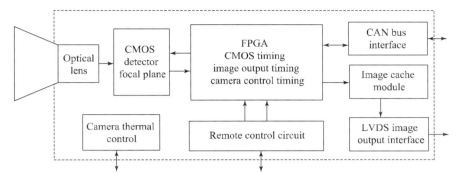

Fig. 5.1 Camera system block diagram

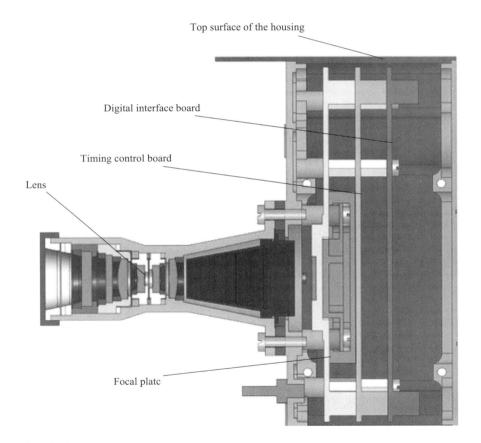

Fig. 5.2 Camera structure

Table 5.7 Relationship between F number and camera focus depth

F number	2	4	6	8	10	12	14	16	20
Depth of focus/mm	0.005	0.019	0.043	0.077	0.12	0.173	0.235	0307	0.48

Table 5.8 Distance of the actual image surface away from best image surface at different distances

Target distance/m	3	4	5	6	7	8	10	15	30	2000
Defocusing amount	0.42	0.21	0.08	0	−0.06	−0.11	−0.17	−0.25	−0.33	−0.42

5.3.4.2 Lightweight Considerations for Optical System Solutions

To achieve the goal of lightweight and miniaturization, the camera structure should use materials with relatively low density as much as possible, and the electronics system should be as simple as possible and use highly integrated devices. In addition, the selection of the optical system should take into account the requirements of lightweight and miniaturization from the beginning. Figure 5.3 is a schematic structural view of the optical system of a panoramic camera.

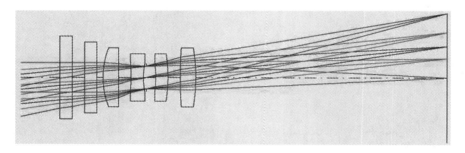

Fig. 5.3 Schematic diagram of the optical system

5.3.4.3 Electronics Design

The panoramic camera circuit mainly consists of three parts: focal plane circuit, camera timing and buffer circuit, and camera power circuit. The main functional modules include CMOS image sensor, power module, crystal oscillator, master FPGA, FPGA configuration circuit, image buffer circuit, CAN bus interface circuit, and LVDS interface circuit, etc. The principle structure is shown in Fig. 5.4.

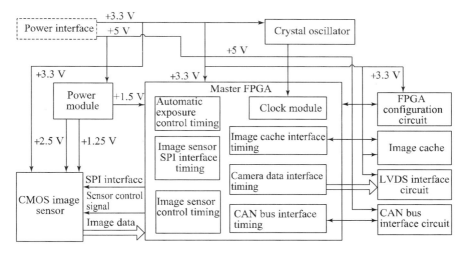

Fig. 5.4 Principle structure of the panoramic camera electronics system

5.3.5 Automatic Exposure Technology

When the panoramic camera is in operation, it is expected that a 360 degree panoramic shot can be taken at different pitch angles of the mast, for which big differences in the solar elevation angles and the target reflections are needed, and there are both front lighting and back lighting images. To improve the adaptability to target and changes of light intensity, the camera is designed with the exposure control function, which can be realized by two methods: automatic and manual. In the lunar and deep space exploration, if the exposure time is adjusted from ground after the test shot, the time resource consumption is large, so the automatic exposure function must be available.

The IA-G3 can be set up to adjust the exposure time. Just by changing the corresponding settings, the integration time of the CMOS sensor can be controlled to achieve the purpose of adjusting the exposure time. The optical response (DN) of the image sensor is linearly related to the exposure time. Based on the exposure range $t_{measured}$ of the current frame of image, the average gray value $DN_{measured}$ and the desired gray value $DN_{desired}$ of the next frame of image, the exposure gear t_{new} of the next frame of image can be calculated, as shown in Eq. (5.1).

$$t_{new} = t_{measured} \times \frac{DN_{desired}}{DN_{measured}}, \tag{5.1}$$

where t_{new} is the exposure gear of the next frame of image; $t_{measured}$ is the exposure gear of the previous frame of image; $DN_{desired}$ is the average gray value of the desired image; $DN_{measured}$ is the average gray value of the previous frame of image.

It should be noted that, considering the fact that the lunar surface has almost no atmospheric scattering, in computing the image mean, the whole image cannot be used for statistics, only the data with the DN value exceeding the dark current for a certain degree is calculated instead.

The first two lines of each frame of the image are dark pixel rows whose DN value represents the size of the dark charge. The dark line filter structure is the same as the pixel line. The dark pixel gray value corresponding to the green filter (GB and GR) in the dark line of the previous frame of image is averaged to obtain the dark pixel average value Dark_pixel_avg; and an offset Dark_offset is added to Dark_pixel_avg to obtain the black background threshold Dark_pixel_dn.

By comparing the gray value corresponding to the green filter (GB and GR) of one frame image with the black background threshold Dark_pixel_dn, and averaging all the pixels larger than the threshold, the gray mean $DN_{measured}$ can be obtained. As the desired image gray mean $DN_{desired}$ is known (uploaded by command), and the exposure gear $t_{measured}$ of the previous frame of image is also known. Therefore, the exposure gear t_{new} of the next frame of image can be obtained by the above formula.

5.3.6 Calibration and Ground Verification Test

Calibration aims to adjust the parameters of the instrument so as to eliminate various errors and inconsistencies of the instrument and make the imaging data of the instrument more accurate. The purpose of the ground verification test is to simulate the ability of the test camera to work on orbit and to verify the imaging performance thereof.

According to the scientific detection tasks of the panoramic camera, the calibration content includes dark current acquisition, flat field calibration of the entire device, mode normalization calibration and color calibration. Ground validation tests include the verification of 360° panoramic imaging and color target imaging capabilities, as shown in Table 5.9.

5.4 Elemental Component Identification Technology

5.4.1 Introduction

Identifying the composition and content of celestial bodies, such as elements, minerals, water, organic matter and isotopes, is an important part of studying the origin of celestial bodies, conducting comparative planetary studies, inferring the evolution process of celestial bodies and identifying the material components required for the origin of life. It can also be applied in exploitation and utilization

5.4 Elemental Component Identification Technology

Table 5.9 Table of calibration and ground verification

No.	Test name	Test content	Desired result
1	Dark current collection	Calibrate the dark current DC component matrix under various states of the panoramic camera (including integration time, gain state)	Obtain the dark current coefficient matrix
2	Flat field calibration of the entire device	Calibrate the relative calibration matrix of various states of the panoramic camera (including exposure time, gain state)	Obtain a relative calibration correction matrix
3	Mode normalization calibration	Calibrate the relative change in output between various states of the panoramic camera (including exposure time, gain state), i.e., the mode normalization coefficient matrix	Obtain the relationship between the camera output DN value, the exposure gear and electronic gain
4	Color calibration	Calibrate color correction matrix and white balance coefficient	Obtain color recovery coefficient matrix
5	360° panoramic imaging	Test camera exposure difference between front lighting and backlighting imaging	Verify camera imaging capability and obtain the relationship between camera output DN value and phase angle
6	Color target imaging	Test camera color imaging performance	Verify camera color imaging capabilities

of celestial resources outside the Earth, and element recognition is the most basic detection in component identification.

This section introduces elemental composition identification technology based on particle-excited X-ray spectrometers, including element identification method, active excitation source selection, sensor technology, system design and calibration test.

5.4.2 Principles of Elemental Composition Identification

The alpha particle X-ray spectrometer (APXS) applies the carried excitation source to bombard the surface of the lunar rocks or soil and detects the generated X-rays that bear the elemental features. The type and content of the primary elements of the lunar surface are obtained through analysis. Nuclear physics methods are adopted for the elemental composition identification and content analysis, and the detection principle is mainly based on the following physical processes.

(1) α-particle-exciting characteristic X-ray mechanism: α particles emitted by the radiation source excite the inner electrons of the elements in the sample to be tested, and the outer electrons perform inward transitions to produce characteristic X-ray fluorescence (Particle-Induced X-ray Emission, PIXE).

(2) X-ray-exciting characteristic X-ray mechanism: Characteristic X-rays of radioactive source decaying sub-nucleus excite the inner electrons of the sample element to be tested and produce characteristic X-ray fluorescence (XIX).

The two excitation mechanisms are shown in Fig. 5.5.

The particle-excited X-ray spectrometer will measure the characteristic X-ray fluorescence produced by the elements under the action of the two mechanisms to determine the type and content of the elements. Considering both factors of the excitation cross section and the detection efficiency, the method is applicable to elements with atomic number $Z = 11$–40.

The characteristic X-ray energies of different elements are shown in Table 5.10. If the energy resolution of the instrument is relatively high, it can distinguish the characteristic lines of different elements and identify the elements and content through data processing.

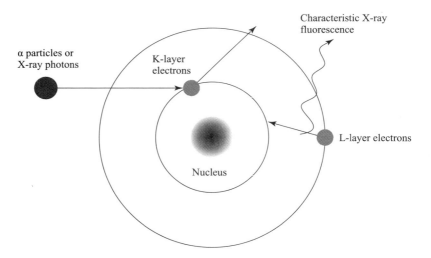

Fig. 5.5 Schematic diagram of PIXE and XIX

5.4 Elemental Component Identification Technology 177

Table 5.10 Characteristic Kα X-ray energy of elements

Element	Li	Be	B	C	N	O	F	Ne
Kα/eV	54.3	108.5	183.3	277	392.4	524.9	676.8	848.6
Element	Na	Mg	Al	Si	P	S	Cl	Ar
Kα/eV	1040.98	1253.60	1486.70	1739.98	2013.70	2307.84	2622.39	2957.70
Element	K	Ca	Sc	Ti	V	Cr	Mn	Fe
Kα/eV	3313.80	3691.68	4090.60	4510.84	4952.20	5414.72	5898.75	6403.84
Element	Co	Ni	Cu	Zn	Ga	Ge	As	Se
Kα/eV	6930.32	7478.15	8047.78	8638.86	9251.74	9886.42	10,543.0	11,222.4
Element	Br	Kr	Rb	Sr	Y	Zr	Nb	Mo
Kα/eV	12,649.0	13,395.0	14,165.0	14,958.0	14,958.0	15,775.0	16,615.1	17,479.3

5.4.3 Selection Strategy of Excitation Source

The particle-excited X-ray spectrometer uses active excitation for detection and is independent of solar activity. Therefore, the selection of the excitation source has an important influence, and it is necessary to attach importance to the half-life period of the source and the energy generated by the decay. The 244 cm source is widely used in the element analysis of deep space exploration in the world. As there is no atmospheric absorption on the Moon, the detection on the lunar surface can also select the combination of ^{55}Fe + ^{109}Cd.

5.4.4 Sensor Selection and Design Techniques

Sensor is a key factor affecting energy resolution and detection efficiency. Table 5.11 compares the advantages and disadvantages of various X-ray detectors. The SDD sensor (Si drift detector) has been successfully used in the Courage and Opportunity Mars rovers and is one of the most advanced X-ray detectors in the world. The SDD sensor is taken as an example to introduce the sensor design technology.

In the SDD, a large-area uniform PN junction is formed on the back surface (incident surface) of the high-purity N-type silicon wafer, a dot-shaped N-type anode is formed in the center of the other surface, and a P-type drift electrode is distributed around the anode. In operation, a reverse voltage is used on the PN junctions on both sides of the device to create a potential well within the device. Applying a voltage

Table 5.11 Comparison of advantages and disadvantages of various X-ray detectors

Detector	Energy resolution	Detector operating temperature	Peak-to-background ratio	Time resolution
Crystal detector (NaI, etc.)	Poor	Normal temperature	Poor	Excellent
CZT	Normal	Normal temperature	Poor	Excellent
X-CCD	High	< − 70 °C	Normal	Very poor
SCD	High	< − 40 °C	Normal	Poor
Si-PIN	High	−40 °C, with increasing temperature, energy resolution becomes significantly worse	High	Excellent
Si (Li)	Very high	< − 100 °C	Very high	Excellent
SDD	Very high	< − 20 °C, the energy resolution does not change much in a certain temperature range	Very high	Excellent

5.4 Elemental Component Identification Technology

to the drift electrode creates a drifting electric field inside the device, the electrons generated by the X-ray acting on the detector will drift to the anode under the action of the drift electric field, and then be collected and read out by the charge-sensitive pre-amplifier to form the electrical signal.

In terms of the manufacturing process, SDD is made into a point-like collecting anode, which significantly reduces the junction capacitance of the detector. For the SDD and Si-PIN with the same area, the junction capacitance of the SDD detector is 1–2 orders of magnitude smaller than that of the Si-PIN. The noise level of the system is not only related to the leakage current of the detector itself, but also directly related to the detector junction capacitance. The smaller the junction capacitance is, the lower the noise level is. The particle-excited X-ray spectrometer uses an SDD silicon drift chamber detector (energy resolution 150 eV@5.9 keV) with a sensitive thickness of 450 μm and a Si wafer area of 10 mm^2.

The detection efficiency depends mainly on the material and thickness of the window and detector. The thickness of the X-ray detector SDD is 450 μm. By simulation, the efficiency of X-ray photons in the energy range of 0.5–20 keV detected by SDD can be calculated for different Be-window thicknesses, as shown in Fig. 5.6.

It can be seen from Fig. 5.7 that the low-energy photons are mainly affected by the absorption of the Be-window, and the detection efficiency of the high-energy photons is independent of the thickness of the Be-window (depending only on the thickness of the SDD).

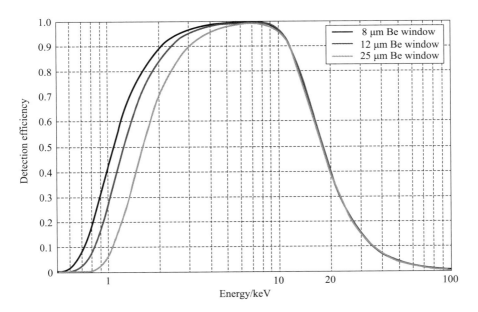

Fig. 5.6 SDD detection efficiency curve (450 μm Si + Be-window of different thickness)

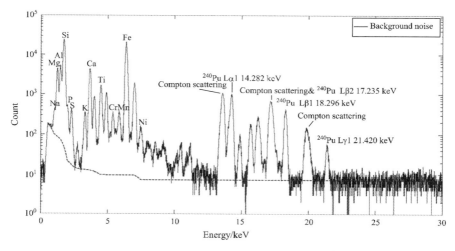

Fig. 5.7 APXS simulation detection spectrum (30 min 150 eV @5.9 keV)

According to the physical design of APXS, the Monte Carlo software Geant-4 is used to establish the simulation model, and the simulated detection energy spectrum of the specified sample is obtained with the X-ray detector (150 eV@5.9 keV) accumulated for 30 min under the ideal measurement condition. Figure 5.7 shows the K fluorescence lines of the 13 elements of Na, Mg, Al, Si, P, S, K, Ca, Ti, Cr, Mn, Fe, Ni as well as the Rayleigh scattering and Compton scattering background of the emission line of the radioactive source 244 cm.

5.4.5 System Design

5.4.5.1 System Composition

The particle-excited X-ray spectrometer of the probe Chang'e 3 includes the detection head, the calibration device, a lunar night survival device, cables and an electronic board with its software portion. The system composition is shown in Fig. 5.8.

The detection head is mounted on the end of the rover arm for scientific detection and distance sensing. It mainly consists of the excitation source, infrared distance sensor, silicon drift X-ray detector (SDD) and pre-amplifier circuit.

The lunar night survival support device is installed on the rover and consists of the active radiant heat source (RHU), external bracket and thermal control coating. In low-temperature environments (such as lunar night), the radiant heat source provides active temperature control for the detection head so as to ensure that the storage temperature of the front-end of the detection head is within the range specified in the design requirements.

5.4 Elemental Component Identification Technology

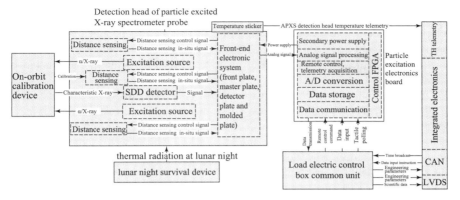

Fig. 5.8 Block diagram of the particle-excited X-ray spectrometer system

The on-orbit calibration device consists of a standard sample and its support, which is the reference for on-orbit measurement calibration to ensure the accuracy of the detection data results.

5.4.5.2 Design of Detection Head

The detection head mainly consists of the front-end electronic component (including SDD detector), the excitation source component at the front-end of the detection head, infrared distance sensor and the supporting structures such as outer cylinder (Fig. 5.9).

Fig. 5.9 APXS detection head

The excitation source component at the front-end of the detection head is composed of eight excitation sources, SDD collimating components, source collimating components, polyimide structures, etc., and provides the functions of isolating lunar thermal radiation, collimating excitation, collimating detection and radiation shielding.

The X-ray enters the SDD detector and forms electron–hole pairs by depositing energy, which are collected by the electric field on the SDD to form an electrical signal whose charge is proportional to the X-ray energy. The front-end amplifier circuit uses the low-noise charge-sensitive amplifier, by which the weak signal generated by the SDD is amplified, filtered by the shaping circuit and then transmitted to and collected the electronic plate to establish the scientific data of the APXS. The front-end electronics block diagram is shown in Fig. 5.10.

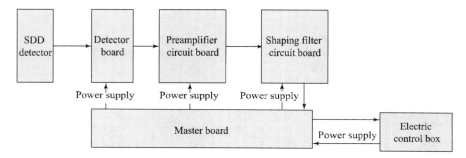

Fig. 5.10 Front-end electronics block diagram

5.4.6 Calibration and Ground Verification Test

The calibration of the particle-excited X-ray spectrometer includes ground calibration and on-orbit calibration, in which ground detailed calibration is mainly used, and the on-orbit calibration is used for data correction. The calibration test items are shown in Table 5.12.

5.5 Lunar-Based Astronomical Observation Technology

5.5.1 Introduction

In deep space exploration, in addition to studies on the origin and evolution of celestial bodies, the origin and evolution of life, human beings also concern about the resources and environment of the celestial bodies being explored for human

5.5 Lunar-Based Astronomical Observation Technology

Table 5.12 Test items of calibration and ground verification test

No.	Test item	Test objective
1	Linear calibration of energy	Obtain the energy–channel relationship and estimate the energy interval based on the linear calibration result
2	Energy-resolution calibration	Obtain the single-point energy resolution of the detector
3	Temperature response test	Obtain energy-channel and energy-resolution relationships of detectors at different temperatures
4	Element content inversion calibration	Obtain the measured energy spectrum of a standard sample with known elemental content
5	Detection capability verification test	Test unknown samples and infer the main element types and contents of unknown samples

use. The period of rotation of the Moon is quite long. Astronomical observation carried out on the Moon enables observation of the same target for a long time with a single telescope. The lunar atmospheric density is 14 orders of magnitude lower than the Earth atmospheric density, so continuous UV observations can be performed under the negligible UV absorption, which is of great advantages over astronomical observations on the Earth.

The lunar-based optical telescope can make full use of the advantages of the Moon to continuously monitor the optical variability of a group of celestial bodies with important scientific value in the near-ultraviolet band for a long period of time and complete the observation tasks that cannot be realized on the ground. The objects to be monitored mainly include interacting double stars with dense stars, active galactic nuclei corresponding to huge black holes, chromospheric active stars dominated by intense magnetic activity and short-period pulsating variable stars. Long-term continuous monitoring can reveal a variety of complex optical variability behaviors more thoroughly.

China's Chang'e 3 is equipped with a lunar-based optical telescope to carry out astronomical observations. This section takes the lunar-based optical telescope as an example to introduce how to use the favorable conditions of the Moon to carry out lunar-based astronomical observations, including the selection of high-output spectral segments and observation sky regions, telescope design, stray light suppression, calibration and ground verification tests [11, 12].

5.5.2 Selection of Spectral Segments and Observation Sky Regions

Lunar-based telescopes make full use of the favorable conditions of the Moon while avoiding unfavorable factors to maximize scientific output. Therefore, the observation bands should favorably be those that are with advantageous for lunar-based observation, namely those cannot be observed in the Earth's atmosphere. As the cost of providing energy for observations at lunar nights is too high, the telescope should be used to make observations in the lunar daytime. To control the stray light, the bands with the strongest solar energy, namely visible light bands, should be avoided as much as possible. Combining these two factors, near-ultraviolet bands are selected for the observations using lunar-based optical telescopes.

Due to the rotation of the Moon, the stars observed from the lunar surface will present apparent diurnal motion. The effect of this apparent motion is most obvious for celestial bodies with low lunar declination (declination in the lunar equatorial coordinate system). The higher the lunar declination is, the slower the apparent diurnal motion of the celestial body is, and the longer the time for continuous monitoring is. The higher the lunar declination of the observed object is, the longer the allowed single-frame exposure time is, and the higher the signal-to-noise ratio of the captured target image is. Meanwhile, to improve the coverage of the observed band, the target sky region selection should also take into account the observation stations on the Earth, so that multi-band coordinated observations in the visible and near-infrared bands can be performed. Based on the above analysis and the consideration of multi-band joint observation, the sky regions with high lunar declination are the best choice.

5.5.3 Telescope Design

The lunar-based optical telescope works in the near-ultraviolet band with a long focal length. The generally available structural forms are catadioptric, Cassegrain-reflective, Ritchey–Chretien-reflective (referred to as R-C system) systems. For catadioptric system, the front and rear correction mirror groups should be added to ensure the relative aperture and the off-axis FOV image quality. Cassegrain-reflective system has the superiorities of no chromatic aberration, small energy loss in the near-ultraviolet band, compact structure and light structure weight. However, its effective FOV is small, and it is difficult to meet the system FOV requirements.

The R-C system itself is a reflective mirror satisfying the aplanatic conditions also with the superiorities, such as no chromatic aberration at all, small energy loss in the near-ultraviolet band, compact structure, light structural weight, and the effective FOV is larger than the that of classical Cassegrain-reflective mirror. However, it is still difficult to achieve excellent image quality in a full FOV (1.92°) with a pure R-C system. It is necessary to add a non-focal power corrector to the system to correct the off-axis aberration. The optical system of the "Chang'e 3" lunar-based optical telescope uses the R-C coude reflector system plus the coude Nyquist system. The materials of the reflective mirror of the optical system are SiC material, glass-ceramic, fused silica glass JGS1. The glass material of the correction group is silica glass JGS1.

5.5.4 Stray Light Suppression

Stray light suppression is a core factor for achieving high-quality detection. The structure type of the optical system should be selected under the consideration of the influence of the Sun on the observation, so that the angle between the pointing direction of the reflecting mirror and the incident direction of sunlight is kept in as large as possible. However, the stray light from the Sun is still the most important source of the background noise of the lunar-based optical telescope. To ensure SNR of the star image observation of the system, the stray light PST (stray light irradiance transmittance) of the lunar-based optical telescope system should be as low as possible. The stray light suppression measures should be so designed that

(1) It should be avoided that the sunlight illuminates directly into the main lens barrel of the lunar-based optical telescope, which causes primary scattered light;
(2) Stray light suppression measures should be applied on the surface of key objects that facilitate stray light transmission in order to suppress secondary scattered light;
(3) Mirror group coating should be adopted to ensure that the transmittance in the non-working band is less than 5% of the transmittance in the working band.

5.5.5 Calibration and Ground Verification Test

The calibration test items for the lunar-based optical telescope include background, dark field, flat field and system spectral response, referring to Table 5.13.

5.6 Prospects

In recent decades, many major scientific problems have been solved in deep space exploration. However, there are still many problems to be solved, and many new problems have been proposed, requiring more and more payloads of new working mechanisms. For example, the future demands for the exploration of extraterrestrial microorganisms require the payload that has the ability to accurately perceive microbial metabolic products, can quickly identify the microorganisms with a low content and is able to accurately identify the isotopes. In addition, sensors are the core

Table 5.13 Ground calibration items of lunar-based optical telescope

No.	Test item	Calibration content
1	Bias field	Obtain the bias field image of the CCD under different working temperatures, gains and readout modes.
2	Dark field	Obtain dark field images of various states of the CCD (different integration time, gain, readout mode, CCD operating temperature).
3	Flat field (Response uniformity)	Take the flat field image in the following three cases: (1) Uniform illumination on the CCD using a near-ultraviolet light source; (2) Uniform illumination on the system using near-ultraviolet light source; (3) Use the LED light of the system itself for lighting.
4	System spectral response	(1) Measure the transmittance/reflectance of each optical device, measure the spectral response of the CCD, and calculate the spectral response of the system; (2) Directly measure the spectral response of the entire system.

elements of payloads. From the previous introduction, it can be seen that whether it is obtaining the morphology of the celestial body, identifying the composition and content of the celestial body, determining the physical characteristics of the celestial body or using the celestial body to carry out astronomical observations, the sensors with high resolution, high detection efficiency and low noise are indispensable.

References and Related Reading

1. Ouyang Z et al (2005) Introduction to Lunar science. China Astronautic Publishing House, Beijing (Chinese vision)
2. Sun H, Dai S, Yang J et al (2005) Scientific objectives and pay-loads of Chang'E-1 Lunar satellite. Earth Syst Sci 114(6):789–794
3. Ouyang Z, Li C, Zou Y et al (2010) The primary science results from the Chang'E-1 probe. Chin J Nat 32(5):249–254 (Chinese vision)
4. Ouyang Z (2013) Chang'E-2 preliminary results. Chin J Nature 35(6):391–395 (Chinese vision)
5. Sun H, Wu J, Zhang X et al (2010) Scientific objectives and payloads of Chang'E-2 Lunar satellite. In: Proceeding of the 23rd National space exploration conference, Xiamen (Chinese vision)
6. Dai S, Jia Y, Zhang B et al (2014) Chang'E-3 scientific payloads and its checkout results. Sci China Technol Sci 44(4):361–368 (Chinese vision)
7. Jia Y, Dai S, Wu J et al (2014) Chang'E-3 Lander's scientific payloads. Chin J Space Sci 34(2):219–225 (Chinese vision)
8. Dai S, Wu J, Sun H et al (2014) Chang'E-3 Lunar Rover's scientific payloads. Chin J Space Sci 34(3):332–340 (Chinese vision)

9. Jia Y, Zou Y, Xue C et al (2018) Scientific objectives and payloads of Chang'e-4 missionormalsize. Chin J Space Sci 38(1):118–130 (Chinese vision)
10. Yang J, Ruan P (2006) Light and miniaturized panorama camera. In: The 8th national symposium on space chemistry and meteoritics, Haikou (Chinese vision)
11. Xu L, Zhao J, Xue X et al (2014) PST research and measurement of Lunar-based optical telescope stray light. Infrared Laser Eng 43(4):1289–1295 (Chinese vision)
12. Xu L, Zhao J, Xue X et al (2012) Detectability calibration of Lunar-based optical telescope on ground. Optics Precis Eng 20(5):972–978 (Chinese vision)
13. Wu M, Wang H, Peng W et al (2012) Temperature effect correction for Chang'E-3 alpha particle X-ray spectrometer. Spectrosc Spectral Anal 32(7):1965–1968 (Chinese vision)
14. Peng W, Wang J, Wu M et al (2009) Physical simulation and analysis of Chang'E-3 alpha particle X-ray spectrometer. In: The 22nd annual meeting of space exploration committee, Chinese Society of Space Science, Dalian (Chinese vision)

Chapter 6
Guidance, Navigation and Control Technology

6.1 Introduction

The guidance, navigation and control (GNC) technology of deep space exploration covers an extensive range of applications, including deep space orbital control, extraterrestrial celestial landing, extraterrestrial celestial patrol and high-velocity reentry in addition to the attitude and orbital control of conventional near-Earth spacecrafts. Taking the typical deep space exploration mission as an example, this chapter mainly discusses the orbital control technology, extraterrestrial celestial landing GNC technology and extraterrestrial celestial patrol GNC technology involved in deep space exploration GNC.

This chapter focuses on the orbital transfer control approaches such as thrust direction angle control and orbital control time-zone control in terms of the orbital control technology through the calculation and analysis. To reduce the orbital transfer error under the large velocity increment, it is necessary to further consider the approaches that improve the orbital transfer precision, such as the pre-calibration of sensors (such as accelerometer), the wheel-controlled establishment of ignition attitude, the pulse width modulation jet based on PID control and the velocity incremental shutdown. In terms of the entry and landing GNC technology, this chapter starts from the requirements for high autonomy and safe control to introduce three typical technical characteristics, namely atmospheric entry control, powered descending control, and obstacle identification and avoidance. Compared with the first two GNC system operations, celestial surface patrol mission is not urgent. However, to cope with the complex terrain environment on the surface, it is necessary to be provided with the capabilities such as autonomous environment perception, position and attitude determination, path planning and motion control. Mission interruptions are implemented by monitoring the motion status in real time.

In each technical field, the GNC subsystems of Chang'e 2 and Chang'e 3 are taken as an example to introduce the mission requirements, system composition, operating mode design and switching as well as core GNC scheme.

Finally, this chapter proposes the prospects for the autonomous GNC technology for deep space exploration.

6.2 Orbital Control Technology

6.2.1 Features of Deep Space Orbital Control

Deep space exploration missions are accompanied with the difficulties such as long flight distance, long mission period, strong environmental uncertainty, long communication delay time and inability to monitor in real time. Therefore, the requirements for reliability, autonomy and control accuracy are higher than those for the near-Earth spacecraft. With the increase of flight distance and mission complexity, the problem of fuel consumption in orbital control and the problem of accurate orbit control are particularly prominent.

Orbital control, also known as orbital maneuver, refers to the operation of the probe to actively change the flight orbit through the thruster action. It is an important difference between the orbital mechanics and the celestial mechanics, that is, the probe is not only flying under the gravitation of a celestial body in space, but also actively changing its own trajectory. The main orbital control requirements for deep space exploration missions are:

Precision orbital control: It is generally used for orbit correction in the orbit transfer phase or for orbit maintenance. Deep space probe is required to compensate for errors and deviations by means of a proper velocity increment due to the existence of injection error, orbital control execution error, navigation error and the drift caused by various types of perturbations so that it can meet the final state requirements after a period of flight. For example, in the lunar transfer phase, the Chang'e 3 probe carried out a number of midcourse orbit corrections. For the missions such as Lagrangian point (libration-point) orbital maintenance and corridor entry control, the precision orbital control (such as the orbit maintenance control of the Chang'e 5 flight bedstead at the Lagrange point) is also required to guarantee the correct execution of the mission. The key to the precision control of small impulses is how to ensure the control accuracy.

Large-impulse orbital control: It is generally used for the implementation of orbital transfer process (from the initial orbit to the transition orbit or final orbit) with large velocity increment by changing the orbital parameters. The transfer missions such as the transfer from Earth orbit to Earth–Moon transfer orbit and from Earth–Moon transfer orbit to lunar orbit are generally achieved by large-impulse orbital control, such as the Lunar Orbit Insertion (LOI) control of the lunar probe and the Mars orbit insertion (MOI) control of the Mars probe. The key to large-impulse orbital control is how to optimize the orbital control strategy to maximize the fuel utilization.

6.2 Orbital Control Technology

The mission of the deep space probe orbital control is to take into account the both demands, reasonably determine the system configuration plan and design the orbital control strategy for the purpose of high precision and less propellant consumption.

6.2.2 Large-Impulse Orbital Control Strategy

For the orbiting missions of deep space exploration, there is generally an orbital control process from the orbit transfer to the target orbiting. Such orbital control process is also called orbit brake control. When the payload is the same, the mass of propellant required by the deep space probe is huge compared to the near-Earth probe. Assuming that the capacity of the launch vehicles is equivalent, the propellant mass will seriously affect the mass of the payload. Therefore, it is important for the orbit brake control process to minimize fuel consumption to achieve the mission target within an acceptable period. From the Tsiolkovsky equation, the formula for calculating the velocity increment actually produced by the orbital maneuver under a finite thrust is derived:

$$\Delta v = -I_{sp} \ln\left(1 - \frac{\Delta m}{m_0}\right) \text{ (m/s)} \tag{6.1}$$

where Δv is the velocity increment in the period from the initial time to the end of braking; I_{sp} is the specific impulse of the orbital control engine, namely the equivalent exhaust velocity of the engine; m_0 is the initial mass of the probe before the start of braking, including the masses of the probe itself, payload and propellant; Δm is the total propellant mass consumed during braking. To be specific, the equation establishes the relationship between the velocity increment and the three parameters (the specific impulse, the initial mass of the probe and the consumed mass).

It can be seen that there are two ways to reduce the propellant consumption Δm, namely increasing the specific impulse of the engine and decreasing the velocity increment required for braking. The former is the thrust characteristic parameter of the engine, that is, the engine thrust is proportional to the specific impulse, namely the equivalent exhaust velocity. The larger the specific impulse is, the smaller the propellant mass is required for the same thrust. The specific impulse depends greatly on the type of propellant and is also proportional to the engine nozzle outlet area and throat area ratio. But this part is the domain of propulsion system design. To reduce fuel consumption, the research on the control system shall mainly proceed from the reduction of the velocity increment Δv. The brake control strategy design can effectively reduce the velocity increment demands.

The use of single-pulse braking is a preferred choice to control the braking. By properly selecting the orbit approaching the extraterrestrial celestial body, the propellant consumption required for the brake control on the approaching orbit may be greatly reduced. Normally, the orbit approaching an extraterrestrial celestial body is a hyperbolic orbit. The total mechanical energy of the orbit is limited by the launch window and the transfer time which are often difficult to be changed. However, the parameters of a braking orbit (mainly pericenter parameters) can be changed by correcting small impulses.

The total mechanical energy for entering the orbit can be expressed as:

$$E = -\frac{\mu}{2a} = -\frac{\mu}{r} + \frac{v^2}{2} \qquad (6.2)$$

where E is the total mechanical energy of the orbit; μ is the gravitational constant; γ is the braking altitude; v is the braking point velocity. And it is obvious that:

$$\frac{\partial E}{\partial v} = v \qquad (6.3)$$

To be specific, suppose the initial total mechanical energy is constant. Then, the greater the probe velocity at the time of braking is, the larger the mechanical energy change caused by the velocity increment will be. As the orbital mechanical energy is conserved, a lower braking attitude will require a smaller velocity increment and less propellant. Therefore, when designing the braking strategies for various deep space probes, it is always chosen to brake near the near-point. As far as the orbital safety is guaranteed, the altitude of the orbit near-point shall be as low as possible through midcourse correction. Except the 6000 km braking was selected for safety reasons by the Sailor 11 probe, all the other Mars-orbiting probes apply the brake at the altitude of less than 600 km.

For the orbital control approaches discussed above, it is assumed that the orbital control engine is pulse-braked, that is, an infinite thrust is implemented in an infinitesimal time. This is just a theoretical ideal case. In engineering, large-impulse orbital maneuver is used to approximate the pulsed orbital maneuver. A large-thrust engine is used to provide the required velocity increment to the probe in a short period of time. Once the engine ignition needs to last for a period of time, there will be gravity loss, namely the fuel consumption difference between the actual orbital maneuver and the ideal pulsed orbital maneuver. If the velocity increment required for the ideal pulsed orbital transfer in one orbital braking is Δv_{pl} and the velocity increment required for the actual finite thrust is Δv, then the gravity loss caused by the orbital transfer will be:

$$\frac{\Delta v - \Delta v_{pl}}{\Delta v_{pl}} \times 100\% \qquad (6.4)$$

In deep space exploration missions, the total mass of the probe is large, the braking amount is big, and the thrust of the orbital engine is limited, so the gravity loss effect is particularly significant. To minimize the gravity loss, it is necessary to establish a corresponding braking strategy. The braking strategy includes a thrust direction angle control strategy and an orbital control time-zone strategy.

6.2 Orbital Control Technology

1. **Thrust direction angle control strategy**

There are three typical strategies to control the thrust direction angle: inertial orientation control, along-track control and uniform rotation control, as shown in Fig. 6.1.

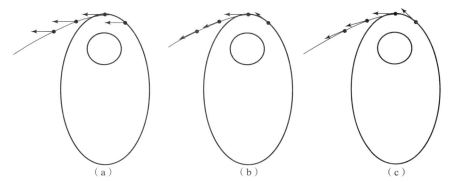

Fig. 6.1 Schematic diagram of thrust direction angle control strategies: **a** inertial orientation control; **b** along-track control; **c** uniform rotation control

Among them, the inertial orientation control refers to the ignition strategy in which the engine thrust direction is always opposite to the near-point velocity direction of the hyperbolic trajectory. This is the most popular control strategy applied nowadays. The along-track control means that the engine thrust direction always follows the opposite direction of the current velocity. The uniform rotation control is that the engine thrust is rotated at a specific constant angular velocity. The comparison of different thrust direction angle control strategies is shown in Table 6.1.

China's Chang'e 1 and Chang'e 2 probes adopt the inertial directional control strategy with the gravity loss of about 5%. This is also the control strategy usually adopted in the orbital braking for lunar near-point capture. The autonomous along-track braking control requires the involvement of inertial navigation technology, which is relatively complicated and rather risky in engineering realization. Therefore, the approximation method of uniform rotation control strategy has emerged. Although it is easier to realize uniform rotation ignition than the along-track ignition, its ground design is still quite difficult. It is necessary to determine the initial thrust direction angle and the rotational angular velocity in the ignition direction according to different ignition time and target orbit parameters. The uniform rotation control has been applied in the Mars probes "Odyssey" and "Cassini" and Chang'e 5 flight tester to greatly reduce the gravity loss.

If the gravity loss caused by the three strategies above is simply compared, the order of strategies from small loss to large loss is: along-track control, uniform rotation control and inertial orientation control. Taking the orbital braking mission of the Mars near-point as an example, the gravity loss of along-track control can be double of that of the inertial orientation control.

Table 6.1 Comparison of thrust direction angle strategies

Strategy	Advantage	Disadvantage	Applications	Engineering application
Inertial orientation control	Good engineering realization and high reliability	The braking efficiency is lowest, and the fuel loss during long-time braking is largest	Short-time ignition or small brake velocity increment, such as lunar capture braking	Chang'e series
Uniform rotation control	Good engineering realization, high reliability and high braking efficiency. The short-time ignition efficiency is similar to that in along-track control	The long-term braking efficiency is poorer than that of along-track control	Long-term ignition braking, such as Mars exploration	MGS MRO "Odyssey" Mars Express
Along-track control	The braking efficiency is highest	The engineering implementation is relatively poor	Long-time ignition braking, such as Mars exploration	Backup solution of Mars Express

2. **Time-zone strategy of orbital control**

As mentioned above, under the same thrust, the orbit arc with higher velocity requires a smaller velocity increment during braking, that is, the ideal position of the pulse maneuver is the near-point of the orbit. However, the ignition of actual orbital engine does not occur in a flash, but may last from a few hundred seconds to a few thousand seconds. Therefore, the selection of the arcs of the orbit for starting and ending ignition may affect the orbital transfer efficiency.

The most basic time-zone strategy for orbital control is a symmetric ignition time strategy that takes the time of the near-point as the midpoint of the ignition times before and after the near-point. To improve this strategy, it is necessary to arrange the ignition time in the orbit segment with higher velocity, that is, if the deceleration braking is performed, the ignition time should be earlier than that of the symmetric ignition strategy; if the acceleration orbital control is performed, the ignition time should be later than that of the symmetric ignition strategy. The optimized strategy should reduce not only the gravity loss but also the ignition duration.

Compared with the thrust direction angle control strategy, the effect of the time-zone strategy on gravity loss is small and closely related to the total ignition duration. Similarly, taking the orbital braking for Mars near-point capture as an example, if the ignition duration is about 6000 s, the optimized ignition time strategy can reduce the gravity loss by about 22%; if the ignition time is about 1800 s, the optimized ignition time strategy can reduce the gravity loss by about 3%.

6.2 Orbital Control Technology

In fact, under the same gravitational system, the most influential factor on gravity loss is engine thrust. When the engine thrust is large, the gravity loss will be greatly reduced. Meanwhile, this also means that the effectiveness of the optimized braking strategy will also be reduced. In the strategy design, it is necessary to decide which brake optimization strategy should be used based on the total braking duration.

3. Comprehensive analysis of braking strategy

In the following, the specific braking strategy for Mars orbit capture is taken as an example for further analysis and explanation. In this example, the probes (2500 kg) carrying 490 N and 1500 N engines will be analyzed and compared, respectively, in light of the braking strategy when transferring the probe from a hyperbolic orbit of $v_\infty = 3.4164$ km/s to an elliptical orbit with the semi-major axis of $\alpha_t = 3.187 \times 10^4$ km. For the thrust direction angle control strategy, the inertial orientation control strategy and the along-track control strategy are mainly analyzed and compared (the uniform rotation control strategy is an approximation of the along-track control strategy). For the time-zone strategy, the basic ignition time strategy (the near-point time is the midpoint) and the optimized ignition time strategy are compared.

1) Comprehensive data comparison

Table 6.2 compares the different orbit braking strategies given the same semi-major axis of final orbit target.

Table 6.2 Comprehensive comparison of orbital braking strategies

Parameters	Basic ignition time strategy		Optimized ignition time strategy	
	Inertial orientation	Along-track control	Inertial orientation	Along-track control
Thrust F/N	490			
Total ignition time t_{ig}/s	7014	6108	6670	5938
Semi-major axis error Δa/km	27.5563	19.257	1.0636	−0.4549
Fuel consumption Δm/kg	1131.129	985	1075.807	957.7419
Thrust F/N	1500			
Total ignition time t_{ig}/s	1820	1758	1813	1755
Semi-major axis error Δa/km	−52.1224	133.7374	7.5704	1.4035
Fuel consumption Δm/kg	898.1238	867.5115	895.1613	866.524

2) Argument of pericenter

The argument of pericenter (ω) changes due to the different velocity increments on both sides of the near-point. Table 6.3 lists the amounts of change in ω under several orbital braking strategies.

Table 6.3 Comparison of changes in argument of pericenter

Parameters	Basic ignition time strategy		Optimized ignition time strategy	
	Inertial orientation	Along-track control	Inertial orientation	Along-track control
Thrust F/N	490			
$\Delta\omega/(°)$	6.2890	6.4717	6.3151	6.1496
Thrust F/N	1500			
$\Delta\omega/(°)$	6.3001	6.3461	6.2634	6.2764

It can be seen that the effects of the orbit braking strategy and the engine thrust on the argument of near-point are limited. The variations of the arguments listed in the table are within the range of (6.1, 6.5), and the difference between them is quite small.

3) Change in the orbital altitude of the near-point

The orbital altitude is correlated to the flight safety of the probe. During the orbital transfer, the altitude of the near-point will change. To clarify the safe altitude, it is necessary to analyze the variation of the orbital altitude of the near-point before and after the orbital transfer.

The inertial orientation control will raise the altitude of the near-point orbit, and the along-track control will reduce the altitude of the near-point orbit. The specific values are shown in Table 6.4.

It can be seen that, with the increase of engine thrust, the amount of change in the near-point orbital altitude decreases accordingly. This effect is particularly remarkable under the inertial orientation control strategy. Inertial orientation control would raise the altitude of the orbit and have no effect on flight safety. This is one of the reasons for which inertial orientation control strategy is selected for most

Table 6.4 Comparison of changes in orbital altitude of near-point

Parameters	Basic ignition time strategy		Optimized ignition time strategy	
	Inertial orientation	Along-track control	Inertial orientation	Along-track control
Thrust F/N	490			
Δh/km	341.1416	−298.7360	678.7962	−549.6520
Thrust F/N	1500			
Δh/km	8.2083	−64.2052	17.1855	−94.1378

orbital capture missions; while the orbital control will reduce the orbital altitude and probably affect the flight safety. If thrust equals to 490 N, the altitude at the lowest point is reduced by up to 549 km, and the altitude of the nominal hyperbolic orbit near-point is about 300 km, indicating that this scheme is no longer available. When designing the safe flight altitude, the problem of reducing the near-point altitude by track control should be considered.

6.2.3 Precision Orbit Control

1. **Application requirements for precision orbital control**

The advantages of the large-thrust orbital braking lie in the facts that the efficiency is relatively high and the generated gravity loss is relatively small. Its disadvantage is that the engine thrust will increase as the orbital control accuracy decreases when other conditions are the same. In actual missions, sometimes the accuracy requirement of the orbital transfer is quite high, and the accuracy of Δv is required to be at the centimeter level. Precision orbit control is applied in the following aspects in deep space exploration.

1) Orbit midcourse correction

In the long-term cruise of the probe, the initial small deviation of the orbit will result in an unacceptable target-aiming error. The midcourse correction of the orbit is to control the target-aiming error within the applicable scope by one or several precision orbit maneuvers during the cruise [1].

Generally, a midcourse correction is difficult to satisfy the aiming accuracy requirements. However, the first correction contributes the most to reducing the deviation. Numerical calculations and engineering practices show that the first correction can reduce the orbit injection error by 99% and leave only 1% of the error uncorrected. The second correction can further reduce the residual deviation by 99%. Therefore, generally the accuracy requirements of the mission can be met by 3–4 corrections.

It is known from the nature of the error divergence that, from the perspective of saving propellant, earlier midcourse correction means a smaller cumulative deviation, a smaller required velocity increment and thus less fuel consumption. From the perspective of correction accuracy, later correction means a smaller control deviation due to the time-related cumulative effect of residual error. Therefore, the rational selection of orbit correction time is correlated to fuel consumption and correction accuracy.

For deep space orbit transfer, orbital errors are mainly from two sources: (1) the mechanical model errors caused by various types of perturbation; (2) the errors caused by orbit insertion, orbit prediction and orbital control. The former error can be reduced by establishing a more accurate mechanical model and can even be neglected in some analyses; the latter is the main source of errors in orbit correction.

The first orbit correction mainly corrects the injection deviation of the probe. The residual error after the first orbit correction does not change greatly with the postponement of the first correction, and the velocity increment required for subsequent correction increases with the delay of the correction. Therefore, the first orbital correction should be performed as early as possible to save the propellant. If the accuracy of the orbit injection is improved, the initial deviation can be reduced, thus causing a sharp decrease in residual error and making it possible to correct earlier.

In addition to the residual error after correction and the required velocity increment, the second orbital correction is also required to consider the navigation error. Though the second correction should also be implemented as early as possible to save the propellant, it should not be too close to the first correction so as not to affect the navigation accuracy.

The correction velocity increment of the subsequent third or even fourth orbit correction is small, and the residual error decreases rapidly with the correction time. The propellant consumed is small compared with the previous two corrections, and the correction time may be selected according to the specific requirements of the orbital accuracy.

2) Libration-point maintenance

Generally, when discussing the problems of deep space orbit transfer correction and orbit maintenance, the two-body model and the conic curve splicing method are adopted, and the concept of influence sphere is introduced. Within the radius of the influence sphere, the gravitational force of the central celestial body dominates. After entering the influence sphere, the probe can be regarded as a perturbed two-body system. In fact, at the boundary of the sphere, the gravity of the central celestial body and that of other large celestial bodies are often similar. For example, in the solar system, the motion of the probe near the influence sphere boundary of a planet is affected by the gravitational forces of both the Sun and that planet. The kinetic model at this time is a restricted three-body model. The so-called restricted three-body model is composed of two large celestial bodies whose movements are completely determined and whose relative motion is a two-body issue, and a small-mass celestial body has no effect on the two celestial bodies. The focus of the study of the restricted three-body issue is in nature the movement of the third small celestial body [2].

Under this model, the motion model has five equilibrium points corresponding to five libration points, which are collectively called Lagrangian points. The resultant force of the gravitational and centripetal forces from the first and second mass bodies that is acted on the spacecraft at the libration point is zero. The spacecraft can circle around the first celestial mass body in the same motion period as the second celestial mass body as long as it has a certain initial velocity. Among the five libration points, the two libration points L1 and L2 which are located on the connection line of the two celestial bodies and are symmetric with respect to the second celestial mass body are also called the collinear libration points. They play an important role in deep space exploration.

6.2 Orbital Control Technology

The orbit around the libration point and perpendicular to the connection line of the two celestial mass bodies is called halo orbit, and the halo orbits around L1 and L2 are good observation positions. L1 and L2 are unstable libration points. Even if the initial state of the probe satisfies the conditions of the collinear libration solution, the probe will be far from the libration point and the corresponding orbital control error will be amplified exponentially in case of just a slight disturbance. Taking the Chang'e 2 probe orbiting around the L2 point between the Sun and the Earth as an example, the orbit period is 6 months and the orbital error is increased by six times per month. The orbital error of 0.01 m/s will result in a velocity error of 2.16 m/s after 3 months. This requires periodic orbital maintenance.

In engineering application, the orbit maintenance control methods are divided into pulse type and continuous small thrust type according to the types of the thrusters. For the pulse type, the change of the velocity increment occurs approximately at a moment. Each maneuver is to pull the spacecraft back to the target orbit to a certain extent. However, there's certainly a correction error, which will accumulate with time. Thus, the correction interval is quite important. If the interval is too long, a large amount of energy is needed to correct the accumulated deviation. If the deviation is too large, it may exceed the capability range of the orbital control engine.

2. Types of errors in orbital control and error countermeasures

After calculating the velocity increment Δv required for the orbital control, an orbital control error, namely the error between final actual control quantity dv and Δv, denoted as δv, will also occur due to the influence of various factors in the process of implementing the orbital control.

The control errors are divided into scalar error and vector error according to the form, and their difference is shown in Fig. 6.2.

Among them, the scalar error is usually caused by the accelerometer error and the shutdown time error, while the vector error is generally caused by the error of the attitude control. That is, there is an error between the actual attitude and the target

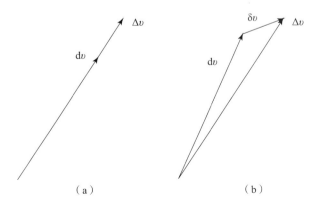

Fig. 6.2 Orbital control errors: **a** scalar error; **b** vector error

attitude, causing the inconsistence between actual ignition direction and desired ignition direction.

1) Method to reduce scalar error

The basis for precision orbit control is to use an accelerometer for velocity increment measurement and use a small thrust engine for control. However, for an accurate orbital control, it is required to start from the pre-orbit-control attitude maneuver to analyze the error source. A typical orbital control process is shown in Fig. 6.3.

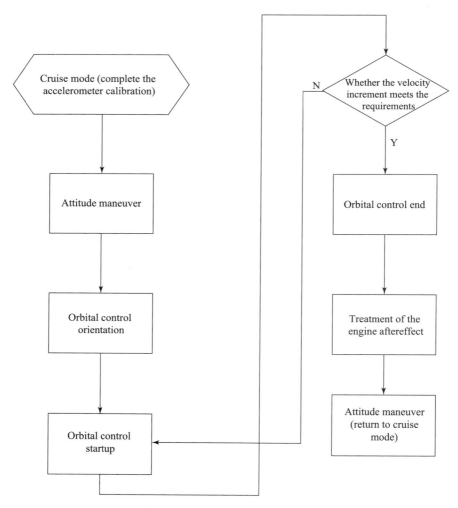

Fig. 6.3 Typical orbital control process

6.2 Orbital Control Technology

The error caused by accelerometers has a relatively significant impact on both orbit control and orbit determination. Therefore, the selection of an accelerometer with higher accuracy is critical to improving orbital control accuracy.

After the orbital control is enabled, the error caused by the accelerometer is mainly a scalar error, which is accumulated over time. Therefore, the cumulative error may be reduced by reducing the ignition time. Obviously, with the fixed value of Δv, a greater thrust means a shorter ignition time, a shorter time of error accumulation and a smaller error caused by the accelerometer.

Accelerometers may result in another type of error—the data truncation error. As the probe control system is in nature a discrete system, the sampling interval cannot be infinitesimal. The velocity increment generated between the two sampling operations may be truncated. Under the same truncation error, the greater the thrust is, the lower the accuracy of the engine orbital control will be accordingly. In addition, the shutdown aftereffect for a large-thrust engine is relatively great. After the engine shutdown, the thrust will not immediately decrease to zero and will still exit for a period of time. If the control system takes Δv as the control target, the final dv will be greater than Δv due to the influence of truncation, aftereffect and other factors. Therefore, the strategy of early shutdown can be adopted in engineering, and the remaining small velocity increment is achieved by the attitude engine modulation with a smaller thrust. The operation using a small thrust engine to achieve the remaining velocity increment is called orbital aftereffect treatment. After adding the aftereffect treatment, the scalar accuracy of the orbital control can be increased by one order of magnitude.

2) Method to reduce vector error

(1) Improve the attitude accuracy during orbital control

After the orbit control begins, attitude control error is the main source of the orbital vector error. Due to the inevitable installation deviation of the orbital control engine, a large disturbance torque will be generated after the orbital control engine is started. The attitude control system using the phase plane control is often in the single-side limit cycle under the action of large disturbance torque, that is, the attitude stays at the edge of the limit cycle and cannot continue to converge, resulting in an attitude offset.

The pseudo-rate modulation attitude controller [1] not only includes the differentiation of the input signal, but also enables the rate damping. In addition, it can introduce an integral link for the input of the rate modulator and has strong constant interference to obtain a stable limit cycle. Thus, it has higher attitude control accuracy. Such controller has long been applied to the "Sailor" probe. A typical block diagram of the attitude control system is shown in Fig. 6.4.

The pseudo-rate modulator is also known as the pseudo-rate increment feedback controller. The illustrated input signal is the attitude angle. The pseudo-rate modulator includes a Schmitt trigger, whose output pulse width and pulse interval are correlated to the input. And θ_D is the dead zone. As can be seen from Fig. 6.4, the average output pulse of the pseudo-rate modulator is:

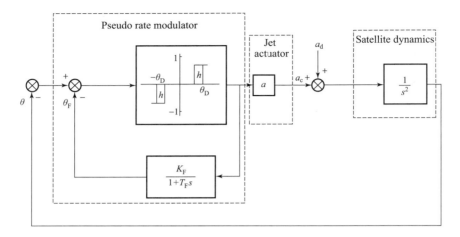

Fig. 6.4 Block diagram of single-axis attitude control system with pseudo-rate controller

$$N_{av} = \frac{\theta - \theta_D + \frac{1}{2}h\theta_D + \dot{\theta}T_F}{K_F} \quad (6.5)$$

It can be seen that the value is proportional to the attitude angle and the angular acceleration of the attitude. To be specific, the pseudo-rate modulator can approximate the proportional differential control, and $\frac{T_F}{K_F}$ can be designed to be relatively small to reduce the sensitivity to measurement noise.

(2) Reasonable arrange the attitude control thrusters to improve orbital control accuracy.

During the orbital control, a large disturbance torque is often generated due to the installation deviation of the orbit control engine, and such disturbance torque is required to be balanced by the attitude control thrusters through the reaction control. Attitude control thruster has a relatively small thrust. To generate a large torque, inclined mounting is generally applied to obtain as great control torque as possible. The schematic diagram of the inclined engine is shown in Fig. 6.5.

The inclined attitude control thruster produces a lateral thrust perpendicular to the direction of orbital maneuver. In case of no installation deviation of the orbit engine, the attitude control thrusters on both sides can work symmetrically, and the lateral thrust can be offset. In case of installation deviation of the orbit engine, the thruster on one side must work more frequently than that on the other side. Finally, the lateral thrust cannot be offset, resulting in a lateral velocity increment. Such lateral velocity increment will introduce an orbital vector error. Therefore, it is necessary to minimize the installation deviation of the orbit control engine and optimize the installation angle of the attitude control thrusters. In addition, along the direction of rotation in the orbital control, a couple attitude control thruster should be configured to minimize the lateral error.

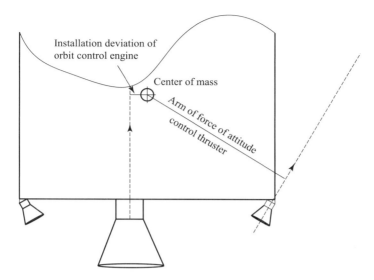

Fig. 6.5 Schematic diagram of inclined engine

(3) Enable non-jet attitude maneuver with an angular momentum exchange device

As can be seen from the orbital control process, the orbital control starts with attitude maneuvering and ends with the attitude maneuver. If the attitude maneuver is driven by jet, unpredictable orbital control errors may occur. Taking the Chang'e 2 probe as an example, such error is generally 50 mm/s. The method of reducing the orbit control error in the attitude maneuvering phase is to use the momentum wheel for attitude maneuver control so as to realize the non-jet attitude maneuver.

6.2.4 Design of Orbit Control System

Take the Chang'e 2 mission as an example to introduce the typical orbital control system design.

1. **Mission requirements**

1) Orbit control

Complete the orbit control missions such as the midcourse Earth–Moon transfer correction, three braking operations at perilune, lunar orbit maintenance, descending to the 100 km × 15 km orbit and recovery, transferring to and maintaining the Earth–Sun L2 libration-point orbit, and rendezvous with and flying over the asteroid Toutatis.

2) Attitude adjustment and maintenance

Realize the establishment of the orbital attitude, the attitude maintenance during the orbital transfer, and the autonomous recovery of the cruising attitude around the Sun or the Moon.

3) Autonomous fault diagnosis and treatment during the orbital transfer

Enable the autonomous diagnosis and treatment of faults during the transition period as well as component-level and system-level fault diagnosis and system reconstruction.

2. System composition

The GNC system mainly consists of the sensors, actuators and controllers, as shown in Fig. 6.6.

Sensors include UV lunar sensor, star sensor, solar sensor, gyros and accelerometer. Except for the UV lunar sensor, other sensors are used for attitude or orbit determination before and after the orbit transfer.

Actuators include 490 N orbit maneuver engine, 10 N thruster (belonging to the propulsion subsystem), solar array drive mechanism and antenna drive mechanism.

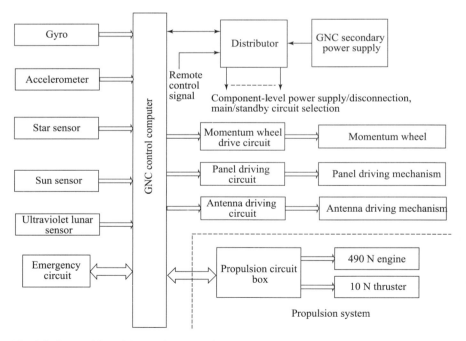

Fig. 6.6 Composition of the GNC system of the Chang'e 2 probe

6.2 Orbital Control Technology

Before and after the orbit transfer, the 490 N orbit maneuver engine, the 10 N thruster (belonging to the propulsion subsystem) and the momentum wheel are mainly used.

Controllers include control computer, emergency circuit, power distributor and GNC secondary power supply.

3. Operating mode

The GNC subsystem is designed with four operating modes for orbital transfer preparation and orbital control [4].

(1) Star capture: Before the probe establishes the orbital control ignition attitude, the inertial attitude of the probe is estimated based on the sensor information, and the probe attitude control is performed;
(2) Inertial attitude adjustment: To establish the probe's orbital control ignition attitude and realize the three-axis large-angle attitude maneuver of the probe;
(3) Star orientation: To ensure the steady-state orientation of the probe after the probe establishes the orbital control ignition attitude;
(4) Orbital control orientation: To conduct the startup and shutdown control of orbit control engine, determine the ignition attitude of the probe and perform the attitude stabilization control during ignition and the recovery after orbit transfer control.

The workflow in the above mode is as follows. Before the probe establishes the orbital control ignition attitude, the probe enters the mode of star capture to calibrate the gyro drift and the zero offset of the accelerometer. At the programmed time point, the probe autonomously shifts into the inertia attitude control phase and performs three-axis large-angle reorientation to establish the orbital control ignition attitude. After the attitude adjustment ends, the probe is autonomously transferred to the attitude stabilization phase before orbital control. The star sensor is used to filter and correct the probe attitude. When the programmed time is up, the probe enters the orbital control orientation mode autonomously, and the onboard autonomous orbit control engine is started. After the engine is shut down according to the guidance law, the probe remains stable for a period of time and then automatically enters the solar orientation mode to restore the cruising attitude.

4. Program design points

1) Orbit control scheme

Navigation: The free flight orbit parameters at each phase are determined by the ground-based measurement and control system; the change of orbit (velocity increment) during ignition is determined jointly by thruster calibration and accelerometer measurement.

Guidance law includes startup time, ignition attitude (rules of changes with time), shutdown conditions (time-based shutdown, velocity-based shutdown). The guidance law is calculated by the ground according to the measured orbital parameters before each orbital transfer and is expressed by the fitting coefficients (the initial direction of the thrust vector, the rotational angular velocity of the thrust vector, the rolling offset angle, the ignition time, the shutdown time and the shutdown velocity increment) and is then input into the GNC computer. To control the engine shutdown, the onboard GNC computer calculates the shutdown equation in real time according to the guidance control parameters and accelerometer output sent by the ground station, and then issues a shutdown command to the engine. It is forbidden to shut down the engine before a certain time interval (shutdown interval) to prevent the velocity increment from being seriously insufficient. In the interval, the engine is turned off based on the velocity increment to achieve high-precision orbit transfer. After the interval, time-based engine shutdown is enabled to prevent a catastrophic failure caused by too large cumulative velocity increment during ignition.

In order to reduce the error of orbital transfer, when the orbital velocity increment is relatively small, the attitude control thruster can be used to change the orbit and the modulation-off control mode can be adopted. During the orbit transfer, the thruster works continuously. When the attitude is controlled, the corresponding thruster is turned off according to the attitude control law to ensure that the attitude is stable. When the shutdown condition is satisfied, the modulation-off mode of the thruster is restored to the modulation-on mode.

2) Attitude control scheme

During the orbit transfer, pulse width modulated (PWM) jet control is used under the control law of PID and filter correction. PWM is functionally equivalent to the pseudo-rate modulator (PRM). Attitude control should overcome the large disturbance torque generated by the off-center thrust of orbit control engine, as well as the effects of liquid sloshing, solar array vibration and flexible antenna vibration. Through the pulse width modulation algorithm and the interface circuit, the control signal is converted into the switching pulse signal of the attitude control thruster, which can effectively ensure the attitude control precision of the probe during the orbit transfer.

The phase plane jet control is used to establish the ignition attitude before the orbit transfer and to restore the orientation to the Sun after the orbit transfer. By using the area divided by the switching lines on the phase plane and according to the actual attitude/angle and desired attitude angle/angular velocity of the probe, the limit cycle around the target attitude is formed by setting the ON/OFF intervals of the thruster, thereby realizing the attitude maneuver and maintenance.

When the ignition attitude is established, the three-axis simultaneous attitude maneuvering method based on the quaternion is used to establish an all-sphere (omnibearing) arbitrary attitude. Meanwhile, to reduce the influence of the three-axis coupling of the probe, the method of setting different angular velocities according to the attitude adjustment target of the probe is applied in the design of the control law within the restricted attitude adjustment time. In addition, the momentum wheel is introduced to establish the ignition attitude and further reduce the impact of jet on the orbit. The control law is zero momentum flywheel control of the classic PID and filter correction.

After the orbital control, solar orientation and solar directional control are restored by solar sensor, and the cruising attitude is autonomously restored.

3) Autonomous fault diagnosis and treatment

Component fault diagnosis: The consistency comparison can locate and isolate the fault in one of the five or six gyros. The star sensor is selected by the consistency of its validity mark and output in time. In case of the failure in the communication between equipment and computer, the autonomous switching of the component communication interface can be realized. Through the three-machine data comparison and arbitration, the control computer on duty can be autonomously switched.

System-level fault diagnosis: If the orbital control ignition cannot be sustained due to a certain fault, it is allowed to shut down the orbit control engine first so that the rate damping mode for onboard autonomous control is activated. Then, in a short time, autonomous system restructuring and mode recovery (propulsion branch switching, on-duty computer changing, et al.) are performed, the orbital control process is reset, and the orbital control engine is restarted again by applying the guidance law that is independently set. In this way, the "Dual Starts" in the key orbital transfer are realized.

6.3 Entry and Landing GNC Technology

6.3.1 *Characteristics of Entry and Landing GNC Technology*

The features of the entry and landing process include short time, fast change of orbit and attitude, failure of real-time control and intervention from the Earth. The landing requirements of entry and landing mission are difficult to be satisfied in terms of accuracy, timeliness, reliability and other aspects via ground station network. And the probes are generally required to have highly autonomous GNC capabilities.

Depending on the different celestial bodies to be entered, the landing GNC technologies can be generally divided into atmospheric landing GNC technology and non-atmospheric landing GNC technology. Presently, typical atmospheric entry and landing missions include deep space high-velocity return and reentry, Mars entry, descending and landing missions. The typical mission of non-atmospheric entry and landing is for lunar powered descending and soft landing.

In case of atmospheric entry control, while guaranteeing the initial atmospheric entry conditions, the GNC system generally has the ability to adjust the trajectory to ensure that the probe meets the required opening point conditions (altitude, dynamic pressure, Mach number, etc.). For the Earth return mission, the GNC system has completed its task at this time. But for the Mars landing mission, the parachute deceleration capability is limited due to thin atmosphere. After the parachuting phase is over, the powered descending control must be continued to further eliminate the final velocity. Thus, the key conditions such as the attitude and velocity of the landing touchdown can be guaranteed to realize safe landing. Therefore, to ensure the continuity of subsequent navigation in the period from the beginning of the parachuting to the powered descending, a reasonable design of the navigation scheme is required.

For the powered descending control, it is necessary to rationally design the guiding law to optimize the propellant consumption. In addition, the navigation method of "inertial navigation + external measurement correction" shall be adopted to improve the navigation accuracy. The pre-touchdown shutdown mode and touchdown shutdown mode shall be reasonably selected to ensure the final safe shutdown of the engine.

To achieve safe landing and avoid the terrain conditions (rocks, craters, slopes, etc.) bad for landing buffer, it is also necessary to image the surface of the landing area by optical or laser imaging sensors to further identify safe landing sites.

6.3.2 Atmospheric Entry Control

Whether in the entry into a celestial body with atmosphere during deep space exploration or in the return to the Earth through near-Earth orbiting, the control of atmospheric entry will be involved. Atmospheric entry control can be divided into two stages: pre-entry control and post-entry control. Pre-entry control is mainly to reach the entering corridor through orbit transfer and meet the key entry parameters such as the initial entry angle and initial position. Post-entry control is mainly to control the lift through the tilt angle adjustment, thereby achieving the entry deceleration ballistic control and finally satisfying the required parachute-opening conditions.

Depending on the lift–drag ratio of the spacecraft, the flight after entering the atmosphere is divided into ballistic entry type, semi-ballistic entry type and lift entry type. The ballistic entry type does not need to control the lift. The principle of the atmospheric entry in deep space exploration is similar to that of Earth-orbiting reentry. Their difference depends mainly on the characteristics of the target atmosphere.

The greatest challenge encountered by the extraterrestrial celestial atmosphere entry, as represented by the Mars entry, is that the atmospheric density of Mars is extremely thin (only 1% of that of the Earth). Meanwhile, the uncertainties of atmospheric density, temperature and wind field of the Mars are relatively great. In addition, due to the large delay between the Earth and the Mars, real-time ground intervention can't be implemented.

Such characteristics require the probe to establish the attitude and navigation reference, implement the deceleration control based on reverse thrust engine and strengthen the autonomous entry and landing control capabilities, in addition to experiencing the pneumatic deceleration and parachute descending. There are two ways to obtain the navigation reference: One is to obtain the horizon information by using the measurements of ranging velocity radar and the difference between different beam measurements; and the other is to start orbital extrapolation and inertial navigation in real time from the orbital operation stage by using the inertial measurement unit (IMU). As affected by terrain, the accuracy of the horizon information obtained by the range and velocity radar is relatively poor with the error generally being 4°–5°; the accuracy of the horizon information obtained by inertial navigation can be within 1°. The current successful Mars probes generally adopt the second method. However, inertial navigation also has its own limitations. Firstly, it has high dependence on orbit measurement. The error of the orbit measurement by the ground is directly reflected in navigation error. Secondly, the inertial navigation has high requirements for IMU, including a large measuring range that can adapt to the parachute-opening shock as well as high precision that can achieve high inertial navigation accuracy under large dynamics. Once the IMU saturation occurs, it will be difficult to guarantee the landing safety. The failure of the lander (Schiaparelli) in the ExoMars 2016 mission was just due to IMU saturation. Therefore, the probe should be equipped with a large-scale, large dynamic IMU assembly to withstand the parachute-opening shock.

After the atmospheric entry, the target of the lift control within the atmosphere is to minimize the landing error. The general design method is to predict the landing point error according to the current angle of heel and the navigation data to generate the heeling angle command, and then predict the new landing point error according to the new flight state. The process should be repeated until the accuracy requirements are satisfied. The control parameter here is the heeling angle sequence (or attitude engine jet command), and the output is the predicted landing point error (see Fig. 6.7).

Take the small lift-to-drag ratio probe such as the Mars Science Laboratory (MSL) [5] as an example. Its flight path in atmospheric entry is controlled by adjusting the heeling angle. Its landing accuracy is greatly improved compared with other successful Mars probes. In the design of the entry guidance law, the dynamic equation is first linearized near the nominal trajectory to obtain the functional relationship between the terminal range deviation and the small disturbance terms of state quantity and control quantity. On this basis, the feedback gain coefficient is optimized to realize the closed-loop guidance.

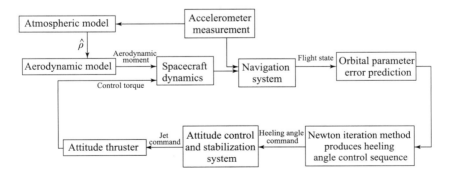

Fig. 6.7 Block diagram of the orbital control system in the atmosphere

6.3.3 Powered Descending Control

Presently, the difficulties in the powered descending control technology using a deceleration engine are: to optimize the guidance law and reduce the propellant consumption in the descending process as much as possible by using a relatively small real-time calculation amount; to use the inertial navigation + external radar to correct the navigation strategy; to formulate the engine shutdown scheme before finally landing on the celestial body.

1. **Optimization of propellant consumption**

For the target objects with no atmosphere, or when the deceleration velocity increment in the pneumatic flight does not satisfy the mission requirements, other power resources are required to provide more velocity increments. Powered descending refers to the mode of landing by means of reverse thrust engine. For the soft landing on the Moon without the atmosphere and the deceleration at the final stage of Mars landing, the method of powered descending deceleration is required. As the process is at the expense of propellant consumption, it is necessary to optimize the propellant consumption as much as possible to effectively improve the landing quality.

The lunar soft-landing mission starts the powered descending from the lunar orbit. The power descending of Martian EDL begins after the parachute separation. Given the same initial velocity and the altitude above the ground, a larger initial thrust-to-weight ratio will result in shorter deceleration time, much smaller gravity loss and less propellant consumption.

2. **Navigation strategy of inertial navigation + external radar correction**

The basic navigation method applied in the powered descending process is inertial navigation. The angular velocity and non-gravitational acceleration measured by the IMU (including the gyro and accelerometer) installed on the probe are used to calculate the attitude, position and velocity of the probe. However, due to the existence of error sources such as initial reference error, IMU error, gravitational field model error and integral step error in the calculation of the inertial navigation system, the external radar data (range and velocity) must be included to correct the error.

3. **Touchdown shutdown strategy**

The last phase of powered descending process is the touchdown phase. In the touchdown phase, it is necessary to shut down the deceleration engine in time to ensure that the vertical and horizontal velocities are confined to a small range when the surface of the celestial body is contacted at the end, so as to meet the working conditions of landing buffer device during the final touchdown. Current shutdown strategies include pre-touchdown shutdown and touchdown shutdown.

The pre-touchdown shutdown is implemented through the gamma shutdown sensor used in the lunar soft-landing mission. The shutdown signal is given by the gamma shutdown sensor at an altitude of 1–2 m before the final landing of the probe. Most of the current successful Mars landing missions have adopted the touchdown mode. The touchdown shutdown is logically simple and requires the use of a touchdown switch mounted on the landing buffer structure. The principle is that when the touchdown switch touches the ground, the mechanical switch will be activated to realize the ON/OFF conversion of the signal circuit, control the computer to collect the signal and send a shutdown command.

6.3.4 Obstacle Identification and Avoidance

The surface of the target celestial body is generally complex in terrain and has the typical features such as craters and the undulating mountains on the Moon and the Mars. To ensure the final stable landing and to rover on the celestial surface, the image acquired by the remote-sensing probe is generally used to analyze the relatively safe landing zone based on engineering constraints. However, due to the limitation of image resolution and coverage, it is not possible to obtain the slope and rock distribution conditions for small-scale baselines. Therefore, most of the current soft-landing missions have a relatively limited ability to adapt to small-scale obstacles. In the actual landing process, the probe should have a certain ability of on-orbit real-time image recognition and safety point selection to achieve the avoidance of small-scale obstacles.

In obstacle avoidance, the probe is required to hover over the landing zone for a short time, acquire the three-dimensional terrain of the landing zone and find the ideal landing point through image recognition. Then, it shall move by translation and slowly descend to the landing site. The core of obstacle avoidance is obstacle recognition. Presently, there are two ways to identify obstacles: One is laser three-dimensional imaging, and the other is binocular vision. With the development of laser radar, laser three-dimensional imaging has become the main approach to realize obstacle recognition in the future. This method has the features such as small calculation load, fast recognition, immunity to sunlight conditions and high recognition accuracy. The binocular vision method uses two cameras with long baselines to take images simultaneously and achieves terrain recognition through image comparison. The accuracy of this method is directly proportional to the baseline length and inversely proportional to the square of the recognized altitude. That is, the lower the imaging altitude and the longer the baseline, the higher the recognition accuracy. In addition, binocular vision requires a large amount of image calculation, so the recognition speed is slow and it takes a relatively long hovering time to achieve binocular vision.

The Chang'e 3 lunar lander has successfully adopted the obstacle recognition function on orbit. By hovering and imaging at an altitude of 100 m, the safe landing point is quickly calculated online, and finally safe landing is achieved. The image taken by the laser three-dimensional imaging sensor of the Chang'e 3 lander is shown in Fig. 6.8. The research on related technologies is being carried out by the Mars 2020 mission based on "Curiosity" platform and the USA's mission of returning to the Moon. And the ground verification of key sensor performance is being carried out for the purpose of engineering applications.

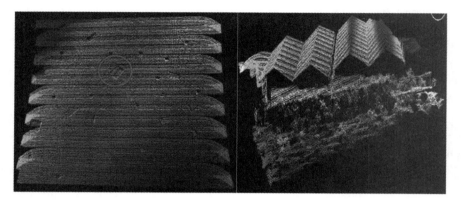

Fig. 6.8 Images taken by laser three-dimensional imaging sensor of the Chang'e 3 lander

6.3.5 Design of Entry and Landing GNC System

The Chang'e 3 lander, which realizes China's first lunar soft landing, is taken as an example to introduce the design of the GNC subsystem.

1. **Mission requirements**

One of the core missions of the Chang'e 3 lander is to achieve highly reliable and safe lunar soft landing, requiring the lander to have the capability of autonomous high-precision GNC. The main missions of the GNC system include:

(1) Apply the deceleration braking from the initial position of the lunar orbit, complete the soft lunar landing and finally satisfy the requirements for the landing area/point accuracy;
(2) During the powered descending, the lander has a certain capability of autonomously choosing an optimum safe landing zone and falls into the designated safe landing zone;
(3) When the lander finally touches the Moon surface, the contact velocity, attitude and angular velocity control accuracy shall meet the requirements.

2. System composition

The CNC subsystem of Chang'e 3 lander is configured as follows (see Fig. 6.9):

1) Controller

The controller includes the control computer, image processing unit and other devices, and is used for the execution of the GNC algorithm and the processing of the safe-point identification image.

2) Sensor

The sensors include star sensor, digital sun sensor, analog sun sensor, inertial measurement unit (including gyro assembly, accelerometer assembly, electronic circuit), laser ranging sensor, microwave ranging and velocity sensor, laser three-dimensional imaging sensor, optical imaging sensor, touchdown sensor (belong to landing buffer subsystem) and gamma shutdown sensor. Except for the sun sensor, all the sensors are used for powered descending control.

3) Actuator

The probe is equipped with a variable thrust engine with a thrust range of 1500-7500 N, a 16×150 N thruster and a 12×10 N thruster. The variable thrust engine is used for orbital control, the 10 N thruster for attitude control, and the 150 N thruster for obstacle avoidance maneuver and attitude control. The 150 N thruster is used for horizontal precise obstacle avoidance. And the coarse obstacle avoidance function is achieved by main engine and attitude maneuver.

3. Operating mode

Seven operating modes of the GNC subsystem are specifically designed for powered descending control.

Ignition preparation mode: The main mission is to correct the ignition timing and the corresponding orbital parameters injected by the ground station before the powered descending ignition, and calculate and adjust the ignition target attitude.

Primary deceleration mode: It is used from the altitude of about 15 km to about 2.7 km above the lunar surface. The main mission is soft-landing braking, which reduces the velocity of the lander (about 1.7 km/s) to a preset value.

Quick adjusting mode: It is used from the altitude of about 2.7 km to about 2.4 km above the actual lunar surface. The main mission is to quickly connect the main deceleration stage and the approaching stage. The fast attitude maneuver is made to achieve the entrance attitude at the approaching stage, and the engine thrust is synchronously reduced to a low thrust level.

Approaching mode: It is used from the altitude of about 2.4 km to about 100 m above the lunar surface. The main mission is coarse obstacle avoidance. To ensure that the field of view of the optical imaging sensor is aligned with the landing area, the trajectory of approaching the target landing area is designed to meet the specific attitude and the descending trajectory. Therefore, the lander is approaching the landing

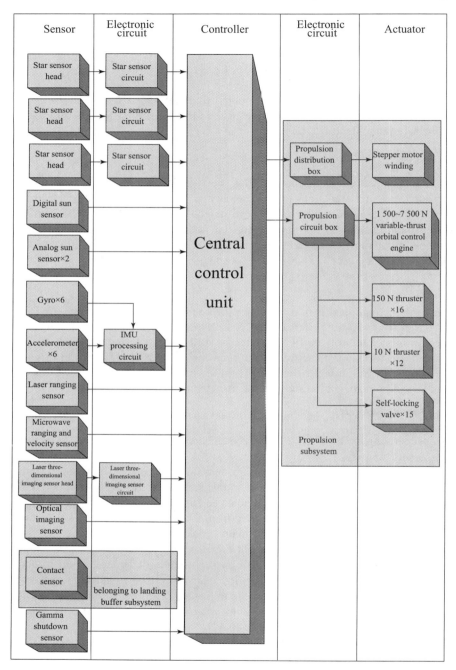

Fig. 6.9 Composition of the GNC system of the Chang'e 3 lander

6.3 Entry and Landing GNC Technology

area at an angle of 45° through almost linear descent. The large obstacles (the rocks or craters larger than 1 m in diameter) are detected by the optical imaging sensor, and the safe landing site is selected. Then, the lander moves and finally reaches the altitude of about 100 m over the landing area.

Hovering mode: It is used at the altitude of about 100 m over the lunar surface. The main mission is the fine obstacle detection for the landing area. The variable thrust engine offsets the lander gravity and keeps the lander in a stable hovering state. The three-dimensional imaging sensor is used to observe the landing area and select a safety landing point.

Obstacle avoidance mode: It is used from the altitude of about 100 m to about 30 m over the lunar surface. The main mission is precise obstacle avoidance and descent. According to the relative position information of the safety landing point given by the hovering segment, the lander descends to the altitude of 30 m above the landing point.

Slow descending mode: It is used from the altitude of about 30 m to the altitude where the signal of the gamma shutdown sensor is effective. The main mission is to ensure that the lander is smoothly and slowly descending to the lunar surface. The velocity and attitude control accuracy for lunar surface landing shall meet the requirements.

4. **Scheme design points**

1) GNC scheme structure for powered descending phase

The general GNC workflow in the powered descending phase is as follows. The navigation sensor performs measurement and sends the measurement data to the control computer. The navigation algorithm processes the navigation measurement data and sends the navigation result to the guidance and attitude control algorithm. The guidance algorithm generates a main engine thrust command and an attitude command based on the navigation result and the guidance parameter, and sends the attitude command to the attitude control algorithm. The attitude control algorithm generates a thruster command according to the navigation result and the attitude command. The control computer sends the main engine thrust command and the thruster command to the main engine and the thruster, respectively. The main engine and thruster complete the execution of the commands.

In the process of obstacle inspection and avoidance, the optical imaging sensor sends the lunar surface image of the landing area to the image processing unit, and the three-dimensional imaging sensor sends the distance data corresponding to the landing area landform to the image processing unit. The algorithm for lunar obstacle recognition and safe landing area selection processes the landing area image and the lunar surface data according to the navigation information sent by the control computer, and determines and sends the information on safe landing point (the line of sight of the safe landing point relative to the lander) to the control computer. Then, this information is further input into the guidance algorithm, and the obstacle avoidance mission is completed.

2) Navigation method for powered descending stage

The powered descending phase adopts the navigation algorithm that combines inertial navigation with ranging and velocity measurement correction (as shown Fig. 6.10). In the initial phase of powered descending, pure inertial navigation is adopted. To overcome the impact of terrain uncertainty and ensure the accurate altitude above the Moon after the completion of main deceleration, the range correction is introduced in the late main deceleration. When the rapid adjusting mode is over, the lander maintains a relatively stable attitude toward the lunar surface. As the velocity accuracy requirement becomes increasingly high, the velocity correction is introduced. In the slow descent mode, the engine plume stirs the lunar dust due to the low altitude of the lander, thus adversely affecting the velocity measuring and ranging sensors. Therefore, pure inertial navigation is restored in this phase.

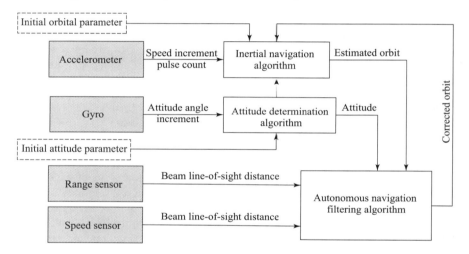

Fig. 6.10 Autonomous navigation algorithm flow

3) Design of guidance law for powered descending phase

Main deceleration guidance: The main mission of this phase is to eliminate the initial horizontal velocity of powered descending phase (about 1.7 km/s), so the optimization of propellant consumption is the main design goal of the guidance law in this phase. In addition, the autonomy and engineering achievability requirements should also be balanced. Based on the linear tangent guidance law, an adaptive dynamic explicit guidance method is adopted. This method not only achieves relatively low propellant consumption but also improves the adaptability of the system to the uncertainties in mass, thrust, specific impulse and other parameters.

Quick adjusting guidance: At the end of the main deceleration phase, the lander attitude is still almost horizontal, and the main engine still operates with the maximum thrust. In the subsequent approaching phase, the lander attitude shall be close to

6.3 Entry and Landing GNC Technology

vertical, the main engine shall operate at low thrust level, and the altitude, velocity and acceleration shall meet a certain relationship. To achieve smooth transition from the main deceleration phase to the approaching phase, the requirements of constant-speed transition of the main engine thrust and the lander attitude shall be observed, and the guidance law of the linear change of the thrust magnitude and direction shall be utilized to meet the requirement of the entry status in the approaching phase.

Approaching guidance: This phase is responsible for connecting the terminal status of the main deceleration phase and the initial condition of the hovering phase to ensure that the optical images of the landing zone are taken and the coarse obstacle avoidance is performed. The quartic polynomial guidance used by the approaching phase of the Apollo mission is selected to ensure that the field of view of the optical imaging sensor can always cover the landing zone. The guidance law allows the landing area to be changed within a certain range, ensuring that the obstacles are evaded in the approaching phase.

Hovering guidance: The main purpose is to finely detect the obstacles in the landing area with the three-dimensional imaging sensor and to give the position information of the safety landing point relative to the sub-satellite point. The initial navigation altitude of the hovering phase is used as the hovering altitude guidance target. The variable thrust engine offsets the lander gravity to ensure that the lander is in a hovering state with zero velocity and stable attitude. The phase plane control mode is applied along horizontal direction, and the PID control mode is adopted along the altitude direction.

Hazard avoidance guidance: The main mission is to avoid obstacles and descent accurately. According to the relative position information of the safety landing point given in the hovering phase, the lander horizontally moves to the altitude above the selected safety landing point, the phase plane control mode is applied along the horizontal direction and the final stage's horizontal velocity is close to zero. The lander descends to about 30 m above the landing point, the PID control mode is applied along the altitude direction and the final stage's descent velocity relative to lunar surface is the preset value (-2 m/s).

Slow descending guidance: It mainly considers the landing safety. To ensure the velocity and attitude control accuracy when touching the lunar surface, the lander descends with a steady attitude and a constant velocity. The horizontal velocity is eliminated until the signal of the shutdown sensor is received. Then, the engine and the thruster are all turned off, and the lander drops to the lunar surface. The phase plane control mode is applied along the horizontal direction, and the PID control mode is applied along the altitude direction to control only the velocity and acceleration.

4) Attitude control method for powered descending phase

The error quaternion is calculated based on the current attitude quaternion, and the target attitude quaternion is given by the guidance law to obtain the angle deviation of the three axes of the body. The thruster jet control is performed using the PID/PI

+ PWM attitude control system. According to the control torque required in PID/PI calculation, the 10 N and 150 N thrusters are reasonably selected for pulse width modulation.

5) Method of obstacle identification and safety landing point selection

Safe-point recognition is enabled by optical imaging and 3D imaging. [6]

Coarse obstacle identification and safe-landing zone selection based on optical images: Two-dimensional optical images are acquired first, and then, the large obstacles are identified by using the images of lunar rocks and craters. The accessibility is evaluated based on propellant consumption. Finally, the safety zone is determined.

Fine obstacle recognition and safety landing zone selection based on 3D images: In the hovering state, the 3D imaging sensor performs 3D imaging of the landing area to obtain high-resolution slant-range data information relative to the lunar landing area. By using the algorithm for fine obstacle recognition and safety landing zone selection, the data is processed to accurately identify the obstacles on lunar surface and select the safety landing zone.

6) Autonomous fault diagnosis and treatment

In addition to the conventional component-level autonomous diagnosis and treatment, a system-level autonomous diagnosis and system reconstruction strategy are designed for different phases of powered descending:

Main deceleration phase: If the attitude in the main deceleration phase is continuously out of tolerance due to the disturbance torque that is greater than the control torque, the strategy of reducing the disturbance torque by switching to the variable thrust of 5000 N or 5000–1500 N is required to be adopted.

Approaching phase: During the approaching and descending process, if the deviation of the actual trajectory from the nominal trajectory is relatively large, the guidance law shall be switch to the safety mode.

Hovering phase and later phase: If the velocity in the *Y-axis* or Z-direction of lander body is out of tolerance, the horizontal position control shall be added to reach the safe landing point.

When the propellant is insufficient: If the propellant is insufficient in the main deceleration phase, the approaching phase shall be cancelled; if the propellant is insufficient in the approaching phase, the time of hovering and obstacle avoidance shall be reduced.

Design principle of the guidance law in safety mode: (1) Priority is given to vertical safety; (2) propellant consumption is as little as possible, and only velocity and acceleration, instead of position, are controlled in the horizontal direction to eliminate horizontal velocity as quickly as possible and reduce the landing risks.

6.4 GNC Technology in Celestial Body Surface Roving

6.4.1 Features of Rover GNC

The navigation of the rover is generally referred to as positioning, that is, determining the position of the rover on the surface of the celestial body. It is divided into global positioning and local positioning. The global positioning is to determine the position of the rover in the geographic coordinate system fixed to the celestial body to be explored. The local positioning is to determine the position of the rover relative to the local reference point (such as the lander or the detecting station).

The guidance of the rover is also generally referred to as path planning. Under the given cost functions and constraints, the path point sequence of the rover is calculated, that is, a path from the starting point to the target point is found. The cost function can be selected as energy function, distance function, etc. The constraints may include velocity constraint, acceleration amplitude constraint as well as the constraints on the turning radius and braking time for obstacle avoidance. Path planning includes global planning and local obstacle avoidance.

The control of the rover is to acquire the motion state of the rover in real time, calculate the control law and give the real-time control command to the rover actuator so that the rover can reach the designated target area from the starting point along the planned path (orbit). In addition, the control of the rover includes the low-level motor servo control of the actuating wheels and steering mechanism, the pointing control of the directional antenna and solar panel, the pointing control of the mast and gimbal for holding the camera and other loads, and the coordinated control of the manipulator motion, etc.

The rover GNC has the following features:

(1) The rover vehicle travels on the surface of an unknown or partially known celestial body to be explored, and the environment is sensed and modeled by onboard sensors to identify obstacles and find a safe route to the target point;
(2) The communication delay caused by the distance between the rover and the Earth requires that the rover GNC must have a certain degree of autonomy;
(3) The trajectory of the rover is unpredictable, and the motion performance of the rover is closely related to the topography and soil characteristics of the surveyed area. There are unpredictable factors such as slipping and skidding, which make it difficult to obtain an accurate dynamic model for the rover and bring challenges to the control of the rover;
(4) The product must adapt to the space environment for launching, flying, entry and landing, and it should also adapt to the environment of the exploration area after reaching the surface of the celestial body to be explored.

Therefore, the navigation and control technology of the rover is a new research topic in the space control technology, and is one of the key technologies for the rover to complete the roving exploration mission. The most complicated and extremely difficult key technology to be addressed by the GNC system is that under the constraint of the onboard computing resources, the rover has the ability to safely travel a certain distance when it independently explores the surface topography with the nominal complexity, while maintaining the position information of the rover with a certain precision (such as 10% of the travel distance) through pose determination, environment perception, path planning, motion control, etc.

6.4.2 Environmental Perception

Vision is the most direct and effective means to sense the unknown environment. Automatic path planning based on visual information is the key to improving the intelligence and autonomy of the rover. So to speak, the level of environmental perception largely determines the autonomy of the rover. The environment-sensing and modeling approaches applicable to the rover mainly include laser ranging method and binocular stereo vision method.

Myron Z. Brown et al. believed that the stereo vision technology had matured in the early 1990s. They comprehensively reviewed three important contents in stereo vision research, namely matching method, occlusion processing and real-time implementation.

The following is a description of the binocular stereo vision algorithm taking MER as an example. Among the MER sensors applied to measure the surrounding terrain, the most powerful sensor is the combination of any available camera pair and the software that processes the camera images. Four pairs of cameras can provide different fields of view. The MER stereo vision software generates the 3D measurements of the images by applying the camera geometry lens correction and the one-dimensional search of the matching metric with SAD (absolute error sum). The results, in the form of 3D point cloud, are delivered to the terrain analysis software for safety performance prediction of the rover. So stereo vision software is an important part of the MER autonomous navigation software.

Stereo vision is mainly used to support the autonomous terrain analysis of the rover. The original image obtained from the camera hardware has 256×1024 pixels and a 12-bit gray value, and is sampled by software into an image with 256×256 pixels and an 8-bit gray value.

The typical stereo vision algorithm steps are as follows:

(1) Input image: The original image obtained from the camera hardware has 1024×256 pixels and a 12-bit gray value, and is sampled by software into an image with 256×256 pixels and an 8-bit gray value image (pixel gray value average). This phenomenon is called the "pyramid" pixel reduction, where each layer is reduced to 1/8 of the original calculation at the cost of reducing the depth resolution of the distance by 1/2;
(2) Epipolar rectification: To reduce the subsequent matching time and correct the left and right images to eliminate the influence of lens distortions (Fig. 6.11b);

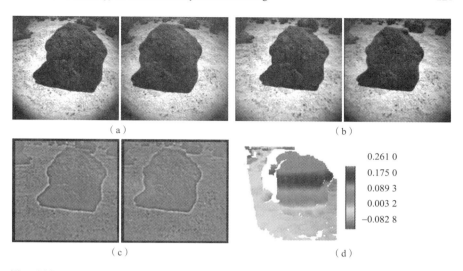

Fig. 6.11 Process of visual algorithm [7]: **a** original image; **b** corrected image; **c** Laplacian transformed image; **d** elevation map result

(3) Perform a Laplace transform on each image to eliminate the luminance difference of the pixel, by actually using a differential Laplace transform with higher computational efficiency, that is, DOG;
(4) Perform a 1D correlation search on the filtered image with the matching window of 7×7, search for each pixel of the left camera image in the right image, and score each possible matching point. The range of the pixel points to be searched is called the disparity range, which is geometrically determined by the input depth value range. The point with the highest score is selected as the matching point, and the corresponding distance estimate is calculated by the camera model;
(5) Match point inspection: Reject unreliable matching points through multiple checks, such as peak filter and reverse matching check. Only the matching correlation curve with a unique minimum value can be judged valid;
(6) Eliminate isolated points: This process is also known as Blob Filter;
(7) Map the disparity values to the X, Y and Z values of the 3D space by using the geometric model of the camera, as shown in Fig. 6.11d.

6.4.3 Position/Attitude Determination and Estimation

For the rover, the positioning capability (i.e., determination of position, attitude and heading) is essential. The positioning capability of the rover determines the size of the exploration range and the accuracy of motion control accordingly. It is also a necessary condition for path planning and autonomous obstacle avoidance.

Presently, the methods of rover positioning are mainly divided into star sensor-based positioning, UHF Doppler tracking-based positioning, natural landmark positioning, "inertial navigation system + odometer" positioning, visual odometry positioning, bundle-adjustment-based positioning, etc. The following focuses on the introduction of visual odometry positioning technology.

It can be seen from the operation mode of MER that in the relatively flat terrain, the use of only IMU and the wheel coder can give a fairly excellent position estimation value. For example, the deviation of the "Spirit" is only 3% after moving for over 2 km. However, in the steep hillsides, the mixed terrain of sand and gravel in the crater, and even the sand dune in the Meridian plain, there is a serious slip. At this time, the only way for MER to detect slip and give the exact location information of the rover is to rely on the navigation camera and visual odometry software.

The visual range of MER applies an orientable navigation camera with a 45° field of view installed on a mast to capture the image pairs, and automatically selects and tracks the topographical features between the two pairs of images (256 × 256 pixels) by in-orbit software. And motion estimation is conducted to obtain the position and attitude of the rover (x, y, z, roll, pitch, yaw). It has been proved that the visual odometry has excellent performance [18]: high convergence rate (Courage: 97%, Opportunity: 95%). With a slope of 31°, it can successfully detect a slip up to 125% and a position change of at least 2 mm. In the actual operation process, the visual odometry has been developed from "the icing on the cake" into the key guarantee for MER safe travel.

The basic visual odometry algorithm can be divided into the following steps: feature detection, feature stereo matching, feature orbiting and robust motion estimation [7].

Feature exploration: Firstly, select, from a pair of images, the feature points that are easy to match and easy to track in the image sequence, and apply image feature operators (such as Forstner, Harris) to one image in the image pair to select the points with high feature evaluation values. To reduce the amount of calculation, a grid with a cell size smaller than the minimum distance of the feature point is superimposed on the left image. Within each grid cell, the point with the most prominent corner feature is selected as an alternative feature point. All the candidate feature points of the image are arranged in a descending order, and the feature points are selected according to their sequence in the list. The constraint on the minimum distance of the feature points is added to ensure the uniformity of the selected feature points in the image.

Feature stereo matching: Calculate the 3D position of the selected feature points by stereo matching. As the camera pair has been calibrated, the feature points can deviate from the polar line by a few pixels for matching purpose. The quasi-normalized correlation coefficient is used as the matching metric to determine the best matching point. To obtain sub-pixel precision, a binary fourth-order polynomial is fitted next to the associated 3×3 pixels, and the extremum of the polynomial is used as the matching correlation value. Ideally, the rays from the left and right images to the same feature point should be intersected in space. They do not necessarily intersect due to image noise or matching error. The distance between the two rays also indicates the quality of the matching. In further processing, the feature points with large distances are removed. In addition, this distance is also used as an error model factor for the calculation of cross-correlation matrices.

Feature tracking: After the rover travels a certain distance, the second pair of images is acquired. Based on the approximate positioning information given by the wheel odometer, the feature points of the previous pair of images can be projected into the second pair of images. Based on the correlation search, the 2D position of the feature point is accurately determined in the second pair of images. The tracked feature points are then stereo-matched on the second pair of images to determine their new 3D position. As the 3D position of these feature points is already known in the previous step, the search range in their stereo matching is greatly reduced.

Robust motion estimation: If the initial motion is known exactly, the difference between the estimated values of the two 3D positions should be within the specified error ellipse. However, when the initial motion deviates, the difference between the two position estimates reflects the error of the initial motion estimate and can be used to determine the change in the position of the rover. Motion estimation can be done in two steps. The first step is to use the least square method with a closed-form solution to obtain a slightly less accurate motion estimate. The advantage of the least square method is simple, fast and robust. Its disadvantage is low accuracy, but its computational cost is small. The second step is to use the nonlinear optimization algorithm to obtain more accurate motion estimates.

6.4.4 Path Planning

In terms of path planning algorithm, the rover is similar to the ground mobile robot. The commonly used algorithms include artificial potential field method, A^* and D^* algorithm, Morphin algorithm, Tangent-Bug and Wedge-Bug algorithm, etc. The MER path planning algorithm uses the GESTALT derived from the Morphin algorithm.

The data source for terrain analysis is the ranging data obtained by stereo vision. The principle is "overriding old data with new data." This is used to solve the following two problems. Firstly, the accuracy of the ranging data for the objects far from the rover is always lower than that of the ranging data when the rover is closer, and overriding old data with new data can eliminate erroneous evaluation arising from the ranging error. Secondly, the new data overriding can enhance the navigation robustness of the rover to the positioning error, as the latest 3D data information is always selected to avoid the possible errors of the old data.

The GESTALT terrain analysis method is to use the ranging data for local plane fitting. The terrain evaluation is based on the difference between each terrain block of rover size and the reference plane. It is necessary to analyze a rover-sized terrain block. All (X, Y, Z) points within this terrain block are fitted into a plane, and the parameters of the fitted plane are used to evaluate the safety of the rover when it is located at any direction. The continuously varying suitability values are derived from several independently calculated filter output values, including the step filter, tilt filter and roughness filter. The value to evaluate the suitability of a unit is the most conservative output value of the above filter. The terrain suitability map for path selection is finally obtained. Therefore, finding a safe travel path is just to find a path with cell width on the suitability map. In Figs. 6.6, 6.7, 6.8, 6.9, 6.10, 6.11 and 6.12, the red area indicates the obstacle and the yellow/orange area indicates the area that the rover can pass.

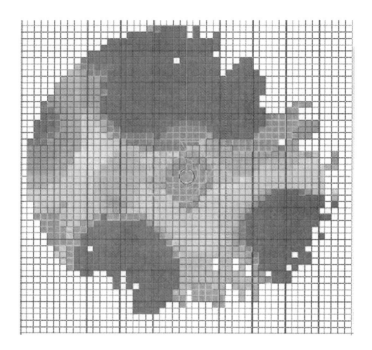

Fig. 6.12 Example of suitability graph [8]

6.4.5 Motion Control

The motion control of the rover is essential to the roving exploration mission and is closely related to the motion performance optimization design of the rover in the natural unstructured environment of the planetary surface, the local planning and safe trajectory generation of the rover as well as the energy optimization of the rover movement, and the realization of scientific goals. Long-distance autonomous navigation on the natural terrain of the surface of the celestial body is still a challenging problem for the rover. To navigate autonomously a few hundred meters without human intervention, the rover is required to have the following capabilities: giving

6.4 GNC Technology in Celestial Body Surface Roving

different representations of the environment, making trajectory planning and execution planning according to different terrains, controlling its own movements and positioning itself.

For motion control, whatever methods are used to generate the motion trajectory, the rover performs the same process to execute the generated trajectory. The command signal of converting the basic geometric motion into a line velocity and angular velocity pair based on the specified velocity is periodically updated and delivered to the servo controller of the low-level motor of the moving mechanism. It determines the commanded velocity of each of the six wheels. The angle and angular velocity of the wheels are servo controlled through the feedback of the goniometer. When the rover travels in less rugged terrain, this simple motion control algorithm can produce satisfactory control effects in most cases. However, if there is rugged terrain, more complex motion control algorithms are required to avoid slipping, skidding and the reduction of wheel grip force. Therefore, it is a quite important issue to monitor incorrect motion behavior when moving on rough terrain. This is related to the travel distance of the rover, the use of external sensory sensors to make estimates and other aspects. Or, some novel mobile mechanism forms may be applied to enhance the adaptability to rugged terrain. However, this also brings challenging problems to motion control.

The rover may suffer serious slipping and skidding when exploring the soft and rugged surface of the celestial body. To adapt to the terrain, the load assigned to each wheel is different. Therefore, to optimally coordinate the driving and steering of the wheels and optimize the driving efficiency, the driving and steering control technology of the rover should be studied to achieve energy optimization control. The issues involved mainly include:

(1) Steering and driving control based on kinematic velocity decomposition;
(2) Decoupling control based on full steering and driving dynamics of the rover;
(3) Sliding rate optimization control based on simplified dynamic model;
(4) Kinematics control based on slipping and skidding estimation.

Major international research institutions, such as JPL, LAAS, CMU, MIT, EPFL and TOHUKU, are conducting in-depth research and testing on the motion control of the rover to develop a control method with better performance in ground test and examination. However, such control method has not been verified by flight tests. According to the published literature, the motion control algorithm currently used on the rover is basically the velocity decomposition and synchronization control method based on kinematic inverse solution. The velocity synchronization algorithm is described below.

When the rover moves, the rocker suspension passively adapts to the terrain, and each wheel stays in contact with the surface as much as possible. Each of the six wheels experiences different loads. In some configurations, the problem of "fighting" between the wheels may occur. Or, it may be described in the following way. Each wheel is in a closed-loop control state to drive the wheel to follow its own command velocity (calculated by the inverse kinematics of the rover). When the load conditions of the wheels are different, the interaction between the wheels will cause one wheel to accelerate to the commanded velocity and the other wheel to accelerate to the given value, so that all six wheels run at the velocities different from the commanded velocity. This results in an increase in the total energy consumed by the rover and excessive slippage of the wheels, as well as a side slip during the obstacle crossing. Therefore, JPL developed a velocity synchronization algorithm for FIDO [10].

The first step in the algorithm is to eliminate the outliers. When running along the nominal velocity curve, it is necessary to determine which of the six wheels is farthest from the closed-loop position. If θ_m is the translational distance of the m-th ($m = 1,\ldots,6$) wheel, the deviation of the distance generated by one wheel from the distance generated by all other wheels will be:

$$\Delta_m = \sum_{i=1, i \neq m}^{6} |\theta_m - \theta_i| \tag{6.6}$$

The average of the deviations is:

$$\Delta_{\text{avg}} = \frac{1}{6} \sum_{i=1}^{6} \Delta_m \tag{6.7}$$

Then, use the voting method to determine the composite scores of the wheels that match each other better. In other words, the composite score attempts to identify the outlier and excludes the outlier when calculating the average distance of the wheel without the outlier. The composite score s is determined by:

$$s = \sum_{m=1}^{6} \begin{cases} \theta_m, & \Delta_m \leq \Delta_{\text{avg}} \\ 0, & \text{other} \end{cases} \tag{6.8}$$

And when calculating the above formula, a counter c is set. Whenever $\Delta_m \leq \Delta_{\text{avg}}$ occurs, 1 is added. Finally, the voted average distance is $\theta_{\text{vote}} = s/c$.

The final step in the velocity synchronization algorithm is to calculate the change in commanded velocity required for the m ($m = 1,\ldots,6$) wheels relative to the nominal velocity setpoint according to the following equation.

$$\delta_m = |\theta_{\text{vote}} - \theta_m| / \theta_{\text{vote}} \tag{6.9}$$

$$V_m = V_{\text{nominal}} + \Delta V \tag{6.10}$$

$$\Delta V = \begin{cases} \delta_m V_{\text{nominal}}, & \delta_m V_{\text{nominal}} \leq \Delta V_{\text{threhold}} \\ \Delta V_{\text{threhold}}, & \text{other} \end{cases} \tag{6.11}$$

where V_{nominal} is the rated velocity setpoint for each wheel; $V_{\text{threshold}}$ represents the dynamic range of velocity change that can be adjusted by the closed-loop control.

The result of using the velocity synchronization algorithm described above is that the required energy is significantly reduced compared to using a strict asynchronous closed-loop control method. Several tests to evaluate the driving performance of the rover on typical terrain have proved that the slippage rate of the driving wheel of the rover is small and the six wheels can start and stop at the same time with higher synchronism.

6.4.6 Design of GNC System for Celestial Body Surface Roving

1. **Mission requirements**

Due to the uncertainty in the surface topography and soil characteristics of the detected extraterrestrial celestial body as well as the constraints of thermal control, energy and communication, the rover should rove and detect the surface of the celestial body in a mode that is changed from "remote operation-oriented" to "all-autonomous-oriented." The main missions of the GNC system are:

1) Position/attitude determination

Determine the three-axis attitude of the rover's rolling, pitching and yaw; determine the position of the rover relative to the fixed-point coordinate system of the celestial body surface, and calculate the moving mileage in real time.

2) Environmental perception

According to the Earth command, the navigation camera can be controlled to perform sequence imaging to obtain a wide range of environmental information around the rover. A self-controlled obstacle avoidance camera is provided to acquire the information of the short-distance obstacles around the rover. The camera can adjust the optimal imaging parameters according to the surrounding environment.

3) Path planning

According to the environmental perception results and other sensor measurement information, an environmental model for planning is established, and then a collision-free safety path from the current location to the target location (area) is planned according to the motion capability of the rover.

4) Motion control

According to the results of the autonomous planning on the rover or the ground-commanded results, make coordinated control of the motion of the suspension, driving wheels and steering wheels.

(1) Wheel coordinated motion control includes the control on blind travel, autonomous obstacle avoidance function, etc;
(2) This control is applicable to the coordinated motion control of the rover wheel and suspension in special situations (seriously submerged and slippery terrain, or under abnormal roving conditions) to achieve the mobile forms such as crab travel, creeping and leg-type walking;
(3) For local attitude adjustment of the rover, the local adjustment of rover attitude and heading is enabled by adjusting the rotating angles of the joints of the active suspension, the wheel craters, the in-situ steering, etc.
5) Safety monitoring

The attitude of the rover is monitored in real time to keep it within a safe range to avoid overturn. In case of an emergency during the rover travel, an emergency response command is issued.

2. **System Composition**

Take the CNC system of the Chang'e 3 rover as an example. The GNC system consists mainly of the sensor, actuator, control computer and navigation control software (as shown in Fig. 6.13).

Sensors: Typical sensors include sun sensors, cameras (navigation cameras, obstacle avoidance cameras), inertial measurement units (IMUs), etc. Meanwhile, it is necessary to obtain the measured values of driving wheel velocity, steering-wheel angle, steering-wheel turning velocity, rocker angle, deck pitch angle and mast yaw angle. The above data is generally collected and converted by the related system and sent to the GNC control computer.

Actuators mainly include wheels, mast mechanisms, etc. Such actuators generally belong to other subsystems of the rover.

Control computer is responsible for the data interaction with other systems. It manages the power on/off and communication of the inertial measurement unit (IMU), solar sensor, navigation camera and obstacle avoidance camera in the GNC system as well as the transmission, processing and storage of data and images, etc.

Navigation control software runs on the control computer and is an important guarantee for the GNC system to achieve intelligent and autonomous operation. The GNC functions such as camera imaging service, position/attitude determination, visual processing and obstacle recognition, planning, autonomous movement control, fault protection design, and software correction and update are all implemented through complex software.

3. **Operating mode**

The communication delay caused by the distance between the rover and the Earth requires the rover GNC to have a certain degree of autonomy in order to improve operational efficiency, reduce the operating costs and improve the mission safety. However, the limited computing power, security requirements and other factors of the GNC computer limit the complexity of the algorithms available for GNC, such as

6.4 GNC Technology in Celestial Body Surface Roving

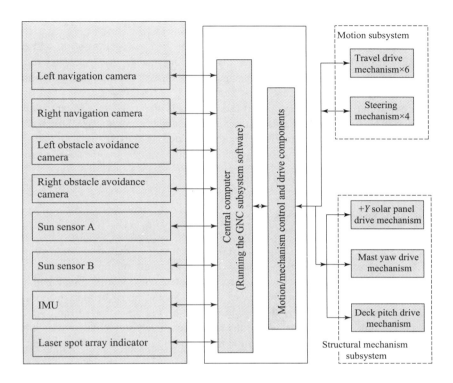

Fig. 6.13 Composition of the GNC system of the Chang'e 3 rover

binocular vision, planning and visual odometry. Therefore, the current control mode of the rover is basically a "teleoperation combined with autonomous operation," which is also called the semi-autonomous mode. "Semi-autonomous mode" refers to that the visualization of remote planning, command sequences, rover activity sequences and related data products is performed by an operation group composed of ground engineers and scientists.

After the rover receives the command sequence from the ground, the following three control levels, namely bottom-level command control, direct driving element and autonomous path selection, are generally applied to control the motion of the rover. The bottom-level command control is to directly control the movement of the actuating wheel motors and the steering mechanism motors. Direct driving element is to control the arc path of the rover (linear and in-situ steering is a special case) and is used by the operator to specify the movement sequence of the rover. Autonomous path selection (automatic navigation) is, according to the current state (including the use of the attitude determination and optional visual odometry positioning of IMU), to effectively select the "direct driving element," that is, to select the local safety path that can control the rover to arrive at the position point designated by the operator in the Descartes space coordinates.

In the three levels of control, the attitude of the rover is monitored in real time to keep it within the safe range and avoid overturning, that is, under reactive hazard detection. In case of an emergency during the movement, an emergency command is issued to implement emergency treatment.

In the design of the GNC system of the rover, it is also necessary to consider the motion control of the mast, the directional antenna and the robot arm on the basis of the mobility control, the realization of functions such as camera imaging and IMU real-time calibration as well as the design of specific modes and the logic of mode switching.

4. **Scheme design points**

1) Attitude determination scheme

Generally, the method of dynamic and static combination is used to determine the attitude, as specifically described below. Before the start of the rover travel, the accelerometer and the sun sensor are used to determine the roll, pitch and yaw attitudes, and the gyro constant bias is calibrated. When the rover is in motion, the gyro output is compensated by the previously calibrated deviation of the gyro and the triaxial attitude is estimated. After the rover movement stops, the accelerometer and the sun sensor are used to determine the roll, pitch and yaw attitudes, and the attitude estimation error is corrected.

2) Position determination scheme

Position determination includes determining the global position of the rover on the surface of the celestial body and its position relative to a certain starting point. The global position determination is generally done by the operators on the Earth and uplinked to the rover. During the travel of the rover, the Earth station periodically updates the position of the rover relative to the lander and uplinks it to the rover.

The local positioning method of the Chang'e 3 rover is the dead reckoning positioning method based on the wheel odometry and introduces the rocker angle, pitch and rolling attitudes to solve the altitude information. The positive kinematics relationship is used to compensate the error caused by the topographic change. The motion mismatch between wheels is reduced by the coordinated control of the wheel system. The slip rate of the rover under various motion conditions is obtained by simulation and experiment and is compensated in the dead reckoning algorithm.

3) Environmental perception scheme

(1) Long-distance panoramic image.

The navigation camera performs pitch and yaw motion along with the mast and images the rover for 360° around with the effective sensing depth of about 10 m. The long-distance panoramic image is used for global path planning on the Earth.

6.4 GNC Technology in Celestial Body Surface Roving

The global path planning generates the target point and path point coordinates or curvature driving instructions and uplinks such information to the GNC system.

(2) Obtain dense terrain information based on binocular stereo vision algorithm.

After receiving the path point and target point commands sent from the ground, a safe path is found between the two path points, that is, local path planning is performed. Dense terrain information is used for the autonomous local obstacle avoidance planning of the rover. Before and during the motion, the left and right obstacle avoidance cameras are used to image the front of the rover, and the dense terrain information is obtained through image rectification, filtering, matching, three-dimensional recovery and other operations.

4) Path planning scheme

The camera image is used to obtain the geometric information of the surrounding terrains, and the environment model in the form of local terrain goodness map is built by evaluating the local terrain. The algorithm evaluates the selected multiple arc paths on the suitability map and selects the path with the highest comprehensive evaluation value as the next safe path. Considering the turning ability of the rover during the travel, there should be no obstacles existing within a certain radius of the target point input from the Earth, so that the GNC system can safely travel to the target point through autonomous planning and motion control.

5) Motion control

Based on the functions of the mobile subsystem, the coordinated control of the wheels is achieved according to different motion requirements. Considering the movement velocity of the rover and the constraints of computing resources on the rover, the wheel coordination control of the Chang'e 3 rover mainly adopts the kinematics model based method, and off-line verification of the coordination control law is made based on dynamic simulation.

The coordinated control of the wheel system motion realized by the Chang'e 3 rover includes the control of rover position and heading angle as well as the coordinated control of driving wheel velocity and steering-wheel turning velocity.

(1) Control of the overall rover velocity and heading angle (yaw angle). According to the current position and attitude information of the rover as well as the motion requirements given by the path planning or Earth command, the desired heading and velocity of the rover are given by the control law design. Due to terrain changes, slipping, skidding and other effects, the rover may deviate from the set heading and trajectory. The GNC subsystem will control the heading angle to the desired value through the closed-loop control of the heading angle, so that the rover will travel along the required trajectory.

(2) Coordinated control of drive wheel velocity and steering-wheel turning velocity. The desired heading and velocity of the rover are decomposed into the desired steering angle of each steering wheel and the desired rotational velocity of the

driving wheel. Meanwhile, the desired steering angular velocity of the steering wheel is given according to the current steering angle of the steering wheel, in order to realize the coordinated control of the driving wheel velocity and the steering-wheel steering velocity is enabled accordingly.

Through the two-layer control of the whole rover and the wheel system, the requirements of the rover's turning in place, the movement along the fixed curvature (including the straight line) and the arrival at the target point along the path point can be satisfied. Each control cycle gives the expected values of the rotating speed of each drive wheel and of the turning speed of each steering wheel as an input to the motor low-level control.

6) Safety monitoring scheme

Perform the safety monitoring of the rover to a certain extent based on available information and issue a stop instruction in case of detection of a hazard or obstacle. Please refer to the Table 6.5 for the safety monitoring scheme of the Chang'e 3 rover design.

Table 6.5 Safety monitoring scheme of Chang'e 3 rover

No.	Stop conditions	Stop method
1	Horizontal tilt out of limit	Emergency stop
2	Anomaly of discrete terrain information extraction	Emergency stop
3	Danger detected by laser	Emergency stop
4	Dense terrain restoration failure	Emergency stop
5	Local obstacle avoidance planning without effective path output	Emergency stop
6	Local obstacle avoidance planning time out of limit	Emergency stop
7	Motion instructions not meeting the control requirements	Emergency stop
8	Continuous deviation of wheel system feedback from GNC command requirements	Emergency stop
9	Continuous movement time out of limit	Emergency stop
10	Movement mileage and angular displacement out of limit	Emergency stop
11	The yaw gyro is faulty, the outputs of the two sun sensors differ greatly during the movement, and the rover cannot make judgment autonomously	Emergency stop

6.5 Outlook

From the perspective of the development trend of deep space exploration, the GNC technology for deep space exploration develops mainly toward the direction of high precision, high reliability and strong autonomy. The autonomous navigation based on optical imaging measurement is the main means to achieve autonomous operation

6.5 Outlook

in deep space exploration missions. With the development of computing technology, there are potential space for the improvement of the image processing accuracy, rapidity and robustness. Meanwhile, with the emergence of new measurement principles, the absolute navigation technologies such as X-ray pulsar navigation are also constantly evolving, and in theory can obtain the performance indicators superior to imaging navigation. On this basis, to adapt to the increasingly diversified and complicated deep space exploration missions, the development of a single navigation method toward the integrated navigation system with multi-source information fusion is an important direction. To cope with the limited system resources in deep space exploration, the products such as navigation sensors are rapidly developing toward miniaturization, modularization, integration and unification. To ensure the reliability and safety of deep space probes and to improve the efficiency of autonomous fault diagnosis of GNC systems, the improvement of autonomous multi-fault/soft-fault diagnosis level as well as self-repairing capability is the main development trend in the future.

In addition, with the rapid development of new propulsion technologies such as electric propulsion and non-working-medium propulsion, the corresponding autonomous control and orbit optimization technologies will play an important role in future deep space exploration missions. It is foreseeable that with the continuous development and implementation of deep space exploration missions, a series of distinctive and forward-looking autonomous GNC technologies oriented to deep space exploration will surely make vigorous development and even greater breakthroughs.

References and Related Reading

1. Kizner W (1961) A method of describing miss distances for Lunar and interplanetary trajectories. Planetary Space Sci 7:125–131
2. Howell KC, Pernicka HJ (1988) Numerical determination of Lissajous trajectories in the restricted three-body problem. Celestial Mech 41:107–124
3. Tu S (2001) Satellite attitude dynamics and control. China Aerospace Publishing House, Beijing (Chinese vision)
4. Huang J, Zong H, Li J et al (2013) High-reliability and high-precision orbit control technology for Chang'e-2 satellite. Scientia Sinica Technologica 43(7):727–732 (Chinese vision)
5. Gavin MF, Craig LE (2011) Entry guidance for the 2011 Mars science laboratory mission. AIAA, Portland
6. Wang D, Li J, Huang X et al (2014) A pinpoint autonomous navigation and hazard avoidance method for Lunar soft landing. J Deep Space Exploration 1(1):44–51 (Chinese vision)
7. Goldberg SB, Maimone MW, Matthies L (2002) Stereo vision and Rover navigation software for planetary exploration. In: IEEE aerospace conference proceedings, March 2002, Big Sky, Montana, USA
8. Biesiadecki JJ, Maimone MW (2006) The Mars exploration Rover surface mobility flight software: driving ambition. In: Proceedings of IEEE aerospace conference, Big Sky, MT
9. Baumgartner ET, Aghazarian H, Trebi-Ollennu A (2001) Rover localization results for the FIDO Rover. In: Proceedings SPIE photonics east conference, October 2001
10. Wang X et al (2009) Spacecraft entry and return technology (Part 1). China Aerospace Publishing House, Beijing (Chinese vision)

11. Xi X, Zeng G, Ren X et al (2016) Orbit design of Lunar probt. National Defense Industry Press, Beijing (Chinese vision)
12. Guo M, Li M, Huang X et al (2017) On guidance algorithm for Martian atmospheric entry in nonconforming terminal constraints. J Deep Space Exploration 4(2):184–189 (Chinese vision)
13. Huang X, Zhang H, Wang D et al (2017) Autonomous navigation and guidance for Chang'e-3 soft landing. J Deep Space Exploration 4(2):184–189 (Chinese vision)
14. Sun Z, Jia Y, Zhang H (2013) Technological advancements and promotion roles of Chang'e-3 lunar probe mission. Scientia Sinica Technologica 43(11):1186–1192 (Chinese vision)
15. Wang D, Guo M (2015) Review of spacecraft entry guidance. J Astronautics 36(1):1–8 (Chinese vision)
16. Zhang H (2015) Theories and methods of spacecraft orbital mechanics. Beijing: National Defense Industry Press (Chinese vision)
17. Peralta F, Flanagan S (1995) Cassini interplanetary trajectory design. Control Eng Pract 3(11):1603–1610
18. Maimone MW, Leger PC, Biesiadecki JJ (2007) Overview of the Mars exploration Rovers autonomous mobility and vision capabilities. In: IEEE international conference on robotics and automation, space robotics workshop, Roma, Italy, 14 April 2007

Chapter 7
Atmospheric Braking Technology

7.1 Introduction

Atmospheric braking plays a key role in the process of entry, descent and landing (EDL) to atmospheric planets. The main functions required for atmospheric braking include braking to a specific terminal descent speed, providing stability (preventing the parachute from rolling or meeting the pointing requirements of instruments and devices) as well as providing differences in the ballistic coefficients of various components for separation and providing the altitude and time scale required for the EDL process.

Safe entry involves multiple disciplines and specialties, including aerodynamics, flight mechanics, thermal structural mechanics, guidance navigation and control, parachute. The overall performance of atmospheric braking depends on the technical development level of each system on one hand and on the other hand, the complex interaction relationship among different systems. It also requires an integrated and integrated design that is well-connected and optimized. In the conceptual/preliminary design phase of extraterrestrial celestial atmospheric braking, the system design mainly addresses three missions:

(1) According to the overall mission objectives, evaluate the ballistic/semi-ballistic/lifting entry schemes, demonstrate the basic aerodynamic shape, clarify the deceleration means and technical approaches for each altitude section, define the key mission point indicators such as parachute-opening point and stable descent speed/altitude and develop an outline of general plan for the entry mission.
(2) Work on the indications of environmental and mission constraints such as atmospheric environment, aerodynamics, aerodynamic thermal environment and location of the landing area and propose preliminary performance requirements for systems such as aerodynamics, structure, propulsion, control and thermal protection, and conduct system analysis and design.

(3) The simulation method is used to comprehensively check and evaluate the indicator research and design results of each system. Through continuous iteration and correction, the comprehensive optimization of the entry index system is completed, the flight control sequence and mission profile of the EDL process are determine, the functional performance requirements on other systems such as thermal control, measurement and control and mechanisms are put forward.

In case of entering the atmospheric celestial body, it is necessary to solve the multi-objective design optimization problem: It must meet multiple the top-level design objectives such as the overall quality, entry overload, aerodynamic thermal protection, landing accuracy, entry sequence and elevation margin. However, different objectives are mutually restricting and conflicting. Therefore, iterations, trade-offs and balancing should be carried out in the whole mission design framework. Generally, compared with the Earth reentry, the aerodynamic braking of deep space mission has some special characteristics or priority issues. For example, the orbit determination accuracy of deep space probe is low, which directly affects the navigation control and the parachute-opening strategy in the entry process. The atmospheres models of extraterrestrial celestial bodies have great uncertainty, and the entire aerodynamic braking capability should be available with sufficient margin. The risk of the braking task is relatively high, and the proper balance should be made between simple and reliable methods and complicated means (such as lift control). The deep space exploration missions are generally costly in terms of launching, and the mass constraints of the probe are relatively high, which limits the application of multiple redundancy and conservative design to a certain extent and adversely affect the inherent reliability of the probe. Different celestial body entry missions require different focuses in aerodynamic braking: When entering the celestial body with dense atmosphere, the aerodynamic thermal environment indication and thermal protection should be the main problems; while in case of entry into a celestial body with think atmosphere, the available altitude and aerodynamic resistance performance of aerodynamic braking should be the first design objective.

There is a problem of multi-system coupling design when entering the atmosphere: Atmospheric entry involves multidisciplinary optimizations, and the mutual constraints among different disciplines are relatively stringent. The aerodynamic shape design largely determines the heat flux distribution on the surface of the probe, and the ablation and back-off of the thermal protection material in turn affect the aerodynamic shape. In case of poor transonic–supersonic dynamic and static stabilities, the problem of relatively low attitude control margin may be avoided by opening the parachute in the supersonic velocity. However, this would result in infinite mass parachute-opening effect, and excessively great interference will lead to control saturation. One of the objectives of the guidance and control is to ensure the state of the parachute-opening point; however, the quality of the parachute-opening conditions will in turn affect the attitude of the probe by the force of the parachute cord. In order to improve braking performance, increasing parachute area may be an option. However, with the increase of the parachute-opening force, the bearing requirements of the structure is further increased, and the ballistic coefficient of the probe may

7.1 Introduction

be reduced. The resistance performance and stability of the aerodynamic shape are contradictory, so are the resistance performance and stability of the parachute. Both relates to the entry control design.

This chapter focuses on the main design processes, elements and methods of aerodynamics, aerodynamic thermal protection, control, parachute systems in atmospheric braking technology.

7.2 Aerodynamics and Aerodynamic Analysis

7.2.1 Basic Concepts of Aerodynamics

1. **Continuous flow and discontinuous flow**

All classical aerodynamic research approaches are based on the assumption of continuous media. From a microscopic point of view, there is a gap between neighboring gas molecules (expressed by molecular free path). However, generally, if an object disturbs the gas, it presents changes in macroscopic properties such as pressure, density, shear stress, velocity, viscosity, temperature, internal energy and thermal conductivity instead of the behavior of individual molecules. At this point, the gas may be regarded as a continuous medium consisting of a continuous mass of particles with no gaps between each other (i.e., the continuous medium assumption), and the corresponding flow may be referred to as the continuous flow.

However, when the entry vehicle is flying in the lower density (thin) celestial atmosphere, with the decrease of the atmosphere density, the gap between the gas molecules gradually increases gradually, which lead to the more prominent properties of the gas molecules as well as the gradual loss of the macroscopic properties of the molecules. Finally, the assumption of continuous media is completely invalid. In order to facilitate the research, non-dimensional Knudsen number Kn is proposed to characterize the flow continuity. Where $Kn = \lambda/L$, that is, the ratio of the mean free path λ of the fluid molecules to the characteristic length L of the spacecraft. According to the value of the Knudsen number, continuous flow and discontinuous flow can be distinguished. The currently accepted criteria are:

(1) $Kn \leq 0.01$—continuous flow zone, where gas flow can be described by the Navier–Stokes equation (N-S equation) without slip boundary;
(2) $0.01 < Kn < 0.1$—slip flow zone with good continuity, where N-S equation is still valid for the whole circumferential flow field. However, consideration should be made to the temperature and velocity jumps, and the slip boundary conditions should be set;
(3) $0.1 \leq Kn \leq 10$—transitional zone, where continuous medium assumption is no longer valid. The analysis of aerodynamic problems in this flow region is usually based on the gas microscopic equation and can be simulated by the molecular dynamics approach based on the Boltzmann equation or the direct simulation

Monte Carlo approach (DSMC, Bird, 1994). Some semiempirical relationships are often used in engineering calculations, such as various bridging function methods.
(4) $Kn > 10$—free molecular flow region. The analysis of aerodynamic problems in this flow region requires the free molecular hypothesis. If there is no interaction between the gas molecules, Boltzmann equation can be used to solve the problem.

It should be noted that the criteria above are not unique, and there are no clear boundary among the flow areas. In addition, Knudsen number varies greatly with the probe scale in simulation. Therefore, if a uniform Knudsen number is used to simulate the circumferential flow field in a certain state, the calculation error may be caused by the inconsistent gas flow states along the surface of the probe.

2. **Incompressible flow and compressible flow**

A flow in which the gas density ρ is constant is referred to as an incompressible flow, and conversely, a flow in which the gas density is variable is referred to as a compressible flow. In fact, the complete incompressible flow does not actually exist. The compressibility of the gas can be characterized by the Mach number $Ma = v/\alpha$, where v is the gas flow rate and α is the local sound velocity. The probe enters the atmosphere of the celestial body at the velocity of several kilometers or even ten kilometers per second. As affected by atmospheric damping, the velocity gradually decreases and the velocity approaches to zero at the time of landing. During such period, according to the compressibility of the bypass gas, it can be divided into several velocity domains as follows:

(1) Hypersonic velocity ($Ma > 5.0$): Due to high velocity, the probe has a strong shock wave, viscous boundary layer effect and even has chemical reactions around the circumferential flow field.
(2) Supersonic velocity ($1.2 < Ma \leq 5.0$): In this velocity domain, there are still shock waves around the circumferential flow field, in which the flow characteristics and flow lines are not continuous. Meanwhile, there are obvious expansion waves in the circumferential flow field.
(3) Transonic velocity ($0.8 < Ma < 1.2$): In this velocity domain, the surface airflow of the probe appears mixed flow of both $Ma < 1$ and $Ma > 1$. With the decrease of the velocity, the shock wave position and intensity change rapidly in the circumferential flow field of the probe, resulting in strong instability in the velocity domain.
(4) Subsonic velocity ($0.3 \leq Ma \leq 0.8$): The flow velocity of the probe around the circumferential flow field is lower than the local sound velocity; there is no shock wave in the circumferential flow field; and the flow is continuous. In this velocity domain, the main factors affecting the aerodynamic characteristics of the probe are viscous and airflow separation.
(5) Low velocity ($Ma < 0.3$): In this velocity domain, the gas flow can be approximately considered as incompressible, and the change in density with pressure can be ignored.

3. Viscosity and viscous flow of gases

From the perspective of molecular motion theory, when the real gas molecules move randomly, molecule exchange may occur between adjacent fluid layers with different velocities to transport mass, momentum and energy from one place to another. Therefore, viscosity is an important transport property of real fluid, and the true flow with transportation activities is called viscous flow. Conversely, a flow without transportation activity is called non-viscous flow. There is no pure non-viscous flow in nature.

The viscosity of the gas is analyzed by the typical flow as follows. When the airflow bypasses an object, the air velocity of the object surface is 0 at a certain point. With the increase of distance to the point, the velocity increases accordingly. The difference of this velocity and the incoming velocity is not significant until at a certain distance δ from the object surface. This flow distribution near the object surface is caused by air viscosity. Viscosity makes no relative velocity between the surface airflow of the object and the surface of the object, which is called no-slip condition. The velocity of the air leaving the object surface δ turns to almost equal to the local free-flow velocity v_∞. This layer of flow where the velocity from the wall surface to airflow is recovered to local velocity is called the boundary layer.

Fluid viscosity is usually expressed by the dynamic viscosity coefficient μ or the kinematic viscosity coefficient, where $\mu = 1/\nu$. The viscosity of the fluid is directly related to temperature and is almost independent of pressure; the viscosity of the gas increases with the increase of the temperature. As proved by experiment, the relationship between gas viscosity coefficient and temperature can be approximated as

$$\frac{\mu}{\mu_0} = \left(\frac{T}{T_0}\right)^{3/2} \left(\frac{T_0 + B}{T + B}\right) \qquad (7.1)$$

where T_0 and μ_0 are the dynamic viscosity coefficients of the reference temperature and the corresponding temperature, respectively; B is the temperature constant related to the gas type, for example $B = 110.4$ K for air and $B = 240$ K for Martian atmosphere.

In order to characterize the comprehensive influence of fluid viscosity on the circumferential flow field, the concept of Reynolds number with dimension 1 is proposed. Reynolds number reflects the ratio of inertial force to viscous force in a moving fluid, which is defined as:

$$Re = \frac{\rho v L}{\mu} \qquad (7.2)$$

where ρ is the fluid density, v is the fluid velocity, and L is the characteristic length. It can be seen from the formula that, with the increase of the Reynolds number, the influence of the viscous force in the fluid becomes increasingly small, and the impacts on friction, heat transfer and diffusion thereof are limited to a quite thin boundary layer. The flow outside the boundary layer is non-viscous. In addition, the frictional drag caused by viscosity is quite small compared with the differential pressure-induced drag and the shock-induced drag, and the non-viscous flow assumption may reasonably predict the overall aerodynamic force around the circumferential flow field.

4. **Characteristics of boundary layer flow**

The boundary layer is a flow thin layer where the viscous force close to the object surface in the high Reynolds number is non-negligible. Friction/shear stress plays a major role in the boundary layer. The thickness of the boundary layer is defined as the distance from the nonslip object surface to the position where the velocity is approximately equal to the local free-flow velocity v (0.99 v or 0.995 v) along the normal direction, which is denoted as δ.

According to the flow state, the boundary layer can mainly fall into three types: laminar boundary layer, turbulent boundary layer and mixed boundary layer (transition). The fluid elements in the laminar boundary layer move in regularly arrangement, the flow between adjacent flow lines is smooth, and there are few lateral fluid movements between adjacent layers; the fluid elements of the turbulent boundary layer are disordered with high pulsation feature, the streamlines are curved, irregular and jagged, and there are lateral movements between different fluid micelles as well as even motion forms that are opposite to the macroscopic flow; boundary layer transition is the intermediate state from the laminar boundary layer to the turbulence boundary layer. There are many factors that lead to this transition, including the increase of the Reynolds number of free incoming flow, the interference of the local structure of the object surface, roughness, etc.

Boundary layer separation refers to the phenomenon in which the boundary layer flow is detached from the surface of the object, which is the result of the interaction of the viscous drag of the moving fluid and the adverse pressure gradient.

5. **High-temperature effect**

During the flight of the probe of high velocity entry into the celestial atmosphere followed by braking, there may be extremely high temperatures around some areas of the circumferential flow field, such as the shock layer and the highly dissipative boundary layer. For example, shock layer temperature of Mars entry vehicle may be up to 10,000 K (Chen, 1992). At such high temperatures, atmospheric molecules are excited by thermo chemical effects such as molecular vibration excitation and electron excitation as well as chemical reactions such as decomposition, recombination, ionization and displacement of gas molecules. Such effect is also called high-temperature effect or high-temperature real gas effect, the occurrence of which significantly affects the atmosphere properties and aerothermodynamics properties. The flow pattern under high-temperature effect can be expressed as chemical equilibrium state, (thermal) chemical non-equilibrium state and frozen state.

Chemical equilibrium state: If the chemical reaction is carried out fast enough compared with the change in state, each chemical reaction can reach equilibrium, and then each chemical reaction in the circumferential flow field has a corresponding equilibrium constant. Then, by solving the chemical reaction equation, the equilibrium component in this state is obtained. This flow state is called chemical equilibrium state.

(Thermal) chemical non-equilibrium state: For the hypersonic velocity real gas flow, due to the great flow rate and short time, the chemical reaction state of the gas component is difficult reach equilibrium by following the state change within a short time, resulting in the chemical non-equilibrium state. When the temperature in the circumferential flow field falls under the excitation threshold of the gas molecular vibration energy and the electron energy or is only partially excitable, the gas molecules can quickly reach the equilibrium between the rotational temperature and the translational temperature with little collision. At this time, the flow state can be regarded as in thermodynamic equilibrium, and a single temperature model is applied to describe the thermodynamic properties of the gas. When the temperature in the circumferential flow field rises so that the vibration energy and the electron energy of the gas molecules are excited in a large amount, the vibration temperature and the electron temperature of the gas molecules are difficult to be balanced within a short time with the rotational temperature by collision and translational motion. At this time, the flow state is not in thermodynamic equilibrium, and it is necessary to describe the thermodynamic properties of the gas by using a dual temperature model representing the vibration temperature or a triple temperature model representing both the vibration temperature and the electron temperature.

Frozen state: For chemical non-equilibrium flow, when the collision frequency of atmospheric molecules decreases or the flow rate increases to a certain extent, the time for gas chemical reaction tends to infinity relative to the state change, and the gas freezes and cannot respond to changes in the flow. This state is called the frozen state.

Different gas molecules have different temperature ranges for vibration excitation, dissociation and ionization. Taking the atmosphere of the Earth as an example, as shown in Fig. 7.1, by maintaining a constant standard atmospheric pressure, if the temperature reaches 800 K, the vibration energy of the molecules will become significant and the vibration excitation will be initiated. If the temperature reaches 2,500 K, O_2 will start to be dissociated, and if to 4000 K, the basic dissociation of O_2 is completed, most of which exists in the form of O atoms, and N_2 begins to be dissociated; if to 9000 K, the basic dissociation of N_2 is completed, and meanwhile, ionization is initiated; if higher than 9000 K, a partially ionized plasma will be formed, and the gas contains N, N^+, O, O^+ and electrons. In addition to the vibration excitation phenomenon which is not affected by the pressure, if the pressure decreases, the starting temperature will drop correspondingly, and vice versa. In addition, a small amount of NO will be formed between 4000 K and 6000 K, some of which will ionize into NO^+ and electrons.

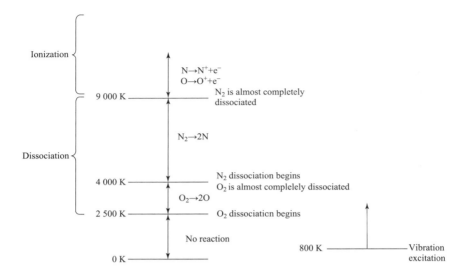

Fig. 7.1 Temperature range at which vibration excitation, dissociation and ionization occur at one standard atmospheric pressure

6. **Circumferential flow field structure**

In the process of the probe entering the planetary atmosphere, the surrounding circumferential flow field structure is very complicated. Figure 7.2 shows the typical circumferential flow field around the circumferential flow field with MSL as an example. From Fig. 7.2, it can be seen that the hypersonic incoming flow forms a strong accurate detected shock wave before the heat shield, and the post-wave gas is compressed. Due to the various high-temperature effects caused by the excited gas in the shock layer, the wave layer is closer to the object surface; the streamline passes through the shock wave to change direction and then evenly bypasses the object surface, and the wall limit flow line spreads from the stagnation point to the periphery and extends to the shoulder; when the gas bypasses the shoulder, a large outer break angle occurs, which produces a series of expansion waves; the backward expansion of flow accelerates and gradually transits to the non-viscous outflow region; the boundary layer of the object surface grows gradually, and to the outflow region at the shoulder, forming a shear layer and resulting in relatively strong vortex motion; at the rear end of the entry vehicle body, the boundary layer flows are separated and the recirculation zone if formed, creating a low pressure zone; the wake bypassing the entry vehicle is concentrated and compressed to form a secondary compression wave parallel to the main flow direction; the compression waves gradually weaken and join the outflow in the far field. The approximate locations of various flow phenomena and structures in the circumferential flow field are shown in Fig. 7.2.

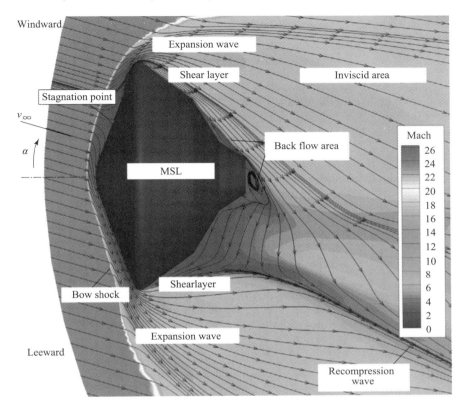

Fig. 7.2 Circumferential flow field structure of MSL under typical hypersonic condition

7.2.2 A Study on Aerodynamic Problems in Atmospheric Entry

1. Mars entry vehicle aerodynamic research

The Viking probe is the first Mars entry vehicle developed and successfully landed on Mars by the USA. In the development process of the Viking, computational fluid dynamics (CFD) technology was not well-developed yet (including software and hardware limitations). Therefore, the aerodynamic data acquisition relied on wind tunnel tests and ballistic range tests considerably, and the impact caused by the differences between atmospheric compositions was corrected. Subsequently, the aerodynamic research processes of the Phoenix Lander, Mars Pathfinder and Mars Exploration Rover mainly relied on CFD technologies, and the ground verification tests were significantly reduced. It was not until the development of the Curiosity when relatively large-scale ground tests were carried out on some special aerodynamic problems such as transition and turbulence.

Based on the aerodynamic test and flight data of Viking, the aerodynamic research of NASA's Phoenix and its subsequent Mars entry vehicles mainly relied on the CFD technology. Table 7.1 shows the velocity domain division and calculation approaches of the Phoenix in the study of the aerodynamic characteristics of the entry vehicle. In addition to the velocity domain $Ma < 1.5$ where Phoenix inherited and adopted the ground test data of the Viking, other velocity ranges are numerically calculated as the ground design basis; the aerodynamic characteristics of the supersonic zone are corrected with the empirical formula obtained after the flight test of the Viking.

Table 7.1 Aerodynamic research velocity domain division and calculation approaches of the Phoenix

Flight area	Applicable rules and research range of attack angle	Input parameter	Calculation approach
Free molecular flow	$Kn > 1000$, $0° < \alpha_t < 180°$	A, β	DACFree
Transition flow	$0.00106 < Kn < 1000$, $0° < \alpha_t < 26°$	Kn, α, β	DSMC
Hypersonic velocity	$Kn < 0.00016$, $Ma > 8.8$, $0° < \alpha_t < 16°$	$v_\infty, \alpha, \beta,$	LAURA (Precursor)
Supersonic velocity	$2 < Ma < 6.3$, $0° < \alpha_t < 16°$	Ma, α, β	LAURA (Precursor + bottom correction)
Transonic velocity	$0.4 < Ma < 1.5$, $0° < \alpha_t < 16°$	Ma, α, β	Ground wind tunnel test data of the Viking

2. **Aerodynamic research of Titan entry vehicle**

The Titan entry vehicle Huygens has a blunt-headed outsole with the half cone angle of 60°. It has a maximum diameter of 2.7 m, a ball nose radius of 1.25 m and a mass of 318.62 kg.

Aerodynamic calculation: Aerodynamic research of the Huygens is mainly focused on the accurate simulation of the (thermal) chemical non-equilibrium flow of Saturn's atmospheric components and the coupling analysis with high-temperature radiation.

3. **Aerodynamic research of Jupiter entry vehicle**

In terms of aerodynamic calculation, the "Galileo" entry vehicle mainly carried out the research on the (thermal) chemical non-equilibrium reaction simulation of different components of Jupiter atmosphere, the simulation of aerodynamic characteristics of thin gas DSMC considering high-temperature effect and the calculation and analysis of high-temperature gas ionization and radiation characteristics. In addition, due to the high entry velocity and the thick atmosphere of Jupiter, the chemical reaction is strong. The thermal ablation of the entry vehicle is very serious. Therefore, the influence of ablation asymmetry on the aerodynamic characteristics is analyzed [1].

7.2.3 Atmospheric Entry Aerodynamic Analysis and Prediction

7.2.3.1 Atmospheric Composition and Its Chemical Reaction Model Selection

The atmospheric components of other celestial bodies with atmospheres in the solar system are more complex, which is different from the simple and stable atmosphere compositions of the Earth. Due to the difficulty of exploration and lack of effective means, there exist discrepancies between the atmospheric composition models obtained by different exploration methods. The chemical reaction models that cause high-temperature effects are also different, which poses a problem for the prediction of high-temperature effects aerodynamic characteristics of high-speed entry vehicles.

Because the high-temperature effect has a relatively great impact on the aerodynamic characteristics of the inlet, in order to effectively carry out the aerodynamic and thermal design of the entry vehicle, it is necessary to first screen the atmospheric components and high-temperature chemical reaction models of the target celestial bodies.

Table 7.2 gives the volumetric mixing ratios of the currently widely recognized components of the Martian atmosphere. Table 7.3 shows the 8-component 12-reaction model for the Martian atmosphere.

Table 7.2 Atmospheric composition of Mars (volume mixing ratio)

Main gas	%	Trace gas	ppm
CO_2	95.3	^{36}Ar	5
N_2	2.7	Ne	2.5
^{40}Ar	1.6	Kr	0.3
O_2	0.13	Xe	0.08
CO	0.07	O_3	0.04–0.2
H_2O	0.03		

The atmospheric composition of Venus is close to that of Mars, so the chemical reaction model may also be basically transplanted.

The atmospheric composition of the Jupiter varies greatly with the temperature of the atmosphere. Table 7.4 gives the three atmospheric composition models of Jupiter. Table 7.5 shows a typical Jupiter atmospheric recreation model.

7.2.3.2 Research on High-Altitude Discontinuous Flow Effect

For a flight process that enters a celestial body with dense atmosphere or travels in the thin atmosphere for a short period of time, the effect of the discontinuous flow effect on the aerodynamic characteristics of the incoming flow is insignificant. However,

Table 7.3 Typical chemical reaction model of Martian atmosphere

Type	Reaction equation	Reactants referred to by M
Dissociation reaction	$CO_2 + M_1 \rightleftharpoons CO + O + M_1$	M_1: CO_2, O_2, CO, N_2, NO, C, N, O
	$CO + M_2 \rightleftharpoons C + O + M_2$	M_2: CO_2, O_2, CO, N_2, NO, C, N, O
	$O_2 + M_3 \rightleftharpoons 2O + M_3$	M_3: $CO2$, O_2, CO, N_2, NO, C, N, O
	$N_2 + M_4 \rightleftharpoons 2N + M_4$	M_4: CO_2, O_2, CO, N_2, NO, C, N, O
	$NO + M_5 \rightleftharpoons N + O + M_5$	M_5: CO_2, O_2, CO, N_2, NO, C, N, O
Exchange reaction	$CO + O \rightleftharpoons O_2 + C$	–
	$CO_2 + O \rightleftharpoons O_2 + CO$	–
	$NO + O \rightleftharpoons N + O_2$	–
	$N_2 + O \rightleftharpoons NO + N$	–
	$CO + N \rightleftharpoons NO + C$	–
	$CO + CO \rightleftharpoons CO_2 + C$	–
	$NO + CO \rightleftharpoons CO_2 + N$	–

Table 7.4 Typical atmospheric composition of the Jupiter (volume mixing ratio)

Main gas	Low-temperature model /%	Medium-temperature model /%	High-temperature model /%
H_2	68.454	86.578	93.754
He	21.057	13.214	6.149
CH_4	0.145	0.062	0.028
NH_3	0.035	0.015	0.007
H_2O	0.24	0.102	0.048
Ne	0.031	0.013	0.006
Others	0.038	0.016	0.008

for an entry process of a celestial body with thin atmosphere such as Mars or which has a thick atmosphere and a relatively long transit time in a thin atmosphere such as Jupiter, it is necessary to focus on the impact of discontinuous flow effects and select the aerodynamic research approaches reasonably in different regions according to the variation of Knudsen number.

Table 7.6 shows the variation of the Knudsen numbers of Mars entry vehicle Pathfinder and the Jupiter entry vehicle Galileo along their respective entry trajectories and gives the corresponding flow patterns.

For free molecular flow regions where Kn_∞ are large, as the molecular free path is large, the change in the velocity distribution of the gas molecules due to the collision may be ignored. Thus, the basic equation of the free molecular flow is the Boltzmann equation without collision terms. For the steady flow problem, the velocity distribution function of the incoming flow is the equilibrium state distribution, the Maxwell distribution, which is expressed as follows:

7.2 Aerodynamics and Aerodynamic Analysis

Table 7.5 Typical atmospheric reaction model of the Jupiter

No.	Reaction formula	Reaction rate/(m$^3 \cdot$s^{-1})	Equilibrium constant K_i/m^{-3}
1	$H_2 + He \rightarrow$ $H + H + He$	$6.93 \times 10^{-12}/T \exp(-52340/T)$	$3.7 \times 10^6 \left[1.0 - \exp(1.50 \times 10^8 T^{-2})\right]$ $\exp(-52340/T)$
2	$H_2 + H_2 \rightarrow$ $\rightarrow H + H + H_2$	$2.5 k_{H_2 + He \rightarrow H+H+He}$	Ditto
3	$H_2 + H \rightarrow$ $H + H + H$	$20 k_{H_2 + He \rightarrow H+H+He}$	Ditto
4	$H_2 + H^+ \rightarrow$ $H + H + H^+$	$20 k_{H_2 + He \rightarrow H+H+He}$	Ditto
5	$H_2 + e \rightarrow$ $H^+ + H + e$	$20 k_{H_2 + He \rightarrow H+H+He}$	Ditto
6	$H + e \rightarrow$ $H^+ + e + e$	$6.09 \times 10^{-23} \sqrt{8kT_e/(\pi \mu_e)} \exp(-15782/T_e)$	–
7	$He + e \rightarrow$ $He^+ + e + e$	$2.56 \times 10^{-23} \sqrt{8kT_e/(\pi \mu_e)} \exp(-285248/T_e)$	$1.62 \times 10^{-2} T^{3/2} \exp(-285248/T)$
8	$H_2^+ + e \rightarrow$ $H + H$	$1.2 \times 10^{-16}(T/300)^{-0.4}$	–
9	$H_{(n=3)} + H$ $H_2^+ + e$	$6.8 \times 10^{-18} T^{0.61} \exp(-13000/T)$	–

Table 7.6 Knudsen number variations and corresponding flow patterns during typical entry processes of Mars and Jupiter

Mars Pathfinder			Galileo		
Altitude/km	Kn_∞	Flow state	Altitude/km	Kn_∞	Flow state
130.9	13.2	Free molecular flow	604	43	Free molecular flow
119	2.76	Transition flow	506	5.55	Transition flow
95	0.11	Transition flow–slip flow	453	1.51	Transition flow
80	0.012	Slip flow	382	0.18	Transition flow–slip flow
65	0.002	Slip flow	353	0.07	Slip flow
55	0.0005	Continuous flow			

$$f_0 = n\left(\frac{m}{2\pi kT}\right)^{3/2} \exp\left[-\frac{m}{2kT}(u'^2 + v'^2 + w'^2)\right] = n\left(\frac{\beta}{\pi^{1/2}}\right)^3 \exp(-\beta^2 c^2) \tag{7.3}$$

where n is the molecular number density of the incoming gas, T is the inflow temperature, m is the molecular mass, c is the molecular velocity value; $k = 1.380658 \times 10^{-23}$ J K^{-1}; $\beta = (2RT)^{-1/2}$. The momentum and energy transfer of the incoming molecules to the surface of the object can be calculated by solving the Maxwell distribution. By assuming a proper molecular surface reflection model (diffuse, specular, finite reflection), the momentum and energy flow of the reflected molecules away from the surface can also be determined.

For the transitional flow region with the molecular free path close to characteristic length of the entry vehicle, the numerical methods include the direct numerical solution based on the Boltzmann equation and DSMC approach combining the molecular dynamics theory with the gas physical model. The advantage of Boltzmann equation is that it can be used to describe the flow in almost all regions. However, the disadvantage is that the equation is extremely complex. Even after simplification, it is difficult to discretize the equation. Therefore, there may be difficulties in improving accuracy. The advantage of DSMC is that it can describe the physical properties of the gas, so a high-temperature (thermal) chemical non-equilibrium model can be introduced. However, the disadvantage is that mesh generation and time step limit the approach. If the Kn number is relatively low, the numerical simulation is difficult to be implemented. At present, the DSMC approach is still mainly used in engineering research. In the actual flight, multi-component mixed gas model and the thermo-chemical non-equilibrium model are often added to simulate the high-temperature effect.

In the slip region with small thickness and Kn_∞ less than 0.1 (which can be extended to 0.2 for some local parts), the discontinuous molecular effect is not obvious, and the gas flow can still be studied by continuous medium model. Starting from the N-S equation and applying the slip boundary conditions, the numerical solution of the flow satisfying the accuracy requirements of the project is finally obtained. It should be noted that since the flow in the slip zone is usually accompanied by high-temperature effect, the conditions of the radiant equilibrium wall of the chemical non-equilibrium flow should be met when dealing with the temperature jump boundary.

7.2.3.3 Heat Radiation Calculation of High-Temperature Gas

When the flight velocity of the entry vehicle reaches certain extent, the high-temperature effect around the circumferential flow field would become quite important, and the high-temperature gas in the nose shock layer would become a strong radiator that heats the entry vehicle by photon radiation. The magnitude of the radiated thermal energy increases with the increase of the nose size of the entry vehicle and the increase of the flight velocity, possibly reaching or exceeding the level of convective heat transfer. According to the previous research experience, for the entry of Martian atmosphere, when the entry velocity reaches 7 km/s or above, high-temperature gas heat radiation must be considered. In the entry aerodynamic analysis of Venus, Jupiter and distant atmospheric celestial bodies, radiation heating effect must be considered.

The heating mechanism of the radiation of high-temperature circumferential flow field is complicated, involving the intersection and fusion of high-temperature gas dynamics, atomic molecular spectroscopy, heat transfer and many other disciplines. The reliable prediction of radiative heating is not only related to the calculation approach and the calculation model itself, but also closely related to the large amount of basic data applied in the approach and model, such as high-temperature gas component spectral data and chemical reaction kinetic data. Presently, related physico-chemical models, calculation approaches and related basic data for radiative heating are still being developed and improved. Common gas radiative heating calculation approaches for high-temperature circumferential flow field are semi-engineering and fully numerical.

The concept of semi-engineering calculation method of radiative heating: Obtain the radiative heating data of the stagnation point along the ballistic trajectory with the stagnation radiant heating quick formula. Work out the radiative heating distribution characteristics with the numerical calculation approach and finally obtain the distribution data of radiative heating of the entry vehicle along the ballistic trajectory.

The quick formula for radiant heating at stagnation point is a simplified functional expression between the stagnation radiant heat flux and the ball nose radius, the local flow density and the flight velocity, which approximates the stagnation radiant heat flux along the ballistic trajectory.

The fast calculation formula for the stagnation point radiant heat flux (W/cm^2) in the entry into the atmosphere of Mars is as follows:

$$q_{r_0} = C r_n^a \rho^b f(v) \tag{7.4}$$

where r_n is the radius of the windward head of the entry vehicle; $C = 2.35 \times 10^4$; $a = 0.526$; $b = 1.19$; the value of the function f(v) is shown in Table 7.7.

Table 7.7 Table of function interpolations in atmospheric radiant heat calculation of Mars $f(v)$

$v/(m \cdot s^{-1})$	$f(v)$	$v/(m \cdot s^{-1})$	$f(v)$
6000	0.2	7600	14.8
6150	1	7800	17.1
6300	1.95	8000	19.2
6500	3.42	8200	21.4
6700	5.1	8400	24.1
6900	7.1	8600	26
7000	8.1	8800	28.9
7200	10.2	9000	32.8
7400	12.5		

The fast calculation formula for the stagnation point radiant heat flux (W/cm^2) in the entry of the atmosphere of Venus is as follows:

$$q_{r_0} = k r_n^{0.497} (\beta \sin \gamma_E)^{1.18} \qquad (7.5)$$

where $k = 1.175$; β is the ballistic coefficient of the entry vehicle; γ_E is the entry angle.

The calculation of the radiative heating distribution mainly includes three parts: the calculation of circumferential flow field, the calculation of radiation characteristics and the calculation of radiation transport. The circumferential flow field calculation is based on the three-dimensional N-S equation containing the chemical reaction source term to simulate the circumferential flow field, and the component number density and temperature at each spatial point in the shock layer are obtained. The calculation of the gas radiation characteristics is mainly based on electronic states needed for spectral calculation through circumferential flow field simulation, and the calculation of frequency-dependent emission and absorption coefficients at each point in space based on different processes and mechanisms such as binding–binding transition (atomic line spectrum), atomic binding–free transition (photoionization), atomic free–free transition (bremsstrahlung) and molecular binding–binding transition (molecular band spectrum). Finally, the radiation intensity formula is established to describe the radiation intensity caused by absorption, emission, scattering, etc., along the ray trajectory, and the radiation intensity distribution across the surface of the entry vehicle is further calculated.

Numerical approaches of radiative heating include the coupled solution method and the non-coupled solution method. The coupled solution method takes into account the coupling interference effect between the circumferential flow field and the radiation field, coupling the radiation transfer equation with the non-equilibrium N-S equation including the radiation source term. The approach has high calculation accuracy. However, the calculation workload is heavy, and the requirements for computing resources are also significant. Currently, it can only be applicable for radiative heating calculation of simple flowing gas. The non-coupled approach is to numerically solve the convection of N-S equation without radiation source term so as to numerically simulate the flowing field and then obtain the radiative heating of the point by solving the radiation transfer equation and integrating all the fluid microelement radiation of a solid angle range of a certain point on the object surface based on the circumferential flow field parameter distribution obtained by numerical simulation. This method does not consider the cooling effect of high-temperature gas radiation on the circumferential flow field. Therefore, the surface radiant heat flux obtained by this method is higher than that obtained by the coupling solution method. However, the calculation workload by this method is relatively small, which is more diversified than the general engineering approach. And it can be used as one of the effective methods to predict the radiative heating.

7.2.3.4 Hypersonic Flow Transition Model Prediction

When the entry vehicle is flying in a dense atmosphere, if it is disturbed by acoustics, vortex, etc., in the incoming flow, or there are factors of increased roughness caused by structural protrusions, pits, large separation and reattachment, and ablation of the heat shield on the surface of the bay, these factors may cause local or even unstable perturbation of the boundary layer around the circumferential flow field. With the exponential growth of the perturbation, the boundary layer flow transits from laminar flow to full turbulent flow.

For ballistic or semi-ballistic entry vehicles generally applying the shape of a large blunt-nose revolving body, the frictional changes caused by the flow transition and the turbulent boundary layer are relatively small compared to the supersonic shock induced drag, so the transition and turbulence have a limited impact on the aerodynamics of the entry vehicle. However, for aerodynamic heating, the peak heat flux density generally appears in the transition zone after the transition, while the peak heat flux density of the laminar flow wall generally appears near the stagnation point. Therefore, the distribution of the heat flux density on the surface of the entry vehicle may change significantly and may have a disruptive effect on the thermal protection design after the hypersonic incoming flow transition and turbulence occurs. According to the previous research experience, it is necessary to carry out research on boundary layer transition prediction and turbulent aerodynamic thermal assessment when one or more characteristics exist: large size (such as windward characteristic length greater than 3 m), large ablation amount, local bulge/pit of the windward surface and incoming flow with high Reynolds number.

The prediction of the transition position of the hypersonic boundary layer flow has been a worldwide highlight topic in the field of aerodynamics. So far, there have been no universal and stable standard prediction methods. At present, there are three methods to determine the location of transition point: empirical method, stability theory method and engineering transition model method.

The empirical method is to analyze a large number of wind tunnel test data and flight test data and then fit to give an empirical relationship. The relationship usually relates the free-flow turbulence, the local pressure gradient and the Reynolds number of the transition momentum thickness. For example, the Abu-Ghannam and Shaw relationship is based on a large number of experimental observations. The advantage of this method is that it is simple and easy. And the disadvantage is that, for hypersonic incoming flow, the test is costly. In addition, as the unstable airflow environment caused by the noise of the cave wall in the wind tunnel is quite different from the actual flight situations, the hypersonic transition conditions obtained by such tests are generally much worse than actual conditions. In addition, since the empirical correlation-based transition simulation requires the comparison of the Reynolds number of actual momentum thickness with the Reynolds number of reciprocal momentum thickness derived from the correlation, and the outer edge of the boundary layer is not clearly defined, the integration along the boundary layer depends on the execution of the search algorithm, bringing challenges to N-S numerical simulation in terms of numerical approaches and programming.

The flow stability theory is applied to determine whether the transition position introduces disturbance in the laminar flow region, and then approaches such as direct numerical simulation, large eddy simulation, solution of parabolic stability equation and linear stability equation are adopted to calculate the development of disturbance and give the transition position. This method can reveal the mechanism of transition by describing the development of disturbances. However, the calculation of stability poses strict demands on computer resources and numerical methods, especially direct numerical simulation methods. In addition, it is currently only possible to make assumptions through experience for giving external disturbances or initial disturbances in the boundary layer, which brings uncertainty to some extent. Therefore, for the complicated problems in transition position prediction such as local flow transition, the direct numerical simulation method is not feasible in engineering.

A major breakthrough in the research of transition predictions is the engineering transition model approach that has evolved since the 1990s. This method is based on the Reynolds average N-S equation by introducing intermittent factors to describe the occurrence and development of the transition. The theoretical system thereof is consistent with that of the turbulence. By this method, the calculations of the transition model and turbulence model are based on a tight coupling relationship within a computational framework. Therefore, various approaches and programs for turbulent model calculation can be applied in the calculation of the engineering transition models. Compared with the direct simulation approach of flow stability, this method has lower requirements on computer resources with easier engineering realization.

In summary, the empirical methods determined by the transition criteria are mainly based on a large number of experiments, and there are conversion problems between the space and the Earth; direct numerical simulation requires intensive computation but the engineering achievability is relatively poor; for engineering transition model, it is required to introduce additional intermittent factor to describe the occurrence and development of transition. And it is also difficult to determine intermittent factors in the non-Earth atmospheric environment or an atmospheric environment with flow characteristics unknown. According to foreign research experience, the hypersonic incoming flow transition models in the process of interplanetary entry are still mainly based on empirical approach analysis and fitting. The Reynolds number of the momentum thickness of the transition boundary is determined after comprehensive ground test and flight measurement data.

7.2.3.5 Plasma Sheath Analysis and "Blackout" Prediction

After the deep space probe enters the interplanetary celestial atmosphere at an extremely high velocity, a strong collision occurs between the probe and the air, resulting in sharp rise of the temperature and pressure in the head shock layer, which further causes the gas molecules to ionize the ablated insulation material, forming a layer of plasma sheath wrapping the spacecraft. This sheath contains a large amount of free electrons and a variety of charged particles, which will absorb and scatter of electromagnetic waves to different extents, resulting in mismatch of the antenna impedance of the antenna and distortion of the pattern as well as reduced antenna efficiency and gain. In addition, it will also result in some difficulties in communication, navigation, measurement, etc., in the process of entry. In severe cases, even the radio contact between the vehicle and the outside would be completely interrupted, which is the so-called blackout phenomenon. To accurately predict the occurrence interval of communication interruption in the process of entry and then take measures accordingly to delay such "blackout" or shorten the communication interruption time, it is necessary to first accurately simulate the plasma distribution characteristics in the plasma sheath.

The numerical simulation approaches of plasma distribution mainly include boundary layer method, viscous shock layer method and full N-S equation capture approach. Among the three methods, the full N-S equation capture method is the closest to the real physical phenomenon. With the great improvement of computing power, the full N-S equation capture method has been widely used to study the chemical non-equilibrium flow with ionization. The main idea of this method is to describe the flow by the complete N-S equation with multi-component chemical source terms, replace the total mass conservation equation with the chemical component mass conservation equations and ignore the work made of interaction force among the body force term in the momentum conservation equation, the radiant energy term in the energy equation and the chemical components. The key to improving the accuracy of plasma simulation is to establish a high-temperature chemical reaction model that is as realistic as possible while giving consideration to the most charged particle reaction as possible, increasing the reaction between the product and the surrounding gas after ablation of the thermal protection material, etc. In addition, the application of reasonable temperature models and boundary conditions is also an important factor to improve prediction accuracy.

7.2.3.6 Transonic/Supersonic Aerodynamic Stability

In order to achieve rapid braking in the process of entry, the entry vehicle of deep space probe is generally aerodynamically shaped with a short blunt body with a large blunt inverted cone. This shape is generally aerodynamically unstable during the flight from supersonic section to transonic section. According to the stability study, the greater the bluntness of the windward head, the smaller the distance from the pressure center of the entry vehicle to the center of mass of the spacecraft is, the worse the stability; the smaller the bluntness of the head, the easier to achieve flight stability.

The stability of a large blunt inverted cone entry vehicle varies with flight altitude and the Mach number. In the hypersonic continuous flow region, although the stable derivative of the large blunt inverted cone is small in magnitude and tends to decrease with the increase of the Mach number and the height, it remains unchanged and stable. In other words, when the flow disturbance causes the spacecraft to deviate from the balance, the aerodynamic momentum can recover it to the equilibrium attitude. However, when the velocity of the entry vehicle is reduced to the supersonic velocity and then the transonic velocity, the pressure center moves forward greatly, and the arm of the pitching moment becomes extremely small and is likely to be negative, making the entry vehicle unstable. Although this dynamic instability is not necessarily completely reversing the entry vehicle, it would cause a limit cycle vibration with the amplitude of about 10°. Therefore, in case of dynamic instability with the angle of attack starting to diverge, attitude control is required. To simulate the flight attitude and control effect and confirm the control margin, it is necessary to calculate the Mach number, the angle of attack range and the corresponding damping derivative values of the dynamic instability of the entry vehicle obtained by calculation in combination with the wind tunnel test.

There are five methods to calculate the damping derivative: (1) dynamic derivative engineering method based on Newton's theoretical approximation method and small disturbance theory; (2) dynamic derivative constraint Euler numerical calculation method based on small perturbation theory; (3) dynamic rotation based on non-inertial system method; (4) free oscillation method based on unsteady fluid dynamics/rigid body dynamics (UCFD/RBD); (5) forced oscillation approach based on UCFD/RBD. The linearity of the entry vehicle with large blunt head and short blunt body at small attack angle aerodynamic force in the supersonic velocity area is poor. And the linear assumption of the forced oscillation method cannot guarantee the accuracy of the transonic ultrasonic simulation of the entry vehicle. Therefore, free oscillation method based on UCFD/RBD is generally applied for dynamic derivative calculation of the entry vehicle with large blunt head and short blunt body, as shown in Fig. 7.3. The fluid dynamics equation is the three-dimensional unsteady compressible N-S equation in the general curvilinear coordinate system, double time step sub-iteration is applied for time discretization, and time difference format is reasonably selected to obtain the high-order time precision. The rigid body dynamics equation is a function of time-dependent variation of a 12-dimensional vector consisting of point position, point velocity, attitude angle and angular velocity and unsteady

7.2 Aerodynamics and Aerodynamic Analysis

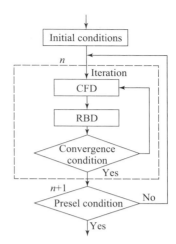

Fig. 7.3 Flowchart of UCFD/RBD coupling approach

aerodynamic force (moment) obtained by flow field integration. After sub-iterative processing, it is advanced simultaneously with the sub-iterations of the fluid dynamics equation. In terms of the turbulence models of the fluid dynamics equation, there are mainly three types: RANS, LES and DES. RANS can efficiently and reliably simulate the turbulent flow in the near-wall region dominated by high-frequency and small-scale motion. However, for large-scale separation flow simulation, it has disadvantages such as large dissipation and easy distortion. LES may accurately calculate the unsteady separation flow area dominated by low-frequency and large-scale motion. However, the total computing resource demands for the circumferential flow field are too large, and it is difficult to be used in engineering development; DES combines the characteristics of RANS near-wall flow efficient simulation and LES unsteady large-scale separation flow area accurate calculation and is applicable for processing separation flow with great Reynolds number with limited computing resources. And it is also applicable for engineering practice.

The main methods for damping derivative test are free vibration test and free flight test. The test model is supported in the incoming flow of the wind tunnel with the support rod in the free vibration test. Under the inflow condition of a certain state, the model is forced to vibrate slightly by the excitation gas source, and the course of attitude variation caused by the free vibration of the model is measured by the dynamic balance so as to identify the test approach for obtaining the aerodynamic damping derivative. This test has the advantage that the variation rule of the oscillation curve of the excitation curve is obvious, the damping derivative symbols so obtained are accurate and the values are reasonable. The disadvantage is that the damping derivative is provided with unavoidable errors due to the interference of the support rod and the vibration limiting device. The free flight test here mainly refers to the ballistic range free flight test, which is carried out in a long closed chamber. The scaled model is accelerated to the required velocity by the light-gas gun launcher and then enters the chamber to fly freely under inertial force and aerodynamic force. Several high-speed photographing and timing stations are set along the flight direction of the model to determine the spatial coordinates, attitude information and time of the model when it flies by each station so as to identify the static aerodynamic forces and damping derivatives during free flight. The advantage of the ballistic target free flight test is that it avoids the interference of the support devices in the conventional wind tunnel test. The flight environment can be set to designed state by adjusting the pressure and gas content in the chamber. However, the identification accuracy of the free flight test depends largely on the accuracy of the model design, the number of measuring stations as well as the accuracy of the measuring equipment. In addition, after the combination of the static data and the second-order damping derivative, it is extremely difficult for the research on the identification approaches. Therefore, if the ballistic range free flight test is applied to identify the damping derivative, data accuracy is also uncertain to some extent. Due to the advantages and disadvantages of the two methods, generally both tests should be carried out simultaneously in order to predict the damper derivative of the inductor across the supersonic velocity accurately if conditions permitting. The damping derivatives obtained by different test approaches are compared and comprehensively analyzed to obtain the final data with greater accuracy. As early as in 1971 when the USA conducted a dynamic stability test on the Viking Mars entry vehicle, as led by the Langley Center, both the ballistic target test and free vibration test were carried out at the Theodore von Karman aerodynamic laboratory. Later on, ballistic range tests were carried out for Titan Huygens entry vehicle, Mars Exploration Rover entry vehicle and Mars Science Laboratory, respectively.

7.3 Aerodynamic Thermal Protection Design

7.3.1 Basic Theory of Thermal Protection Technology

During the aerodynamic braking of the atmosphere into the atmosphere at hypersonic velocity, due to the intense friction between the probe and the surrounding gas and the intense compression of the atmosphere ahead of the probe, most of the kinetic energy of the probe is dissipated in the form of shock waves and wake vortices into the surrounding environment. The remaining part of the kinetic energy is converted into the thermal energy of the gas and is unevenly applied onto the probe by means of convection heating and radiative heating. Therefore, aerodynamic thermal protection is closely related to aerodynamic thermodynamics under hypersonic velocity conditions. Hypersonic velocity thermodynamics is the theoretical basis of aerodynamic thermal protection, and aerodynamic thermal protection is the main research object of hypersonic velocity aerodynamics. It is an important integrated interdisciplinary subject developed based on basic disciplines such as aerodynamics, heat and mass transfer, physical chemistry, materials science, mechanics and solid mechanics.

7.3 Aerodynamic Thermal Protection Design

There are two ways to solve the problem of thermal protection during the process of probe entry: one is to reduce the amount of aerodynamic heating of the incoming flow; the other is to absorb or dissipate such aerodynamic heating as possible. The former depends on the actual conditions of the orbit, aerodynamic shape and probe quality characteristics and is often highly coupled with other constraints of the exploration task; the latter is the research object of aerodynamic thermal protection, where constraints such as material development, design level and process technology are considered to complete the design and verification of the thermal protection system. Its purpose is to provide complete protection for the probe, maintain the integrity of the probe structure, control the aerodynamic shape and maintain the inside of the probe within the ranges of allowable temperature and pressure.

The development of aerodynamic thermal protection technology not only depends on the in-depth knowledge and understanding of the aerodynamic thermal environment, but also depends on the development of technical fields such as thermal protection materials, design and manufacturing, and test verification.

Aerodynamic thermal environment is the starting point of aerodynamic thermal protection research. In other words, aerodynamic thermal protection is carried out in a certain aerodynamic thermal environment. Therefore, the aerodynamic thermal environment is the primary research object of aerodynamic thermal protection, which includes appendage flow and separated flow thermal environment; simple shape and complex shape thermal environment; laminar and turbulent thermal environment; single-phase and multi-phase thermal environment; freezing, balanced and unbalanced gas thermal environment, the wall surface mass addition ejection and the chemical reaction thermal environment.

Different missions have different entry ballistic trajectory and atmospheric components, and the thermal environments often vary greatly. Therefore, the requirements for aerodynamic thermal protection and the thermal protection technologies used are also different. Aerodynamic thermal protection research started from the warhead of missile and later on developed into recoverable satellites, manned spacecraft, space shuttle, probes of moon and other celestial bodies, etc. Various forms of thermal protection have been adopted, among which heat sinking, radiative heating and ablative thermal protection are the most widely used [2].

7.3.1.1 Mechanism of Heat Sinking Thermal Absorption

The thermal protection mechanism of heat sink is a thermal protection method that utilizes the heat capacity of the material to absorb heat during temperature rise to enable thermal protection. The outer layer of the spacecraft structure is covered with a layer of material with a large heat capacity, which will absorb most of the aerodynamic heat, thus reducing the heat entering the structure. As long as the heat capacity of this material is large enough, the internal temperature raise can be maintained below the allowable value. The layer of material for thermal absorption is the heat shield of the endothermic thermal protection structure, and the thermal protection mechanism and main characteristics thereof can be described by an energy balance relationship. Specifically, the net heat flux of the incoming material on the surface of the heat shield is

$$q_n = q_e \left(1 - \frac{h_w}{h_s}\right) - \sigma \varepsilon T_w^4 \quad (7.6)$$

where q_n is the net heat flux from the surface into the heat shield; q_e is the heat flux density introduced into the surface of the heat shield when the surface is assumed at a thermodynamic temperature of zero; h_w is the gas specific enthalpy on the surface of the heat shield; h_s is the stagnation specific enthalpy of the surrounding gas; ε is the total emissivity of the surface of the heat shield; σ is the Stefan Boltzmann constant, $\sigma = 5.67 \times 10^{-8}$ W/(m^2 K^4); T_w is the thermodynamic temperature of the heat shield surface.

For most heat absorbing materials, the allowable temperatures of the materials are unlikely to be quite high. Therefore, the allowable surface temperature T_w is also unlikely to be high. As a result, the radiation term at the right end of Formula (7.6) can be generally omitted. In addition, as long as the thermal protection material is an excellent heat conductor, the heat applied on the surface can quickly diffuse to the entire heat shield. Therefore, theoretically, the maximum amount of heat Q that can be absorbed by the heat shield per unit area is expressed as

$$Q = \rho d c_p (T_w - T_0) = W c_p (T_w - T_0) \quad (7.7)$$

where Q is the heat absorbed by the heat shield per unit area; ρ is the density of the heat absorbing material; d is the thickness of the heat absorbing material; c_p is the specific heat capacity of the heat absorbing material; w is the areal density of the heat absorbing material; T_w is the temperature of the wall of the thermal material; T_0 is the initial temperature of the heat absorbing material.

In Formula (7.7), T_w is the highest temperature of the heat shield after being heated, and its maximum value is either the melting point of the material or the temperature at which the material begins to intensively oxidize or initiate other harmful reactions. As the heat transferred from the surface cannot be conducted to the entire heat shield in time, the heat absorption calculated by Formula (7.7) is only the maximum value when the thermal conductivity of the heat absorbing material tends to be infinite.

The simple mechanism analysis above shows the following basic characteristics of endothermic thermal protection:

(1) The total mass of the heat shield is proportional to the total heat input to the surface. Therefore, this thermal protection method is generally only used in the case of short heating time and low heat flux density; otherwise, the heat shield will be excessively bulky.

7.3 Aerodynamic Thermal Protection Design

(2) In the process of thermal protection, the surface shape and physical state of the heat shield remain unchanged. Some spacecraft are required to maintain their shape in return or entry. Such thermal protection approach has the superiority of maintaining shape and being reusable.
(3) The allowable surface temperature of this thermal protection method is low, and the maximum temperature cannot exceed the melting point of the material. Sometimes, the allowable temperature is even lower, as many materials would start intense oxidation at a temperature far below the melting point. As for general practical heat absorbing materials, the allowable surface temperature of the endothermic heat shield is 600–700 °C, which is much lower than the surface working temperatures of general carbonized ablative materials. Therefore, the radiation heat dissipation hardly works in the heat absorption and thermal protection mechanism.

7.3.1.2 Mechanism of Radiation Heat Dissipation

The mechanism of radiant heat-dissipating thermal protection is to use certain materials with high radiation characteristics at high temperatures to dissipate the surface heat by radiation. This technique is applied to some partial surfaces of the spacecraft, and the radiant thermal protection structure consists of the outer skin directly contacting the high-temperature environment, the internal structure as well as the heat insulating material between both. The outer skin is made of high-temperature-resistant material, and the surface is also covered with a high radiation coating. Therefore, if heated to a relatively high temperature, a large amount of heat energy will be radiated in the form of radiation. Reasonable design can dissipate most of the aerodynamic heat, so that the residual heat entering the internal structure is very small. From the nature of the process, the heat dissipation process of the radiant thermal protection structure is only a physical phenomenon and is not accompanied by the consumption of the heat shield material. As long as the heat flux density of the aerodynamic heating is less than a certain value, the temperature of the outer skin can be lower than its allowable temperature. Therefore, the thermal protection effect will not decrease with the heating time. Based on the energy balance of the skin surface, the net heat Q transferred from the skin to the structure may be expressed as

$$Q = Q_1 + Q_2 + Q_3 - Q_4 = Q_5 + Q_6 + Q_7 + Q_8 \tag{7.8}$$

where Q is the net heat of the incoming structure; Q_1 is the convection blocking factor multiplied by the convective heat; Q_2 is the radiation blocking factor multiplied by the radiation heat; Q_3 is the solar radiation heat; Q_4 is the heat dissipating capacity of the outer surface of the skin; Q_5 is the temperature rise of the structural material; Q_6 is the heat transferred into the probe; Q_7 is the phase change heat absorption of the heat shield material; Q_8 is the heat absorbed by the additional cooling device of the structure.

Q_1 in Formula (7.8) is the convective aerodynamic heat received by the surface skin. The convection blocking effect is caused by the reduction of convective heating when the surface of the heat shield has mass injected into the boundary layer. In the absorbing heat shield and radiant heat shield, there is no mass injection into the thermal layer. Therefore, the convection blocking factor is equal to 1. Q_2 in the formula represents the radiant heat of the high-temperature gas surrounding the probe, which is generally negligible in case of atmosphere entry at low velocity. Q_3 is the solar radiation heat, which is always negligible compared to Q_1 and Q_2. Q_4 in the formula is the heat radiated from the outer surface of the skin to the surrounding environment. The higher the amount is, the smaller the net heat of the incoming surface is. In other words, the burden of various thermal protection and absorption measures indicated by Q_5 to Q_8 are smaller accordingly.

The basic principle of radiation thermal protection is to increase the radiative heat dissipation of the surface of the thermal protection layer. The optimal structure for radiant thermal protection is that the emissivity of the inner surface of the skin equals to zero, or the thermal conductivity of the insulating material behind the skin equals to zero. In this optimal configuration, the external heating of the surface is completely offset by the radiant heat of the surface, which is, the net heat entering the surface is zero. The skin temperature in this case is defined as the radiant equilibrium temperature $T_{r,\max}$. If the solar radiant heat and the high-temperature gas radiant heat are neglected, the radiant equilibrium temperature is determined by the following formula

$$q_{\max} = \sigma \varepsilon_{ou} T_{r,\max}^4 \tag{7.9}$$

where $q_{\max}$ is the maximum heat flux density of convective aerodynamic heat; σ is the Stefan Boltzmann constant, $\sigma = 5.67 \times 10^{-8}$ W/(m^2 K^4); ε_{ou} is the hemispherical total emissivity of the outer surface of the skin; $T_{r,\max}$ is the radiant equilibrium temperature.

The radiant equilibrium temperature is an ideal temperature, which is, the highest temperature that the skin may reach at a certain heat flux density. For radiant thermal protection structures, the closer the skin surface is to this temperature, the higher the efficiency of thermal protection is. In addition, the radiant equilibrium temperature corresponds to the maximum heat flux density. Therefore, regardless of the thermal protection approach actually used, the radiant equilibrium temperature can be taken to express the thermal environment of the spacecraft entering the atmosphere and the distribution of heat flux density at various locations. If the thermal environment is expressed by the radiant equilibrium temperature, the surface emissivity is generally taken as 0.8. This method is more intuitive than by directly giving the magnitude and distribution of the heat flux density, and it enables quick determination of which parts are not able to adopt the radiant thermal protection structure—the radiant equilibrium temperature of such parts has exceeded the allowable range by the practical materials. In other parts, if a radiant protection structure is applied, it is also a simple and reliable approach to select the skin material according to the radiant equilibrium temperature.

7.3.1.3 Mechanism of Ablative Thermal Protection

Ablative thermal protection dissipates the heat by means of melting, vaporizing and pyrolysis on the surface layer of the thermal protection material and loses part of the thermal protection material to meet the internal environment requirements of the probe. Most ablative thermal protection materials are composites with organic resin as the continuous phase. After heating, the gas product (mostly hydrocarbons) by the decomposition of the resin penetrates into the hot surface and is then injected into the boundary layer; the carbonaceous residue is deposited on the surface of the reinforcement and forms a carbon layer. The gas penetrates into the hot surface through the porous carbon and provides part of the convective cooling. The gas is further injected into the adjacent boundary layer located on the surface to provide part of the evaporative cooling. The surface carbon layer may be consumed by chemical reaction with the boundary layer gas, causing the retreating of surface ablation. A carbonaceous surface carbon layer is a preferred choice as it can withstand quite high temperatures. Obviously, compared with heat sinking thermal absorption and radiant thermal dissipation, the interaction between ablative thermal protection material and surrounding ambient gas is much more complicated, and many mechanisms work together in the process of aerodynamic heating.

Taking the typical ablative material as an example, set T_{P1} as the temperature at which the material starts to pyrolyze after being heated; T_{P2} is the temperature at which the material is completely pyrolyzed into a carbon layer. With the heating of the surface of the spacecraft, the surface temperature starts to rise. During the heating process, a part of the heat is absorbed by the heat capacity of the material itself, and meanwhile, a part of the heat is introduced into the material through the solid conduction. In addition, through surface radiation, a part of the heat is scattered into the atmosphere. As long as the surface temperature continues to be below the pyrolysis temperature of the material, the above state will continue. At this time, the entire ablative thermal protection structure will work similarly to a heat capacity endothermic structure. As the surface temperature continues to rise, the material starts to pyrolyze and carbonize, meanwhile producing a very large temperature field along the thickness direction. As the materials have different states in different temperature ranges, three different layered zones are created within the material: the original material zone, the pyrolysis zone and the carbonization zone.

Raw material zone: In this zone, the material temperature is lower than T_{P1}. Therefore, the physical state of the material remains unchanged except for a significant change in temperature distribution. There are only two thermal phenomena occurring within this area: one is the heat absorption of the material itself, and the other is the heat conduction to the internal material.

Pyrolysis zone: The pyrolysis zone is unstable with an inner boundary temperature of T_{P1} and an outer boundary temperature of T_{P2}, both of which move inward at a certain velocity. The material within this area is in a continuous pyrolysis process: on one hand, it gives out many mixed gases composed of high molecular weight and low molecular weight gas, and on the other, the foamy solid carbon layer is residual, so the area is a multi-phase coexistence area. The thermal phenomena occurring in this area mainly consist of three parts: one is thermal desorption of materials; one is heat absorption of solid carbon and pyrolysis gases; and the third one is solid heat conduction.

Carbonization zone: If the temperature of the material exceeds T_{P2}, the organic matter in the material has been basically decomposed, leaving a stable carbon layer. And the pyrolysis gas generated by the pyrolysis zone flows to the surface through the carbon layer. As the main component of the carbon layer is carbon, the melting point thereof is high. In addition, the carbon layer is porous, and its thermal conductivity is small. Therefore, a high-temperature thermal isolation layer is naturally formed on the surface of the material. Thermal phenomena occurring in this region include heat capacity endotherm of carbon layer and pyrolysis gas, solid conduction as well as possible pyrolysis gas secondary cracking and carbon–silicon reaction.

Material surface: The thermal phenomena that occur on the surface of the material are extremely complicated with both heating and heat dissipation. The convection and radiation from the aerodynamic heating may directly affect the surface of the material; the hot carbon layer will chemically react with the surrounding atmosphere to generate a part of the reaction gas and absorb/release part of the heat; a large amount of pyrolysis gas and combustion product are injected into boundary layer, changing the original velocity, temperature and concentration of each component gas in the boundary layer. Therefore, the boundary layer is thickened, and the convective heating of the boundary layer gas to the surface is reduced. The carbon layer in the high-temperature state radiates a large amount of heat to the surrounding space. If the surface temperature is high, the carbon in the carbon layer will directly sublime into a gas injected into the boundary layer and absorb the sublimation heat.

According to the principle of energy conservation, all thermal phenomena occurring in the above three regions and the boundary of the material surface should satisfy the energy balance conditions, and their mathematical expressions are as follows.

$$\varphi Q_e + \varepsilon Q_r + \sum_{i=1}^{7} Q_i = 0 \qquad (7.10)$$

where Q_e is the total convection heating amount; φ is the convection blocking coefficient, $\varphi \leq 1$; Q_r is the shock radiative heating amount; ε is the surface gray body coefficient; Q_1 is the carbon layer oxidation combustion thermal protection; Q_2 is the surface radiation heat dissipation; Q_3 is the thermal absorption of the material heat capacity; Q_4 is the thermal decomposition heat of the material; Q_5 is the absorption endothermic heat of the pyrolysis gas; Q_6 is the heat absorption of the carbon sublimation; and Q_7 is the heat introduced into the structure.

7.3.2 Aerodynamic Thermal Protection Technology

Since the beginning of the deep space exploration of the humans, the complex environmental conditions, high development costs and stringent weight indicators that are inherent in such mission have continuously challenged the aerodynamic thermal protection technology.

Within the solar system, all planets except the Mercury have the atmosphere. Therefore, in the field of interplanetary deep space explorations, aerodynamic thermal protection technology is of great important practical significance and has been widely applied for deep space exploration missions of celestial bodies with atmosphere.

The energy absorbed or dissipated by the probe is significantly different with different entry modes for different exploration targets. Regardless of the actual gas effect, at subsonic and low Mach, the temperature of surrounding gas will rise due to compression and friction of the probe. The probe mainly absorbs heat by convection heat transfer. In contrast, in hypersonic velocity conditions, the stagnation temperature of the front edge of the probe may be as high as 10,000 K or even higher, and the radiative heating of the high-temperature gas in the shock layer may also reach the order of convection heating. In such a high-temperature environment, atmospheric composition has become an important factor. The atmosphere itself may undergo dissociation and ionization reactions. In addition, the high-temperature components will react with the surface materials of the probe. In most planetary entry missions, the thermal protection design should also consider the phase change effect of material, wall catalytic effect, and particle erosion effect, the coupling effect of the aerodynamic shape and the aerodynamic environment as well as the structural thermal response characteristics of the probe.

Deep space probes generally require a huge amount of development funds, a huge investment of human and material resources, and a long development process to achieve success. At present, researches on extraterrestrial planets by the humans are not quite comprehensive and sufficient, and there are still quite uncertain factors in various environmental conditions during the process of planetary entry. Therefore, improving the design redundancy of the thermal protection of the probe seems to be a reasonable response. However, the total mass of the probe depends on the carrying capacity of the rocket. The total mass of the probe in deep space exploration mission is always strictly restrained. This is also the problem that the designers should face objectively.

Thermal protection technology, which is lightweight, efficiency and reliable, has always been the requirement of deep space exploration. However, there is no "universal" thermal protection measure in this field, and each thermal protection design for a specific entry mission should be adequate. Considering the specific aerodynamic environmental conditions, the thermal protection performance and heat insulation efficiency should be well-balanced, and the test verification should be carried out on the ground as far as possible. For example, in low heat flow conditions, it is best to use thermal protection materials that can start carbonization at a lower temperature for ablation and heat protection. Under the condition of long heating time, the materials often require relatively low thermal conductivity and relatively low density; for relatively moderate total heating conditions featuring high heat flux and high stagnation pressure. The carbon phenolic material with general thermal insulation properties but excellent ablation resistance is usually the first choice [3].

Generally, the thermal protection design for deep space probes is always restricted by many factors, and the reasonable thermal protection design should be able to reliably cope with the aerodynamic thermal environment expected by the mission plan. In addition, the mass ratio of the heat shield should be minimized to improve the performance of the whole system.

At present, all planetary entry and exploration missions have been designed for specific aerodynamic thermal environments. And ablative thermal protection or a combination of ablative and other thermal protection approaches such as ablative radiant thermal protection and ablative heat and thermal protection are applied. The greatest advantage of ablation thermal protection is safety and reliability. It can adapt to the change of the circumferential flow field and the uncertainty of atmospheric parameters or task environment. It is an efficient and reliable thermal protection technology suitable for deep space exploration missions.

1. **Thermal Protection Technology in Mars Exploration** [4]

In the late 1960s, based on the research of Apollo thermal protection, in order to adapt to the aerodynamic thermal protection environment of Mars probe, the mass of the heat shield of Mars entry vehicle was further reduced and the payload of Mars entry was increased. The USA launched the Super Lightweight Ablators (SLA) program and developed SLA-561 V thermal protection materials for Mars probes. After entering the thermal environment of Mars, SLA-561 V has 50% higher ablation efficiency than the original low-density carbonized ablative materials, such as nylon phenolic ablative materials and silicone-filled ablative materials. In 1975, SLA-561 V succeeded in the thermal protection of the Viking probe, making it the preferred thermal protection material for Mars entry and exploration. It also obtained great success in the subsequent probes such as Mars Pathfinder, Courage, Opportunity and Phoenix.

In the 1990s, after 20 years of development, the USA once again attached importance to exploration programs for Mars and other celestial bodies. To further improve the performance of thermal protection materials and reduce the total mass of thermal protection systems, we began to find alternative thermal protection materials with advanced performance to replace Avcoat 5026-39 and SLA-561 V. NASA Ames Research Center launched the Lightweight Ceramic Ablators (LCA) Project, which is dedicated to the synthesis of high-temperature, high-strength, heat-resistant materials with low-density fiber substrates and organic resins. Unlike conventional ablative materials, LCA adopts a porous ceramic or carbon fiber precursor as a substrate to provide the support structure and adopts the polymer resin as the filler. The substrate mainly includes Reusable Surface Insulation (RSI), such as Lockheed Insulation's LI-900, Ames Center's Alumina Enhanced Thermal Barrier and carbon fiber body heat insulating materials by Fiber Materials Inc. (FMI). And the polymethyl methacrylate resin, the epoxy resin and the phenol resin are selected as fillers. By utilizing

7.3 Aerodynamic Thermal Protection Design   265

the different combinations of the above substrates and fillers, in combination with the corresponding impregnation process, 14 lightweight thermal protection materials including phenolic impregnated carbon ablator (PICA) were developed.

MSL was the first Mars probe that encountered the turbulent thermal environment. In ground test, due to the existence of the turbulent thermal environment, the SLA-561 V experienced an unexpected catastrophic failure in the aerodynamic thermal shear test, and the thermal protection design was shifted to the PICA with the equivalent density, and finally the test succeeded. The process of MSL thermal protection from failure to success indicates that even the materials applied in the previous applications may encounter new problems in new missions. In addition, the gas medium for previous aerodynamic thermal test is air, the atmospheric composition of Mars is mainly CO_2 and the change of gas composition will result in difference between the catalytic effect of the material and that in the Earth's atmosphere. To faithfully evaluate the aerodynamic environment adaptability of heat-resistant materials and heat-resistant components, NASA has begun to use the CO_2 as a medium for experimental research of the environment and to simulate the orbital simulation test technology of Mars orbital heating in the arc wind tunnel of JSC. Germany has applied CO_2 as a medium on high-frequency induction heaters and also carried out related experimental research on entry into the thermal environment of Mars.

For the future thermal protection requirements in Mars exploration, the thermal protection material being researched and tested in the USA is the Reusable Ceramic Ablation Material (SIRCA). This material is slightly lighter than the SLA-561 V, and initial test results show that it has higher compressive capacity and higher heat flux rate than SLA-561 V. According to the previous experience of thermal protection design and aerodynamic heating analysis of Mars entry mission, MAV with SIRCA advanced has a mass ratio of thermal ablative material of about 6%.

2. **Thermal protection technology in Venus exploration** [5]

At present, only the Soviet Union and the USA have completed Venus landing mission, while most of others are failed or partially completed. Due to the dense atmosphere, high temperature and strong pressure of Venus, the entry environment of Venus exploration is very harsh, and the requirements for braking and thermal protection of the probe are very high. In addition, for a dense atmosphere environment such as Venus, the current probe has a high ballistic coefficient, which averages 200 kg/cm^2 or higher, resulting in poor aerodynamic braking, large acceleration overload (200–450 g) as well as high heat flux peak density (2000–7000 W/cm^2). To withstand such a large heat flux density, high-density carbon phenolic materials are selected for the thermal protection materials. This is the only ablative material that has been characterized in detail to cope with such high heat flux and result in higher mass ratio of the heat-resistant materials of Venus probes, thus limiting the transport capacity of the probe load.

In the future, if Venus probe enters the atmosphere by means of aerodynamic capture, the peak heat flux will be relatively low. However, the total heating will increase significantly. In such cases, high-density carbon phenolic is not the optimal material. Medium-density heat-insulating materials with better thermal insulation properties may be a better choice. In addition, solid ablated body back lined with high-temperature, low-density insulation is also a consideration. Compared to carbon phenolic materials, such new solutions can reduce the proportion of thermal protection by 50%.

3. **Thermal protection technology in Jupiter exploration** [5]

The Galileo Program for Jupiter exploration is quite challenging in the planetary entry missions. The probe applied a 45° blunt-conical heat shield and entered the Jupiter atmosphere at a rate of 47.4 km/s. The entry environment was extremely harsh, with an expected peak heat flux (including conduction and radiant heat flux) of approximately 35 kW/cm^2 and a total heating capacity of approximately 200 kJ/cm^2.

High-density carbon phenolic materials thermally protected the first half of the Galileo probe. For the heat-resistant design, the most advanced engineering techniques of the 1970s were applied. Several groups developed the design independently and their approaches were used for analysis and design evaluation. Such models addressed the correlation between the boundary layer and the impact layer that undergo a chemical reaction in the case of thermochemical ablation and a small amount of ablation. However, due to the limitations of ground testing, some models could not be verified (such as impact layer radiation). The final thermal protection design, especially the thickness distribution, was constructed by adding a margin to the various conservative boundaries. Such margin is determined by the teams and the ablation of the Galileo probe was installed. The sensor provided a basis for evaluating the accuracy of the design approach. As indicated by the ablation data derived from the flight data, that the amount of retreat of the stagnation point was smaller than expected, and the amount of retreat of the shoulder was significantly larger than expected (almost burning through).

In the future Jupiter's entry and exploration programs, some probes will enter higher latitudes, and their entry velocity will also be higher. As the amount of heating increases in quadratic velocity, such a high heating amount is too severe for high-density carbon phenolic aldehyde, and the thermal protection quality ratio will exceed 70%, so that the payload may be quite small. Therefore, higher requirements have been proposed for thermal protection materials, and it is necessary to further improve existing mature materials, develop new materials, propose new structural design concepts, build higher-level test equipment and establish more accurate and efficient analytical models. New challenges have emerged in all aspects of technological development.

7.3.3 Aerodynamic Thermal Protection Design

The thermal protection system (TPS) is a subsystem of the probe that can protect the probe from the harsh aerodynamic heating environment when it passes through the planetary atmosphere at high hypersonic velocity and enters the planet safely. Because TPS is the single fault point of the probe, it is the key to the success of

7.3 Aerodynamic Thermal Protection Design

the exploration mission. It can be said that in most deep space exploration missions requiring solve the problem of aerodynamic thermal protection, the structure and materials of the probe mainly depend on the requirements of the design of the protection system, and the performance should be analyzed and verified as possible through simulation calculations and ground tests.

In summary, the thermal protection system design to be solved for the probe is as follows:

(1) Different missions are subject to different aerodynamic thermal environments. Even for the same probe, the heating conditions of different parts are quite different. Therefore, it is necessary to correctly analyze the aerodynamic thermal environment faced by the probe, reasonably develop a thermal protection measure and specify alternative thermal protection materials.
(2) Select suitable thermal protection materials based on the comprehensive comparison of the thermal protection, thermal insulation characteristics, structural compatibility, process feasibility, etc., and carry out structural design accordingly.
(3) Establish a reasonable analysis model based on engineering algorithm or numerical algorithm through material-level ground wind tunnel tests and performance tests and make necessary model correction with the experimental data and physical parameters to determine the optimal global design for thermal prevention layer thickness.
(4) Reasonably handle the local thermal protection problems on the probe surface such as protrusions, cracks and steps.

7.3.3.1 Design Factor Analysis

There are two major parameters that affect the choice of probe thermal protection measure: one is environmental parameter and the other is the constraint condition. The former reflects the parameters that the probe will face such as the launch section environment, the track environment, the heat flux density and the total heating amount. The latter includes the overall design and technical requirements of the thermal protection system for each subsystem, such as aerodynamic shape, connection separation surface design, instrument surface opening, the maximum allowable inner wall temperature rise and the quality characteristics of the distribution.

Peak heat flux density and total heating capacity are important environmental parameters that affect the design of the thermal protection measure. For example, the peak heat flux density directly reflects the radiant equilibrium temperature that the probe may reach during the process of entering the atmosphere, while the excessive radiant equilibrium temperature would limit the use of the endothermic and radiant thermal protection measures. The total heating amount is a comprehensive parameter reflecting the heat flux density and the length of heating time, which directly affects the selection of thermal prevention and insulation materials and thickness design.

The maximum allowable inner wall temperature rise by the design is an important standard to determine the thickness of heat shield. Obviously, the smaller the wall temperature rise is, the thicker the thickness of the required heat shield is, and the greater of the mass of corresponding heat shield is. However, due to low wall temperature, the bearing structure does not cause decreased strength performance due to the temperature rise. The required structural quality is also relatively small. Conversely, if the allowable temperature rise of the wall is increased, the thickness of the heat shield may be appropriately decreased and the mass of the heat shield can be reduced accordingly. However, due to the increase in the working temperature of the bearing structure, the strength of the material may be lowered, and the required structural mass would increase. Therefore, the maximum inner wall surface temperature rise index can be reasonably determined from the perspective of global design.

After solving the global heat shield material layout and thickness design issues, special attention should be paid to local thermal protection problems. Due to functional and application requirements, it is often necessary to design some protrusions, cavities and gap structures on the surface of the probe. Ablative heat transfer effects different from the large-area heat shield are formed around such local structures.

As confirmed by both theory and practice, whether at the subsonic or hypersonic conditions, and whether the incoming flow is laminar or turbulent, the protrusion or surface discontinuity will cause the increase in the heat flux density in local ranges. More seriously, when the incoming flow is laminar, the convexity and discontinuity of the surface will cause the fluid properties to transit into turbulence, further resulting in sharp rise of local heat flux density, thereby causing local heat failure. In order to realize the reliable design of the local thermal protection structures, it is necessary to judge the aerodynamic coupling and the ablation heat transfer effect between the local structure and the incoming flow on both practical and theoretical basis. However, this effect is generally accompanied by complicated flow phenomena, such as flowing shock, separation, reattachment and vortices. Therefore, for the key parts, it is also necessary to design local structural specimens with the same size ratio for wind tunnel test evaluation to ensure the reliability and safety of the thermal protection design.

7.3.3.2 Anti-heat Material Screening

After the preliminary thermal protection plan is formulated, appropriate thermal protection materials should be selected. There is no thermal protection material suitable for all environments. Generally, materials with high heat flux have high thermal conductivity, but the thermal protection effect is not satisfying when the heating time is too long. Low-density material has low strengths. When the stagnation point pressure is too high, it is prone to denudation and the thermal protection fails. A certain kind of thermal protection material may perform well in a specific peak heat flux environment. However, it would fail under the same peak heat flux when the

7.3 Aerodynamic Thermal Protection Design

stagnation pressure is significantly increased. It is sometimes necessary to improve the existing thermal protection in order to cope with the difficult task environment. The material has been improved, and even new heat-resistant material systems have been developed.

1) Heat sink absorbing thermal protection material

Heat sink thermal absorption is an inefficient thermal protection method, but it is simple and easy to use. Therefore, it has been widely applied in missiles and early loaded spacecraft. Therefore, tantalum, beryllium oxy chloride and graphite are the preferred materials, which all have the following common characteristics: large specific heat capacity, high melting point and high thermal conductivity. Among them, graphite has the best heat absorption performance. However, the oxidation begins at a temperature far below the melting point, so graphite without antioxidation measures cannot give full play to the endothermic potential of the material.

2) Radiant dissipating thermal protection material

For radiant heat dissipating, the maximum heat flux allowed by the heat shield depends on the maximum temperature that the skin material can withstand. According to the current conditions of materials and processes, titanium germanium alloy is generally used when the radiant equilibrium temperature is lower than 500 °C; and high-temperature alloy based on iron, cobalt and nickel is applied when the temperature is lower than 500 °C and lower than 950 °C; refractory metal with antioxidation coating is applied when the temperature is within the range of 1000–1650 °C. If the temperature is even higher, temperature-resistant ceramic and specially treated carbon fiber or graphite fiber composite is applied. To solve the problem of rapid oxidation of the skin material under high-temperature conditions, it is necessary to apply an antioxidation coating on the surface and avoid failure after being exposed to long-term space high and low temperature and vacuum environment. In addition, high-temperature insulation materials are part of the main components in radiant thermal protection structures. Inorganic nonmetallic materials have the advantages of larger specific heat capacity and lower thermal conductivity, which are the major components of most high-temperature insulation materials.

3) Ablative thermal protection material

Ablative heat-resistant materials can adapt to a wide range of variations and mutations of external heat flux and are provided with high safety and reliability. Such materials are currently the most widely applied thermal protection materials in deep space exploration. The principle of ablation thermal protection is that ablative materials produce a series of physical and chemical reactions in a heated environment (i.e., melting, pyrolysis, pyrolysis gas into the boundary layer, sublimation, in-phase and out-of-phase chemical reactions, etc.). In this physical and chemical process, on one hand, ablative materials are consumed; on the other hand, the heat imparted to the materials by the environment is dissipated in different ways to ensure that the internal structure can operate at the allowable temperature.

There are many ways to classify ablative materials. Generally, ablative materials can be classified into sublimation, melting and carbonization according to different ablation mechanisms, of which carbonized ablative materials are widely applied in deep space probes. Depending on the material density, the ablative materials are further subdivided into high-density ablative materials, medium-density ablative materials and low-density ablative materials.

The sublimation materials are characterized in that the materials are directly injected into the boundary layer without being heated through the liquid phase. Such materials rely primarily on sublimation heat absorption and convective blocking effects of gaseous products to prevent heat. Typical sublimation ablative materials include polytetrafluoroethylene, graphite, etc.

The characteristic of the molten ablative materials lies in that, except the pyrolysis and carbonization of the resin, the components of glass, high silicon oxide, quartz, etc., are melted into the liquid, and the liquid permeates through the carbon layer to form a viscous liquid protective layer on material surface. On one hand, this layer of liquid makes it impossible for the boundary layer gas to directly heat the material, and on the other hand, it absorbs and takes away a large amount of heat during the process of evaporation and loss. Such materials include inorganic fiber reinforced plastics such as glass phenolic and high siloxane phenolic.

The most important feature of carbonized ablative materials is low-temperature pyrolysis and high-temperature operation. The material starts to undergo pyrolysis and phase transformation at a relatively low temperature, such as 250–300 °C, and then forms a thick carbon-based carbon layer on the surface of the material. This layer of carbon may withstand high temperatures, effective radiative heat dissipation, and may also act as a high-temperature insulation material to protect the internal structure. Typical carbonized ablative materials include nylon phenolic, carbon phenolic, silicone resins, thermoplastic epoxy, etc.

4) Material screening

The screening of heat-resistant materials is affected by multiple factors, of which the most important ones involve two aspects: one is the choice of materials for thermal prevention and insulation, and the other is the consideration of structure and process performance of the material. In addition, the ability of materials to work under vacuum, ultraviolet radiation as well as electron and proton irradiation should also be considered.

If materials with low technical maturity are selected, or if the mission is faced with a new aerodynamic environment, it is usually necessary to study the ablation heat transfer characteristics of materials under specific thermal conditions through material-level wind tunnel tests, especially for the failure thresholds and failure mechanisms of new materials. Figure 7.4 shows the circumferential flow field of the wind tunnel test.

7.3 Aerodynamic Thermal Protection Design

Fig. 7.4 Circumferential flow field of wind tunnel test

Thermal prevention materials are always moving toward more reliable and efficient directions. "Reliable" means that the performance of the material must be consistent with expectations without failure, while "efficient" means that the material is used in the most efficient area, that is, to generate carbon at expected temperature is carbon and forms a reradiating surface and to effectively crack without excessive ablation. Therefore, it is most suitable for a thermal protection material that can function efficiently on a reliable basis to minimize the mass of the thermal protection system; that is, lightweight is always the objective for the design of the thermal protection system. To achieve this objective, it is necessary to establish a numerical analysis model based on the appropriate theoretical basis of aerodynamic thermal protection. And experimental data and physical parameters should be used to make the necessary model corrections, so as to determine the optimal global thermal insulation material layout and thickness design.

7.3.3.3 Ablation and Heat Transfer Analysis

The hypersonic chemical boundary layer is the basis of the aerodynamic thermal protection theory of missiles and other spacecraft. It is also the mechanism and performance prediction for studying the interaction between the thermal environment of the warhead or spacecraft and the thermal protection material. Because this interaction is often related to the mass loss of materials, it is usually called the boundary layer. It has mass injection and chemical reaction, which is one of the theoretical bases for the establishment of ablation heat transfer analysis model.

The ablation mechanism of thermal protection materials mainly refers to chemical reaction, sublimation and liquefaction loss of thermal protection material in the ablation process, and finally, the ablation morphology of the surface of the thermal protection structure is obtained. Different adiabatic materials are provided with different ablative heat transfer effects and mechanisms, and the corresponding analysis models have different focuses. The study of ablation mechanism mainly depends on theoretical analysis and experimental research. In theoretical analysis, the ablation chemical reaction of the material is taken as the research object, and the reaction rate in the chemical equilibrium state is obtained through the main chemical reaction. The static test method is mainly used to obtain the thermal weight loss curve of the material under high-temperature environment. The results of theoretical analysis are corrected.

According to the different ablation mechanisms, ablative thermal protection materials can be mainly divided into three types: silicon-based composite, carbon-based composite and carbonized composite. Generally, the silicon-based composite belongs to the molten ablative material, the carbonized composite belongs to the sublimation type ablative material, and the carbonized composite belongs to the cracked ablative material.

Taking Mars exploration mission as an example, the factors to be considered or involved in the ablation heat transfer analysis model of Mars probe are shown in Fig. 7.5.

The aerodynamic heating environment of Mars is highly correlated with the surface catalytic properties of the thermal protection material. The non-catalytic surface peak heat flux is 20%–30% of the total catalytic surface heat flux.

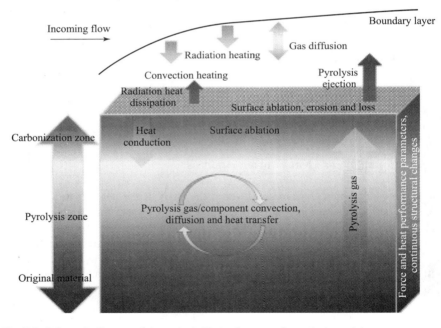

Fig. 7.5 Schematic diagram of the typical ablation heat transfer analysis model

7.3 Aerodynamic Thermal Protection Design

If a cellular-reinforced low-density ablative material, which belongs to the carbon–silicon hybrid system material, is applied. At high temperature, it is necessary to consider the liquefaction loss of silicon-based materials and the chemical reaction with gas; while PICA materials are mainly carbon-based material, which are on the hot Mars. Under environmental conditions, the most important factor to consider is the chemical reaction with gases. In the air environment, the chemical reaction mainly considers the oxidation of oxygen. In the process of Mars entry, the oxidation of CO_2 should be considered.

Generally, in the design and analysis process, it is necessary to pay attention to the consequences and mechanisms of airflow and material interaction. There are two main connotations. One is the destruction of the material by the airflow, where the ablative thermal protection mechanism is to scarify the mass of the material so as to absorb the heat of the airflow acting on the material and the radiant heat of the outside to achieve the purpose of thermal protection. Therefore, while discussing the destructive effect of the gas flow on the material, it is also necessary to study the heat absorption and heat dissipation principles of the material. The other is the reaction of the material to the airflow after ablation. An important example of this reaction is the effect of ablation on the aerodynamic force coefficient, and the effect of surface ablation on aerodynamic shear and even laminar transition.

Wind tunnel ablation test

Wind tunnel ablation test is based on the test and calculation results of the thermal environment, establishes the test conditions with simulation significance, tests the thermal protection material or the local structure model, examines the thermal response characteristics or predicts and evaluates the effect of the thermal protection design. In addition, it belongs to the field of experimental aerodynamics. Due to the multifaceted mechanism of the airflow on the thermal protection material, it is determined that the thermal environment simulation involves relatively great number of parameters. These parameters include the heat flux, enthalpy and boundary layer flow conditions that affect the thermal effect; heating time and total heating amount that affect heat accumulation effects; pressure and velocity affecting force; temperature, pressure, and gas flow components that affect chemical action. Because the absolute values of these parameters are applied to the thermal protection material, the thermal environment simulation mainly refers to the simulation of the absolute values of these parameters.

The main characterization parameter of the aerodynamic heating environment is heat flux, that is, the heat per unit time and unit area. Therefore, the preferred simulation parameter for any wind tunnel ablation test is heat flux. There are many factors affecting the size of the heat flux, such as airflow enthalpy, pressure, velocity and flow state, of which the most important one is the enthalpy value. Any heat transfer phenomenon is caused by the temperature difference, and the aerodynamic heating depends mainly on the difference between the boundary layer air temperature and the solid wall temperature. The initial wall temperature is normal temperature. Therefore, the determining factor is the air temperature. As the heating process progresses, the wall temperature increases and the corresponding heat flux changes.

The varying wall temperature poses difficulties in the experimental measurement of heat flux, so the concept of "cold wall heat flux" is introduced. That is, in the actual measurement, the wall temperature of the heat flux sensor is controlled to the initial temperature so that the heat flux can have a certain value. Compared with the cold wall heat flux, the corresponding heat flux when the wall surface temperature increases is called "hot wall heat flux." It can be easily seen that the real heat transfer actually acts on the hot wall heat flux. Therefore, the experimentally determined cold wall heat flux must be corrected before being applied.

The total enthalpy (or total temperature) of the airflow not only dominates the heat flux. In addition, it directly affects the physical and chemical reactions of the thermal protection material. The airflow pressure not only affects the heat flux but also is the dominant factor for the mechanical ablation of the thermal protection material. It also affects related chemical reactions. Therefore, total enthalpy and pressure are also important simulation parameters for wind tunnel ablation tests.

Due to the heat accumulation effect of the external heating response of the probe, the wind tunnel ablation test should generally consider the simulation of heating time and total heating. In cases where the heat flux along the orbit does not change greatly, consider the approach of determining the test state by multiplying the average heat flux by the test time equal to the total heating amount. Several typical heat fluxes can also be selected in the range of heat flux variation to give a variation range of the thermal response characteristics of the test piece.

In case of long flight time and large parameter changes, orbit simulation technology should be considered. Specifically, in flight conditions, the curve of the heat flux with time is divided into several small segments, so that the heat flux in each segment does not change greatly, and the average value of the heat flux in each segment and the corresponding time period are taken as objects, and the airflow is adjusted in the same test. The parameter is segmented to simulate the heating process. Orbital simulation has high requirements on experimental equipment and technology. It requires enough test time and adjustment of airflow parameters over a wide range during the test. This is also an important research topic in the current wind tunnel ablation test technology.

Generally, the wind tunnel ablation test does not simulate incoming Mach numbers. If the absolute value of the velocity is applied to the heating or shearing of the model, velocity simulation may be considered if necessary. For test projects where the Mach number plays a leading role, such as the local thermal environment with shock interference characteristics, the simulation of the local Mach number should also be considered.

At present, the wind tunnel ablation test of the thermal protection structure is generally divided into four categories:

(1) By material ablation performance research test, the ablation performance and thermal insulation performance of the ablated material are compared and analyzed to provide basis in material selection and thermal protection design for the ablative thermal protection structure;

(2) By the shear ablation test, the ability of ablative materials is assessed to resist high-temperature airflow, research and determine the ablation matching performance of the joints of two or more ablative materials on the heat shield;
(3) By the ablative performance acceptance test, the ablation performance of the material is tested in the specified thermal environment, so as to determine whether the manufacturing quality of the material can satisfy the acceptance requirements;
(4) By the thermal ablation test, the temperature distribution and structural characteristics of the heat-resistant structural component are tested under the specified thermal environment, and the thermal protection and structural properties of the component are studied and determined.

7.4 Atmospheric Entry Guidance and Control Design

7.4.1 Atmospheric Entry Guidance and Control Technology

The research on atmospheric entry guidance process is mainly aimed at a series of problems brought about by the high-velocity entry into the atmosphere, including the entry into the orbit design, the entry into the guidance and control approaches, etc. The basic mission guidance and control in the process of entry is to disengage the entry vehicle from the original orbit, establish and transfer into the orbit and allow the entry vehicle to enter and pass through the atmosphere and safely land in the intended zone without any loss.

In case of Earth entry, the ballistic-lift-type entry is generally adopted in the return processes of satellite and manned spacecraft (which can be regarded as an improvement for the ballistic entry). The spacecraft with lift–drag ratio for Earth entry is designed with its center of mass deviated from the central axis by a certain distance. After entering the atmosphere, it will produce a small angle of attack and generate a certain lift, which can increase the width of the entry corridor, reduce the overload peak in the process of entry, thereby increasing the success rate of entry. The ballistic-lifting entry trajectory can be divided into a hopping entry trajectory and a direct entry trajectory. The hopping entry is generally used for entry missions when the deep space probe returns at high velocity, such as the entry guidance process in which the Apollo spacecraft returned.

When the deep space probe returns to the Earth, the initial velocity of entering the atmosphere can reach 11 km/s, and the initial kinetic energy is about twice that of the near-Earth orbit spacecraft. It brings a series of difficulties and challenges to the entrance guidance, such as entry corridor. The limits and guidance laws are more sensitive to the initial entry velocity and initial entry angle deviation, the peak heat flux density and the excessive peak of the incoming overload, which would seriously affect the accuracy, safety and reliability of the return of the deep space probe.

For spacecraft entering the atmosphere at a velocity close to the second cosmic space, generally the hopping form is applied to reduce the entry overloading and adjust the landing point over a wide range. Table 7.8 compares the key performance parameters of the hopping entry and one-time entry of the lunar probe. The data in the table is typical data. It can be seen that, for small lifting bodies such as lunar probes, the hopping track plays an important role in expanding the range, reducing the overload and peak heat flux density, etc. However, due to the large range and the long entry time, the deviation of the landing points because of the deviation of atmospheric density, the deviation of the aerodynamic coefficient, the navigation error and the aerodynamic shape deformation caused by the ablation. Therefore, higher requirements are raised on the guidance system of high-precision landing.

Table 7.8 Comparison of performance parameters between hopping entry and one-off entry

Parameters	Hopping entry	One-off entry	Problems in engineering application
Entry point velocity/(km·s^{-1})	11.03	11.03	
Maximum range/km	>10,000	3500	Ability to adapt to entry error
Maximum overload/g	5	12	Endurance of astronauts
Peak heat flux/(MW·m^{-2})	3.0	4.0	High-temperature resistance of thermal protection materials
Total heat absorption/(MW·m^{-2})	400	250	Thermal protection system design
Entry time/s	1200	500	Navigation and control system design difficulty

The atmospheric entry process of Mars refers to the atmospheric flight from the upper boundary of the Martian atmosphere about 120 km away from the surface of Mars to the point before the parachute-opening point. This phase generally lasts 4–5 min. In the braking control, the value of the bank angle is mainly adjusted in real time to change the direction of the lift, thereby adjusting the flight path of the probe. However, it requires very high capability of braking due to the short period and the rapid change of the state. The atmospheric entry process of Mars has a certain similarity with the recoverable satellite or spaceship. Since the atmosphere of Mars is quite thin, the density is about 1/100 of that of the Earth, the atmospheric density on Mars varies greatly with the Martian years, which is mainly affected by the seasons. In addition, the densities of the atmosphere in different latitudes are also very different, directly affecting the choice of the height of the parachute-opening point. The composition and physical properties of the Martian atmosphere are quite different from those of the Earth. The temperature varies greatly from place to place, so the wind velocity is relatively high. The minimum wind velocity is 1.1 m/s, and the average wind velocity is 4.3 m/s. There are often storms with an average wind velocity of 50 m/s and a maximum wind velocity of 150 m/s. It can be seen that

7.4 Atmospheric Entry Guidance and Control Design

the atmospheric condition of Mars, especially in the dense atmosphere areas with the altitude below 20 km, has greater uncertainty compared with the Earth, which seriously challenges the robustness of the entry guidance algorithm.

Mars Science Laboratory (MSL) is a small lift–drag ratio probe in which the flight path in the atmospheric entry process was controlled by adjusting the bank angle. The landing accuracy is greatly improved compared to other successful Mars landing probes. In the design MSL entry guidance law, the dynamic equation near the nominal trajectory is first linearized, and the functional relationship between the terminal trajectory deviation and the small disturbance terms of the state quantity and small quantity is obtained. On this basis, the design feedback gain coefficient is optimized and the closed-loop guidance is formed [6]. The atmospheric entry into the guidance process at MSL is the first successful example of a Martian ballistic-lift entry.

7.4.2 Atmospheric Entry Trajectory Design

7.4.2.1 Atmospheric Entry Trajectory Classification

In the process of the probe deorbiting, entry into the atmosphere and landing on the surface of the celestial body, the orbit of the centroid movement is called the atmospheric entry trajectory or entry trajectory [7]. The study of atmospheric entry trajectory design and guidance approaches is an important part of the overall design and control system design of the probe. By shapes, the atmospheric entry trajectory is divided into ballistic trajectory, lifting trajectory, hopping trajectory and elliptical attenuating trajectory.

1) Ballistic trajectory

The ballistic entry process trajectory is also called a ballistic trajectory. After the ingress enters the atmosphere, the state of flight is maintained at zero lift, or although there is lift, the direction of lift is not controlled. Thus, the entry vehicle will land on the surface of the planet along a monotonically decreasing flight path. This kind of entry technology is relatively simple and easy to implement. However, the overload caused by aerodynamics is large in the process of entry, and the landing accuracy is also relatively poor. The first generation of manned spacecraft by the USA and the Soviet Union (the "Mercury" in the USA, the "Oriental" in the Soviet Union) applied ballistic trajectories.

2) Lift trajectory

The trajectory-lift-type entry and lift-type entry processes are collectively referred to as lift-type trajectories. After the entry vehicle enters the atmosphere, it uses the lift generated during its movement in the atmosphere to descend in a relatively gentle trajectory. Compared with the ballistic trajectory, the lift trajectory features a long braking time, and thus, the overload is greatly reduced. By controlling the magnitude and direction of the lift, both the longitudinal and lateral directions may be properly orbited to improve the accuracy of the descending. Compared to the ballistic trajectory, the entry corridor of the lift trajectory is also wider. Lift trajectories were applied for the Gemini spacecraft of the USA, Symbol spacecraft of the Soviet Union and the space shuttles of the USA.

3) Hopping trajectory

After entering the atmosphere at a small entry angle, the entry vehicle enters and then dashes out of the atmosphere by the lift to complete a ballistic flight. Then, it enters the atmosphere. The probe may also enter and exit the atmosphere for multiple times. Each time it enters the atmosphere, the velocity is reduced by the atmosphere. This height of the entry trajectory fluctuates greatly, so it is called a hopping trajectory. In case of the trajectory without hopping out of the atmosphere after entering it, but the trajectory still changes a lot in the trajectory by the lift, it is also called the hopping trajectory. The probe entering the atmosphere at a velocity close to the second cosmic velocity often takes a hopping into the trajectory to reduce the entry overload and adjust the landing point within a relatively large range. After completing the lunar mission, the US Apollo spacecraft also entered the orbit in a hopping manner. The spacecraft hopped when the flight altitude dropped to about 55 km. However, the highest point of the hopping was only 67 km, which was still within the atmosphere. The general hopping trajectory is shown in Fig. 7.6.

4) Elliptical attenuated trajectory

This trajectory is also known as the "brake ellipse" trajectory. For a spacecraft returning to the Earth in near the second cosmic velocity, assuming that there's no atmosphere on the Earth, it will moves along a Kepler orbit. The perigee of the Kepler orbit is called the virtual near-location. If the virtual perigee is too high from the ground, the spacecraft is only weakly resisted by the thin atmosphere, which is not sufficient to make the spacecraft entry into the Earth. Due to poor braking effect, the spacecraft will penetrate out of the atmosphere and form a large elliptical orbit around the Earth. After one circle, it will enter the atmosphere again, and then slow down a bit, reenter the elliptical orbit with a smaller size and slightly changed position. The long axis is rotated through an angle due to a deviation between the directions of penetrating into and out of the atmosphere. In principle, such "brake ellipse" may be repeated to reduce the large initial entry velocity. The disadvantage of the "brake ellipse" entry is that the landing point cannot be preselected and a lot of braking time is required.

7.4.2.2 Analysis of Atmospheric Entry Corridors

Entry corridor analysis is a prerequisite for standard ballistic design. The goal is to ensure that the maximum braking overload of the probe through the atmosphere and limit the elapsed time to a certain range. In addition, the heat generated should

7.4 Atmospheric Entry Guidance and Control Design

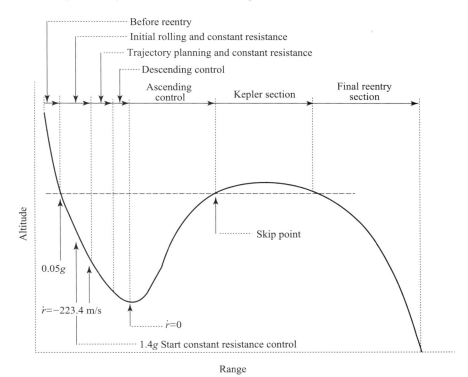

Fig. 7.6 Jump into the guided flight path

not damage the probe and the probe will land in the designated area. This section takes Mars exploration mission as the background and analyzes the atmospheric entry corridor of Mars, so as to provide a reference for the preliminary design of the lift–drag ratio and initial entry angle of the probe.

Entry corridor is defined as the initial range of entry angles necessary to enable the probe to successfully enter the atmosphere of Mars. When entering the atmosphere of Mars, the maximum overload, maximum heat flux and total heat absorption of the probe must meet certain constraints, and the constraints of the terminal on the parachute-opening state are also quite strict. Such process quantities and terminal state quantities are mainly affected by the entry angle. If the entry angle is excessively large, the trajectory will be excessively steep, the overload peak will be excessively large, and the parachute-opening height will be excessively low. On the contrary, if the entry angle is excessively small, the flight time will be relatively long, the total heat absorption will exceed the allowable value, and the probe may passes over the edge of the dense atmosphere without getting into the atmosphere, the scheduled entry becomes impossible. The entry corridor of the ballistic-lift-type entry vehicle can be determined by mathematical simulation during the design of the trajectory.

In a given mission, the different initial entry angles and bank angles are processed by ergodic transformation, and the corresponding process quantity and terminal state quantity are obtained by the integral dynamics equation. The feasibility of the entry angle selection may be judged according to the constraints. In the general dynamics integration process, when the overload is less than 0.05 g, the probe keeps flying in the trim angle of attack attitude by attitude control; and if the overload is greater than 0.05 g, the pitching moment coefficient, the axial force coefficient, the normal force coefficient and the position of the pressure core are all functions of height, Mach number, angle of attack and angle of sideslip. In the calculation of aerodynamic force, multi-dimensional linear interpolation is applied to obtain various aerodynamic coefficients, thereby calculating aerodynamic lift and resistance, aerodynamic moment, etc.; the atmospheric density and sound velocity are obtained through the interpolation of the atmospheric data table.

The bank angle is used as the controlled variable and is set to a fixed constant value during the ergodic transformation. The analysis range of the bank angle is generally selected as $[0°, 90°]$ during the simulation, and the entry angle ergodic transformation range can be selected as $[-17°, -10°]$. The bank angle is taken every $2°$, and the entry angle is taken every $0.1°$. After the ergodic transformation of the simulation, the analysis of entry corridor mainly includes the following aspects:

(1) Determining the selectable range of the initial entry angle based on the constraints of successful entry satisfying the parachute-opening conditions;
(2) Analysis of the influence of different entry angles on the range;
(3) Analysis of the influence of different entry angles on the height of the terminal parachute opening;
(4) Analyze the influence of different entry angles on the dynamic pressure of the terminal;
(5) Obtain the width of the entry angle for a given range and overload constraints.

Under different bank angles, the results of entry corridors considering the overload and range constraints and satisfying the parachute-opening conditions (opening dynamic pressure [250, 800 Pa], open parachute height greater than 4.5 km, Mach number less than 1.8) are calculated as $[-17°, -10.8°]$, as shown in Fig. 7.7.

7.4.2.3 Atmospheric Entry Trajectory Design

In the optimal design of the atmospheric entry trajectory, it is required to propose optimized indicators and design schemes according to specific missions. In addition to satisfying some restraints, such as angle of attack constraints and normal overload constraints, the atmospheric entry can also provide the best requirements for some performance indicators of the ballistic design. For Mars exploration mission, the guarantee of the parachute-opening state is the most important, so the optimal atmospheric entry trajectory should be the trajectory with optimized parachute-opening state that can satisfy the robust requirements.

7.4 Atmospheric Entry Guidance and Control Design

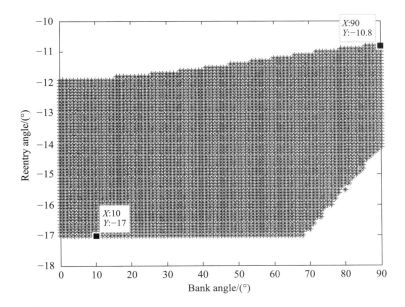

Fig. 7.7 Determination of the selectable range of initial entry angle by restraints satisfying the parachute-opening conditions

First, the optimal range of variation of the bank angle of the atmospheric entry process is analyzed. The designed reference profile is required to provide enough margins, so as to cope with various environmental dispersion problems, and the maneuverability of the entry vehicle should also be considered; that is, in an uncertain environment, the guidance law does not saturate the bank angle to lose the ability to reduce the range deviation to converge. The limit of the nominal dip angle profile is estimated from the dispersion of atmospheric density ρ and aerodynamic coefficient C_d.

$$\sigma_{minnom} = \arccos(100\% - \rho\% - C_d\%) \quad (7.11)$$

where the typical atmospheric density spread is 30%, the aerodynamic coefficient distribution is 12% and the minimum allowable initial bank angle σ_{minnom} is 55°. Similarly, the maximum bank angle is 125°. If the nominal angle of the bank angle does not satisfy the range constraints, it is necessary to increase the lift–drag ratio or reduce the ballistic coefficient.

According to the maneuverable range of the bank angle, it is further considered that, to maximize the height of the parachute opening, and the influence of the different entry angles analyzed by the previous section on the height of the opening of the terminal, the optimal initial angle value can be determined as −14.4°, as shown in Fig. 7.13.

Finally, the bank angle reference profile is determined, and the optimized design of the nominal entry trajectory is completed by integral dynamics. Mars probe is guided by controlling the bank angle during the flight, and the controllable variable is single. Therefore, the bank angle profile should also be relatively simple. The engineering bank angle profile is usually designed in a simplified manner such as piecewise constant value or piecewise linearization. Investigations have shown that in most cases, a feasible variable value reference bank angle profile is used, generally in the initial segment of atmospheric entry, so that the lift is smaller in the vertical plane component and the bank angle is chosen to be 70° or 80°. The bank angle value then decreases linearly till the minimum value analyzed previously of about 55°, as shown in Fig. 7.8.

The reference trajectory is designed in the form of a "linear + constant value" with a tilt angle profile.

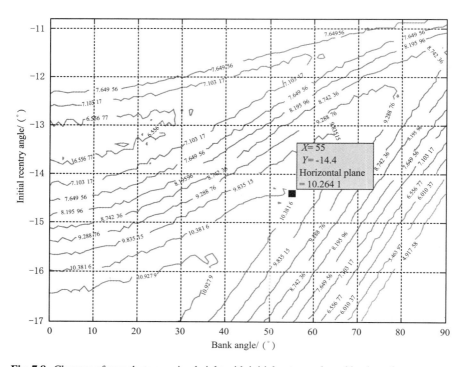

Fig. 7.8 Changes of parachute-opening height with initial entry angle and bank angle

7.4.3 *Atmospheric Entry Guidance and Control Design*

7.4.3.1 Classification of Atmospheric Entry Guidance Methods

The entry guidance problem of probe with small lift–drag ratio is a single-variable control problem; that is, control is made only by changing the bank angle. As determined by the special design of the probe mass, inertia, shape, etc., the entry attitude is maintained near the trim angle of attack, and the total lift force cannot be changed. The entry process control system changes the rolling angle according to certain control logic, that is, change the bank angle to adjust the components of the lift in the horizontal and vertical directions, so as to achieve the purpose of controlling the probe overload, heat flux density and landing point position within a certain range.

The entry guidance method can be classified into two kinds: one is the guidance method to predict the range of the landing point, which is called the prediction point approach (or the range prediction approach); and the other is the guidance method to track the standard trajectory, which is called the standard trajectory method.

The standard trajectory guidance method has no quite demanding requirements on the probe compute. However, the processing power is limited for many emergencies. The method is to preload several sets of standard entry trajectory parameters in the computer of the entry vehicle entry vehicle, which can be the functions of time, velocity, etc. After atmospheric entry, it deviates from the standard design trajectory due to factors such as initial conditional errors, changes in atmospheric density and aerodynamic characteristics. At this time, the navigation system measures the attitude parameters and velocity increments of the entry vehicle, and the computer calculates the orbital parameters such as position and velocity. The measured parameters are compared with the standard trajectory parameters to generate an error signal. Taking the error signal as input, the required attitude angle and attitude angular velocity are calculated by the guidance equation, and a control command is issued to the attitude control system to adjust the attitude angle of the entry vehicle, thereby changing the direction of the lift and realizing the guidance control of entry trajectory.

Compared with the standard trajectory method, the former focuses on eliminating the error of the corresponding falling point and the theoretical design point of the actual entry into the trajectory at any time. The predicted falling point approach is to store the characteristic parameters corresponding to the theoretical falling point in the input computer. According to the incoming state parameter measured by the navigation platform, the falling point calculation is performed in real time, and the calculated result is compared with the theoretical falling point to issue an error control signal into the computer guidance equation. The attitude angle of the entry vehicle is controlled according to the specified guidance law, and the magnitude and direction of the lift is adjusted to achieve accurate landing.

7.4.3.2 Atmospheric Entry Nominal Trajectory Approach

Under the corresponding technical levels and conditions, most of them adopt standard trajectory guidance approaches, such as the Mercury and Apollo by the USA. The key to the standard trajectory approach is to plan a standard entry trajectory satisfying various constraints and accurately track it in the entry process.

Trajectory planning: The problem of standard entry trajectory optimization design can be considered as the optimal control problem with constraints under different performance indicators. The solution approaches can be divided into two categories: direct method and indirect method. The indirect method finally transforms the optimal control problem into a two-point boundary value problem. The necessary conditional equations are written according to the Lev Pontryagin minimum value principle, namely the regular equation, the cross-sectional condition equation and the governing equation, and then numerical methods such as the gradient method, conjugate gradient method, adjacent extremum method, boundary value shooting method, algebraic function method and Newton method are applied to solve the problem. Although the indirect method may better analyze and solve the unconstrained optimal control problem, various constraints existing in the process make it excessively complicated for the derivation process of the indirect method, and the solution process is highly sensitive to the initial value of the covariate variable, making it difficult to converge. Compared with the indirect method, the direct method has advantages in the robustness of convergence and the applicability of solving practical problems.

The direct method transforms the optimal control problem into a nonlinear programming problem and overcomes the disadvantages of the indirect method to find the analytical solution by using multiple numerical solutions. The direct collocation method generally uses Chebyshev, Cubic and other polynomials to approximate the time history of states and control variables. However, the number of design variables of this method is huge. A recently developed improved Gaussian pseudo-spectral approach for solving optimal control problems is a direct collocation method based on global and local interpolation polynomial mixing. Its advantage over the general direct collocation approach lies in that relatively high precision may be obtained with as few nodes as possible.

7.4.3.3 Atmospheric Entry Prediction Correction Method

The standard trajectory method has limited processing capacity for many emergencies, and the high-velocity entry of the Moon returns is greatly affected by uncertainty factors. Therefore, a more flexible guidance method is required. The predictive correction guidance method is preferred as it can predict and correct the trajectory online in real time, and with the continuous improvement of computing power of the computer, the numerical prediction correction guidance method is easier to be implemented than before. The prediction correction method is divided into two major parts: prediction and correction. This prediction method can be further divided into two types: fast prediction (numerical prediction) and approximate prediction (analytical prediction). Accordingly, the prediction correction guidance may also be divided into two categories: numerical prediction correction guidance and analytical prediction correction guidance.

The principle of the numerical prediction method is to apply the ability of the computer for quick application, the current state is set as the initial value and the numerical integral motion equation predicts the future trajectory and the falling point. The principle of the analytical prediction guidance method is to approximate the future trajectory with a simplified analytical expression.

7.5 Parachute Deceleration System Design

7.5.1 *Overview of Parachute Deceleration Technology*

Parachute deceleration technology is a comprehensive technology generally consisting of parachute deceleration system design technology, parachute technology, parachute deployment technology, parachute-opening point control technology. The object to be decelerated may be the aeronautic vehicle (such as spacecraft and UAV), space reentry vehicle (such as a recoverable satellite return cabin, a manned spacecraft return module) or an entry vehicle (such as a Mars entry module), airdrop equipment and materials (such as combat equipment, relief supplies), weapons (such as bombs, mines, missile warheads). For deep space exploration, the object to be decelerated is typically an entry probe (a probe entering the atmosphere of another celestial body) or a reentry vehicle (a vehicle returning to the atmosphere of the Earth).

After the recovery subsystem (or control subsystem) is initiated, it is judged whether the condition can satisfy the parachute-opening requirements. If these conditions are met, the parachute is ejected out of the parachute bay by the pyrotechnics, and the parachute is pulled out from the pack. Then it is inflated under the action of a gas stream, forming a drag area, decelerating the fore body, separating from the fore body under specified conditions or landing on land or splashing on water.

The structure of the parachute is shown in Fig. 7.9. A parachute generally consists of the following functional components:

(1) Decelerator: parachute;
(2) Load-bearing components: metal joints, bridles, risers, swivels;
(3) Inflation control component: apex parachute, sacrifice panel, reefing cord, etc.;
(4) Pack and assembly components: parachute pack, locking loop, pack fixing tape, etc.;
(5) Other components: cutting knife, cutting knife tape, etc.

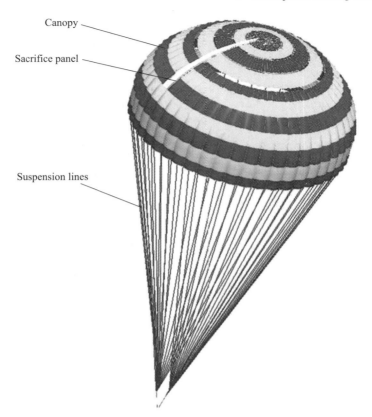

Fig. 7.9 Schematic diagram of the parachute

The parachute deployment and inflation process generally includes stretch (the stretch of the suspension lines and canopy), pre-inflating, first filling, over-inflating, adjusting and stable inflation, as shown in Fig. 7.10.

The stretch process of the parachute begins with the parachute pack being pulled out of the parachute bay till the parachute top is separated from the parachute pack. The stretch method includes the cord-first method and the canopy-first method. Cord-first method is also called the pull-down method, in which the parachute cord is first pulled out from the parachute pack, and then the parachute canopy is pulled out. In such stretch method, the stretch maximum load generally appears at the bottom of the parachute canopy pulls out the parachute pack. At the moment, the stretch load is generally less than the maximum load of the entire parachute-opening process, and the aerospace recycling parachute generally adopts such type of pulling parachute. The canopy-first method is also called sequential pulling method. The parachute canopy is first pulled out from the parachute pack, and then the parachute suspension

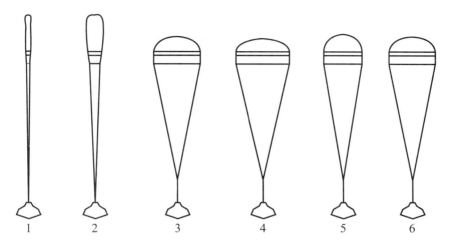

Fig. 7.10 Typical parachute inflation process. 1-Straight; 2-Pre-inflated; 3-First inflation to full; 4-Over-inflated; 5-Adjusted; 6-Stable inflation

lines are pulled out. In such stretch method, the maximum load of stretch generally occurs when the parachute system pulls out the parachute pack. The stretch load is generally greater than the maximum load of the entire parachute-opening process, and the parachute is generally used by the human parachute.

The pre-inflation process generally starts with the separation of the parachute top and the parachute pack, and finally, the first air mass is charged to the parachute top. For parachutes with poor stretch control, pre-inflation will also start during the stretch of the canopy.

The first full inflation process generally starts with the pre-inflation, and finally, the parachute for the first time reaches a steady state in full shape. For limited mass inflation, the maximum load of the parachute opening generally occurs during the gradual filling of the parachute.

Over-inflation generally starts at the end of first full inflation, and finally, the bottom of the vest is tensioned. For infinite mass inflation, the maximum load of the parachute opening generally occurs at this stage.

During the adjustment phase, the inflatable shape of the parachute gradually reaches a stable process. For limited mass inflation, it is generally accompanied by partial collapse of the canopy top. For supersonic parachute opening, it is generally accompanied by a relatively long period of breathing.

After the air is stabilized, the inflatable shape of the parachute reaches a steady state.

7.5.2 Atmospheric Entry Parachute Technology

There are seven celestial bodies suitable for aerodynamic deceleration using parachutes, namely Venus, Earth, Mars, Jupiter, Saturn, Titan and Neptune. As of 2017, in addition to Saturn and Neptune, humans have applied parachute deceleration technology to the remaining five celestial bodies. The characteristics of parachute deceleration system of each celestial body except the Earth are shown in Table 7.9.

For different planetary entry missions, the parachute type should be selected based on the characteristics of the mission and the parachute as well as the functional performance requirements.

For the exploration of Venus, Jupiter, Saturn, Titan and Neptune, the velocity of parachute opening is subsonic, and the dynamic pressure is about 1000 Pa. Solid textile or slotted parachute with less air permeability may be used such as

Table 7.9 Deceleration characteristics of deep space exploration system

No.	Celestial body	Parachute deceleration system characteristics
1	Venus	(1) The height of the parachute opening is about 70 km, the subsonic parachute opening ($M\alpha < 0.5$), and the parachute-opening dynamic pressure is low (about 500 Pa). The parachute-opening condition is similar to the Earth deceleration condition (2) There is a large amount of sulfuric acid cloud in the atmosphere of Venus at an altitude of about 60 km. The parachute material should have good acid resistance and chemical stability (3) The temperature of the atmosphere of Venus is about 80 °C at an altitude of 50 km and about 475 °C near the ground. The parachute material should have good high-temperature resistance. In most cases, the parachute fails due to severe heat loss at high temperatures before landing onto the surface of Venus
2	Mars	(1) Supersonic velocity ($M\alpha \approx 2.0$), low dynamic pressure (about 600 Pa), low density (0.005 kg/m^3) (2) Under supersonic conditions, parachute flutter and breathing are severe (3) Under the condition of low density and low dynamic pressure, the process of opening the parachute is infinite mass opening, the opening load is large, the stability is poor, the deceleration efficiency is low, the deceleration time is long, and the height loss is much (4) Low temperature: ambient temperature about -100 °C (5) Generally, the Martian parachute system decelerates and should have good supersonic low dynamic pressure opening performance
3	Jupiter	(1) The velocity of opening the parachute and the conditions of dynamic pressure are similar to those of the Earth returning to the parachute (2) The ambient temperature is low (about -150 °C) (3) Generally, Jupiter, Saturn and Titan parachute systems are decelerating, and the main challenge is to withstand low-temperature environments
4	Jupiter	
5	Titan	
6	Neptune	Neptune's opening velocity and dynamic pressure conditions are similar to those of Venus, but the ambient temperature is extremely low, and there are severe weather conditions such as strong storms

7.5 Parachute Deceleration System Design

extended skirt, conical, guide surface, parasailing, cross, ring-slot and disk-gap-band parachute.

For Mars exploration, due to the condition of supersonic and low dynamic pressure, the solid textile parachute or a slotted parachute with small air permeability is generally selected, such as disk-gap-band and ring-slot parachute. It is further considered that if the probe is landed by airbag, the horizontal velocity of the probe should be as small as possible. The stable parachute type should be preferred; if the probe is landed by buffer engine, the selection of parachute type should focus on good drag performance.

7.5.2.1 US Mars Exploration

A total of seven probes in the USA have landed softly on the surface of Mars. Their process of entry and deceleration are basically the same; that is, the probe initially is braked by the aerodynamic shape of the entry vehicle and then further decelerated by the parachute. The main difference is the final deceleration and landing buffering process. Please refer to Table 7.10 for the relevant parameters for entering the Entry Decelerate Landing (EDL).

By summarizing Mars exploration of the parachute system deceleration process of the USA, several technical features are concluded:

(1) The first-stage supersonic disk-gap-band parachute that pops the parachute with a bomb tube is applied;
(2) The range of parachute-opening Mach number of the parachute is 1.1–2.0;
(3) The dynamic pressure range of the parachute is 350–750 Pa;
(4) The steady descending velocity range of the parachute is 60–100 m/s;
(5) In Viking, the radar altimeter is used for open control; in the MSL, the parachute-opening control method with the open Mach number as the control target, and the rest apply the open control approach with dynamic pressure as the control target.

7.5.2.2 European Mars Exploration

The ESA has totally three Mars Landing Exploration Projects, Mars Express mission in 2003, the 2016 Mars Biology (ExoMars 2016) mission in 2006, and the 2020 Mars Biology (ExoMars 2020) mission in 2020.

1) Mars Express

The Beagle lander in Mars Express mission entered the atmosphere of Mars in a ballistic manner. The deceleration system consists of a deceleration parachute and a main parachute. The disk-gap-band parachute was used to decelerate the Mach number of the entry vehicle from 1.5 to 0.4–0.6. The main parachute is a parasailing parachute for decelerating the entry vehicle to 16 m/s.

Table 7.10 EDL related parameters of the US Mars exploration mission

Parameters	Viking 1/2	Mars Pathfinder	Opportunity/spirit	Phoenix	MSL
Entry mass/kg	992	584	827/832	600	3257
Lift control	Yes	No	No	No	Yes
Entry navigation	No	No	No	No	Yes
Boundary diameter	3.5	2.65	2.65	2.65	4.5
Disk-gap-band parachute type	Viking type	MPF	MPF improved type	Viking type	Viking type
Parachute-opening velocity (Nα)	1.1	1.57	1.77	1.65	1.75
Parachute-opening dynamic pressure	350	585	725	490	493.6
Parachute-opening altitude/km	5.79	9.4	7.4	12.9	7.1
Parachute-opening control model	Radar altimeter	Based on dynamic pressure	Based on dynamic pressure	Based on dynamic pressure	Based on Mach number
Parachute stable ascending velocity/(m s^{-1})	~64	~63	~72	~60	~100
Deceleration before landing	Variable thrust engine	Solid rocket	Solid rocket	Variable thrust engine	Variable thrust Engine
Attitude control before landing	Yes	No	No	Yes	Yes
Landing buffer	3 landing legs	Omnidirectional air bag	Omnidirectional air bag	3 landing legs	Aerial crane
Landing point altitude	−3.5	−2.5	−1.9	−3.5	−3.2

2) ExoMars 2016

The ExoMars 2016 mission of EAS with a similar deceleration scheme as the US mission adopted a one-stage reduction scheme (nominal area 114 m^2) using a disk-gap-band parachute.

3) ExoMars 2020

The ExoMars 2020 mission is a follow-up mission to the ExoMars 2016 mission, which primarily achieves a soft landing on Mars and releases a rover. ExoMars 2020 adopts a two-stage parachute deceleration method. The first parachute, a supersonic parachute, is opened by a pilot shute. The second-stage parachute is a subsonic parachute, which is pulled out by a pilot shute.

7.5.3 *Deep Space Probe-Parachute Design*

7.5.3.1 Mission Analysis

The parachute deceleration system works in the astrophysics environment to be explored; in this respect, the physical environment and mission requirements of different celestial bodies must be fully researched. Take the analysis of Mars parachute mission as an example to introduce the factors that should be prioritized for the outline design of the parachute.

There is great environmental uncertainty: The atmospheric composition of Mars is mainly CO_2, and the surface atmospheric density is about 1% of the Earth. The atmospheric parameters of Mars vary greatly with seasons and geographical locations, and there are severe meteorological conditions such as dust storms. There are large uncertainties in various atmospheric parameters forming the major basis of the design.

There are two major effects of supersonic conditions on the parachute: (1) Under supersonic conditions, the aerodynamic interference is complicated; (2) during the supersonic flight, the parachute will appear "breathing" and high-frequency chattering phenomenon, which is easy to cause the damage of the parachute.

Impact of low-density environment on parachute opening in three major aspects: (1) As a result of low density, the deceleration efficiency of the parachute is reduced. To achieve the same steady-state velocity, a larger parachute area is required; (2) In the low-density atmosphere, the effective permeability of the parachute material is sharply reduced compared with Earth environment, and the decrease in the air permeability will deteriorate the stability of the parachute; (3) due to the low atmospheric density, the aerodynamic deceleration process of the entry vehicle itself is relatively long. At parachute opening, the entry vehicle has not reached the steady-down phase and has a negative acceleration of 3 times the gravity of Mars.

Effect of low dynamic pressure environment on the parachute opening: During the parachute opening, the low dynamic pressure and insufficient aerodynamic damping will cause phenomena such as swaying and "breathing" and the relatively large amplitude of swaying or "breathing" so that the recovery is relatively difficult.

According to this special parachute-opening conditions, the following points should be focused on the development of Mars parachute systems:

(1) Select the appropriate parachute type based on the overall task characteristics of the entry vehicle;
(2) The parachute should be so deployed that the parachute passes through the wake zone reliably and reaches the straightened state;
(3) In the design of the parachute structure, measures to prevent flutters and damages should be taken.

7.5.3.2 Scheme Selection

1) Selection of parachute deceleration method

There are two types of parachute deployment methods: one-stage deceleration and multistage deceleration. The advantage of the one-stage deceleration method lies in that the system is simple, the process is streamlined and the flight time is short. Considering the complexity of the system, the one-stage deceleration method has fewer links and high reliability; according to the time of the landing patrol EDL process, relatively long flight time can provide more space for subsequent procedures. The disadvantage of the one-stage deceleration method lies in that the parachute-opening load is great, and the entry structure should be provided with a high load carrying capacity; and the multistage deceleration method has the advantages of small parachute-opening load and high space utilization.

In the Earth return mission, due to the relatively large atmospheric density and large air pressure for parachute opening, the multistage parachute deceleration method is generally used to reduce the opening load. The spacecraft Shenzhou is equipped with a deceleration parachute and a main parachute, in which the deceleration former is used to reduce the velocity of the return module from 200 to 90 m/s, and the latter is used to reduce the velocity of the return module from around 90 to 7 m/s; all current Mars landing exploration missions in the USA adopt a one-stage deceleration method; in the ExoMars 2020 mission by the Europe, the deceleration of the entry module adopts a two-stage deceleration method of "deceleration parachute + main parachute".

2) Parachute deployment design

There are two types of parachute deployment: one is using pyrotechnics, one is using pilot parachute.

Figure 7.11 shows the relative movement velocity of the parachute pack and the fore body as a function of time when the parachute is ejected in different open conditions. Figure 7.12 shows the relative movement velocity of the parachute pack and the fore body as a function of time when the parachute is pulled out by the pilot shute under different parachute-opening conditions.

By comparing the pyrotechnics and the pilot shute methods in parachute opening, the following characteristics can be obtained:

(1) When the parachute is ejected, the movement velocity of the parachute pack becomes increasing slow during the stretch of the parachute system. When the parachute canopy are pulled out from the parachute pack, the relative velocity is 30–40 m/s; When the parachute is opened, the movement velocity of the parachute pack is getting increasingly fast, and the relative velocity is 40–120 m/s when the parachute canopy is pulled out from the parachute pack. If the parachute canopy is pulled out from the parachute pack too fast, it would easily cause friction damage of the parachute canopy, and vent whip would occur for the parachute.

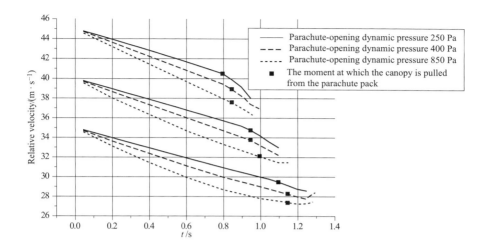

Fig. 7.11 Relative velocity curves of the parachute pack and the fore body when the parachute is ejected under different opening conditions

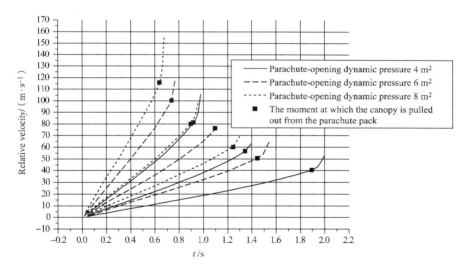

Fig. 7.12 Relative velocity curves of the parachute pack and the fore body when the parachute is pulled out by pilot shute under different opening conditions

(2) Compared with the approach of projecting the parachute to open the parachute, the pilot shute is more sensitive to the opening conditions. Under the high dynamic pressure, the velocity of the parachute would be too fast to pull out from the parachute pack, while under low dynamic pressure, the pullout velocity of the parachute canopy from the parachute pack would be too slow, and it is necessary to make relatively accurate control of the parachute-opening conditions.

(3) The advantage of the pilot shute + main parachute method lies in that the ejection force is small and the space utilization rate is high; the advantage of the parachute ejection parachute method is that it is less affected by the opening condition, the uniformity of the parachute stretch process is excellent, and the work link is small. Considering the reliability of the process of stretch the parachute, the parachute-elevating parachute method has a relatively low relative velocity and the consistency of the stretch process is excellent.

There is a limit to the ejection of a parachute. With the increase of the parachute area, the pyrotechnics will become larger and larger, and the development of the large-diameter ammunition tube will be more difficult, and the ejection force will be greatly increased. At present, the largest one-shot parachute is used by the US MSL probe, with a nominal area of 358 m².

7.5.3.3 Drag Area Design

The determination of the drag area of the parachute is an iterative process, and the determined process is shown in Fig. 7.13.

1) Parachute drag area determined by steady descending velocity

Considering the constraints of steady velocity, flight path angle and atmospheric uncertainty, the minimum drag area of the parachute may be calculated by the following formula:

$$C_d A \geq \frac{2mg \cdot \sin(-\theta)}{k\rho v^2} - C_w A_w \tag{7.12}$$

where m is the mass of the parachute system at steady state; g is the gravitational acceleration; and for Mars, the surface gravitational acceleration is 3.73 m/s²; θ flight trajectory angle; k is the reduction factor after considering the uncertainty of atmospheric density; ρ is the nominal atmospheric density of steady descending altitude; v is the steady velocity; C_w is the drag coefficient of the entry vehicle; A_w is the reference area of the entry vehicle; C_d is the drag coefficient of the parachute; and A is the drag area of the parachute.

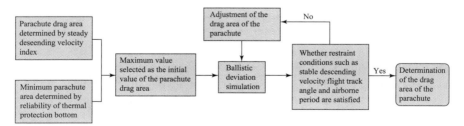

Fig. 7.13 Parachute drag area determination process

7.5 Parachute Deceleration System Design

2) The minimum area of the parachute determined by the separation reliability of the heat shield

Whether it is Earth Returner or Mars Probe, there is a large cone-shaped thermal shield to withstand the high temperatures generated by aerodynamic heating during the entry process. Before landing, it is generally necessary to throw away the thermal shield to expose the landing platform. The action of throwing the thermal shield is generally carried out during the deceleration of the parachute. The thermal shield itself has a high drag area and a small mass. Defining the ballistic coefficient as the ratio of the mass of the flying object to the area of the drag, the thermal shield has a small ballistic coefficient. In the process of preventing the heat out of the outsole, the parachute and the entry vehicle that throws away the thermal shield are a flying body, and the thermal shield is a flying body, the both must not collide with a safety accident. The thermal shield must move faster than the parachute and the entry vehicle assembly. For this reason, the ballistic coefficient of the parachute and the entry vehicle assembly is less than the ballistic coefficient of the thermal shield.

The minimum parachute drag area determined by the ballistic coefficient ratio constraint can be calculated by the following formula:

$$C_d A \geqslant \frac{k_r}{R_w} \cdot m - C_w A_w \tag{7.13}$$

where R_w is the ballistic coefficient of the thermal shield; k_r is the ballistic coefficient ratio of the thermal shield and the bay parachute combination, which may generally be specified as 1.2.

3) Deviation ballistic analysis

Between the parachute drag areas determined from (1) and (2), the largest one is the initial parachute drag area. Whether the selected parachute drag area is appropriate or not should be checked by ballistic deviation simulation.

The ballistic deviation simulation can be performed by three-degree-of-freedom mass-point ballistic Monte Carlo shooting simulation. The deviation factors to be considered should include the altitude of the parachute opening, the velocity of the parachute opening, the angle of the flight path of the parachute-opening point, the atmospheric density, the mass of the entry vehicle, the aerodynamic characteristics of the entry vehicle and the parachute drag area. The constraint conditions for judging whether the drag area is suitable for mainly include the steady-state velocity of the parachute system, the flight path angle and the hovering time when the parachute is separated. Step iterative method is applied for ballistic deviation simulation. Under the acceptable probability, if all the constraints are satisfied, the drag area of the parachute can be determined.

7.5.3.4 Parachute-Opening Point Control Design

The parachute-opening point is generally the starting point of the parachute deceleration section. The parachute cannot be opened under all conditions, and some constraints should be satisfied. In the selection of the parachute-opening point of the parachute, it is required to comprehensively consider multiple factors, such as the ballistic characteristics of the entry vehicle, the actual capacity of the parachute, and the subsequent demand conditions of the descending power. And then determination should be made by multiple rounds of system optimization.

In the process of probe entry, the parachute-opening command should be accurately given at the specified parachute-opening point, and the parachute opening should be guaranteed by a set of control methods. There are several control methods available for selection:

Pure time control method: Pure time control method is widely used in uncontrolled sounding rocket recovery system. According to the requirements of the design flight trajectory and flight mission of the rocket flight, the time required for rocket head/body separation, deceleration parachute opening and main parachute opening is determined. The parachute-opening point parameters obtained by the pure time parachute-opening control method are highly discrete, and the range of the deviation trajectory that may be covered is quite small.

Overload-time control: This control method is commonly used in Earth return missions. Compared with the pure time control method, the overload-time control method is to change the overload value experienced by the returner in the returning process to the time zero point. The parachute-opening parameter range of the parachute is greatly improved compared with the pure time control method. However, it is still relatively large.

Air pressure altitude control method: The relationship between atmospheric static pressure and altitude is used to control the opening point. This method adopts a barometric altimeter to sense the local atmospheric static pressure and derives the local altitude accordingly. Therefore, the control method should be provided with a corresponding relationship between atmospheric static pressure and high accuracy. This method is suitable for the global environment, because there are more accurate atmospheric data. China's Shenzhou spacecraft and Russia's Soyuz spacecraft have both adopted this method to control the parachute opening. For deep space exploration, it is difficult to obtain accurate correspondence between atmospheric static pressure and altitude of celestial bodies, such as Mars. Therefore, this method is not available in the field of deep space exploration.

Radar height control method: Based on the altitude and velocity information provided by the landing radar, the guidance system on the lander sequentially issues control commands for the various program actions of the landing process. The advantage of this method lies in that the principle is relatively simple, the altitude of the parachute opening is directly measured and the accuracy is high; the disadvantage is that the radar transmitting and receiving antenna must directly face the landing

7.5 Parachute Deceleration System Design

surface, and the antenna should be opened in the thermal shield, increasing the complexity of the thermal protection. The USA has only used this method in the Viking mission and suspended the use later.

Adaptive parachute-opening control method: Adaptive parachute-opening control takes some parameters such as the dynamic pressure, Mach number, altitude and velocity as the control targets, the acceleration value measured by the accelerometer and the preset parachute-opening control law (relationship function of delay time and acceleration) are used to recognize the ballistic characteristics, and to dynamically determine the parachute-opening time, thereby ensuring that the parachute's open Mach number, dynamic pressure and height are within the required ranges. In Mars Pathfinder mission in the USA, this method has been used for parachute control. The advantage of this method lies in that the hardware requirements are simple and can adapt to the excessively large ballistic deviation caused by the excessive uncertainty of the atmospheric parameters to a certain extent. In addition, it is suitable for the ballistic entry of Mars probe.

7.5.3.5 Parachute Structure Design

The main function of the parachute is to provide aerodynamic drag, so the drag is the most important performance indicator of the parachute. The drag of the parachute may be calculated by the following formula:

$$F = q \cdot C_d A \tag{7.14}$$

where F is the aerodynamic drag of the parachute; q is the dynamic pressure; C_d is the drag coefficient; A is the characteristic area for calculating the drag coefficient, which is generally the nominal area.

Main factors affecting the drag coefficient of the parachute include number of cords, length of the cord, air permeability, velocity, fore body wake, etc.

For parachutes of the same type, the inflatable shape of the parachute may also be affected by structural parameters, such as the number of cords and the length of the cord, thus affecting the drag coefficient. By appropriately increasing the number of cords and the length of the cord, the projected area of the parachute may be increased accordingly. However, if the number and length of the parachute cords continue to be increased, the projected area of the parachute canopy will not be increased significantly.

The airflow of the parachute is expressed in terms of total air permeability, including structural air permeability and fabric air permeability. The structural air permeability indicates the size of the opening area of the parachute canopy and is the ratio of the total area of the parachute-opening canopy to the nominal area of the parachute canopy. The amount of fabric air permeability indicates the size of the air permeability of the fabric, which may be expressed by two methods: the one is the dimension method, also known as effective air permeability, which is defined as the ratio of the average velocity of the airflow through the fabric to the free airflow velocity; and the other is the flow representation method, which is defined as the volume of air passing through a unit area of fabric per unit time under a certain pressure differential. The greater the total air permeability of the parachute is, the lower the drag coefficient is.

The drag coefficient of the parachute varies with the Mach number.

The drag performance of the parachute is also affected by the wake of the fore body. The higher the Mach number, the closer the distance between the parachute and the fore body and the more obvious the influence of the wake of the fore body on the aerodynamic performance of the parachute will be. The main structure of the parachute includes parachute canopy, parachute cord, radial tape, zonal tape, connecting tape, etc. In addition, the structure also involves parachute clothing protection cloth, closing cord, closing cord cutter fixing device, closing cord cutter and closing ring as well as entry vehicles such as rings, closing ring fixing tapes, edge tapes and parachute cord rings. The functions of each major component are shown in Table 7.11.

When selecting the parachute material, the various environments of use experienced during the entire life cycle should be considered, including temperature, humidity, salt spray, damp heat, space irradiation, ultraviolet radiation, natural aging, storage environment, etc. Materials suitable for environmental adaptability should be used. In addition, it is necessary to consider the processability of the material, and the materials with high sewing difficulty and large loss of strength after sewing should be carefully selected.

For Venus parachutes, the parachute material is required to have good acid drag due to the large amount of sulfuric acid cloud at an altitude of about 60 km. Due to the high atmospheric temperature in Venus, the parachute material should be of good heat drag and high thermal strength.

For Mars parachutes, in addition to meeting the strength requirements, the parachute material should be selected on the basis of the material strength loss caused by conditions, such as long-term vacuum, thermal environment and high-density packaging.

1. **Parachute-opening load analysis**

The parachute-opening process may generally be divided into the infinite mass parachute opening and the limited mass parachute opening. If the infinite mass is used, it is approximated that the velocity of the fore body remains unchanged during the inflation of the parachute, and the maximum load of the parachute appears in the instant of full inflation of the parachute. If the finite mass is used, the velocity of the system will change significantly during the inflation process. The maximum load of the parachute will appear during the inflation of the parachute.

For parachutes with no closing design, if the infinite mass is used, the parachute-opening load of the parachute should be calculated by the dynamic load factor method. The calculation formula is as follows:

$$F_{\max} = q \cdot C_d A \cdot K \tag{7.15}$$

7.5 Parachute Deceleration System Design

Table 7.11 Main components and functions of the parachute

No.	Composition	Function	Remark
1	Canopy	Provide drag area and withstand aerodynamic loads	–
2	Parachute cord	The parachute and the connecting tape are connected to form a stable inflatable shape, and the aerodynamic load on the parachute is transmitted to the connecting tape	–
3	Radial tape	The reinforced tape along the radial direction of the awning is the main bearing tape on the awning, which is used to transfer the aerodynamic force on the awning to the cord	Usually a single layer, located on the outside of the parachute coat; it can also be a double layer, located on the upper and lower sides of the parachute
4	Zonal tape	The reinforced tape along the latitude of the awning is the main bearing tape on the awning, which is used to withstand the lateral load during the filling of the vest and limit the expansion of the ripper tear	Including top hole reinforcement tape, bottom edge reinforcement tape and intermediate weft reinforcement tape
5	Connecting tape	Used to bundle the parachute cords and connect them to other connecting devices to transfer the cord load to other connecting devices	–
6	Top hole cord	Connect the opposite two radial tapes together with the top hole reinforcement tape to bear the aerodynamic load on the radial tape and the top width	–
7	Canopy Protection cloth	The parachute protective cloth is a piece of cloth sewn on the radial tape of the parachute. In the process of wrapping the parachute, it is used to wrap the parachute to prevent the high-velocity friction between the parachutes during the high-velocity pulling of the parachute. damage	–

(continued)

Table 7.11 (continued)

No.	Composition	Function	Remark
8	Closing cord	Used to limit the drag area of the parachute and reduce the parachute-opening load	–
9	Closing cord cutter fixing device	It is used for fixing the closing cord cutter, protecting the cutter flame burning parachute clothing, fixing the cutter to hit the buckle for the ear cord, and providing the connection and fixing structure for the firing of the cord cutter	Including the closing cord cutter fixing sleeve, the parachute clothing protective cloth, the buckle for pulling the ear cord, etc.
10	Closing cord cutter	Used to delay the cutting the cord	Initiating explosive device
11	Closing ring	Provide positioning and guidance of the closing cord at the bottom of the parachute	–
12	Closing ring fixing strap	It is used to fix the closing ring on the bottom of the parachute and apply the restraint of the closing cord to the bottom of the parachute	–
13	Edge closing tape	It is used to tighten the bottom edge of the parachute canopy, which is a measure to help improve the inflation symmetry of the parachute canopy.	–
14	Parachute cord loop	It is used to connect the parachute cord and the connecting tape to transfer the load on the cord to the connecting tape	–

where F_{max} is the maximum parachute-opening load; q is parachute-opening dynamic pressure; C_dA is the drag area; and K is the dynamic load coefficient.

For parachutes with closing design, when the infinite mass is opened, the parachute-opening load of the parachute should be calculated by the dynamic load factor method. The calculation formula is as follows:

$$F_{maxR} = q \cdot (C_dA)_R \cdot K_R \quad (7.16)$$

where F_{maxR} is the maximum parachute-opening load in the closing state; q is the open state dynamic pressure; $(C_dA)_R$ is the closing drag area; K_R is the closing dynamic load coefficient, which is generally higher than the dynamic load coefficient of parachutes with no closing design. The values of the closing dynamic load factor of various parachute types are measured by means of testing. Figure 7.14 shows the change curve of the dynamic load factor of the closing ring with the area ratio of the closing drag.

Limited mass parachute-opening method: In engineering design, the inflation time method or the inflation distance method is generally used.

2. Parachute deployment design

Before the probe enters the planetary atmosphere, the parachute is required be securely fixed in the bay. To deploy the parachute, the restraint on the parachute pack in the parachute bay should be first removed, and then the parachute pack out of the bay is pushed by the method of ejection or traction and leaves from the

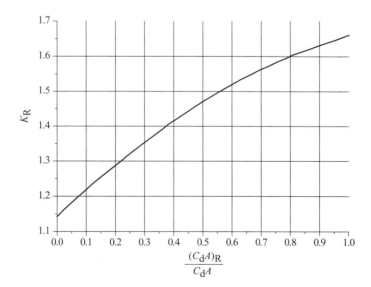

Fig. 7.14 Relationship between the dynamic load factor of the closing ring and the area ratio of the closing drag

probe. Then the parachute system is pulled out from the parachute pack, inflated and deployed under the action of the airflow, and the deceleration action is executed. The parachute deployment refers to the process of removing the parachute pack from the parachute bay and the driving the parachute out of the bay and out of the probe. Before opening the parachute, it is first necessary to eject the rear heat shield (bottom cover) or the parachute hatch of the entry vehicle to the rear of the flight, so as to clear the parachute-opening channel. The separated heat shield will be moved in the wake of the probe to guarantee that the bottom cover is separated from the probe and escapes from the wake region, which is one of the key deceleration actions of the parachute system. After the bottom cover is ejected, the parachute is generally ejected by the bomb tube. In the parachute deployment process, problems similar to the ejection bottom cover will also be encountered.

To enable parachute deployment, it is necessary to apply the pyrotechnic device, such as the catapult, bomb tube, etc.

The catapult is the most widely used in the pyrotechnic device type. The bomb tube generally consists of an inner and an outer cylinder, and the inner cylinder serves as a piston of the outer cylinder and has a stroke of a certain length. The inner cylinder performs an acceleration motion in the outer cylinder under the pressure of the pyrotechnic gas. Eventually, the inner and outer cylinders are completely disengaged at a certain velocity, which is also known as the exit velocity. The object to be ejected (such as a canopy cover) is either fixed in the inner cylinder or the outer cylinder. Thus, the ejected object is separated from the probe at a certain exit velocity along the axis of the bomb tube, which has a piston disk in its outer cylinder. Place the wrapped parachute pack on the piston disk. The piston disk is fixed in the outer cylinder by a shear pin. The piston disk is pushed, and the shear pin is cut by the pyrotechnic pressure of the chamber. The piston disk holds the parachute pack under the pressure of the pyrotechnic gas, accelerates it in the outer cylinder and finally breaks from the bomb tube at a certain exit velocity. The parachute relies on its own kinetic energy at the time of exit and moves to a predetermined position to open the parachute.

7.5.4 *Simulation Analysis of Parachute Design*

In the engineering design of the parachute system, the main functions of the simulation include three points: (1) Conduct preliminary evaluation and optimization design of the design parameters through simulation, such as static analysis of structural parts, modal analysis, vibration analysis, internal ballistic analysis of the canister and dynamic analysis of the parachute process; (2) make up for the deficiencies of the physical test through simulation. The physical test of the parachute system deceleration system for deep space exploration cannot fully simulate the actual working conditions, and the number of physical tests is limited. Furthermore, it is impossible to

7.5 Parachute Deceleration System Design

traverse the possible working conditions; (3) through simulation, it can provide technical support for large-scale physical experiments such as high-altitude parachute-opening test and airdrop test, optimize control parameters, ensure the realization of specified test conditions, reconstruct physical processes with test data and identify important parameters.

In the process of the development of the deceleration parachute deceleration system, the typical simulation projects include the parachute ejection stretch process, the parachute-opening inflation process and the dynamic simulation of the probe-parachute assembly.

Parachute ejection and stretch process simulation mainly focuses on three aspects, including determination the ejection separation velocity to ensure that the stretch process is reliable; determination of the maximum stretch force to verify the design strength of the parachute, connection and separation mechanism and the entry vehicle structure; analysis of the stretch process to confirm the possibility of "Line Sail" or even "Vent Whip". The ejection and stretch process of the parachute are a kinetic problem of variable mass multi-body and coupled continuous elastomers. Presently, the multi-body dynamics method is mainly used for analysis, and the parachute canopy and the parachute cord are discretized into a series of mass points, and the spring damping model is set between mass points.

Simulation of open air inflation process: The inflation process is the most critical part for the parachute. The load generated by the inflation, the inflation time, the inflation distance, the aerodynamic force generated during and after the inflation process are the parameters most concerned by the parachute system. The inflation of the parachute is the most complicated stage in the process of parachute operation. The unsteady flow and nonlinear fluid–solid coupling problems involved in the parachute have posed great research difficulty. Presently, there are no fully developed theoretical models available at home and abroad. Based on the data obtained from wind tunnel tests, airdrop tests, and based on semiempirical and semi-theoretical methods, researchers established analytical models for various parachute inflation processes, such as the aeration distance, the inflation time and the momentum method. With the development of CAE technology, the simulation analysis of the parachute inflation process has gradually deepened from the loose coupling analysis based on the single circumferential flow field model (CFD) or the circumferential flow field model and the parachute canopy structure model to strong fluid–solid coupling analysis for simultaneous solution (ALE) with the parachute structure equation or the fluid mechanics equation.

Dynamics simulation of the probe-parachute assembly is based on the parachute and the combination of the entry vehicle, and the various kinematics and dynamics of the probe-parachute assembly from the parachute to the landing are analyzed. The dynamics simulation of the probe-parachute assembly provides the basis for determining the reasonability of parachute design parameters, the motion characteristics of the probe-parachute assembly, and the stability of the analyzer-parachute assembly. The simulation method is mainly based on the simulation model of the probe-parachute assembly dynamics, and Monte Carlo target is executed under the consideration of the influence of the parachute drag coefficient deviation, the atmospheric density deviation as well as the probe aerodynamic coefficient deviation.

7.6 Prospects

Looking into the future, to meet the demands of deep space exploration missions with more targets and more forms of exploration, as for the deep space exploration reentry and returning and atmospheric celestial entry technologies, the main technical development demands in the future will be:

(1) **Aerodynamic analysis calculation and simulation technology.** Continuously develop and improve the methods and means of aerodynamic analysis to improve the accuracy of atmospheric aerodynamic analysis and prediction, and provide supports for the new aerodynamic shape optimization design and atmospheric entry exploration missions.
(2) **New thermal protection technology.** Investigate the inflatable-expansion-type multi-functional thermal protection structure technology and advanced thermal protection material system to solve the difficulties in design, analysis, process, test verification and the thermal protection problems under long-term space environment and extreme reentry conditions.
(3) **Navigation control technology.** Utilize a variety of navigation means to achieve high-precision navigation, such as the drag to derive highly assisted navigation technology and visual aided navigation technology; fully utilize the onboard computer capabilities to improve the intelligence level of guidance and control technology.
(4) **Group parachute deceleration technology.** Utilize the newly developed parachute fluid–solid coupling simulation method to research and develop a highly reliable group parachute deceleration system to adapt to the demands of large-mass probes.

References and Related Reading

1. Haas BL, Milos FS (1995) Simulated rarefied entry of the Galileo probe into the atmosphere of jupiter. NASA, USA
2. Guiqing J, Lianyuan L (2003) Heat trans for of hypersonic gas and ablation thermal protection. National Defense Industry Press, Beijing (Chinese vision)
3. Bower A, Gary J, Lorenz R, Krzanovic V (2010) Planetary lander and entry detector. China Aerospace Publishing House, Beijing (Chinese vision)
4. Yun W, Yunhua Y, Zhihai F (2013) Current status and further trend of thermal protection materials for deep space exploration. Aerospace Mater Technol 5:1–10 (Chinese vision)
5. Blockley R, Shyy W (2016) Fluid dynamics and aerothermodynamics. Beijing Institute of Technology Press, Beijing (Chinese vision)

6. Dayi W, Minwen G (2015) Review of spacecraft entry guidance. J Astronaut 36(1):1–8 (Chinese vision)
7. Hanyuan Z (1997) Spacecraft entry dynamics and guidance. Changsha: National University of Defense Technology Press (Chinese vision)
8. Xi W (1991) Spacecraft entry and return technology. China Aerospace Publishing House, Beijing (Chinese vision)
9. Siyuan C, Wenbo M, Xiaoli C (2013) Study on radiative heating calculation method for reentry vehicles. Acta Aerodynamica Sinica 6 31(3): 333–343 (Chinese vision)
10. Chunping Z, Jun H, Chengqi S (2006) Application of feedback linearization method to reentry guidance. Guilin, Guangxi, The 12th national conference on space and moving body control technology (Chinese vision)
11. Jun H (1998) A kind of mixed reentry guidance method for manned spacecraft. The eighth national conference on space and moving body control technology, Huangshan, Chinese Society of Automation (Chinese vision)

Chapter 8
TT&C and Communication Technology

8.1 Introduction

Similar to the TT&C and communication system of a near-Earth spacecraft, the deep space TT&C and communication system has also three basic missions. The first mission is to measure and predict the position of the spacecraft. The second mission is to obtain the health status and scientific exploration data of the probe through the downlink. The third mission is to make corresponding controls on the probe based on the remote commands transmitted via the uplink. The three basic missions are also commonly referred to as TT&C, namely tracking, telemetry and command.

To accomplish such three basic missions, the principles and basic composition of the deep space TT&C and communication system are generally similar to those of near-Earth spacecraft communication system. Through the basic external measurement methods such as ranging, velocity measurement and angle measurement, the real-time measurement of the probe position is enabled. In addition, according to the orbit model, the flight path of the probe for a certain period in the future may be predicted to complete the mission of probe tracking. The health status and scientific exploration data of the probe is collected and processed through framing and other techniques to form the data information. Then, the data information is finally modulated by appropriate modulation methods to the radio frequency (RF) signal convenient for wireless transmission and then transmitted to the Earth station for storage and processing. Thus, the telemetry of the probe is completed. As for the remote control of the probe, the Earth station controls the probe motion through the control data information modulated to the radio frequency signal. As for the basic composition of the TT&C and communication system, the early near-Earth spacecrafts were equipped with an independent tracking subsystem, a telemetry subsystem and a telecommand subsystem. Later on, the RF carriers of the three subsystems were unified into one common TT&C transponder. The basic composition of this unified carrier is also the composition of the TT&C and communication system used by most of the deep space spacecrafts.

Though the principles and basic compositions of the two systems (deep space communication system and near-Earth communication system) are roughly the same, the establishment of the TT&C and communication between the deep space probe and the Earth is quite different from that between the near-Earth spacecraft and the Earth that we are familiar with as a result of vast distance in deep space. The long distance in deep space has brought two difficulties in TT&C and communication, namely huge delay and huge path loss. The huge delay greatly reduces the accuracy of conventional external measurement methods. Some conventional external measurement methods are even not suitable for deep space exploration, and real-time telecommand is almost impossible. Due to huge path loss, the signal level is almost too weak to tolerate. As a result, it is even impossible to establish a normal communication link. Such difficulties and "bottlenecks" are exactly the problems to be researched and solved both practically and theoretically in the field of deep space TT&C communication.

This chapter focuses on the specific theoretical and engineering implementation methods as the solutions to the above-mentioned difficulties in deep space TT&C and communication. Generally, the conventional near-Earth TT&C and communication methods are not discussed in detail here. In this chapter, some deep space radio measurement technologies such as innovative ranging, velocity measurement and angle measurement are firstly introduced for the tracking missions. Secondly, for the telemetry and telecommand missions, the deep space RF system technology and deep space data communication technology for resolving the problems of large path loss and delay are introduced from two aspects: radio frequency and data. Finally, a design example of deep space TT&C and communication system is presented.

8.2 Deep Space Radio Measurement Technology

8.2.1 *Deep Space Ranging*

The basic principle of obtaining the radial distance between the spacecraft and the Earth station is as follows. The Earth station transmits the uplink RF signals modulated with the ranging information. After the coherent forwarding by the spacecraft transponder, the Earth station receives the demodulated ranging information and obtains the time delay difference between the transmission and reception of the ranging information by comparison. The product of this delay difference and the velocity of light is just the distance between the spacecraft and the Earth station.

8.2.1.1 Method of Ranging Transmission and Reception

The ranging methods can be divided into two-way ranging and three-way ranging based on whether or not the same Earth station is used for both ranging signal transmission and receiving. Whether in a near-Earth mission or a deep space mission, two-way ranging is a commonly used measurement method. However, in some cases, only three-way ranging is applied to deep space ranging due to the long space delay in deep space TT&C as well as the impact of the rotation of the Earth. For example, in Neptune exploration missions, the two-way communication delay is over 8 h, so the receiving station is not able to receive the ranging signals sent by the station itself. Therefore, it is impossible for the same station to carry out either single-station two-way coherent measurement or two-way ranging. Instead, three-way ranging must be performed through the transmission and reception at a different station.

The specific implementation of the three-way ranging is explicated as follows. The Earth station A transmits a signal at the time t_1. After the signal is coherently forwarded by the transponder, the Earth station B receives the signal at the time t_2 and compares its phase with the phase at t_1, thereby completing the measurement of the radial distance from the Earth station to the spacecraft. Depending on whether the transmitted signal and the received signal are homologous, the three-way ranging methods may be further divided into homologous three-way ranging and non-homologous three-way ranging. If the distance between the transmitting station and the receiving station is relatively small, the two stations are connected by a microwave link or a fiber link to ensure the frequency coherence between the two stations. This is the homologous three-way ranging. If the distance between the transmitting station and the receiving station is quite large, the inter-station frequency coherence cannot be guaranteed. This is non-homologous three-way ranging. In the deep space exploration missions, non-homologous three-way ranging is mainly applied.

From the perspective of measurement principle and ranging information composition, the two-way ranging and the three-way ranging are basically the same. Their essential difference is that the information is transmitted and received by the same Earth station in the two-way ranging, but by different Earth stations in the three-way ranging. Due to the difference of transmitting and receiving sources, the three-way ranging accuracy is generally lower than the two-way ranging accuracy. The key technology to improve the accuracy of three-way ranging is to realize the clock synchronization between the transmitting station and the receiving station and to avoid the zero range between the transmitting link and the receiving link as much as possible so that the three-way ranging error introduced by the equipment calibration can be reduced.

8.2.1.2 Ranging Signal System

According to the type of ranging signal, the ranging methods may be further divided into side-tone ranging, pseudo-code ranging and tone-code ranging.

The side-tone ranging signal, generally called single-tone ranging signal or side-tone ranging signal, consists of a series of sine waves or square waves. The pure side-tone ranging signal can facilitate the formation of unified TT&C system by sharing the TT&C channel with the telemetry sub-carrier and the remote sub-carrier. This technology is well developed and easy to implement. It is widely applied in the field of aerospace TT&C in China and in the TT&C systems of Earth orbit spacecrafts developed by the NASA and ESA. For deep space TT&C, the NASA and ESA have maintained such system. And China's deep space TT&C stations also list such ranging system as an important option among a variety of ranging systems. However, the shortcomings of pure side-tone ranging system in deep space ranging are also obvious. Firstly, when capturing the distance, a deep space TT&C station uses the traditional side-tone ranging method, in which the transmission of side tones in sequence requires a longer time. Therefore, in deep space TT&C, after the primary tone is sent, the secondary tone is generally sent without waiting for the return of the primary tone. This requires a strong logical relationship between the transmitted and received signals, thus increasing the complexity of the system operation. Secondly, due to the limitation of transmitted energy and the influence of space path loss, the signal received by the TT&C station is extremely weak. To track and receive highly dynamic and weak signals, the deep space TT&C stations must adopt a high order narrow-band phase-locked loop with stable conditions and high Doppler compensation to extract the ranging signals. Therefore, the system is complicated.

The ranging signal in the pseudo-code ranging system consists of a set of pseudorandom codes. By utilizing the two-value nature of the pseudorandom sequence autocorrelation function, the delay of the received code relative to the transmitted code can be determined, so that the radial distance between the probe and the Earth station may be obtained [1]. As limited by the means of implementation, the minimum side tone used for pure side-tone ranging can only be 0.5 Hz in general, and the corresponding maximum unambiguous distance is 300,000 km, which is a far cry from the actual demands of deep space TT&C. The greatest superiority of pseudo-code ranging is that it can solve the problem of range ambiguity. In pseudo-code ranging, the code length determines the maximum unambiguous distance, and the capability of range ambiguity resolving may be further extended by appropriately increasing the code length. Depending on whether the simple coherent forwarding or the ranging signal regeneration is adopted by the responder, the pseudo-code ranging methods can be further divided into non-regenerative pseudo-code ranging and regenerative pseudo-code ranging. The superiority of regenerative pseudo-code ranging method is that the uplink noise does not affect the downlink signal. This is greatly favorable for the detection of deep space weak signals and thus is mostly widely applied in deep space ranging.

The methods of side-tone ranging and pseudo-code ranging have their own advantages and disadvantages. For pure side-tone ranging, the side-tone frequency may be improved to obtain high-precision ranging. In addition, the occupied bandwidth is narrow, the acquisition is fast, but the process of resolving phase ambiguity is relatively complicated. However, the pseudo-code ranging is featured with long unambiguous distance. To improve the accuracy, the code element width must be reduced, resulting in increased occupied bandwidth, more complicated capturing and extended occupation time. The tone-code ranging system tries to combine the advantages of the two methods and avoids their shortcomings. The tone-code ranging utilizes the high resolution and excellent signal-to-noise ratio of side-tone ranging as well as the merits of code ambiguity resolving. In the tone-code ranging method, although the resistance to the interference with telemetry signals is not as good as that in the pseudo-code ranging method, the interference may be avoided in most cases by selecting the side-tone frequency [2].

Presently, there is no unified deep space ranging system in the world. Side-tone ranging, pseudo-code ranging and tone-code mixed ranging are all applied in differently deep space TT&C missions.

8.2.2 Deep Space Velocity Measurement

Like the velocity measurement of the low-Earth orbit (LEO) spacecraft, the velocity measurement of the deep space probe is also realized by measuring the Doppler frequency offset of the carrier. When the transmitter and the receiver get close to each other in relative movement, the signal frequency f_R received by the receiver will be higher than the signal frequency f_T transmitted by the transmitter. But when they are moving apart from each other, the received signal frequency f_R will be lower than the transmitted signal frequency f_T. Such phenomenon, in which the received frequency is different from the transmitted frequency due to the relative motion, is called Doppler effect [3]. Depending on the tracking mode, the velocity of deep space probes can be measured by three methods: one-way Doppler measurement, two-way Doppler measurement and three-way Doppler measurement.

8.2.2.1 One-Way Doppler Measurement

In the one-way tracking mode, the deep space probe utilizes its high-stability frequency oscillator as a reference to generate the nominal downlink transmitted signals, which are then received by the Earth station after spatial transmission and delay. The Earth station measures the frequency difference of the received signals relative to the nominal downlink transmitted signals (using the high-stability frequency oscillator in the Earth station as a reference). The one-way velocity measurement system generally consists of the beacon on the probe and the ground receiving device, as shown in Fig. 8.1.

Suppose the frequency of the signal transmitted by the probe is f_T. Then, the frequency of the signal received by the ground is

$$f_R = \sqrt{\frac{c+v}{c-v}} f_T \tag{8.1}$$

where f_R is the frequency of the signal received by the Earth station, c is the speed of light, and v is the radial velocity of the probe.

If the Earth station can accurately replicate the frequency f_T, the Doppler frequency can be measured as

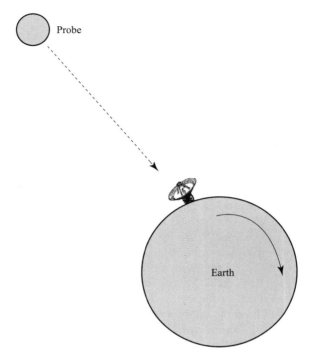

Fig. 8.1 Schematic diagram of one-way Doppler measurement

$$f_d = f_R - f_T = \left(\sqrt{\frac{c+v}{c-v}}\right)f_T \approx \frac{v}{c}f_T \tag{8.2}$$

where f_d is the frequency of the received signal.

Therefore, the radial velocity v of the probe relative to the Earth station can be calculated.

The accuracy of one-way Doppler velocity measurement is directly affected by the frequency stability of the probe. Due to the limitation of onboard conditions, it is difficult to manufacture a crystal oscillator with high frequency stability, which will directly affect the accuracy of Doppler velocity measurement. The unknown fixed frequency variation Δf with respect to the nominal frequency f is converted into the velocity error $\Delta v = c\frac{\Delta f}{f}$.

8.2.2.2 Two-Way Doppler Measurement

The two-way velocity measurement system typically consists of a responder (coherent) on the probe and a ground transmitting/receiving device, as shown in Fig. 8.2.

8.2 Deep Space Radio Measurement Technology

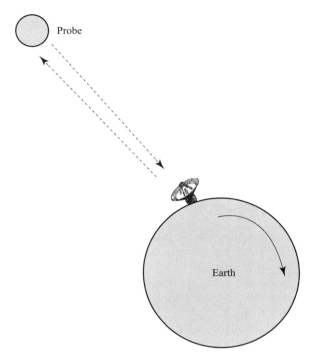

Fig. 8.2 Schematic diagram of two-way Doppler measurement

If the transmission frequency of the Earth station is f_T and the radial velocity of the probe relative to the Earth station is v, then the frequency of the signal received by the probe will be

$$f'_R = \sqrt{\frac{c+v}{c-v}} f_T \tag{8.3}$$

If the probe performs a coherent forwarding of the received signal to the Earth station, for which the forwarding ratio is q, then the frequency of the signal transmitted by the probe will be

$$f'_T = q f'_R = \sqrt{\frac{c+v}{c-v}} q f_T \tag{8.4}$$

The frequency of the signal received by the Earth station is

$$f_R = \sqrt{\frac{c+v}{c-v}} f'_T = \frac{c+v}{c-v} q f_T \tag{8.5}$$

Therefore, the measured Doppler frequency offset is expressed as

$$f_d = f_R - qf_T = \left(\frac{c+v}{c-v} - 1\right)qf_T \approx \frac{2v}{c}qf_T \qquad (8.6)$$

If v is much smaller than the speed of light, Eq. (8.6) is approximately valid.

In two-way Doppler measurement, the same ground frequency standard is generally adopted as the reference signal for the uplink signal detector and downlink signal detector, and a highly stable hydrogen clock is generally used to obtain high-precision Doppler data.

8.2.2.3 Three-Way Doppler Measurement

When the probe is far away from the Earth station, the two-way Doppler measurement may not be realized due to the influence of the rotation of the Earth. In this case, three-way Doppler measurement must be applied, as shown in Fig. 8.3. The three-way velocity measurement system generally consists of the transmitting device of the Earth station G1, the coherent transponder on the spacecraft and the receiving device of the Earth station G2.

If the transmitting frequency of the Earth station G1 is f_{G1_T}, and the radial velocity of the probe relative to the Earth station is v_{G1}, then the frequency of the signal received by the probe will be

$$f'_R = \sqrt{\frac{c + v_{G1}}{c - v_{G1}}} f_{G1\_T} \qquad (8.7)$$

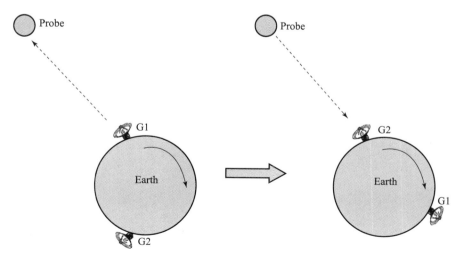

Fig. 8.3 Schematic diagram of three-way Doppler measurement

8.2 Deep Space Radio Measurement Technology

If the forwarding ratio of the received signal that is coherently forwarded by the probe is q, then the frequency of the signal transmitted by the probe will be

$$f'_T = q f'_R = q \sqrt{\frac{c + v_{G1}}{c - v_{G1}}} f_{G1\_T} \tag{8.8}$$

If the radial velocity of the probe relative to the Earth station G2 is v_{G2}, then the frequency of the signal received by the Earth station G2 will be

$$f_R = \sqrt{\frac{c + v_{G2}}{c - v_{G2}}} f'_T = \sqrt{\frac{c + v_{G2}}{c - v_{G2}} \cdot \frac{c + v_{G1}}{c - v_{G1}}} q f_T \tag{8.9}$$

Considering the fact that the probe is quite far away from the Earth station, the equation $v_{G1} = v_{G2} = v$ can be considered true, so:

$$f_R = \frac{c + v}{c - v} q f_T. \tag{8.10}$$

If the Earth station G2 can accurately replicate the transmitting signal frequency f_T of the Earth station G1, the Doppler frequency can be measured as

$$f_d = f_R - q f_T = \left(\frac{c + v}{c - v} - 1\right) q f_T \approx \frac{2v}{c} q f_T. \tag{8.11}$$

If v is much smaller than the speed of light, Eq. (8.11) is approximately valid.

Therefore, three-way Doppler measurement is similar to two-way Doppler measurement. The only difference between them is the frequency reference source used by the receiving device and the transmitting device. The position fixation deviation between them will be directly converted into the velocity measurement error.

Considering the fact that high-stability hydrogen clocks are generally applied in deep space Earth stations, the unknown fixation deviation between the Earth stations can be controlled to have no great influence on the accuracy of the velocity measurement.

8.2.3 Deep Space Angle Measurement

Angle measurement is one of the important basic elements for determining the orbit of the probe. The first and foremost condition for tracking the probe is to ensure that the ground antenna is able to point to the probe at any time. Angle tracking information can generally be obtained by two approaches—the amplitude of the incoming signal and the phase of the incoming signal.

The principle of angle measurement by amplitude comparison is to measure the amplitude of the incoming RF signals through the comparison with other signals received by the same Earth station antenna to reflect the deviation of the probe from the RF axis of the Earth station antenna, thus obtaining the pitch and azimuth angles of the spacecraft relative to the Earth station antenna. This scheme is well developed, simple in principle and relatively easy to implement in engineering. It has been widely applied in the spacecrafts on near-Earth orbits. The premise of the amplitude comparison is that the Earth station receiver must first reliably capture the downlink carrier signals of the spacecraft. However, a probe in deep space is remote. Compared with the near-Earth spacecraft, the RF signal of deep space spacecraft arriving at the Earth station has an extremely low signal-to-noise ratio, resulting in a larger amplitude comparison error and a larger angle measurement error. In addition, due to the long distance, the positioning error of the deep space spacecraft is much greater than that of the near-Earth spacecraft given the same angular error. Therefore, the angular accuracy in the amplitude comparison scheme cannot satisfy the requirements of orbit measurement and determination in deep space missions.

The angle measurement by phase comparison is based on the principle of phase interference. The phase difference of the incoming RF signals is measured through the comparison with the signals received by different Earth station antennas to obtain the direction cosine of the line vector between the Earth station and spacecraft relative to the coordinate axis of the Earth station, thereby determining the pitch and azimuth angles of the spacecraft relative to the Earth station antenna. Such phase interferometry scheme can achieve relatively high angle measurement accuracy and is the main method used for current deep space angle measurement. The engineering implementation techniques in common use include VLBI as well as the improved technologies based on VLBI such as ΔVLBI and SBI.

8.2.3.1 VLBI

The VLBI technology was developed in the 1960s and was originally applied to astronomical observations for accurately measuring the time difference of the signal waves from the same Extra-Galactic Radio Resource (EGRS) arriving at two (or more) stations. Its basic principle is shown in Fig. 8.4.

The two stations simultaneously observe the same EGRS. At each station, the instantaneous phase of the received signal (a random Gaussian process) is recorded on each channel. This process is the VLBI data acquisition. The data recorded by these two stations is sent to the same processing center, where the data from the matching channels of the two stations is cross-correlated to determine the geometric time delay τ_g. This process is the cross-correlation of the VLBI data. By correcting the time delay while accounting for the influence of the factors such as other media, the angle between the signal direction and the baseline direction can be determined. Based on sufficient observations and a certain solution model, the address coordinates of each tracking station, the angular position of the observed source as well as other parameters related to the signal propagation path can be obtained.

8.2 Deep Space Radio Measurement Technology

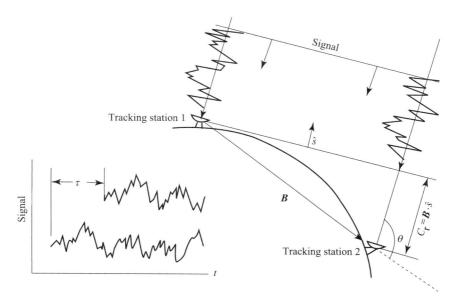

Fig. 8.4 Schematic diagram of the basic principle of VLBI

Currently, VLBI technology is the astronomical observation technology with the highest angular resolution. When the baseline length reaches tens of thousands of kilometers, the angle measurement accuracy can be up to 20–30 nrad. By using the VLBI technology, the angle data of the probe can be directly measured to apply a rational constraint on the lateral position and velocity of the probe. At the same time, by using traditional Doppler velocity measurement and ranging method, the radial distance and velocity of the probe can be directly measured to apply a rational constraint on the line-of-sight (LOS) position and velocity of the probe. By combining these two kinds of data for joint orbit determination, the orbit determination accuracy can be effectively improved, and the requirements of deep space exploration can be satisfied [4].

The application of VLBI technology in deep space probes can be dated back to the late 1970s, when the NASA used VLBI technology to measure the phase delay of the Mars probe "Viking" and the Venus probe "Pioneer". With the continuous development of science and technology, the VLBI technology has also made continuous progress. In the Chang'e 1 mission, China applied VLBI in space missions for the first time.

In the VLBI technique, the radio frequency signals from the same source are recorded at two stations and then cross-correlated to obtain the time delay. Therefore, after the signals are transmitted and before they are cross-correlated, any factors that may affect the signals (such as solar plasma) will have some effects on the time delay obtained from the final cross-correlation. Specifically, these effects are applied on tropospheric delay, ionospheric delay, the clock difference and clock difference rate between the two stations, device phase jitter, device phase delay, as well as Earth orientation parameters and station position error. These systematic errors may severely affect the accuracy of the VLBI geometric delay. If they are not appropriately treated, the measurement accuracy may not be guaranteed. For example, the nominal ionospheric delay correction in the zenith direction is between 0.1 and 2.0 ns for X-band and between 1.3 and 26.7 ns for the S-band. At the elevation angle of $20°$, the ionospheric delay corrections will be 0.3–5.9 ns and 3.9–78 ns for the X-band and S-band, respectively.

If the empirical model is used to calculate the systematic error, the effect is often less than ideal. The reason is that the variation of these systematic errors is generally very complicated, and the empirical model can only be based on the average regularity. This problem can be better solved by the differential measurement method, namely ΔVLBI technology.

8.2.3.2 ΔVLBI

The ΔVLBI technology can effectively eliminate the influence of various system errors in VLBI time delay, such as instrument delay, clock difference and atmosphere delay. Therefore, it is widely applied in VLBI astronomical observation and deep space probe navigation. ΔVLBI refers to the technology that a reference source (strong dense source) is observed simultaneously or almost simultaneously in a small region near the measured target, and then, its system error is used to determine the time delay of the source to be measured.

If the Doppler shift of the downlink probe signal (such as carrier) is obtained by the VLBI technique and then subtracted from the Doppler shift of the radio source signal obtained in the same manner within a similar angular distance, the resulting quantity of observation will be double-difference Doppler shift. This technology is called narrow-band ΔVLBI technique, also known as ΔDOD technique. If the group delay of the probe is generated from the phase information of the probe widespectrum signal obtained by the VLBI technique and then subtracted from the group delay of the radio star obtained in the same manner within close angular distance, the resulting quantity of observation will be differential delay. This technology is called wide-band ΔVLBI technology, also known as ΔDOR technology. Since the ΔDOR technology can provide a more accurate probe angular position, it is more widely applied in deep space exploration navigation [5].

Similar to the amount of information provided by the VLBI time delay, the ΔDOR technology provides a direct geometric angular measurement of the probe in the sky plane in a coordinate system defined using the radio source. Figure 8.5 shows the principle of ΔDOR measurement.

As the double difference greatly eliminates the systematic error in the measurement, the ΔDOR has high angular measurement accuracy. The angular measurement accuracy achieved by the DSN with ΔDOR is 2nrad. Compared with the orbital solution derived only from the LOS measurement data, the orbital solution derived from both the LOS data and ΔDOR data is insensitive to systematic errors. In addition,

8.2 Deep Space Radio Measurement Technology

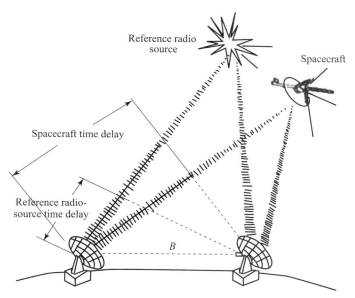

Fig. 8.5 Schematic diagram of ΔDOR measurement geometry

since ΔDOR is a direct geometric measurement, the solution obtained by joint orbit determination using ΔDOR observation data has no singularity in low declination and other special geometric configurations. Another significant advantage of ΔDOR is that it performs well in short-arc orbit determination and positioning and can measure the deep space probe state rapidly after a maneuver. This technique is especially suitable for the scenario in which only a few observation opportunities are available but long-term navigation accuracy is still required.

The ΔDOD and ΔDOR technologies applied to deep space exploration have followed the basic idea of VLBI technology, that is, through the continuous extension of the baseline, various measurement errors are evenly spread on the baseline, thereby improving the measurement accuracy. Meanwhile, by determining the differences between the Doppler shift and time delay of the probe and the corresponding measurements of the radio source, the influence of the system errors is further reduced. A significant difference between the ΔDOD/ΔDOR techniques and the astronomical VLBI observation is that for astronomical VLBI observations, since the radio source is a broad-spectrum signal similar to white noise, the radio signal should be recorded and cross-correlated in a certain spectral sub-bandwidth to form a certain equivalent extended bandwidth so as to obtain the time delay. However, to obtain a similar time delay for the probe, it is only required to modulate two side-tone (sine wave or square wave) signals with an interval equal to the required extended bandwidth on the downlink carrier spectrum, or to use the higher harmonics of the telemetry signals. Moreover, as the probe signal is deterministic, the time delay corresponding to the equivalent bandwidth can be obtained without cross-correlation processing.

8.3 Deep Space RF System Technology

8.3.1 Radio Frequency Modulation

Presently, the TT&C adopts a well-developed unified carrier system. The system has several advantages. The main advantage is that it uses the sub-carrier method that is compatible with both data transmission and side-tone ranging, thus greatly reducing the spaceborne and ground-based devices. However, it also has certain disadvantages. Under this system, multiple sub-carriers are modulated on one carrier and occupy a large total bandwidth, and the nonlinear effect inevitably generates the intermodulation spectrum and the remodulation of the uplink noise in the downlink return channel. Therefore, the power utilization rate is not high. In addition, since the power is distributed to multiple sub-carriers, the actual power obtained by each sub-carrier is not high. For example, when the satellite TT&C channel is used for telemetry, telecommand and ranging at the same time, if the modulation indexes for telecommand and uplink ranging tones are 0.9 rad, and the modulation indexes for telemetry and downlink ranging tones are 0.8 and 0.3 rad, then the modulation loss of the telecommand sub-carrier will be about 5 dB, the modulation loss of the telemetry sub-carrier will be about 7 dB, and the modulation loss of the downlink ranging tone will be about 16 dB.

If a PSK-based digital direct modulation system is applied, these modulation losses will not exist theoretically. Generally, QPSK, OQPSK, MSK, GMSK, CPM, TCM and other systems are relatively satisfying. When the carrier is directly modulated by the data, the way to effectively insert the ranging information should be studied. International space stations adopt the approach to modulate the carrier wave by the data and ranging pseudo-code through UQPSK or OQPSK modulation method. In some current studies, ranging information is treated as a service of AOS that is included in the virtual channel scheduling, mainly to solve the problem of time delay uncertainty.

System noise may also be equivalently reduced by the modulation system, for which an example is the "Phase Inversion Symmetric Modulation" (PISM) scheme. PISM is a new type of digital signal composite modulation. "Phase Inversion" is to simultaneously transmit two signals that have mutually inverted phases at the transmitting end. "Symmetric" means that the processing methods of the two signals are as uniform as possible so that the interference they receive is basically the same, that is, the two signals are in the same disturbed channel reference system. The interference signal is used as the reference signal of the reference system and can be canceled during the processing by subtracting any two signals in the reference frame to obtain the difference signal of the two useful signals. When the two useful signals have inverted phases and a correlation coefficient of -1, the useful signal amplitude is doubled, and the noise is greatly attenuated. Presently, the researches similar to PISM are far from the practical application in engineering. However, it is still necessary to strengthen similar theoretical investigations in the field of deep space communication.

8.3 Deep Space RF System Technology

In addition to adopting a new modulation system, the development of an adaptive communication system in deep space communication is also a new direction to be highlighted. For example, the adaptive system with variable bit rate and variable frame length and structure is a breakthrough of the concept of traditional fixed length and code rate and can better adapt to the situation in which the signal channel and signal source vary greatly during flight.

8.3.2 High-Sensitivity Reception

In the TT&C and communication system, it is necessary to first perform acquisition analysis on the received signals in order to complete the demodulation and measurement of the received signals. In deep space communication, the signal-to-noise ratio (SNR) of the received signals is extremely low due to the restriction of various conditions. Meanwhile, the velocity of the deep space target reaches the second cosmic velocity, resulting in high Doppler dynamics. The acquisition of such highly dynamic low-SNR signals is a key technology.

There are many approaches to achieve low-threshold demodulation. The common methods include negative feedback demodulation, tracking filter, phase-locked loop, etc. The threshold of the phase-locked loop depends on the loop signal-to-noise ratio (SNR) instead of the input SNR. The loop SNR is generally higher than the loop input SNR.

In practical engineering, if the loop SNR is $\geqslant 6$ dB, the loop can function well. Therefore, as the passband of the loop is usually much narrower than the pre-passband of the loop input end, the loop SNR is significantly higher than the input SNR, and the loop may work under low input SNR conditions, namely the loop has the excellent feature of low threshold. Therefore, as long as the loop is designed as a narrow band, the weak signal submerged in the noise may be extracted. By using the low-threshold characteristic of the phase-locked loop and applying it to the demodulation of frequency modulation and phase modulation signals, the weak signal may be demodulated, and the demodulation threshold may be lowered.

In the phase-locked loop, regardless of the type of the loop filter of the second-order loop, the closed-loop frequency response exhibits a low-pass characteristic, while the error frequency response exhibits a high-pass characteristic. To be specific, as long as the phase modulation frequency Ω of the input signal is lower than the natural frequency ω_n of the loop (strictly speaking, the cutoff frequency), the loop can appropriately transmit the phase modulation, and the output phase $\theta_o(t)$ of the voltage controlled oscillator can appropriately track the change of the input phase $\theta_i(t)$ to obtain a very small error phase $\theta_e(t)$ of the loop. When the phase modulation frequency Ω is much higher than the natural frequency ω_n of the loop, the loop cannot transmit the phase modulation, and the output phase $\theta_o(t)$ of the voltage controlled oscillator can no longer track the change of the input phase $\theta_i(t)$. At this time, the error phase $\theta_e(t)$ of the loop will change with the input phase $\theta_i(t)$.

According to the above two characteristics of the phase-locked loop (i.e., carrier tracking and modulation tracking), two types of phase-locked demodulators may be formed. The first phase-locked demodulator is to use the carrier tracking characteristic of the loop to extract the carrier information of the modulated signals by means of the phase-locked loop, then detect the phase of the modulated signals through the phase detector, from which the demodulated signal is output. The second phase-locked demodulator is to use the modulation tracking characteristic for phase-locked demodulation. To improve the ability of tracking highly dynamic low-SNR signals, the third-order phase-locked loop has been used for carrier phase tracking as the traditional second-order phase-locked loop cannot satisfy the related requirements any longer.

When the bandwidth of the phase-locked loop is narrow, it is generally difficult to lock the loop. To shorten the loop lock time and improve the loop tracking capability, the variable bandwidth tracking technology is applied, that is, the phase-locked loop is locked in a relatively wide acquisition band, and then, the loop bandwidth is gradually reduced until the required loop signal-to-noise ratio is achieved to track extremely narrow band.

For the residual carrier PM modulation signals, the acquisition can be realized by improving the conventional frequency FFT acquisition method. However, if the received signals are modulated through complete carrier suppression or complete sub-carrier suppression, an improved algorithm such as dual-branch FFT can be adopted. The fast Fourier transform with dual branches exhibits two advantages. On one hand, the in-phase and quadrature channels of the carrier are simultaneously utilized, so the signal-to-noise ratio during FFT acquisition can be improved, the FFT period can be shortened, and the instability of the carrier can be significantly improved. On the other hand, the application of the dual-branch FFT enables not only the sub-carrier and code element acquisition before carrier acquisition, but also the simultaneous acquisition of sub-carrier, code element and carrier. If carrier acquisition is performed after the acquisition of sub-carrier and code element, the FFT period required for carrier acquisition can be greatly reduced.

Upon the completion of the FFT acquisition of the carrier or sub-carrier, the captured carrier and sub-carrier should be phase-locked before normal data demodulation and measurement. Since the received signals are extremely weak, the loop bandwidth should be narrowed down to Hz level or even less than 1 Hz to improve the sensitivity of the receiver and ensure normal operation of the carrier and sub-carrier phase-locked loop.

8.3.3 High EIRP Emission

Equivalent isotropic radiated power (EIRP) is determined by the transmitter power and the antenna gain. Therefore, the power of the probe transmitter and the gain of the spacecraft antenna should be increased to improve the EIRP of the probe.

8.3 Deep Space RF System Technology

Due to the limitation by various factors such as size, weight and power consumption, the RF emission power of deep space probe is unlikely to increase significantly. In addition, when the probe is operating in the vacuum environment, if the transmitted power of the device is too high, the microdischarge phenomenon may easily occur to cause adverse consequences such as instantaneous power drop and even equipment damage. Meanwhile, the greater the transmitter power is, the greater the absolute value of the power loss caused by cable loss and the power demand for the probe will be. This also brings great difficulties to the thermal control design. Generally, the TT&C transmission power of the near-Earth orbit spacecraft is less than 1 W; the TT&C transmission power of the synchronous orbit spacecraft is generally about 10 W; the TT&C transmission power of the lunar probe is generally about 20 W; and the TT&C transmission power of the Mars probe is generally around 100 W.

The gain of the antenna is proportional to the square of its aperture. Essentially, the antenna gain represents the directionality of the antenna. With the increase of the antenna aperture, the ability of the antenna to gather the microwave energy of a certain wavelength will improve. By tightening the antenna beam, the antenna gain is improved, thereby equivalently reducing the spatial loss of the radio frequency. For deep space probes, large-aperture and high-gain antennas are generally applied. Meanwhile, to make the antenna beam center point to the ground station/data receiving station, the corresponding rotating mechanism should be provided. Presently, the high-gain antennas used in deep space probes have an aperture of about 2–4 m, and their gains in the X-band are generally about 45 dB. In the future, the deep space exploration missions for exploring farther targets will require larger flexible antennas, whose aperture may be up to ten meters or even tens of meters. Since the antenna gain is proportional to the square of the carrier frequency, the high gain of the antenna can be obtained by increasing the carrier frequency. Presently, the main frequency band for deep space exploration is the X frequency band, which will be increased to the Ka frequency band or even higher frequency bands in the future. However, as the frequency band increases, the requirement for antenna profile accuracy becomes higher, and the antenna design and manufacture become more difficult accordingly. Meanwhile, as the frequency band increases, the beam of the antenna will become narrower, and the requirements for the attitude control precision of the probe and for the accuracy of the antenna pointing control will become higher.

8.3.4 Laser Communication

Laser communication refers to the communication using a laser beam as information carrier. Compared with traditional microwave communication, the wavelength used for laser communication is significantly shorter than microwave wavelength and has high coherence and spatial directionality. Therefore, laser communication has the advantages such as large communication capacity, small equipment weight, low power consumption, small size and high confidentiality. Since the wavelength of the laser is short, the apertures of both the transmitting and receiving telescopes can be reduced under the same requirements for divergence angle and receiving field of view. As a result, the huge dish-shaped antenna of the RF system can be eliminated, and the high efficiency of energy utilization can be achieved with a small mass and volume. Such superiority may well solve the problem of huge path loss in deep space TT&C and communication [6].

The space laser communication system is mainly composed of a laser light source subsystem, a transmitting/receiving subsystem and a subsystem for acquisition, tracking and aiming (ATP). The communication light source is quite important, as it directly affects the selection of the detection components and the communication distance, and the aperture and gain of the optical antenna. The commonly used laser is the AlGaAs laser with a wavelength range of 800–850 nm. The APD detection components in this wavelength range operate at the peak value with high quantum efficiency and high gain. The frequency-doubled Nd:YAG lasers or argon-ion lasers with the wavelengths between 514 and 532 nm are the preferred choices for onboard laser sources. The transmitting/receiving subsystem mainly involves the selection of transmitting and receiving antennas, filtering components and signal detection and processing component, as well as modulation and demodulation systems. The transmitting and receiving antennas are actually optical telescopes. As the case may be, a Cassegrain-type reflective antenna or a transmissive antenna can be used. The aperture of the antenna is generally 10–30 cm. Filters and detectors are the important parts of the receiving system. Interference filters are generally applied. And avalanche photodiodes (APDs) are usually used as detectors.

The modulated digital baseband signals pass through the power driving circuit and make the laser to generate a laser beam so that the laser carrying the communication signals is transmitted through the optical transmitting antenna. When receiving the signals, the laser communication machine on the other end converges the collected optical signals onto the photo-electric detector through the optical receiving antenna. The optical signals are converted by the photo-electric detector into electrical signals and amplified, and the useful signals are extracted by threshold detection method. Then, the baseband signal is recovered by the demodulation circuit. In such process, the tracking and aiming (ATP) subsystem enables the capture, tracking and aiming of the laser beams at both ends.

The key technologies of laser communication terminals include the short-time acquisition technology with narrow beam and high probability; the tracking and aiming technology at the microradian level; vibration suppression technology; the optical amplification transmitting technology with high modulation rate, high power and high energy conversion efficiency and the optical amplification receiving technology with high sensitivity and anti-jamming performance.

8.4 Deep Space Telemetry and Telecommand and Data Communication Technology

8.4.1 Data Format

8.4.1.1 PCM Telemetry and Telecommand

The traditional telemetry and telecommand of spacecrafts is in PCM format, which is applicable to the point-to-point closed link between spacecraft and Earth station.

The working principle of the PCM telemetry system is as follows. At the transmitting end of the spacecraft, through the selection of the multi-way switch, the analog parameters obtained by the sensors at various physical telemetry points are digitized by analog-to-digital conversion and then integrated with other digital quantities, formatted as scheduled and processed with sub-carrier modulation after the addition of synchronization words. Finally, they are transmitted to the Earth through RF modulation. At the receiving terminal on the Earth, the radio frequency and sub-carrier are demodulated in turn. Then, the telemetry data is processed, stored, recorded and distributed to each user.

Like PCM telemetry, the PCM telecommand system is also suitable for point-to-point closed links. The key of its implementation is the encoding and decoding of telecommands and data to satisfy the requirements of reliability and security. A fixed data stream format is also applied to PCM telecommand. The working principle of the PCM telecommand system is as follows. In the ground TT&C station, the telecommand command and the uplink data are input into the telecommand encoder for formatting and encoding to form a digital telecommand frame. Then, sub-carrier modulation is performed. After the completion of radio frequency (RF) modulation, the frame is transmitted through the radio channel. At the receiving end of the spacecraft, the received telecommand RF signals are processed with RF demodulation, followed by sub-carrier demodulation. Then, the telecommand PCM code is obtained. The telecommand and data are distinguished by decoding and format identification and then are output to the corresponding individual users on the spacecraft.

With the increasing complexity of spacecraft, many autonomous tasks have been proposed, such as programmed command, time delay command, payload operation management, energy management, system level security management, system level reorganization, spacecraft timing systems as well as data exchange and sharing among the subsystems. These tasks are needed for almost every spacecraft. Initially, such tasks can be implemented by expanding the functions of the telemetry and telecommand subsystem, such as by combining telecommand with program control and telemetry with timing system. However, the tasks of autonomous management have become increasingly intensive and complicated. The traditional PCM telemetry and telecommand system has been unable to adapt to the space missions requiring increasingly high autonomy, such as deep space exploration missions. As a result, space data systems have emerged.

8.4.1.2 Packet Telemetry and Telecommand

The packet telemetry and telecommand was first proposed by the Consultative Committee for Space Data Systems (CCSDS).

The telemetry and telecommand technology recommendations developed by CCSDS provide the format and method of data transmission. The spacecraft-to-ground telemetry information data transmission is managed by packet telemetry recommendations, and the ground-to-spacecraft command information is managed by telecommand recommendations. The Advanced Orbiting System (AOS) provides the standards of data transmission between advanced space system (such as space station) and the Earth. The widely applied CCSDS telemetry and telecommand technology recommendations mainly involve packet telemetry, telemetry signal encoding, telecommand data management services, telecommand data routing and telecommand channel services. Among them, two types of standard data are mainly used for data transmission: space packet and transmission frame. Space package is a data packet generated by a spaceborne application and can be transmitted at the application layer, transport layer and network layer of the OSI model. The transmission frame corresponds to the data link layer of the OSI model and is used for the communication between the spacecraft with the radio frequency (RF) link and the ground. Each protocol is a communication protocol based on a single layer of the OSI model and can be shared with the protocols of other layers.

The two most important concepts for packet telemetry are source packet and virtual channel. Source packet is the data packet generated by a spaceborne information source (subsystem or device) during a certain application process. The data source packets generated by different application processes have different data generation rates and packet lengths. Such source packets are autonomously and randomly generated. And different source packets are asynchronous. Such source packets generated by different application processes and with different probabilities of occurrence will be transmitted back to the Earth by the same space-to-ground link and then distributed to different data sinks on the ground according to different application processes. The virtual channel is a dynamic channel management mechanism for multiple data streams. In fact, multiple information sources virtually monopolize the physical channels in a dynamic time-division manner so that multiple data streams can share the same physical channel for data transmission. The CCSDS specifies that up to eight virtual channels can be utilized in packet telemetry. The source packets generated in different application processes may share the same virtual channel. The allocation of source packets to virtual channels is usually organized according to the data characteristics. Therefore, the virtual channel does not limit the source and sink of the data. The superiority of a virtual channel is that each virtual channel can be managed independently, and different virtual channels may have different priorities so as to provide different levels of service for different types of data. The entry of different source packets into the virtual channel is asynchronous in terms of timing sequence. However, the source packets entering the same virtual channel are transmitted in chronological order, that is, each source packet is queued in the virtual channel. The physical concept of virtual channel is similar to a street with

8.4 Deep Space Telemetry and Telecommand and Data Communication Technology

several bus lines. Each bus is independently scheduled, but all the buses are mixed and running sequentially on the street. The occupation of the physical channel by different virtual channels is dynamically scheduled. The system designer devises a reasonable scheduling algorithm and system parameter configuration according to the task requirements to satisfy the requirements of different users and ensure the timeliness and integrity of data transmission.

Packet telemetry has significant advantages over traditional telemetry. As the packet telemetry has realized dynamic multiplexing on a single physical channel, the channel utilization rate is improved. As each application process can independently generate the source packets of different lengths according to its own needs without being limited by a fixed sampling rate, the autonomy of the packet telemetry source can be guaranteed to fully support the different data requirements of each application process. Furthermore, as the virtual channel mechanism is available to provide different levels of services for the data with different transmission requirements, packet telemetry may be compatible with different characteristic and real-time requirements of each information source. Since there is no limitation by fixed telemetry format, the data system can be designed in parallel with the spacecraft system and user systems, with no need to wait for the user to propose the sampling rate and for the system to complete the frame format design. Meanwhile, in the system integration test, the multiplexing of asynchronous data stream is quite convenient. The sources may be directly added, removed and adjusted without changing the original system configuration, thus greatly simplifying the system integration test. The system using packet telemetry can provide the user-oriented and application-oriented transparent services with a high degree of standardization and strong self-adaptability.

Similar to packet telemetry, packet telecommand is also based on source packets and virtual channels and adopts the layered system. The complex spacecraft control process is boiled down to a series of simple standard operations at each layer that are equally implemented. The interface between two layers is defined by a standard data format according to a certain protocol. The hierarchical model of packet telecommand is shown in Fig. 8.6. From top to bottom, it is divided into an application process layer, a system management layer, a packet layer, a segmentation layer, a transmission layer, a channel encoding layer and a physical layer. The application process layer and the packet layer are used to implement the data management service; the segmentation layer and the transmission layer are used to implement the data routing service; and the lower two layers are used to implement the channel service so as to complete the data transmission function.

The CCSDS has also attached great importance to the projects related to deep space exploration, among which two of the most compelling studies are the proximity-space links and the interplanetary Internet (IPN Internet). There is a problem of proximity-space link communication and operation between the orbiter and lander in deep space explorations. For example, the communication between the Mars rover and the Earth is quite limited so that the data is relayed by the orbiter flying around Mars; while proximity-space link communication is required between the Mars rover and the Mars orbiter. Meanwhile, when the explorations of multiple planets have started, it is necessary to construct a backbone network consisting of

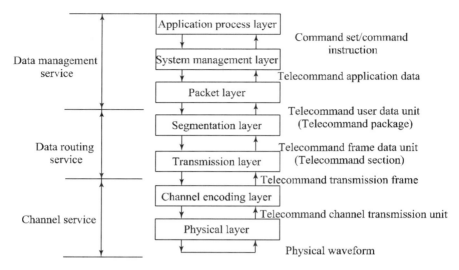

Fig. 8.6 Layered model of packet telecommand

special gateways and ultra-long-distance transmission links in the interplanetary space to connect various probes and planetary station facilities located on and around the planets. Thus, the transmission, exchange and sharing of multimedia information can be achieved in the vast interplanetary space.

The object of proximity link protocol research is the fixed or mobile close-range two-way radio communication link with the characteristics of short delay, moderate signal strength and short independent dialogues. Proximity link protocols can support the communication and navigation demands among multiple types of spacecrafts. As different spacecrafts have different communication capabilities and mission requirements, their communication modes can be simplex, half-duplex or full-duplex. At the beginning of the communication process, a request or negotiation process is first completed between the communicating parties at a lower code rate over a contact channel to determine various channel parameters and establish a communication channel. The contact process can be initiated by either party of the communication. When the communication process uses only one physical channel, the contact channel can be the same as the communication channel; when the communication process uses multiple physical channels, the contact channel occupies a separate physical channel.

The proximity link protocol supports both asynchronous and synchronous communication modes. For asynchronous data links, the frame length varies with the data contents. The maintenance of the link is achieved by inserting an idle sequence between frames that is transmitted only under necessary circumstances. The data field of the frame may contain an integer number of undivided standard packets, single data segments or user-defined data. The proximity link protocol consists of a physical layer and a data link layer. Its main functions are completed in the data link layer. Of course, the physical layer is also required to complete the input and output of RF signals. The main functions of each layer are shown in Fig. 8.7.

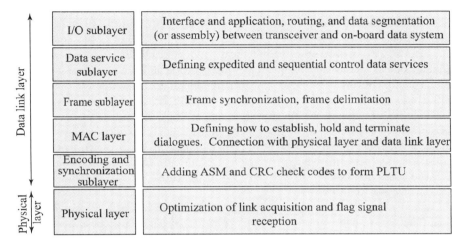

Fig. 8.7 Schematic diagram of hierarchical structure of proximity link protocol

8.4.2 Channel Encoding

Another method to improve the reception of weak signals is channel encoding and decoding. In channel encoding, a particular control method is adopted for introducing an appropriate number of redundant bits to overcome the effects of noise and interference on the information during transmission. According to the channel encoding theorem proposed by Shannon, any stationary discrete memoryless noisy information source has a fixed quantity called channel capacity, which is denoted as C. As long as the transmission rate of information is lower than the channel capacity, an encoding method is surely available so that the probability of occurrence of information error tends to be arbitrarily small as the code length increases. Conversely, when the information transmission rate exceeds the channel capacity, such encoding method is unavailable. This is the well-known channel encoding theorem, which gives the upper limit of the information transmission rate on a particular channel.

A classic channel encoding scheme is a concatenated code (a convolution code concatenating RS code) recommended by the CCSDS. When the bit error rate is 1×10^{-7}, a channel gain of about 7 dB can generally be obtained in engineering. However, in many cases, especially in the systems where bandwidth and power are both limited, this approach still fails to satisfy the application requirements. Among the encoding schemes that have been proposed so far, both the Turbo code and the LDPC code have the encoding gains high enough to obtain the encoding performance close to the Shannon limit under the extremely low F_b/N_0. This is a hot research topic. The CCSDS has adopted the Turbo code as the standard for deep space communication and has gradually standardized the application of LDPC codes in space communication.

Code word cascading is a method of constructing excellent code words within the controllable decoding complexity. The two cascading code words are called inner code and outer code, respectively. The "RS code + convolution code" cascading scheme is applied in deep space communication, in which the inner code is a convolutional code of Viterbi decoding and the outer code is an RS code. Thus, the inner code, namely convolutional code, can correct enough errors so that the outer code with a high code rate can control the error rate within the range of specification.

The basic principle of the Turbo code is that multiple sub-codes are cascaded in parallel or serially through an interleaver and then iteratively decoded by a mechanism similar to the reuse of exhaust gas of internal combustion engine so as to obtain the error correction performance close to the Shannon limit. The implementation forms of Turbo codes are roughly divided into two types: Turbo Product Code (TPC) and Turbo Convolutional Code (TCC).

Currently, LDPC is the encoding method that is the closest to the Shannon limit. The LDPC code belongs to a linear block code and can be described by (N, i, j), where N represents the code length, and $i(j)$ represents the number of non-zero elements in each row (column) of the check matrix H, called the row (column) weight. Constructing an LDPC code is equivalent to constructing a sparse check matrix H, mainly by the methods such as random construction and algebraic construction. Compared with the random construction method, algebraic construction has a rigorous mathematical structure, which is more conducive to performance analysis and may be provided with a cyclic or quasi-cyclic structure to reduce the encoding complexity. In the LDPC decoding, the algorithm based on belief propagation (BP), such as sum product algorithm (SPA), is commonly used. The LDPC channel encoding technology was applied in the China Chang'e 2 mission to obtain the encoding gain of over 8 dB in practical applications.

8.5 Design of Deep Space TT&C and Communication System

This section takes the Mars exploration mission as an example to explicate the design ideas and solutions of a general deep space TT&C and communication system.

8.5 Design of Deep Space TT&C and Communication System

8.5.1 Mission Analysis

In general, the mission analysis of the deep space TT&C and communication system mainly includes the selection of TT&C system based on the condition of ground station support and equipment maturity, the calculation of power and sensitivity based on the flight orbit input as well as the analysis of Doppler frequency and antenna coverage.

8.5.1.1 TT&C System

1) TT&C frequency band

The radio frequency band to be adopted is determined according to the ITU requirements for deep space exploration frequencies, the support condition of the ground-based deep space station and the equipment maturity and other factors. At present, in the Mars exploration missions, X-band is mostly adopted by the Mars probes as the band for TT&C and data transmission.

2) TT&C system

(1) Carrier modulation method

The unified carrier TT&C system and the PSK data modulation system are both well developed and reliable carrier modulation methods and are also widely used in near-Earth and deep space exploration missions. The advantages and disadvantages of both methods are described in Sect. 8.3. Unified carrier TT&C is required in deep space orbit measurement, and PSK modulation system may be applied in data transmission.

(2) Ranging method

Presently, the ranging based on the coherent forwarding of pure side tones is commonly used in near-Earth missions but is not applicable in deep space exploration missions. Considering the mission requirements and technical capabilities, regenerative pseudo-code ranging + side-tone ranging method may be used for deep space ranging.

(3) Angle measurement method

In deep space exploration, angle measurement is generally based on the VLBI principle to improve accuracy. The Mars probes may adopt the ΔDOR method for angle measurement.

(4) Velocity measurement method

Under current conditions, deep space Doppler measurement is generally implemented by two-way tracking due to the limitation by the stability of the crystal oscillator. The velocity of a Mars probe is measured by the two-way Doppler velocity measurement method.

3) Channel encoding method

The R-S+ convolutional concatenated code is currently a relatively matured encoding method, in which the convolutional code is used as the inner code and the R-S code as the outer code, so the demodulation threshold is reduced from 9.6 to 2.6 dB, resulting in a channel encoding gain of 7 dB. LDPC encoding is also applicable to deep space explorations. In the Mars exploration missions, R-S+ convolutional concatenated coding or LDPC coding can be used.

8.5.1.2 Power and Sensitivity Analysis

During the mission, the distance between the probe and the Earth varies with the mission phase, as shown in Fig. 8.8.

If the probe is supposed to orbit Mars for 3 years, the distance between the probe orbiting Mars and the Earth during the entire mission can be as far as 3.8×10^8 km and as near as 7.5×10^7 km. According to the different distances of the probe from the Earth in each flight phase, the received power and sensitivity of the probe are analyzed.

The total power of the signal arriving at the probe receiver is calculated based on the distance from the probe to the Earth in different flight phases as well as the EIRP of the Earth station:

$$[P_r] = [\text{EIRP}] - [L_{\text{tap}}] - [L_p] - [L_s] - [L_a] - [L_{\text{rap}}] + [G_r] + [L_{\text{rf}}]. \qquad (8.12)$$

where

$[\text{EIRP}] = [P_t] - [L_{\text{tf}}] + [G_t]$, effective radiated power (dBW);

$[L_{\text{tap}}]$ is the pointing error loss of the transmitting antenna (dB);

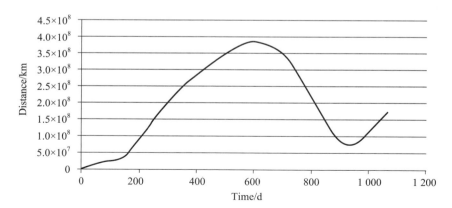

Fig. 8.8 Variation of the distance between probe and the Earth over time

8.5 Design of Deep Space TT&C and Communication System

$[L_p]$ is the polarization loss (dB);

$[L_s]$ is the free space loss (dB);

$[L_a]$ is the atmospheric loss (dB);

$[L_{rap}]$ is the pointing error loss of the receiving antenna (dB);

$[G_r]$ is the gain of the receiving antenna (dB);

$[L_{rf}]$ is the feeder loss of the receiving system (dB).

From the above formula, the signal power that arrives at the receiver of the transponder from the low-gain antenna and the directional antenna, respectively, during the exploration process and changes with the distance can be calculated.

(1) Earth station EIRP: 104 dBW;
(2) Transmitting antenna pointing loss + polarization loss + atmospheric loss, etc.: −1 dB;
(3) X-band free space loss at the altitude of 380 million kilometers: −282 dB;
(4) Receiving antenna gain: 3 dBi;
(5) Receiving system feeder loss: −3 dB;

The calculation based on the above parameters gives $P_r = -179$ dBW $= -149$ dBm.

The total signal power arriving at the receiver of the X-band transponder of the probe is −149 dBm. Considering the 1 dB margin, the carrier acquisition sensitivity should be better than −150 dBm.

8.5.1.3 Doppler Analysis

To determine the performance specifications of the receiver of the X-band deep space transponder, the radial velocity and acceleration variations during the mission of the probe are simulated.

Figure 8.9 shows the distribution of radial velocity during the mission of the probe. The range of the probe velocity during the mission is mainly between −18.88 and 14.66 km/s, of which the maximum velocity in the positive radial direction is 14.66 km/s and the maximum velocity in the negative radial direction is 18.88 km/s.

Figure 8.10 shows the distribution of radial acceleration during the mission. The acceleration varies from −1.3 to 3.1 m/s² with a maximum value of 3.1 m/s².

The variation of the Doppler Shift can be calculated from the velocity and acceleration distributions.

According to the velocity and acceleration of the Mars probe during the mission as well as the resulting Doppler shift and Doppler shift rate, the scan bandwidth and scan rate necessary for the transponder receiver to acquire the uplink carrier can be analyzed.

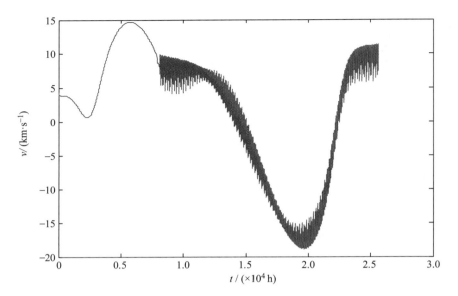

Fig. 8.9 Distribution of radial probe velocity with respect to the Earth

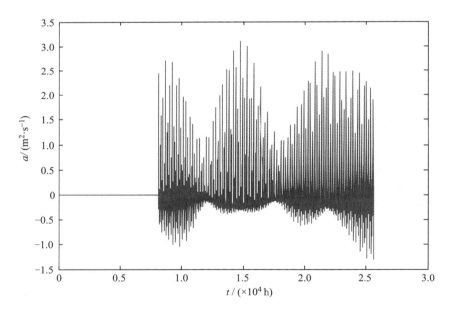

Fig. 8.10 Distribution of radial probe acceleration with respect to the Earth

8.5 Design of Deep Space TT&C and Communication System

8.5.1.4 Antenna Coverage Analysis

Considering the probe flight attitude, the antenna installation form and the probe TT&C working mode, the ground coverage of each spaceborne antenna at different working stages is analyzed.

1) Definition and description of related angles

(1) Definition of SPE angle

The angle between the "Probe-Sun Vector" and the "Probe-Earth Vector" is defined as the SPE angle. The SPE angle is used to describe the relationship between the variations of the Sun-pointing direction and Earth-pointing direction of the probe during the flight process. When the probe is turned into a Sun-pointing cruise attitude or Sun-pointing emergency attitude, the change in the SPE angle helps determine the installation position of the onboard fixed antenna, the antenna beam width, as well as the Earth-pointing direction.

(2) Definition of γ angle

The angle between $+X$-axis of the probe and the vector from the center of mass of the probe to the Earth is defined as the γ angle. Since the antenna is mounted along the X-axis of the probe, the variation range of γ angle can reflect the Earth station coverage of the probe antenna. When the γ angle is between $90°$ and $180°$, the Earth station is within the coverage of the $-X$ TT&C antenna of the probe; when the γ angle is between $0°$ and $90°$, the Earth station is within the coverage of the $+X$ TT&C antenna of the probe.

(3) Relationship between the γ angle and the SPE angle of the probe

In the phase from the probe launch to the probe-rocket separation, the attitude of the probe is determined by the rocket's flight trajectory, and the probe is not pointing to the Sun. In this phase, γ angle is used for antenna coverage analysis.

In the period after the probe-rocket separation and before the Mars orbiting, exclusive of orbital maneuver stage and attitude adjustment stage, the normal flight attitude of the probe during the near-Earth stage and the Earth–Mars transfer stage is always Sun-pointing, that is, the $+X$-axis of the probe points to the Sun. At this time, the γ angle of the probe coincides with the SPE angle. The SPE angle can be applied when analyzing the Earth coverage of the antenna at this stage.

After the Mars orbiting, the probe in normal mode is turned to Mars, that is, its $+Z$-axis is pointing to Mars. At this time, the γ angle changes periodically with the Mars-orbiting flight of the probe. So, this angle is used for analyzing the antenna coverage at this stage.

In the emergency mode, the probe is turned to the Sun. At whether the cruise stage or the Mars-orbiting stage, the $+X$-axis of the probe is pointing to the Sun, and the SPE angle and the γ angle are always coincident. The SPE angle can be applied when analyzing the Earth coverage of the antenna at this stage.

2) Change of SPE angle during the probe mission

After being separated from the rocket, the probe flies in the Sun-pointing attitude at the cruise stage, that is, its +X-axis is pointing to the Sun.

Figure 8.11 shows the simulation curve of SPE angle variation with time, and Fig. 8.12 shows the simulation curve of SPE angle variation with distance.

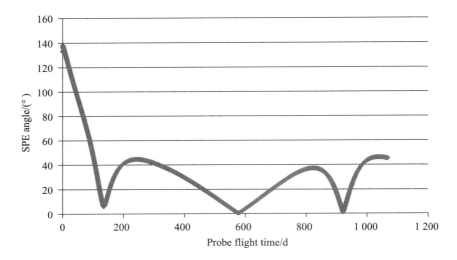

Fig. 8.11 Curve of SPE angle variation with time

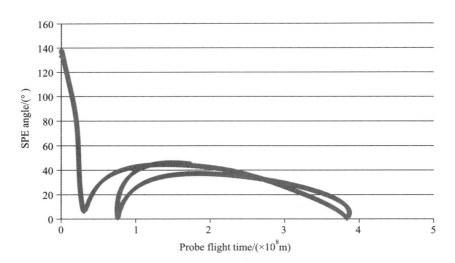

Fig. 8.12 Curve of SPE angle variation with distance

8.5 Design of Deep Space TT&C and Communication System

It can be seen that the SPE angle of the probe gradually decreases from 140° and then increases during the Earth-to-Mars transfer phase. After the Mars orbit insertion, the SPE angle oscillates up to 46° in a Martian year.

3) Working mode of probe antenna

According to the flight attitude and TT&C working mode of the probe, low-gain antenna is used for X-band TT&C to achieve uplink/downlink half-space coverage when the probe is under the normal condition and within the distance of 2×10^6 km from the Earth. Beyond 2×10^6 km, the directional antenna is unfolded to realize the TT&C and data transmission, while the low-gain antenna is used for emergency TT&C.

In the case of an emergency, the code rate of uplink telecommand is 7.8125 b/s, and the X-band low-gain antenna can support the full mission cycle till the distance of 3.8×10^8 km from the Earth.

4) Analysis of the coverage of X-band low-gain TT&C antenna

The working period of X-band low-gain TT&C antenna starts from the flight arc visible to China's X-band station to the end of the whole Mars-orbiting exploration process. Within the distance of 2×10^6 km from the Earth, the antenna works in normal working mode; while within the distance of 2×10^6–3.8×10^8 km, it works in emergency mode.

In the near-Earth flight stage, the attitude of the probe changes greatly, and the probe changes its orbit and attitude quite frequently. To ensure reliable reception of the commands and uplink data, the beam coverage of the X-band low-gain antenna should be designed into full-space coverage. To achieve this goal, four X-band low-gain TT&C antennas, two receiving antennas and two transmitting antennas are used, and the transmitting and receiving antennas are functionally separated. The time-sharing half-space coverage is achieved for uplink and downlink, respectively, by using a switch. The X-band low-gain antenna is mounted along the X-axis of the probe, and one receiving antenna and one transmitting antenna are mounted on the $+X$ plane and the $-X$ plane, respectively.

At the cruise stage after probe-rocket separation and before Mars orbiting, the probe is always oriented to the Sun, and the γ angle coincides with the SPE angle of the probe. So, the SPE angle is used in the following analysis.

Based on the SPE angle and distance variation curves in combination with the channel budget, the relationship between the gain and angle of X-band low-gain antenna (i.e., the antenna pattern) is calculated, and the curve of the required antenna gain with respect to the changing SPE angle is obtained.

When the distance between the probe and the Earth is within 2×10^6 km, the low-gain antenna is required to work in the normal TT&C mode, and the uplink telecommand shall adopt the high code rate (1000 b/s). The maximum gain requirement for a low-gain antenna is -20 dBi at a distance of 2×10^6 km and an SPE angle of about 46°.

When the distance between the probe and the Earth is 2×10^6–3.8×10^8 km, the low-gain antenna is required to work in the emergency TT&C mode, and the uplink telecommand shall adopt the low code rate (7.8125 b/s). Based on the uplink channel budget, the required maximum antenna gain is 3 dBi when the distance is 380 million kilometers (at the uplink code rate of 7.8125 b/s). After the probe enters the Mars orbit, the maximum SPE angle is 46° in the Sun-oriented mode. Considering the attitude deviation, the low-gain antenna should have a gain greater than 3 dB within ±50°.

5) Analysis of the coverage of directional antenna

In the launch and near-Earth phases, the X-band low-gain transmitting and receiving antennas are used by the Mars probe to achieve normal TT&C toward the Earth. After flying to the distance of 2×10^6 km and beyond from the Earth, the directional antenna is used for the uplink reception and downlink transmission of the TT&C signals. The following is an analysis of the Earth coverage of the directional antenna, from which the specification of the directional antenna beam is obtained.

Figure 8.13 shows the Earth coverage of the directional antenna.

In Fig. 8.13, the angle θ is the beam range requirement of the directional antenna of the probe, R_1 is the distance of the probe from the ground, and R_2 is the Earth radius.

Assume the pointing accuracy of directional antenna is σ_1, the probe control error is σ_2, and the position error of the probe with respect to the Earth is σ_3. From the pointing accuracy of the directional antenna and the probe control accuracy, it is known that σ_1 is better than ±0.3°, and σ_2 is better than ±0.1°. Moreover, it is known from the orbit determination precision that, when the directional antenna is used at a distance of 2×10^6 km or beyond, σ_3 is better than 0.18°. As a result, the synthesis of these three errors gives $\theta = \sigma_1 + \sigma_2 + \sigma_3 = \pm 0.58°$.

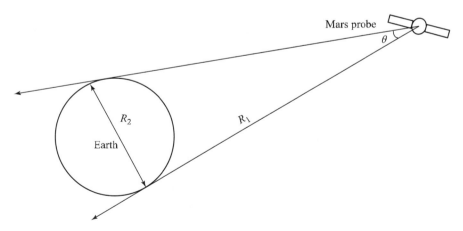

Fig. 8.13 Earth coverage of directional antenna

8.5 Design of Deep Space TT&C and Communication System

Therefore, as long as the half beam index of the directional antenna is greater than 0.58°, it can meet the demand of full coverage of the Earth at a distance of 2 million kilometers from the Earth. Considering a certain engineering margin, the half beam specification of the directional antenna is designed to be 1°.

8.5.2 System Scheme

8.5.2.1 System Scheme Composition

The principle of the system scheme is shown in Fig. 8.14.

The X-band TT&C is realized by two half-space coverage arrays, which jointly provide the time-sharing full-space TT&C signal coverage. The design of the upper and lower half-space coverage eliminates the TT&C blind area near the waist of the probe and realizes the half-space coverage during the near-Earth flight of the probe. The X-band TT&C consists of two channels: TT&C channel A (X-band deep space transponder A and traveling wave tube amplifier A) and TT&C channel B (X-band deep space transponder B and traveling wave tube amplifier B). The two channels serve as the backup for each other.

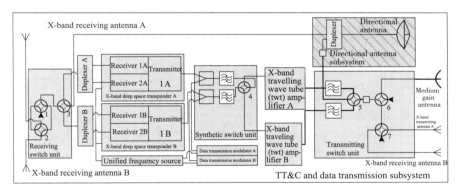

Fig. 8.14 Block diagram of TT&C and communication system of the probe

In the launch phase and the near-Earth flight phase (within the distance of 2×10^6 km from the Earth), if the antenna in the upper half space is accessible to the Earth, the uplink signal is received by the X-band receiving antenna A and then sent to the X-band transponder A for reception and demodulation. The downlink signals from the transmitter are amplified by the X-band traveling wave tube amplifier A and then transmitted by the X-band transmitting antenna A via the transmitting switch unit. If the antenna in the lower half space is accessible to the Earth, the uplink signal is received by the X-band receiving antenna B and then sent to the X-band transponder B for reception and demodulation. The downlink signals from the transmitter are amplified by the X-band traveling wave tube amplifier B and then sent by the X-band transmitting antenna B via the transmitting switch unit.

At the distance beyond 2×10^6 km, the TT&C uplink/downlink signals in the normal mode are both received and transmitted through the directional antenna. The low-gain antennas and medium gain antennas are used for emergency TT&C. Meanwhile, four code rates, namely 7.8125, 31.25, 125 and 2000 b/s, are designed so that the telecommand demodulation can satisfy the threshold requirements at different distances and different antenna gains.

Data transmission is implemented by two channels—main channel and standby channel. The signals of data transmission modulator A and TT&C transponder A are integrated through the synthesizer and then share the X-band traveling wave tube amplifier A to form the data transmission channel A. The signals of data transmission modulator B and TT&C transponder B are integrated through the synthesizer and then share the X-band traveling wave tube amplifier B to form the data transmission channel B. When data transmission is required, the signal modulated by the data transmission modulator A and amplified by the X-band traveling wave tube amplifier A is transmitted by the directional antenna. The data transmission channel B is used as a backup of the data transmission channel A. In case of the failure of the data transmission channel A, the data transmission channel B is selected by the microwave switch to transmit the data transmission radio frequency signals to the directional antenna. The data transmission downlink and TT&C downlink work in a time-sharing manner.

8.5.2.2 Analysis of Key System Metrics

1) High-sensitivity carrier acquisition

The carrier acquisition performance of the probe is achieved in the baseband of the X-band deep space transponder. According to the mission analysis, the carrier acquisition sensitivity requirement of Mars probe transponder is -150 dBm. Considering the noise factor of the channel and the implementation loss of the system, the signal-to-noise ratio of the input signal is required to be 19 dBHz. To obtain the required frequency variation range ± 100 kHz and the frequency change rate

8.5 Design of Deep Space TT&C and Communication System

100 Hz/s, the X-band deep space transponder adopts the FFT estimation-assisted large-frequency-offset acquisition scheme and the pre-sweep assisted acquisition scheme.

(1) Analysis of FFT-assisted acquisition scheme

If FFT algorithm is used for frequency offset acquisition analysis, the effect of swept frequency on FFT analysis is that in an FFT analysis period, the input signal is not a single-tone signal with an infinitesimal bandwidth, but a narrow band signal with a certain frequency range. The signal bandwidth is equal to the product of the FFT analysis period and the sweep rate. If the signal bandwidth is greater than the FFT resolution bandwidth, the signal-to-noise ratio (SNR) of FFT analysis will deteriorate to affect the performance of FFT analysis and even cause the failure in determining the correct frequency position through analysis.

The implementation block diagram of the FFT-assisted carrier acquisition scheme is shown in Fig. 8.15.

Fig. 8.15 Block diagram of FFT-assisted carrier acquisition scheme

The FFT length is considered as 65,536 points, and the period of a single FFT estimation is 328 ms. The relationship between sweep rate and signal-to-noise ratio shown in Table 8.1 may be obtained from the analysis.

As can be seen from Table 8.1, when the frequency variation rate is 100 Hz/s, the SNR threshold value required for FFT calculation is 25.2 dBHz, which is greater than the required value 19 dBHz. Therefore, it is necessary to improve the signal-to-noise ratio required for FFT calculation at the frequency variation rate of 100 Hz/s. Therefore, the pre-sweep assisted acquisition scheme is adopted to reduce the frequency variation rate to below 20 Hz/s so as to ensure the reliability of the FFT estimation results.

(2) Analysis of pre-sweep assisted acquisition scheme

Table 8.1 Relationship between sweep rate and signal-to-noise ratio

Frequency variation rate/ $(Hz \cdot s^{-1})$	328 ms frequency variation range/Hz	Baseband threshold signal-to-noise ratio /dBHz
20	6.56	18.2
50	16.4	22.2
100	32.8	25.2
500	164	32.2
5000	1640	42.2

The block diagram of the pre-sweep assisted acquisition technology is shown in Fig. 8.16. Through the down-conversion and low-pass filtering processing, the input intermediate frequency signal becomes a quasi-baseband complex signal with a ±100 kHz frequency offset and a slow swept frequency characteristic. The complex signal is, respectively, output to multiple parallel preset frequency sweep circuits for frequency sweep pre-calibration followed by FFT processing. Finally, the correct frequency estimation output is obtained by the frequency estimation effective detection logic and then sent to the carrier loop circuit at the back end.

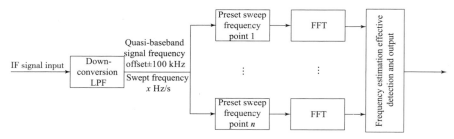

Fig. 8.16 Block diagram of pre-sweep assisted acquisition scheme

Assuming that the slow sweep rate is arbitrarily distributed between ±100 Hz/s, three calibration sweep frequencies can be set to pre-correct the slow sweep signal. Let the correction frequency be ±80, ±40 and 0 Hz/s. Regardless of the sweep frequency of the input signal, the frequency offset of the output slow sweep signal can be less than 20 Hz/s. A direct FFT analysis of this signal can be carried out to obtain reliable analysis results.

2) Low-SNR telecommand demodulation

The telecommand demodulation module is mainly to synchronize the carrier of the telecommand signal separated out of the carrier acquisition and tracking module with the code clock and to demodulate the telecommand data. To achieve the goal of low-SNR telecommand demodulation, the following measures should be taken:

(1) Improve the algorithms of sub-carrier tracking module, sub-carrier bit synchronization module and other modules to reduce the demodulation loss.
(2) Optimize the low noise amplifier (LNA), reduce its noise factor, and provide a larger margin for telecommand demodulation.
(3) Improve the phase noise performance of the receiving local oscillator, especially the phase noise at 10 and 100 Hz, and reduce the extra jitter introduced by the local oscillator and the SNR deterioration caused by it.

Before the telecommand demodulation, the acquisition and tracking of the primary carrier is completed. During the telecommand demodulation, the frequency offset of the telecommand sub-carrier can be corrected by using the tracking result of the primary carrier, leaving only a small frequency offset and phase offset in the corrected telecommand sub-carrier. At this time, the traditional second-order phase-locked loop

8.5 Design of Deep Space TT&C and Communication System

can be used to stably track the frequency and phase of the telecommand sub-carrier. The loop bandwidth of the phase-locked loop can be set to be quite narrow, so as to reduce the demodulation loss as much as possible. Due to the existence of the prior information, its locking speed is also quite fast.

The transponder needs to adaptively demodulate the telecommand data of different code rates. It is necessary to judge the telecommand code rate before entering the bit synchronization. By judging the telecommand code rate, the current telecommand code rate information and the data conversion time may be obtained as a priori information of the subsequent bit synchronization loop to accelerate the convergence speed of the bit synchronization loop. Therefore, the bit synchronization and judgment of the telecommand data can be completed by using a conventional data conversion and tracking loop.

When the carrier tracking is initially performed, the prior information such as code rate is unknown. As mentioned above, the transponder needs to adaptively demodulate four different telecommand code rates ranging from 7.8125 to 2000 b/s. The carrier tracking bandwidths required for different code rates are different. Therefore, four loops of different parameters are used to track the telecommand sub-carriers in parallel. The code rate detection unit receives the in-phase branch integration result of the four-channel carrier tracking and then adds it to the local boot sequence correlatively. Then, the maximum accumulated result is compared with the threshold. If the threshold is exceeded, the telecommand sub-carrier tracking is considered to be successful, and the telecommand code rate information and data conversion time are obtained. When the data rate of local boot sequence is 2000 b/s, the sliding correlation results at the telecommand boot sequence rates of 2000, 125, 31.25 and 7.8125 b/s are input, as shown in Fig. 8.17.

When the input telecommand booting head rate is the same as the local booting head rate, the sliding correlation achieves the maximum correlation value. Therefore, the telecommand code rate can be judged by the magnitude of the correlation value. It can be seen from Fig. 8.17a that when the input telecommand code rate is the same as the local telecommand code rate, the sliding correlation value is jagged. When their phases are the same, the maximum correlation value is obtained; when their phases are opposite, the minimum correlation value is obtained. Therefore, the phase of the telecommand code can also be judged by the magnitude of the correlation value. With such method, the telecommand code rate information and the data conversion time is obtained. Then, such information is sent to the subsequent bit synchronization module to greatly accelerate the convergence speed of the bit synchronization loop.

8.5.3 Simulation and Verification

In addition to the general circuit design and electrical performance test and environmental test, the characteristic simulation and verification items of the TT&C and communication system also include the simulation and test of antenna gain pattern, space-ground integrating test, etc.

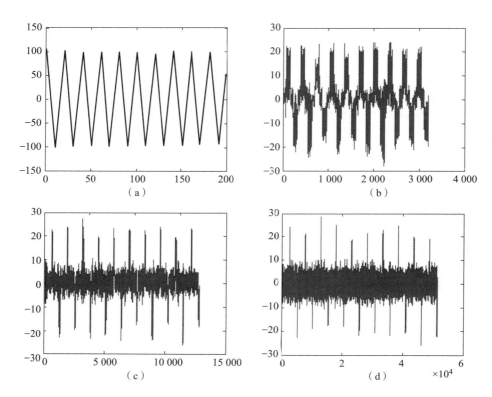

Fig. 8.17 Simulation of sliding correlation characteristic. **a** sliding correlation value at the telecommand code rate of 2000 b/s; **b** sliding correlation value at the telecommand code rate of 125 b/s; **c** sliding correlation value at the telecommand code rate of 31.25 b/s; **d** sliding correlation value at the telecommand code rate of 7.812 5 b/s

8.5.3.1 Antenna Gain Pattern Simulation

The omnidirectional antenna gain pattern is greatly affected by the probe surface status. With the increase of probe functions, the probe surface status becomes more complicated and makes it very difficult to directly calculate the omnidirectional antenna pattern. The antenna pattern is usually obtained using some matured electromagnetic field simulation software for simulation. The core of the simulation software is to solve the boundary value problem based on Maxwell's equation. According to the degree of approximation of the calculation results, the calculation methods may be divided into two types: analytical method and approximation method.

The advantage of the analytic method is that a rigorous solution can be obtained. This method is appropriate when the electric dimension to be solved is not quite large. However, when the solution domain is much larger than the wavelength, this method is no longer suitable as it occupies intensive computer resources. Instead, the superiority of the approximation method becomes evident. The commonly used analytical algorithms include the method of moment (MoM), finite element method (FEM), finite difference method (FDM), finite integral method (FIM), boundary element method (BEM), finite volume method (FVM), transmission line theory (TLM), etc. The commonly used approximation algorithms include geometric optics method (GO), physical optics method (PO), geometrical theory of diffraction (GTD), uniform theory of diffraction (UTD), physical theory of diffraction (PTD), magnetic equivalent circuit (MEC), Gaussian beam method (GB), etc.

8.5.3.2 Antenna Gain Pattern Test

Antenna gain pattern is one of the key indexes of the deep space TT&C and communication system and directly determines whether or not the probe-Earth communication capability is superior. Therefore, it is of great significance to obtain the test results of the antenna gain pattern.

The antenna gain pattern tests are generally divided into near field test, far-field test and compact field test. The near field test is to place a test probe in the near-area (near the antenna) of the antenna to be tested and sample the field data and then obtain the far-field parameters through the data conversion. This test method is not so accurate, but relatively simple, fast and low-cost, so it is suitable for the tests without high-precision requirements. In contrast, the far-field test is more accurate, but its test distance is required to meet the far-field free space condition. Therefore, it is very difficult for far-field test method to match the requirement of testing the deep space large-aperture antennas or omnidirectional antennas on the probe.

The most common test method for deep space TT&C antennas is the compact field test, which is a test method using an intermediate distance between near field and far field for measurement. Its basic principle is to use a large-aperture hyperboloid and a parabolic reflector to convert the spherical wave emitted by the test feed source into a plane wave, which produces an ideal plane wave test environment in a short distance. In an indoor environment, the radiation characteristics of the entire probe can be measured. This test is intuitive and quick and has relatively high accuracy.

8.5.3.3 Space-Ground Integrating Test

The space-ground integrating test refers to the matching verification test of the interface between the TT&C and communication equipment of the probe and the Earth station equipment. It generally includes wired integrating test and wireless integrating test. The wired integrating test focuses on the verification of the interfaces such as uplink and downlink radio signal characteristics and ranging zero value, while the wireless integrating test focuses on the interface verification such as antenna polarization matching.

The inspection of uplink and downlink radio signal characteristics includes testing the modulation characteristics of uplink and downlink signals such as modulation modes and modulation degrees; checking the format consistency of the uplink telecommand and injection data and the downlink telemetry and communication data; and testing the demodulation performance. In addition, when the ranging signals are transmitted from the Earth station, forwarded by the probe and then received by the ground TT&C equipment, a fixed system time delay deviation, also known as ranging zero value, will be introduced. The zero value is to be calibrated in the space-ground integrating process and will be used as one of the bases for the on-orbit ranging correction of the probe.

To avoid the influence of the Earth's atmosphere on the radio propagation channel, circularly polarized antennas are generally used as spacecraft TT&C and communication antennas. The methods of circular polarization include left-handed and right-handed circular polarization. The probe system and the Earth station must share the same antenna polarization mode, so the polarization matching verification of the spaceborne and ground-based antennas is one of the important tasks in the wireless integrating test.

8.6 Prospects

With the continuous improvement of space technology, the breadth and depth of deep space exploration activities are constantly expanding. Accordingly, higher requirements have been proposed for the TT&C and communication system. The development trends of TT&C and communication technologies for future deep space exploration missions are as follows.

1) High-precision radio measurement technology. With the continuous expansion of deep space explorations, the accuracy of classical pseudo-code/side-tone ranging, two-way Doppler velocity measurement and amplitude comparison angle measurement has become less able to satisfy the demands of orbit measurement and determination in deep space missions. Especially, as the distance increases, higher angle measurement accuracy is demanded. As such, the space VLBI with larger baseline, the connected element interferometry (CEI) with multi-station high-precision timing, the phase reference interferometry with weak reference source phase correction and other high-precision angle measurement technologies are expected to represent the development trend of high-precision radio measurements for the next generation of deep space probes.

2) Reception and demodulation technology under an extremely low signal-to-noise ratio. The main problem of deep space exploration is that the RF signal at the receiving end is very weak and the signal-to-noise ratio is extremely low. Therefore, the reception and demodulation technology under an extremely low signal-to-noise ratio is the key technology of deep space communication. For some new types of noise-reduction modulation systems (such as reverse symmetric modulation), the in-depth study of extremely narrow-band phase-locked tracking loop,

adaptive filtering, higher-gain channel encoding and decoding and other technologies will make breakthroughs in the reception and demodulation under an extremely low signal-to-noise ratio.

3) Space Internet technology. To overcome the difficulty in provide continuous real-time TT&C and communication in deep space due to long distance, a space Internet may be established with reference to the ground microwave station relay and Internet technology. The information from each deep space probe is transmitted to the Earth station through step-by-step transmission. The space Internet consists of a number of near-space networks close to the planets and the solar system interplanetary backbone network that interconnects such near-space networks. The key technologies to be conquered include 3D high-accuracy acquisition and tracking, power self-adjustment and distribution, long-time-delay interplanetary routing protocol algorithm, multi-link protocol compatibility technology, etc.

References and Related Reading

1. Ding S, Dong G, Zhu K (2003) The experiment of verifying pseudo code ranging by insert service of AOS. Aerosp Shanghai 20(6):45–48 (Chinese vision)
2. Dong G (2014) Development of new technology in deep space TT&C. J Deep Space Explor 1(4):243–249 (Chinese vision)
3. Fan M, Wang H, Li H, Zhao H (2013) Analysis of errors of deep space X-band range-rate measurement. J Spacecr TT&C Technol 32(2):168–172 (Chinese vision)
4. Zhu X, Li C, Zhang H (2010) A survey of VLBI technique for deep space exploration and trend in China current situation and development. J Astronaut 31(8):1894–1899 (Chinese vision)
5. Li H, Zhou H, Hao W, Dong G (2013) Development and prospect of radio interferometry for deep space navigation. J Aircr Meas Control 32(6):471–477 (Chinese vision)
6. Rong J, Zhu B, Zhong X (2004) Satellite laser communication technology and prospects. J Guizhou Univ (Nat Sci Ed) 21(3):293–296 (Chinese vision)
7. Wu X, Tan W (2009) The difficulty and countermeasure of the deep space communication. J Telem Track Command 30(4): 1–5 (Chinese vision)
8. Wu W, Dong G, Li H et al (2013) Engineering and technology of deep space TT&C system. Science Press, Beijing (Chinese vision)
9. Chen Y, Yin L et al (2007) Satellite radio measurement and control technology. China Aerospace Publishing House, Beijing (Chinese vision)
10. Jiang C (2000) Fundamental issues of deep space communication and tracking, international solutions and China's countermeasures. In: Proceedings of the 2000 symposium on aerospace measurement and control technology. Chinese Astronautical Society, Beijing, pp 37–43 (Chinese vision)
11. Tan W, Gu Y (2004) Space data system. China Science and Technology Press, Beijing (Chinese vision)
12. Thornton CL, Border JS (2005) Radio tracking measurement technology for deep space navigation (trans: Li H). Tsinghua University Press, Beijing (Chinese vision)
13. Gao W (2005) Preliminary exploration of the application of space laser communication technology in China's deep space exploration. In: The 2nd academic conference of China aerospace society's deep space exploration technology professional committee, Beijing (Chinese vision)
14. Rogstad DH, Mileant A (2003) Antenna arraying techniques in the deep space network. Wiley, Hoboken

15. Yuen JH (1983) Deep space telecommunications systems engineering. Plenum Press, New York
16. Border JS, Lanyi GE, Shin DK (2008) Radiometric tracking for deep space navigation. In: 31st annual AAS guidance and control conference, Breckenridge

Chapter 9
Thermal Control Technology

9.1 Introduction

The design of thermal control system for a deep space probe is to incorporate advanced active temperature control technology and advanced system design method into the thermal control design of low Earth orbit (LEO) spacecraft fully used for reference, to ensure that the temperature of equipments and structural elements of the deep space probe is within the required range. Compared to LEO spacecraft, the main characteristics of thermal control system of deep space probe are as follows: (1) The weight and electric power of deep space probe are very precious, so it is necessary to integrate the thermal control system design into the probe's overall design, optimize the system design and reuse the product function when necessary. (2) The deep space probe has a complex mission process and undergoes great changes in the thermal environment, so more active thermal control technologies with strong control ability are needed to adapt to the complex and changeable external thermal environment [1]. In this chapter, the characteristics of the thermal environment that the deep space probe is faced with are given at first. On this basis, the functions, performance and applications of the thermal control products commonly used in deep space probes are introduced. Finally, the design method of deep space thermal control system is introduced in consideration of the characteristics of thermal environment and thermal control products in deep space exploration.

9.2 Characteristics of Thermal Environment in Deep Space

During the interplanetary flight, most of the environmental heating comes from the Sun's thermal radiation. During the flight over a planet, the probe will also be heated by the planetary infrared radiation and the reflection radiation. The geometric data and basic orbital parameters of the planets and the Moon are presented in Chap. 2.

The average distance between the Earth and the Sun is defined as an astronomical unit (AU), and the average solar irradiation intensity near the top of the Earth's atmosphere is defined as a solar constant (S). The solar irradiance intensity at different distances from the Sun can be calculated by the following formula:

$$\text{Solar Irradiation Intensity} = \frac{1367.5}{\text{AU}^2} \frac{\text{W}}{\text{m}^2} \quad (9.1)$$

In order to intuitively understand the thermal environment encountered in interplanetary flight, the concept of "reference sphere" is used here. The reference sphere is an isothermal sphere. Its surface solar absorption ratio and hemispheric infrared emissivity are both 1.0. It orbits a planet at an attitude of 1/10 of the planet radius. The reference sphere temperature can roughly indicate its degree of coldness or hotness. Table 9.1 gives the equilibrium temperature of the reference sphere for the solar system planets.

Table 9.1 Temperature of planetary reference sphere

Planet	Equilibrium temperature of reference sphere/K
Mercury	447
Venus	324
Earth	279
Mars	226
Jupiter	123
Saturn	90
Uranus	64
Neptune	51

9.2.1 Mercury's Thermal Environment

Mercury is the planet nearest to the Sun. Mercury's orbit period is about 88 Earth days, and its rotation period is about 58 Earth days. One Mercury "day" is 176 Earth days. Due to such a slow rotation, the surface of Mercury facing the Sun is very hot, while the back side is very cold. The temperature of the Mercury surface from the sub-solar point to the terminator can be expressed by Hanson's cosine function [2]. Taking the angle φ from the surface location to the sub-solar point as the variable, the surface temperature of Mercury is expressed as follows:

$$T = T_{\text{subsolar}}(\cos \varphi)^{\frac{1}{4}} + T_{\text{terminator}}\left(\frac{\varphi}{90}\right)^3, \quad \varphi \leq 90° \quad (9.2)$$

$$T = T_{\text{terminator}}, \quad \varphi > 90° \quad (9.3)$$

9.2 Characteristics of Thermal Environment in Deep Space

$$T_{subsolar} = 407 \pm \frac{8}{r^{0.5}} K \quad (9.4)$$

$$T_{terminator} = 110 K \quad (9.5)$$

where the angle φ is calculated from the sub-solar point, in degree (°); r is the distance from Mercury to the Sun, in AU; Tsubsolar is the sub-solar temperature, in K; and T terminator is the temperature of the terminator, in K.

Table 9.2 gives the thermal environment of Mercury's orbit. The surface albedo and emissivity of Mercury are 0.12 and 0.77, respectively.

Place a reference sphere in Mercury's circular orbit at an attitude of 1/10 of Mercury radius. When the angle β between sunlight and the orbital plane of the reference sphere is 0° and 90°, the temperature of the reference sphere is as shown in Table 9.3.

Table 9.2 External heat flow in Mercury's orbit (unit: W/m²)

	Solar irradiation heat flow	Reflection heat flow	Infrared radiation heat flow (Max)	Infrared radiation heat flow (Min)
Aphelion	6278	753	5500	6
Average	9126	1095	8000	
Perihelion	14,462	1735	12,700	

Table 9.3 Reference sphere temperature at the orbit altitude of 1/10 of mercury radium

Calculating conditions		Temperature /K		
		Maximum value	Average	Minimum value
$\beta = 0°$	Aphelion	495	300	76
	Perihelion	602	362	76
$\beta = 90°$	Aphelion	420		
	Perihelion	518		

9.2.2 Venus's Thermal Environment

The thermal environment of Venus orbit is not only colder than that of Mercury due to a farther distance from the Sun, but also has different relative distributions of solar radiation and infrared radiation. Mercury's albedo ratio is very small, so most of the incoming solar energy can be absorbed by Mercury surface and then turned into infrared radiation. But Venus is completely covered by clouds, with a very high albedo (about 0.8). Higher albedo results in lower cloud top temperatures, so Venus's infrared radiation is even smaller than that of the Earth.

Venus's clouds cause the directional scattering effect of solar radiation when the angle of sub-solar point is large. This effect brightens the terminator. For low-altitude orbits, a relatively accurate result can be obtained by treating the reflection radiation of Venus as diffuse reflection that is supposed to decrease from the sub-solar point according to the cosine laws. For higher orbits, the reverse irradiation at the edge of the dawn-dusk point cannot be neglected. Because the reflection intensity is relatively small compared to the direct solar radiation, these deviations are not important. Of course, Venus's reflection direction characteristics cannot be ignored for some sensors. The thermal environment of Venus's orbit is shown in Table 9.4, with the albedo value of 0.8, and the temperature of the reference sphere is shown in Table 9.5.

Table 9.4 External heat flow in Venus's orbit (W/m^2)

	Solar irradiation heat flow	Reflection irradiation heat flow	Infrared reflection heat flow (Max)
Aphelion	2650	2120	153
Average	2614	2091	
Perihelion	2759	2207	

Table 9.5 Reference sphere temperature in Venus's orbit

Calculating conditions		Temperature /K		
		Maximum value	Average	Minimum value
$\beta = 0°$	Aphelion	392	285	168
	Perihelion	395	287	168
$\beta = 90°$	Aphelion	337		
	Perihelion	340		

9.2.3 Lunar Thermal Environment

Similarly to Mercury, the Moon has no atmosphere and has a long rotation period. Lunar infrared radiation is also calculated by cosine function (up to about 70°), i.e., it decreases from the sub-solar point as the angle increases. Table 9.6 shows the radiation heat flow on the lunar equatorial surface, as sourced from "Apollo 11" (*Note* the emissivity of the lunar surface is 0.92 and the solar albedo is 0.073). The temperature on the dark side of the Moon is in the order of -170 °C.

Different lunar regions have different albedos. The average lunar albedo is about 0.073, even the maximum lunar albedo is less than 0.13. Therefore, the surface temperature of the sunlight area is relatively high.

9.2 Characteristics of Thermal Environment in Deep Space

Table 9.6 External heat flow in lunar orbit (W/m²)

	Solar irradiation heat flow	Reflection irradiation heat flow	Infrared reflection heat flow (Max)	Infrared reflection heat flow (Min)
Aphelion	1323	97	1226	5.2
Average	1367.5	100	1268	
Perihelion	1414	103	1314	

The reference sphere temperature in orbit at the attitude of 1/10 of the Moon radius is shown in Table 9.7. It can be seen from the table that, like Mercury, the temperature at the sub-solar point of lunar orbit is high, while the temperature in the shadow area is very low. This shows that the Moon has a long lunar day and has no atmosphere to maintain its surface heat, leading to very low surface temperature and very small infrared radiation in the shadow area.

Table 9.7 Reference sphere temperature in lunar orbit

Calculating conditions		Temperature /K		
		Maximum value	Average	Minimum value
$\beta = 0°$	Aphelion	334	214	74
	Perihelion	340	217	74
$\beta = 90°$	Aphelion	290		
	Perihelion	295		

According to the above characteristics of thermal environment, some design states should be noticed in the thermal design for the lunar probes flying around or landing on the Moon, such as:

(1) For lunar probes orbiting the Moon, the effect of lunar infrared radiation on them is obviously greater than that on the Earth orbit. Therefore, the surface with low absorption and high emissivity should be oriented toward the sky whenever possible.
(2) For the equipments landing on the lunar surface, the infrared radiation effects coming from the nearby mountains cannot be ignored.
(3) The influence of lunar dust pollution on the lunar surface will significantly increase the solar absorptivity of the probe surface.
(4) The thermal conductivity of lunar soil on the lunar surface is very low, so the temperature of the occluded area decreases rapidly to cause local low temperature easily.

9.2.4 Mars Thermal Environment

Mars is the last planet where the probes experience significant ambient heating while traveling in the solar system. Its orbital thermal environment is shown in Table 9.8. (*Note* The solar albedo on Mars is 0.29.) The average solar radiation heat flux near Mars is 589 W/m², which is about 42% of the average solar radiation heat flux near the Earth. The albedo of Mars is similar to that of the Earth, ranging from 0.25 to 0.28 near the equator and increasing with latitude up to 0.5. Mars is called "Red Planet," because the spectrum it reflects is red, with a peak of 0.7 μm (0.47 μm for the Earth).

The reference sphere temperatures in the orbit with the altitude of 1/10 of the Mars radius are shown in Table 9.9. As can be seen from the table, the temperature at night is much lower than that at daytime. Because Mars has a thin atmosphere and the transmittance of cloudless infrared radiation is very high, the temperature of the reference sphere in the shadows is very low when $\beta = 0°$. This is different from the Earth and Venus, which have thick atmospheres that block planetary infrared radiation and thus have relatively uniform temperatures during the day and night.

Table 9.8 External heat flow in Mars orbit (W/m²)

	Solar irradiation heat flow	Reflection irradiation heat flow	Infrared reflection heat flow (Max)	Infrared reflection heat flow (Min)
Aphelion	493	143	315	30
Average	589	171	390	
Perihelion	717	208	470	

Table 9.9 Reference Sphere temperature in martian orbit

Calculating conditions		Temperature/K		
		Maximum value	Average	Minimum value
$\beta = 0°$	Aphelion	257	191	110
	Perihelion	284	210	111
$\beta = 90°$	Aphelion	247	230	238
	Perihelion	273	251	241

9.2.5 Thermal Environment of Exoplanets

For the exoplanets, their solar irradiation intensity and the corresponding solar reflection heat flow and infrared radiation heat flow will decrease as their distance from the Sun increases. The spacecrafts operating near the exoplanets are facing an extremely cold environment. The thermal load generated by the thermal environment of an exoplanet only affects those sensitive detection devices and cryogenic radiation coolers, while having a negligible impact on the equipments operating at room temperature.

Tables 9.10 shows the orbital thermal environment of these exoplanets.

Table 9.10 Orbital external heat flow of exoplanet (unit: W/m^2)

	Solar irradiation heat flow	Reflection irradiation heat flow	Infrared radiation heat flow (Max)
Jupiter			
Aphelion	46	15.8	13.4
Average	51	17.5	15.1
Perihelion	56	19.2	16.8
Saturn			
Aphelion	13.4	4.6	4.5
Average	13.6	4.7	4.6
Perihelion	13.7	4.7	4.7
Uranus			
Aphelion	3.39	1.2	0.55
Average	3.71	1.3	0.63
Perihelion	4.09	1.4	0.72
Neptune			
Aphelion	1.49	0.42	0.52
Average	1.51	0.43	
Perihelion	1.54	0.43	

9.3 Key Technologies of Thermal Control S

This section mainly introduces four key thermal control technologies, including aerogel technology, gravity-assisted two-phase fluid loop technology, water sublimator technology and variable conduction heat pipe technology, and presents their hardware product characteristics. These four key technologies occupy an important position in the thermal control systems of China's lunar and Mars probes and play an indispensable role in realizing the function and performance of thermal control system.

9.3.1 Gravity-Assisted Two-Phase Fluid Loop Technology

Gravity-assisted two-phase fluid loop technology is a heat transfer technology which is driven by heat and routed by gravity and does not need to consume electric power resources. This technology is used to transfer the heat of radioisotope heat unit to the probe's thermal insulation cabin in the absence of electric power during the period of power-off and dormancy, so as to realize the thermal insulation and survival of the onboard equipments at lunar night.

The gravity-assisted two-phase fluid loop consists of evaporator components (including evaporator, vapor confluence ring, liquid diversion ring), vapor pipeline, condenser, liquid pipeline, liquid reservoir and control valve, as shown in Fig. 9.1. The evaporator is coupled with the heat source and the condenser is installed where the heat is needed. When the loop works, the evaporator is heated by the heat source so that the liquid working medium inside the evaporator absorbs the heat and becomes gaseous and the steam pipeline is filled with steam. Because the density of the working medium in steam pipeline is different from that in liquid pipeline, a driving force will be formed in the gravity field to cause the drop of the working medium in the liquid pipeline and the rise of the working medium in the steam pipeline and enable

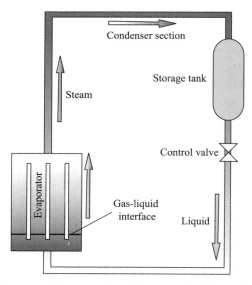

Fig. 9.1 Schematic diagram of the composition of gravity-assisted two-phase fluid loop system

9.3 Key Technologies of Thermal Control S

the working medium to circulate in pipeline. Because of the existence of the gravity field, the gaseous working substance flows in a specific direction without backflow.

The gravity-assisted two-phase fluid loop technology has been used in the thermal control systems of both Chang'e 3 lander and rover to overcome the great difficulty in controllably transferring the heat from radioisotope heat unit to the cabin to heat the equipments at lunar night. The layout of the two-phase fluid system is detailed in Figs. 9.2 and 9.3. In this fluid loop system, the inner diameter of the evaporator is 8 mm. Its inner structure is a composite structure of a light tube and a metal wire mesh barrel with 400 meshes, which can meet the requirements of working fluid flow and evaporation during the high-temperature start-up process in day/night conversion. The vapor pipeline has a saddle-shaped structure to prevent condensate from flowing back to the evaporator (in order to adapt to the inclined attitude of the probe). The inner diameter and outer diameter of liquid pipeline, steam pipeline and

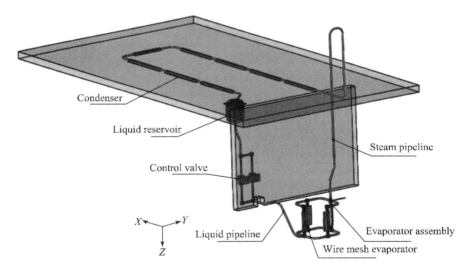

Fig. 9.2 Schematic diagram of the gravity-assisted two-phase fluid loop system in Chang'e 3 rover

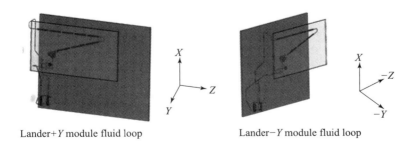

Fig. 9.3 Schematic diagram of the gravity-assisted two-phase fluid loop system in Chang'e 3 lander

condenser pipeline are 4.4 mm and 6 mm, respectively. The working medium used for heat transfer is ammonia. The system material is 316L stainless steel, which is compatible with ammonia.

In the design of gravity-assisted two-phase fluid loop, the driving height, namely the difference between the liquid levels of the reservoir and evaporator, is a key parameter, which determines the maximum driving force provided by the fluid loop system. In order to enhance the heat transfer capability of the lunar gravity-driven two-phase fluid loop as much as possible, leave enough engineering margin and reduce the risk, the height difference between the liquid reservoir and the evaporator assembly should be maximized as the layout permits, and the transverse spacing should be reduced to enhance the adaptability to the inclined attitude on the lunar surface. Another factor affecting the system performance is system resistance. The main source of the system resistance is on-way resistance. Increasing the pipe diameter is an effective method to reduce the on-way resistance. A saddle-shaped transition pipeline higher than the condenser is formed between the steam pipeline and the condenser in order to prevent the probe's uncertain attitude from causing some liquid working medium to flow back from the condenser to the steam pipeline and evaporator to result in temperature fluctuation and heat transfer instability during the working period of two-phase fluid loop at lunar night.

9.3.2 Water Sublimator Technology

Water sublimator technology refers to the direct sublimation of solid ice into steam after the heat absorption without melting in an environment where the ambient pressure is lower than the three-phase pressure of water. The latent heat absorbed during sublimation is much higher than the heat released by liquid water condensation, so heat dissipation is achieved. Water sublimator is a kind of medium-consuming heat sink, applying to the heat dissipation of a large heat-consuming device working for a short time in vacuum environment or low pressure environment. The water sublimator can be used alone or as a supplement to the main heat dissipation system.

Figure 9.4 shows the composition of water sublimator system, which mainly includes working medium storage tank, drain valve, liquid pressure reducing valve, self-locking valve, pressure sensor, sublimation heat exchanger, pipeline, working medium, etc. The sublimation heat exchanger is the core component of the water sublimator system, which is generally composed of a plate-fin gas–liquid heat exchanger and a porous plate.

The working principle of sublimator assembly is shown in Fig. 9.5. The external surface of porous plate in the sublimator assembly is exposed to vacuum environment. For the sublimator using water as the working medium, the feed water enters the feed water gap under certain pressure, and heat is transferred from one side to the water in the feed water gap. When the water entering the feed water gap or continuously infiltrating into the porous plate reaches (or is below) the three-phase point pressure, most of the water will be solidified into ice. Meanwhile, a small part of the water

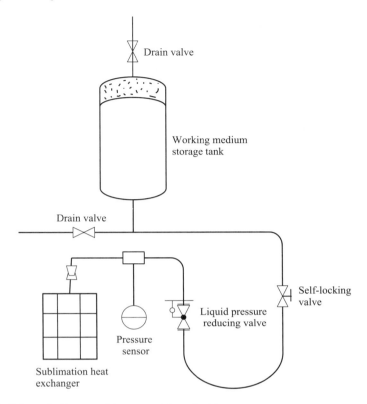

Fig. 9.4 Schematic diagram of the composition of sublimator system

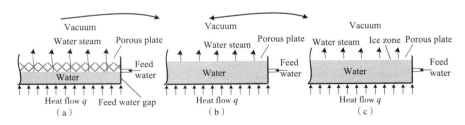

Fig. 9.5 Schematic diagram of the working principle of water sublimator

will sublimate into vapor to dissipate to the vacuum environment, thus taking away the heat of the heat source.

According to different conditions (such as heat load), the water sublimator has four working modes:

(1) Sublimation mode: The ice layer is formed on the inner surface of the porous plate (on the feed water side) and sublimates at the interface between the ice and the porous plate. When the heat load is low or the external ambient pressure is low

enough, the pressure drop generated by the sublimation steam flowing through the porous plate is small. As a result, the vapor pressure on the sublimating surface is lower than the three-phase pressure of water, and the ice continuously sublimates inside the porous plate. The ice thickness depends on the heat load and the pressure drop of the vapor generated by sublimation and flowing through the porous plate. Amid the ice sublimation, the feed water condenses on the water–ice interface at the same rate. Driven by feed water pressure, the ice layer will slide continuously toward the porous plate to ensure that the physical position of sublimation interface is relatively stable. Sublimation is the most desirable working mode of water sublimator.

(2) Evaporation mode: When the heat load and environmental pressure are large, the pressure drop of the vapor generated by sublimation and flowing through the porous plate may be higher than the three-phase point pressure. So, no ice will be formed, and water can enter the porous plate directly. Water will produce surface tension in the hole of the porous plate, so water can stay in the porous plate. At this time, the sublimator will dissipate heat through the evaporation of water. However, the surface tension on the water-pore interface is inversely proportional to the size of the pore. If the pore size of the porous plate is too large or the feed water pressure is too high, the "breakdown" phenomenon will occur, that is, water directly passes through the porous plate and enters the vacuum environment without undergoing the phase transformation, thus causing the loss of the heat dissipation capability of the sublimator.

(3) Mixing mode: Because the shape and size of the holes in the actual porous plate differ a lot and randomly distributed, no ice will be formed in some holes when the heat load increases. Instead, the water phase is changed by evaporation. But ice is formed in other holes, where the water phase is changed by sublimation. This phenomenon is called mixing mode. However, the research shows that this mode can occur only when the porous plate material is not wetted by water, but most of the porous plate materials can be wetted by water.

(4) Periodic mode: When the pressure drop of the vapor flowing through the porous plate is greater than the three-phase point pressure, ice cannot be formed in the feed water gap, and water will enter the capillary hole of the porous plate. Water cannot condense into ice until the pressure drop of the vapor flowing through the remaining holes is less than the three-phase point pressure. At this time, the ice interface is in the porous plate, and the heat and mass transfer is realized by sublimation. When the ice sublimation causes the sublimation interface to retreat to the point where the vapor pressure is higher than the three-phase point pressure of water, the ice will disappear, and water will flow again toward the direction of lower pressure in the hole, and an instantaneous evaporation effect will occur. When the vapor pressure reaches the three-phase point pressure again, the interface temperature will drop to the freezing point and then reenter the state of heat and mass transfer via sublimation to start a new cycle. Generally speaking, due to the hindrance of the microholes in the porous plate, ice will

9.3 Key Technologies of Thermal Control S 361

not slip in the porous plate under the water supply pressure; only the retreat of the ice interface will occur instead. When the water supply pressure is high, the slip process of the ice layer may occur.

In the design of water sublimator system, the design of sublimation component is very important. The objective of sublimation component design is to achieve a relatively stable ice sublimation interface in the porous plate. The core design parameters of sublimation module include the pore size and thickness of the porous plate, as well as water supply pressure control. The pore size of the porous plate affects the water vapor emission pressure. The location of ice sublimation interface can be controlled by adjusting the thickness of porous plate and water supply pressure when the pore size of the porous plate is fixed. In addition, the heat flux density transferred from heat exchanger to sublimation interface needs to be designed to ensure that it matches the heat dissipation capacity of the water sublimation interface.

9.3.3 Variable Conductivity Heat Pipe Technology

Heat pipe is a heat transfer element with high thermal conductivity. Because its heat transfer coefficient of evaporation and condensation changes little in a certain range of heat flow and temperature, its thermal conductivity can be basically considered unchanged. In order to achieve variable thermal conductivity, the method of condensation section blocking by gas is most commonly applied. Its principle is as follows. The heat pipe is filled with a certain amount of non-condensable gas such as nitrogen, helium or argon. When the heat pipe is not working, the gas and working steam would be mixed and distributed in the steam space of the heat pipe. After the heat pipe starts operation, the gas flows to the condensation section along with the steam. The steam condenses into liquid on the inner wall of the condensation section and returns from the core of the tube to the evaporation section. The gas is not condensable. It stays at the end of the condensation section. All the non-condensable gases gather in the condensation section and form a gas plug. This type of heat pipe with variable thermal conductivity is just variable conductivity heat pipe, as shown in Fig. 9.6. In order to improve the control performance, a gas storage chamber is connected to the end of the condensation section to increase the length-changing rate of the gas blocking section. The gas storage chamber may have a mesh core connected with the heat pipe core or may have no mesh core.

The schematic diagram of the working principle of the variable conductivity heat pipe is shown in Fig. 9.7. When the heat load in the evaporation section is large, the steam temperature and pressure in the evaporation section rise rapidly. The steam drives the vapor/controlled gas interface to move to the side of the storage chamber, so that the volume of the controlled gas is reduced and a longer condensation section is opened to increase the heat dissipation capacity. Conversely, when the heat load of

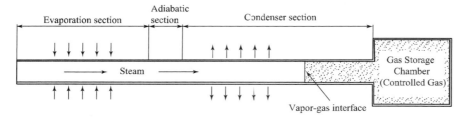

Fig. 9.6 Schematic diagram of the structure of variable conductivity heat pipe

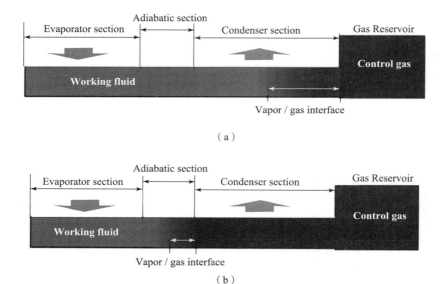

Fig. 9.7 Schematic diagram of the working principle of variable conductivity heat pipe **a** High-power operation; **b** Low-power operation

the evaporation section is smaller, the steam temperature in the evaporation section decreases, and the steam pressure decreases accordingly. At this time, the volume of the control gas will expand and push the vapor/controlled gas interface to move away from the storage chamber. As a result, a longer condensation section is blocked by the controlled gas and the heat dissipation capacity is reduced. From the working principle of variable conductivity heat pipe, it can be seen that the heat pipe can adapt its heat transfer capacity autonomously to the changing heat load and keep the temperature of evaporation section relative stable.

In order to study the working characteristics of the variable conductivity heat pipe, two models are currently used: One is diffusion interface theory model and the other is planar interface theory model. In analyzing the performance of variable

9.3 Key Technologies of Thermal Control S

conductivity heat pipe, the theoretical model of planar interface is simple and clear, and its design accuracy can meet the engineering requirements, so it is widely applied in engineering design.

Variable conductivity heat pipes are usually used in the occasions where the demand for heat dissipation capability varies greatly, for example, in the lunar probe compartment requiring heat dissipation during the lunar day and heat preservation at lunar night. The use of variable conductivity heat pipes can meet the needs of both heat dissipation during the lunar day and heat preservation at lunar night.

9.3.4 Aerogel Technology

The term aerogel was first proposed by Kistler in 1932. It refers to the well-preserved solid structure after the liquid in gels is replaced by gas. Aerogel is a light, amorphous and porous solid material composed of nano-colloidal particles or polymer molecules. It has very low density, high specific surface area and high porosity. The pore size (<50 nm) of aerogel is less than the mean free path of air molecules (about 70 nm). In the aerogel pores, air convection cannot occur, and only, the heat exchange between gas molecules and solid wall can exist. At the same time, owing to very low thermal conductivity, solid aerogel is considered to be the best insulation material—super insulating material—that has been found so far. Figure 9.8 gives a physical picture of pure silica aerogel, and Fig. 9.9 gives a physical picture of fiber-reinforced silica aerogel. Nano-aerogel is commonly used in the thermal insulation systems

Fig. 9.8 Photo of pure silica aerogel

Fig. 9.9 Photo of fiber-reinforced silica aerogel

under atmospheric pressure to replace multi-layer insulation modules for isolating radiation, heat conduction and convection, for example, in the heat insulation system of a Mars surface probe.

The thermal insulation properties of aerogels are related to ambient pressure and atmosphere. When the pressure is greater than 1000 Pa, the thermal conductivity of aerogel will increase with the pressure. When the pressure is less than 1000 Pa, the change of gas pressure has little effect on the aerogel's thermal conductivity, as shown in Fig. 9.10. For different atmospheric environmental conditions, the higher the thermal conductivity of the gas, the greater the aerogel's thermal conductivity, as shown in Fig. 9.11.

Aerogel is a kind of brittle material. Moreover, the silica dioxide aerogel particles on the product surface can fall off easily and form redundant substances. Therefore, in the engineering application of aerogel materials, the encapsulation of aerogel is a problem that needs special attention. Encapsulation will increase the weight of the insulation system and degrade the insulation performance of the thermal insulation system. The detailed encapsulation design is needed to minimize the degradation of the thermal insulation performance caused by the addition of encapsulation structure.

9.3 Key Technologies of Thermal Control S

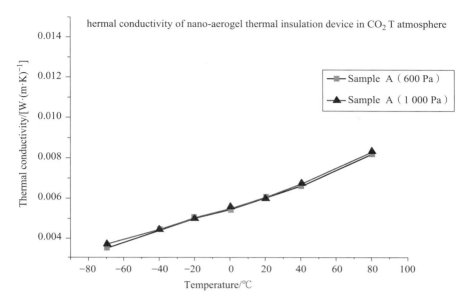

Fig. 9.10 Thermal conductivity variation of aerogels with pressure (provided by china academy of space technology)

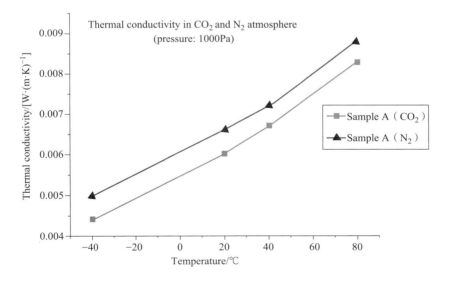

Fig. 9.11 Thermal conductivity variation of aerogels with temperature under different atmospheres (provided by China academy of space technology)

9.4 Thermal Control System Design for Deep Space Probe

The probe's thermal design is to organize the heat exchange processes inside and outside the probe according to the requirements of the probe's mission and the heat loads inside and outside the working probe, so as to ensure that the temperature level of all the instruments and structures in the probe is kept within the specified range during the whole operation period. Compared to the near-Earth spacecraft, the thermal control system in deep space probe usually faces severe situations such as complex thermal environment, complex mission process and resources shortage (such as mass allocation). This requires that the thermal control system in deep space probe must be more closely linked with other systems and even needs to realize the integration and reuse of some functions. Thermal control system also needs strong environmental adaptability and regulating ability. With the development of deep space probes and thermal control technology, multi-disciplinary integrated design will become an important part of thermal control system design. In this section, the thermal control system designs for typical international deep space probes are introduced at first. On this basis, the thermal control design principles, design methods, thermal analysis methods and ground test verification methods of deep space probes are discussed.

9.4.1 Introduction of Typical Thermal Control Systems for Deep Space Probes

9.4.1.1 Design Method of Thermal Control System for Lunar Probe

The temperature of the lunar surface varies greatly from 93 to 395 K. The rotation period of about 27 days leads to the long duration of the high- and low-temperature cycle, including about 240 h at the temperature of above 300 K and 290 h at the temperature of less than 100 K. The harsh thermal environment requires a specially designed thermal control system.

From 1966 to 1968, the USA launched seven "Surveyor" lunar probes to the Moon. The goal was to explore one lunar day and part of the lunar night to prepare for manned landing. The "Surveyor" probe firstly adopted the design idea of "Open System." That is, the equipments are divided into several groups according to the working mode and temperature requirement, and thermal isolation is set up between the equipment groups and between an equipment group and the main probe structure. The thermal control system is designed for the compartments of the grouped equipments, not for the whole probe. In the thermal control design of each compartment, besides conventional passive thermal control technology, mechanical thermal switch technology has also been used for the first time.

9.4 Thermal Control System Design for Deep Space Probe

The Lunokhod 1 from the Soviet Union is a relatively successful lunar probe, which embodies the design idea of the Soviet deep space probe. The Lunokhod 1 used a sealed cabin structure, which was filled with the gas under a certain pressure. The service equipment and payload equipment were placed in a bowl-shaped sealed cabin, where thermal control was achieved by using convection ventilation heat transfer technology. Radioisotope heat source was placed outside the sealed cabin and connected with the cabin through the gas channel. During the lunar day, the gas passage to the radioisotope heat source was closed. The gaseous working medium circulated inside the sealed cabin to collect waste heat from the equipment and transfer it to the radiation surface for heat dissipation. During the lunar night, the radiation surface was covered by a closed solar array. The gas passage to radioisotope heat source was opened, so the circulating gas carried the heat from the radioisotope heat source to the cabin to maintain the temperature of the equipments in the cabin. In addition, in order to adapt to the low-temperature environment at light, most of the equipments went into dormancy during lunar light, and only, a few equipments that had to work were on standby to awaken the lunar rover at dawn.

The China's Chang'e 3 lander adopted the independent thermal control design for compartments and the targeted thermal design according to the functions of each compartment, as shown in Fig. 9.12. After landing on the lunar surface, the lander's central module and $-Z$ module have completed their functions and have no need for survival during lunar night. Therefore, their thermal control functions are similar to those of traditional spacecrafts, so they adopted the thermal control measures that are similar to those of spacecrafts on low Earth orbit. In contrast, the $+Y$ module and $-Y$ module had the need for survival during lunar night. Therefore, the variable conductivity heat pipe was used to dissipate heat during lunar day, while isotopic heat source and gravity-assisted two-phase fluid loop were used for the autonomous transmission and utilization of radioisotope heat during lunar night, so as to achieve the balance of heat dissipation during lunar day and heat preservation at lunar night.

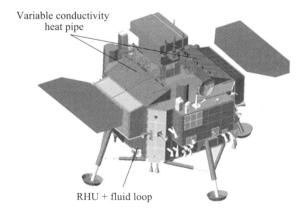

Fig. 9.12 Schematic diagram of the composition of thermal control system in Chang'e 3 lander

At lunar night, with the decrease of cabin temperature, the heat dissipation section of the variable conductivity heat pipe was blocked, so the heat dissipation stopped. Meanwhile, the valve of the gravity-assisted two-phase fluid loop was closed, and the fluid loop operated independently under the action of gravity to transfer heat from the radioisotope heat source to the cabin for equipment insulation. During the lunar day, with the increase of equipment temperature in the cabin, the heat dissipation section of the variable conductivity heat pipe was opened to dissipate heat to the outer space, and the valve of the gravity-assisted two-phase fluid loop was disconnected. As a result, the loop stopped working, so as to prevent the heat of radioisotope heat source from entering the cabin and keep the temperature of the cabin equipment within the working temperature range.

The thermal control system in Chang'e 3 rover should also resolve the contradiction between heat dissipation on lunar day and heat preservation at night. The radiating surface of the rover is directly arranged on the upper surface of the roof to dissipate heat to space, and the equipment that needs to dissipate heat is directly installed on the lower surface of the roof to meet the heat dissipation demand of the equipment during lunar day. During lunar night, the mast is retracted and the $+Y$ solar wing is closed to cover the radiating surface. Meanwhile, the gravity-assisted fluid loop starts working, transferring heat from the radioisotope heat source to the rover compartment to achieve the survival temperature of the equipments, as shown in Fig. 9.13 [3].

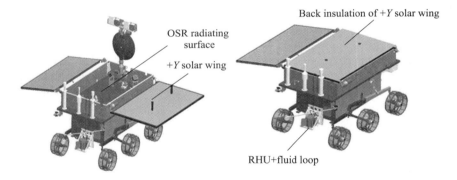

Fig. 9.13 Schematic diagram of the composition of thermal control system in Chang'e 3 rover

9.4.1.2 Thermal Control Design for Mars Probes

Mars is 1.52 AU away from the Sun. Because the day and night durations, topography and chemical composition on Mars are very similar to those on the Earth, Mars has become the preferred place for human beings to search for extraterrestrial life. There is an atmosphere on the Marian surface. On Mars, the temperature difference between day and night is 100–150 °C. At night, the lowest temperature in the polar region is

9.4 Thermal Control System Design for Deep Space Probe

−128 °C, and at noon, the highest temperature near the equator is about + 27 °C. The design of thermal control system in Mars probes mainly focuses on the thermal insulation design for Mars probes under the condition of low temperature at night.

The USA has advantages in the development of Mars probes. "Mars Pathfinder" is a representative Mars probe that has been successfully launched. During the cruise phase, pump-driven fluid loop (named Heat Rejection System, HRS for short) is adopted as the main structure for the thermal management of the whole probe. The working medium is R-11 and the maximum working pressure is 690 kPa. In order to adjust the temperature inside the probe, the main route-bypass configuration is adopted in the fluid loop, and the flow distribution between the outer loop and the bypass is regulated by a paraffin-driven thermal control valve. "Mars Pathfinder" is the first probe to use a single-phase fluid cooling loop in interplanetary exploration [4].

As part of the "Mars Exploration Rover" (MER) project, the "Spirit" and "Opportunity" Mars rovers inherited and improved the mature fluid loop technology of the "Mars Pathfinder." Two paraffin-driven thermal switches and six radioisotope heating units (RHUs) were used for thermal control in the battery packs of "Spirit" and "Opportunity" Mars rovers. In addition, the external equipments of these two Mars rovers adopted the sleep–wake work mode [5, 6].

The "Mars Science Laboratory" (MSL) probe is a large Mars vehicle developed by the USA and launched in 2011. In this probe, a multi-mission radioisotope thermoelectric generator was used to supply power, and the waste heat of about 2000 W produced by radioisotope thermoelectric generator was used to compensate for the heating of internal instruments and equipments. Battery power was used for the temperature control of the external equipment such as fluid pipeline, which is difficult to arrange. The MSL consisted of two single-phase fluid loops: one for the thermal control of the whole probe during cruise phase, with a heat transfer capacity of 2150 W; the other for the thermal control of the Mars rover working on the Mars surface. Both loops used the working fluid CFC-11. Owing to the use of radioisotope thermoelectric generator and fluid loop, the whole vehicle has strong environmental adaptability and can work in any season within the 60° southern and northern latitudes of the Martian surface [7, 8].

The Mars rover of the "ExoMars" mission will explore the areas near the Martian equator. The Mars rover has a box structure and is driven by six wheels. The basic idea of thermal design for the Mars rover is to place temperature-sensitive equipments such as electronic equipments and batteries in the box, design the thermal insulation between the equipments and the box body and increase the thermal time constant (the product of thermal resistance and heat capacity) of the box. The equipment inside the box will be heated by RHUs, and the excess heat will be transferred through the heat switch and the loop heat pipe (LHP) to the external radiator for heat dissipation. Separate thermal control measures and electric resistance heater are used by the equipments outside the box [9].

9.4.1.3 Thermal Control Design for Venus Probes

The characteristic thermal control measure of the "Magellan" Venus probe is the thermal control shutter driven by bimetallic sheets. During the detection process, the temperature of the equipments in some compartments rises due to the performance degradation of the thermal control coating. The probe would reduce the equipment temperature by adjusting the on-orbit operation attitude and using high-gain antennas to block the solar radiation.

9.4.1.4 Thermal Control System Design for Saturn Probes

The most representative Saturn probes are European "Huygens" probe and American "Cassini" probe.

Like most of the spacecrafts, "Huygens" mainly relies on thermal insulation, heaters and surface treatment for equipment temperature control. However, the thermal control design of "Huygens" probe also has unique characteristics. For example, the thermal insulation measures in "Huygens" probe can adapt to vacuum, atmosphere and the environments aerodynamically heated with high heat flux. "Huygens" uses radioisotope heating unit (RHU) for thermal compensation. In addition, in the environment with strong solar radiation, the high-gain antenna in "Cassini" probe is used as a sunshade to avoid the overheating of "Cassini" and "Huygens."

9.4.1.5 Thermal Control Design for Other Planetary Probes

During the flight mission of the European "Rosetta" probe, the solar radiation intensity it can receive varies by a factor up to 25, that is, the minimum solar radiation intensity it can receive is only 4% of that on the low Earth orbit. "Rosetta" does not use a radioisotope electric heater. It maintains the probe's normal temperature by strictly controlling the heat leak of the entire spacecraft. It uses 14 thermal shutter radiators, each with an area of 0.17 m^2. During the long journey of 10 years, the whole probe is in a dormant state. The equipment temperature is maintained by electric heater totaling 132 W [10].

The key to the design of thermal control system of the American "Messenger" Mercury probe is how to adapt to Mercury's high-temperature environment. Therefore, the sunshade of the Messenger probe is the most critical component. It is made of high-temperature resistant ceramic material, which protects the probe from direct sunlight.

9.4.2 Basic Principles for Thermal Design

After referring to the design principle of thermal control system for near-Earth spacecraft, summarizing the development experience of thermal control system for the internationally launched deep space probes and fully considering the mission characteristics of deep space probe and the particularity of thermal environment, the following basic principles have been proposed for the design of thermal control system in deep space probes.

(1) In thermal design, the thermal control requirements of ground operation phase, launch phase, orbital operation phase (including transfer phase or cruise phase), entry phase (reentry phase), planetary surface operation phase and other phases should be considered comprehensively, the thermal control measures adopted in different stages should be coordinated, and the general thermal control technology for deep space probes should be fully utilized to satisfy the thermal control requirements of various mission stages. In the design of a system, it is necessary to consider the full coordination with the whole-probe system and other subsystems and properly deal with the interface relationship between them through the design of gas emission channel in an enclosed space, the control of material exhaust pollution, the control of optical interference in the view field of optical equipments, the control of microvibration interference of mechanical pumps and so on.

(2) The thermal control system should have strong adaptability. This adaptability includes not only the adaptability to the design modification caused by the deviation of input conditions, but also the adaptability to the change of uncertainty in the space environment conditions.

(3) The processes of heat exchange, heat transfer, heat utilization and heat dissipation should be properly organized and the energy (electric or thermal energy) in the probe should be controllably used to simplify the design of thermal control system. In general, the total amount of energy in the probe is relatively fixed. Thus, when designing the thermal control system, the designer should consider the energy balance of the probe system and rationally use two heat sources, namely the heat consumption of the working device and the electric heating compensation for the device not working, so as to achieve the total heat stability in the probe.

(4) Reduce the weight of the thermal control system as much as possible. Saving the weight resources is one of the general principles in the design of deep space probes. Therefore, in the design of thermal control system, it is necessary to optimize the system configuration and the design of thermal control components so as to minimize the weight of the system. The thermal control system of the deep space probe generally weighs 7%–10% of the total weight of the probe.

(5) When designing the thermal control system, the feasibility and convenience of thermal balance test verification should be considered. In the thermal control design, consideration shall be given to the thermal control technology that is

significantly affected by gravity field and microgravity field conditions, such as the channel heat pipe whose heat transfer performance is greatly affected by gravity field.

9.4.3 Thermal Design

The main contents of the thermal design of the probe are as follows: (1) selection of the thermal control system scheme; (2) detailed design of the thermal control system, including the selection of working conditions and the design of heat dissipation surface and thermal control measures; (3) thermal design of the key stand-alone equipments; (4) evaluation of key technologies; (5) analysis and verification of the thermal control system design; (6) reliability and safety analysis of the thermal control system; (7) thermal balance test scheme; (8) development procedure of the thermal control system. In the following sections, only a few of them are discussed [11].

9.4.3.1 Selection of Thermal Control System Scheme

The selection of thermal control system scheme is usually carried out in the scheme development stage of the project. As mentioned above, the deep space probes have very strict requirements for mass and electric power resources. Therefore, in the design of thermal control systems, the designer must pay special attention to the rational use of mass and electric power resources when choosing thermal control schemes, make full use of the favorable opportunity of scheme feasibility demonstration and actively coordinate with the probe system and other subsystems. The relevant requirements for thermal control design shall be integrated into the system configuration and layout design of the probes as early as possible and into the thermal design of stand-alone equipment to hierarchically control and implement thermal control functions.

In the scheme development stage, when choosing the scheme of thermal control system, the system scheme and the subsystem schemes are usually in continuous iteration, and the input conditions needed for the scheme design of thermal control system are usually not specific and variable greatly. In view of this characteristic, the designer should pay attention to the following aspects when choosing the scheme of thermal control system:

(1) The thermal requirements for all instruments and equipments during the whole process from launch to the end of mission should be analyzed in detail, especially those of peripheral, remote and small equipments and the equipments without signing the interface data sheets (such as unplugging). Moreover, the designer must fully refer to the development experience of thermal control system for similar probes in the world and compare multiple schemes to determine the

9.4 Thermal Control System Design for Deep Space Probe

best theme control scheme. At the same time, the ground test phase should be taken into account as much as possible. Especially, the temperature control of the instruments and equipments during the launch zone test at the launching site should be considered.

(2) The recessive requirements related to thermal control system, such as pollution control requirements, anti-static requirements, gas emission or inflation requirements in enclosed space, dust-proof requirements, anti-stray light requirements for optical equipment and vibration isolation requirements, except for thermal requirements, should be analyzed in detail for the whole process from launch to the end of mission.

(3) Major efforts should be focused on the thermal control of the modules or key equipments with high requirements and complex working conditions and affecting the completion of probe mission, for example, the thermal insulation of Chang'e 3 lander ($\pm Y$ modules) that needs to survive at lunar night, the heat dissipation of high-power electronic equipment and the thermal control schemes of high-resolution cameras requiring constant temperature or isothermal property.

(4) The adopted thermal control measures should leave enough margin or room for adjustment to deal with the design deviation caused by the inaccurate design input of the total heat generated by the instruments and equipments.

9.4.3.2 Selection of Design Conditions

The first step of the thermal design for the probe is the selection of design working conditions. The so-called working condition is a combination of internal heat source and external heat flow. In principle, the high-temperature working condition occurs when both the external orbital heat flow and the internal heat source are the maximum; while the low-temperature working condition occurs when both the external orbital heat flow and the internal heat source are the minimum. According to the rates of change of external heat flow and of internal heat source, the working conditions can be divided into transient and steady working conditions. Generally speaking, the transient working condition is chosen for the lunar orbiter, and the steady working condition is chosen for the lunar surface probe. Depending on different thermal statuses and mission stages, the working conditions selected in thermal design are as follows.

(1) High-temperature and low-temperature working conditions of the whole probe during the mission. For the lunar orbiter, its high-temperature working condition is usually the condition with a small β angle because of the large infrared radiation of the Moon; while for the Mars probe, its high-temperature working condition is usually the condition with long sunshine time and vice versa because of the small infrared radiation on Mars.

(2) High-temperature and low-temperature working conditions of key equipments during the mission. Generally speaking, consideration should be focused on the high-temperature and low-temperature conditions of the battery pack, power

supply controller, as well as TT&C equipment. For example, for the Chang'e 1 probe during the lunar orbiting flight, the high-temperature condition of the battery pack occurs in the orbit with the β angle of 45°. However, it is not recommended to verify the thermal design of stand-alone equipment in the probe-level test. Instead, the thermal design verification of key stand-alone equipment should be carried out in the equipment-level thermal balance test.

(3) Other special conditions that need to be considered. These conditions include the working conditions for verifying the release rate of the pressure inside the module during ascent phase, the high-temperature working conditions when reentering the atmosphere, the working conditions when entering the Mars atmosphere and other conditions. These conditions are usually closely related to the special mission process.

9.4.3.3 Design of Heat Dissipating Surface

Based on the analysis of the external heat flow in the whole mission process, the surface with the smallest external heat flow and the smallest fluctuation range is selected as the heat dissipating surface to maximize the heat dissipation capacity and minimize the temperature fluctuation. It should be noted that the range of external heat flow spectrum should also be considered when choosing the heat dissipating surface. At present, the coatings applied on the heat dissipating surface of the probe mainly target the solar spectrum with smaller absorptivity and larger infrared emissivity. If the external heat flow is mainly infrared, the coatings on the heat dissipating surface will no longer have the ability to dissipate heat. For example, affected by the strong infrared radiation, heat dissipating surface is unfit for the side of lunar orbiter facing the lunar surface. In addition, the selection of a heat dissipating surface should also consider the avoidance of dust fall and accumulation.

According to the above principles, the heat dissipating surfaces selected for the probes in different orbits are also different. The heat dissipating surface of the lunar orbiter is generally located on the side not facing the Moon, while the heat dissipating surface of the lunar surface probe is generally oriented toward the sky to avoid the influence of heat flow of strong lunar infrared radiation on the heat dissipating surface. For the Mars-orbiting probe, because the infrared radiation and reflection of Mars are both relatively small, the location of the heat dissipating surface is relatively free, usually on the surface not exposed to sunlight. Moreover, the heat dissipating surface of the Martian surface probe is usually set on the vertical plane to avoid the accumulation of Martian dust.

9.4.3.4 Selection of Thermal Control Measures

Thermal design is achieved through various thermal control measures. The selection of thermal control measures is directly related to the thermal control performance, safety and reliability and economical efficiency.

9.4 Thermal Control System Design for Deep Space Probe

The selection of thermal control measures is closely related to the design idea of thermal control system and needs to serve the thermal control design objectives. For example, when the temperature of a device on orbit is too high, the general thermal control system cannot handle this problem. However, the problem of low temperature can often be solved by electric heating. Therefore, the designer will consider the adoption of low-temperature design to leave more design margin for high-temperature condition, while using electric heating to ensure not exceeding the low-temperature limit. The basic principle for choosing thermal control measures is to choose at first the global and external thermal control measures and then the local and internal thermal control measures, and to choose at first the passive thermal control measures and then the active thermal control measures.

It should be pointed out that for a general spacecraft, some factors related to thermal control measures often exist without being "selected" by thermal control system, such as probe structure geometry, component connection method, component material and instrument layout. For thermal control design, the adjustable range is limited. However, deep space probes are very sensitive to resource occupancy. Thermal control subsystem engineers should actively participate in even the demonstration of preliminary thermal control scheme and analyze the above-mentioned items that are not "selected" by thermal control. Also, some items that seriously affect the design of thermal control systems should be properly selected, and their design status should be modified after coordinating with the probe system and other subsystems when necessary, so as to simplify the design measures of thermal control system and improve the comprehensive performance of the probe system.

9.4.4 Thermal Analysis

The thermal analysis for probes mainly includes three aspects: orbit calculation, external heat flow calculation and temperature calculation. The main purpose of thermal analysis is to confirm whether the temperature of all the equipments and components on the probe is within the required range by analysis and calculation under the specified input conditions. Thermal analysis penetrates the whole development process and mission execution process of the probe and is an important part of the development of thermal control subsystem. This section focuses on the selection of analysis conditions and the establishment of lunar thermal model in the thermal analysis of deep space probes.

9.4.4.1 Selection of Thermal Analysis Conditions

The probe has many thermal states (working conditions) when flying in space. Any change of orbit, attitude, equipment working mode and thermal power consumption, as well as any degradation of coating, will bring about a new working condition. At the same time, the thermal state of the probe is constantly changing when it flies in space.

The changes of thermal state include the changes of sunlight conditions, equipment working mode and heat consumption, attitude, thermal state of the celestial body, etc. Based on the above situation, thermal analysis conditions can be divided into high-temperature analysis condition and low-temperature analysis condition according to the temperature level, or into transient analysis condition and steady-state analysis condition according to the type of working conditions.

When choosing thermal analysis conditions, we should comprehensively analyze the environmental conditions and equipment working modes in the whole process of the mission and determine the thermal analysis conditions based on experience. Generally, this analysis should cover the extreme temperature conditions (extreme working conditions) that the probe may encounter during the whole mission. The extreme working conditions should be selected and determined according to the selected objects for analysis (the whole probe or target location). The extreme working conditions can be selected and determined by the following methods.

(1) Indirect screening method. When the external heat flow absorption in space or the heat consumption in the cabin reaches the extreme value respectively, they are used as the candidate working conditions to be combined, and the possible extreme value of their combinations is the extreme working condition. The time of heat consumption extremum in cabin is usually determined according to the designed working mode of probe system. There are two methods to determine the time of external heat flow extremum in space: (a) directly determining, by experience, the time when the external heat flow absorbed from space reaches the extremum at acertain solar illumination angle; (b) calculating external heat flow in space at multiple solar illumination angles and multiple probe attitudes through theoretical analysis or enumeration; and determining the time of extremum occurrence according to the numerical value of the external heat flow absorbed from space.

(2) Direct screening method. When it is difficult to determine extreme working conditions by indirect screening, the temperature of probe or target location should be calculated under various conditions, and the extreme working conditions could be determined directly according to the temperature results.

In selecting the type of thermal analysis condition, the selected analysis object should be observed under whether steady-state or transient analysis condition. The thermal analysis results should basically reflect the real state of the selected object in orbit, without producing unacceptable deviations. Generally, when the following factors are present, the analysis of transient working conditions should be carried out:

(a) Large fluctuation of external heat flow received by the heat dissipating surface;
(b) Multiple exposed parts;
(c) Multiple equipments with large heat consumption fluctuation;
(d) Multiple heating loops for automatic temperature control.

9.4 Thermal Control System Design for Deep Space Probe

In the following text, Chang'e 3 rover is used as an example to illustrate the selection of thermal analysis conditions.

Chang'e 3 rover is a lunar surface probe, which is carried by the lander to the lunar surface to carry out scientific exploration activities. In determining the thermal analysis condition of the rover, the environmental conditions and working modes that may be encountered during the whole process of the rover's mission should be firstly analyzed, including launch window, equipment working mode, lunar topography, landing attitude, the working mode and heat transfer capability of the two-phase fluid loop in the chosen thermal control product, lunar dust, the thermal properties degradation of thermal control coatings, solar irradiation intensity variation and so on. According to the above analysis factors in combination with the rover's task planning, the typical analysis conditions in the mission process could be determined, including 23 working conditions such as the Earth–Moon transfer, orbit maneuver during Earth–Moon transfer, Moon orbiting, powered descent, separation of lander and rover, high temperature on the lunar surface, low temperature on the lunar surface, day–night shift, lunar night, night–day shift, lunar dusk influence and so on. Among them, orbit maneuver, lunar orbiting, powered descent, separation of lander and rover, day–night shift and night–day shift are transient working conditions, while the others are steady-state working conditions. Figure 9.14 presents the lander-rover joint thermal analysis model before the rover separation, and Fig. 9.15 shows the thermal analysis model for the rover alone after separation.

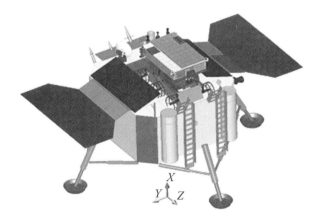

Fig. 9.14 Thermal analysis model for the combination of the rover and lander

9.4.4.2 Lunar Surface Temperature Field for Lunar Orbiter

The calculation of lunar infrared radiation heat flux is the basis of thermal analysis of lunar orbiter. Zhong Qi and his colleagues in China Academy of Space Technology have constructed the temperature field variation of the lunar surface on the basis of

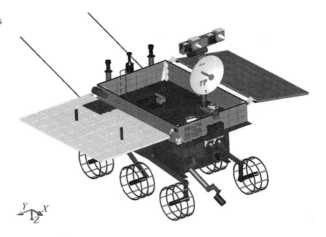

Fig. 9.15 Thermal analysis models for the rover

references from the international exploration data, in order to calculate the infrared radiation heat flux of lunar orbiter [12].

The lunar surface temperature field is defined as follows:

(1) Firstly, the definitions of lunar longitude and latitude are clarified. The longitudinal and latitude network on the lunar surface is established with the zero latitude line defined as the lunar equator, and the zero longitude line defined as the longitude line located on the intersection point of the lunar probe's orbit and the lunar equator. The density of latitude and longitude grid can be dynamically adjusted according to the users' needs.

(2) The temperatures at various points (φ, φ) on the lunar surface are calculated by the following formula:

$$T(\phi, \varphi) = \cos^{\frac{1}{4}} \varphi \cdot T(\phi, 0) \quad (9.6)$$

where

$$T(\phi, 0) = \begin{cases} [T_{ss}^9 \cos \phi + T_{270}^9]^{\frac{1}{9}}, & \phi \in [-90, 0) \\ [T_{ss}^7 \cos \phi + T_{90}^7]^{\frac{1}{7}}, & \phi \in [0, 90) \\ [\frac{3\sigma\varepsilon}{C}(\phi - 270) + T_{270}^{-3}]^{-\frac{1}{3}}, & \phi \in [90, 270) \end{cases}$$

$$T_{ss} = \left[\frac{S(1-\rho)}{\sigma\varepsilon}\right]^{\frac{1}{4}};$$

$$T_{90} = \left[-180 \times \frac{3\sigma\varepsilon}{C} + T_{270}^{-3}\right]^{-\frac{1}{3}};$$

$$T_{270} = 89.05 K;$$

9.4 Thermal Control System Design for Deep Space Probe

S is the solar constant (W/m^2);
ρ is the average albedo of the Moon to the Sun;
ε is the lunar infrared emissivity;
σ is Stephen–Boltzmann constant;
ϕ is the longitude away from the sub-solar point (°);
φ is the latitude (°);
T is the lunar surface temperature (K);
C is the fitting constant, $C = 25.21$.

According to the temperature of the lunar surface points (ϕ, φ), the infrared radiation intensities at the lunar surface points can be obtained.

9.4.4.3 Lunar Thermal Model for Lunar Probe

The analysis on external heat flow of a lunar probe needs to consider the local state of the lunar surface properties. When analyzing external heat flow of a lunar probe, a common practice is to build a lunar surface model and use it as part of the thermal analysis model to analyze the infrared radiation heat flow and albedo heat flow from lunar surface to the lunar probe, and the mutual heat effect between the probe and the lunar surface, such as the shadow cast by the probe on the lunar surface.

With the increase of the depth below the lunar surface, the range of temperature fluctuation between day and night becomes smaller. At the depth of 35 cm below the lunar surface, the temperature fluctuation between day and night is less than 6 K. At the depth of 80 cm below the lunar surface, the temperature fluctuation between day and night stops completely and enters the constant temperature state (about 252 K). Therefore, the thickness scale in the lunar model should be no less than 80 cm. The thermal conductivity of lunar soil varies with depth, which is related to the material state and properties at different depths. Generally speaking, the lunar soil nearer to the lunar surface is looser and has a lower thermal conductivity.

The lunar surface model should be able to approximate the real thermal conditions of the lunar surface, such as surface physical properties, temperature distribution and so on. According to the accuracy requirement for the analysis results, the lunar surface models with different refinement degrees but the basically same structure can be established, and their difference is mainly reflected in the refinement degree of parameter setting. The state of the lunar thermal model is described as follows:

(1) The thickness dimension in the lunar surface model should be no less than 80 cm, and the surface dimension (generally expressed by circular diameter) in the model should be no less than 5 times of the probe's characteristic size. The difference between the calculation results of the radiation heat transfer between the lunar surface model and the probe under this condition and the results under the real condition would be less than 5%.
(2) Set the surface radiation properties in the lunar surface model. According to the literature, the infrared emissivity and albedo are 0.92 and 0.073, respectively.
(3) Set the thermal conductivity, density and specific heat capacity of lunar soil using the following formulas. The formula for thermal conductivity is $K = 1.66 \times$

$10^{-2} \times (Z/100)^{3/5} + 8.4 \times 10^{-11} \times T^3$ (W/(m·K)), where Z is the depth from the lunar surface (in cm) and T is temperature (in K). The formula for specific heat capacity is $c = -23.173 + 2.127\,0\,T + 1.500\,9 \times 10^{-2}T^2 - 7.369\,9 \times 10^{-5}T^3 + 9.655\,2 \times 10^{-8}T^4$ (J/(kg·K)), to which the applicable temperature range is 90–50 K. The formula for density is $\rho = 1.92\,(Z + 12.2)/(Z + 18) \times 1\,000$ (kg/m³).

(4) Discretization of lunar surface model. As shown in Fig. 9.16, the probe's location is taken as the center to divide the lunar surface model into many grids. The lunar surface grids near the probe are dense, while the grids far from the probe are sparse. The grids near the lunar surface are sub-divided.

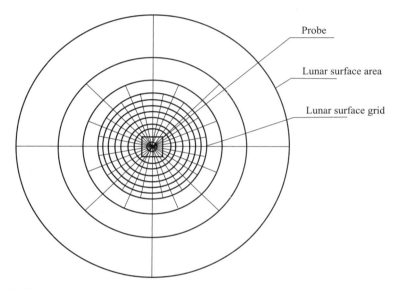

Fig. 9.16 Schematic diagram of the thermal model structure of lunar surface

When the above lunar surface model is input into the thermal analysis software as part of the probe model, the heat flux from the lunar surface to the probe can be calculated. It should be noted that the radiation heat flow from the lunar surface model to the probe is not output as external heat flow, so manual processing is needed.

9.4.5 Ground Simulation Test

In order to verify the function and performance of the probe in space environment and ensure the reliable operation of the probe during the mission, it is necessary to carry out adequate environmental simulation tests on the ground. In particular, the thermal balance test under the simulated spatial thermal environment is an important test item. The purpose of thermal balance test is to verify the correctness of thermal

9.4 Thermal Control System Design for Deep Space Probe

design and thermal analysis model for the probe. In this section, only a few issues related to the thermal balance test of deep space probes are discussed. For other contents, the relevant literatures can be referred to.

9.4.5.1 External Heat Flow Simulator

According to the characteristics of the simulated heat flow, the simulation methods of external heat flow can be divided into two categories. One method is called incident heat flow simulation method, which is to simulate the irradiance, direction and spectral characteristics of the spatial external heat flow. The other method, called absorbed heat flow simulation method, is to simulate only the intensity of the absorbed heat flow on the probe surface to obtain the same thermal effect. Solar simulator is just an external heat flow simulator using the incident heat flow simulation method. The external heat flow simulators based on absorbed heat flow simulation method are far-infrared heating cage, radiation heating plate, infrared heating rod, infrared lamp array, resistance heater and others.

Different external heat flow simulators have different characteristics. When developing the probe's thermal balance test, an appropriate external heat flow simulator should be selected according to the characteristics of external heat flow that need to be simulated. The characteristics of different external heat flow simulators are compared in Table 9.11. In the thermal balance test of deep space probes, the general principles for selecting the external heat flow simulator are as follows.

(1) The maximum and minimum heat flows required by the simulation should be taken into account, such as the large heat flow during lunar day and the small heat flow at lunar night. Solar simulator, infrared lamp array, infrared heating rod and resistance heater have a smaller occlusion coefficient to the test model and can simulate large heat flow and small heat flow. It is usually difficult for the far-infrared heating cage, infrared heating plate and other external heat flow simulators to simulate small heat flow.

(2) The transient characteristic of external heat flow can be simulated according to the simulation test requirement. For the lunar orbiter, the transient characteristics of external heat flow are obvious. The simulation of the transient characteristics of external heat flow in the test is very important for ensuring the effectiveness of thermal balance test. Solar simulator, infrared lamp array and resistance heater usually have better transient heat flow characteristics.

(3) The simulator should adapt to the shape of thermal test model. Solar simulator and resistance heater can adapt to the complex shape of thermal test model, while infrared lamp array, infrared heating rod and radiation heating plate are suitable for the heating of regular large planes.

(4) The simulator should match the characteristics of heat flow measurement device, which are usually related to the spectral characteristics of the measured heat flow. The selected external heat flow simulation device should match the characteristics of heat flow measurement device.

Table 9.11 Characteristic comparison of external heat flow simulators

Name	Spectral characteristics of heat flow	Transient characteristics of heat flow	Occlusion coefficient	Applications	Effect on thermal test model	Operational costs	Remarks
Solar simulator	Solar spectrum	Best	Unoccluded	Various complex surfaces	None	Highest	Radiation simulation
Infrared lamp array	Spectrum related to the applied voltage	Good	Small	Relatively regular surface with fewer protrusions	None	Relatively high	Radiation simulation
Far-infrared heating cage	Far-infrared spectrum	Poor	Relatively large	Relatively regular surface with fewer protrusions	None	Low	Radiation simulation
Radiation heating plate	Far-infrared spectrum	Relatively poor	Largest	Regular surface without protrusions	None	Relatively high	Radiation simulation
Infrared heating rod	Far-infrared spectrum	Relatively poor	Large	Relatively regular surface with fewer protrusions	None	Relatively high	Radiation simulation
Resistance heater	-	Relatively good	Unshielded	Various complex surfaces	Needing to destroy the mounted surface	Lowest	Contact simulation

(5) The selection of external heat flow simulator should consider the requirements of thermal balance tests in both the initial prototype design phase and flight model design phase.
(6) The simulator should be easy to implement and affordable to operate.

9.4.5.2 External Heat Flow Measurement

In the probe's thermal balance tests, all the external heat flow simulation methods—except for the simulation tests of surface contact electric heating and radiation heating plate—shall measure the heat flow reaching the surface of the probes being tested. In the probe's infrared heat flow simulation test, the radiation heat flow meter is used most frequently. The schematic principle and structure of the radiation heat flow meter are shown in Fig. 9.17. The sensitive sheets for absorbing the heat flow are usually round metal sheets coated with high absorption ratio coatings, with thermocouples installed in the center. The sensitive sheet is connected with the base plate through multi-layer insulation material to form a heat flow meter. The heat flow meter in use is suspended to a predetermined position on the probe surface. The properties of sensitive sheet coating and the spectral distribution of radiation heat flow are very important to the accuracy of heat flow measurement.

If the sensor coating is the same as the coating of the tested probe surface, the sensor temperature would depend only on the external space heat flux absorbed by the probe surface and the infrared emissivity of the probe surface coating. Using this type of heat flow meter to measure the external heat flow would eliminate the influence of the change of radiation spectrum of simulated heat flow on the determination of the test heat flow and greatly simplifies the work of heat flow measurement and analysis.

If a high absorption ratio coating, such as black paint, is applied on the sensitive sheet (called "black sheet") and is different from that on the tested probe surface, the sensor temperature is related to not only the external space heat flux absorbed by the probe surface, but also the absorption ratio of the probe surface coating to the simulated heat flux.

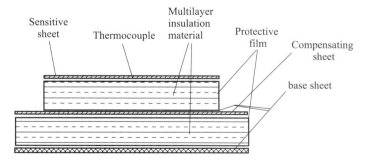

Fig. 9.17 Schematic diagram of the structure of radiation heat flow meter

Therefore, when simulating the heat flow in the far-infrared spectrum (such as far-infrared heating cage and infrared heating rod), accurate measurement results can be obtained by using a "black sheet" heat flow meter or a heat flow meter whose sensitive sheet coating is consistent with the measured surface coating. To simulate the heat flow whose spectrum contains incomplete infrared spectra (such as visible light) generated by solar simulator, infrared lamp array and other devices, it is necessary to use a heat flow meter with the same sensitive sheet coating as the measured surface coating to measure the heat flow. On the other hand, using a "black sheet" heat flow meter will lead to a smaller absorbed heat flow on the probe surface.

9.4.5.3 Selection of Working Conditions for Thermal Balance Test

The thermal balance test of deep space probes should verify the combinations of different equipment working modes, orbit insertion conditions, solar angle, maximum and minimum heat consumptions of equipments, lunar eclipse, Mars night and other conditions during different seasons and all the mission phases, that is, all the harshest hot environments and the harshest cold environments for the probes. Both the sufficiency of thermal control design verification and the affordability of test cost should be taken into account in the selection of thermal balance test conditions. While ensuring sufficient thermal design verification, the test conditions can be reduced or the duration of test conditions can be shortened.

The general principles for determining the thermal balance test conditions of deep space probes are as follows:

(1) The key parameters needed to support thermal analysis model validation, and flight mission prediction should be obtained.
(2) Generally, both high-temperature and low-temperature working conditions should be included. When it is difficult for the ground equipment to simulate high-temperature or low-temperature working conditions, the design of test conditions should meet the requirements of (1).
(3) According to the need of thermal analysis model modification, the working conditions of quasi-steady state, periodic transient state or transient state for thermal analysis model modification can be included.
(4) General consideration is given to verifying the power demand and control capability of the heating loop.
(5) Generally, the possible failure conditions which have a significant impact on the function of probe products should be considered.
(6) Upon request, the calibration conditions can be added to determine the relationship between the heat flow and the electric current applied on the heat flow simulator.
(7) The function and performance of thermal control system and products should be fully validated in the prototype development stage. The system validation requirements, the limitation of thermal test model and the implementability of ground equipments should be considered in the flight model development stage.

9.4.5.4 Thermal Analysis Model Modification

Thermal analysis model modification is the continuation of thermal balance test conditions and an integral part of thermal balance test. In view of the special and complex thermal environment of deep space probes, sometimes its thermal effects (such as Martian atmospheric convection environment) are hard to simulate on the ground. So, verifying and modifying the thermal analysis model of deep space probes is particularly important and could even become the main purpose of the probe's thermal balance test verification.

The modification of thermal analysis model is implemented as follows. The obtained test data such as heat sink temperature, external heat flow measurement, thermal boundary measurement and equipment temperature is used to modify the relevant parameters of the thermal analysis model (i.e., thermal analysis test model) of the probe products in thermal testing state and test environment. The analysis results should be consistent with the obtained test results. Moreover, the modified thermal analysis test model is used to predict the test results under the next working condition. If there is any deviation, the model needs to be further modified. The modifications of the thermal analysis test model should be applicable to all the test conditions, and the revised parameters should be within a reasonable range. Among them, the modified results of the temperature points of intravehicular components and of the temperature points of inner heat dissipating surface should satisfy the Formulas (9.7)–(9.10) simultaneously.

$$\Delta T = |T_{Mi} - T_{Pi}| \leq 5K \text{ (extravehicular components)}, \tag{9.7}$$

$$\Delta T = |T_{Mi} - T_{Pi}| \leq 10K \text{ (extravehicular components)}, \tag{9.8}$$

$$\Delta T_{\text{mean}} = \frac{1}{N} \sum_{i=1}^{N} (T_{Mi} - T_{Pi}) \leq 2K \tag{9.9}$$

$$\sigma = \sqrt{\frac{\sum_{i=1}^{N} \{(T_{Mi} - T_{Pi}) - \Delta T\}^2}{N - 1}} \leq 3K \tag{9.10}$$

where

- ΔT Temperature deviation, K
- T_{Mi} Measured temperature value, K
- T_{Pi} Calculated temperature value, K
- ΔT_{mean} Average temperature deviation, K
- N Number of temperature measurement points, usually ≥ 25 (generally excluding active temperature control points)
- σ Standard deviation, K.

After the modification of thermal analysis test model is completed, the relevant modification parameters should be transplanted into the thermal analysis model (i.e., flight thermal analysis model) that describes the probe product in the mission-specific working state and spatial operation environment conditions, and the flight thermal analysis model should be used to predict the on-orbit flight state.

The thermal analysis test model constructed in the thermal balance test of Chang'e 3 rover is shown in Fig. 9.18. In addition to the test parts of the rover, the thermal analysis model also includes the ground equipments such as space environment simulator, moving cold plate, far-infrared heater, test bracket and ground test cables. The test data of model modification conditions is used to modify the thermal analysis test model, and then, other steady-state test data is used to verify and confirm the correctness of the modification results. After revising the ground thermal analysis model, the standard deviations of thermal analysis results and test results are less than 3 °C, so that the purpose of thermal analysis model modification is achieved. In the process of correlation correction of the thermal analysis model, it is also found that the deviation of installation position of far-infrared heater and the thermal impact of some ground equipments on the test pieces are too significant to ignore. For example, the existence of test brackets in the thermal analysis model of the rover test increases the temperature of the device on the rover test piece by 2–3 °C. Initially, when the on-orbit thermal analysis model is modified directly with test data, this effect is often "forcibly" corrected to the thermal conductivity of the device or the thermal performance parameters of thermal control products, thus resulting in the fake consistency between the test data and the analysis data [13].

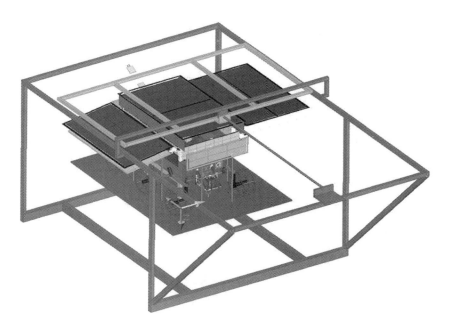

Fig. 9.18 Thermal analysis test model of Chang'e 3 rover

9.5 Prospect

The development of thermal control technology for deep space probes is drawn by mission demand and driven by technology development. However, the feature of traction-induced development is remarkable. The thermal control technology for deep space exploration is greatly influenced by the top-level system design of the probe mission. Complex missions and harsh and changeable space environment have put forward many new and high requirements for the thermal control system of deep space probe and also greatly promoted the development of thermal control technology. This can be seen from the traction effect of the lunar exploration project on China's thermal control technology. The implementation of the lunar exploration project has enabled China to successfully develop a number of new technologies, such as gravity-assisted two-phase fluid loop technology, variable conductivity heat pipe technology, water sublimator technology, phase change material—composite heat pipe technology, engine thermal protection technology, which have been successfully applied in on-orbit flight.

As far as the future deep space missions such as lunar scientific research station, Mars sample return, Jupiter exploration and asteroid exploration are concerned, their complexity and difficulty are unprecedented. The mission environment covers the extreme conditions of high and low temperature, which poses new and higher requirements for the design of thermal control system and puts forward more urgent needs for the application of the following new thermal control technologies:

(1) Lightweight and efficient thermal insulation technologies: including the thermal insulation technologies applying in the vacuum and non-vacuum environments;
(2) High-efficiency heat dissipation technologies: including heat pump technology, lightweight and efficient radiator technology, high-emissivity coating technology with high-temperature resistance, as well as intelligent controllable coating technology, etc;
(3) High-reliability heat transfer technologies: including pump-driven two-phase fluid loop technology, capillary pump-driven two-phase fluid loop technology, as well as single-phase high-reliability long-life mechanical pump technology, high-efficiency piezoelectric pump technology and electrohydrodynamic pump technology (efficiency > 90%);
(4) High-temperature thermal protection technologies: including reusable lightweight high-temperature thermal protection technology and multi-functional thermal protection technology integrating micrometeor prevention, space debris prevention and thermal radiation protection.

References and Related Reading

1. Xiang Y, Wu Y, Shao X (2007) Investigation on thermal control design methods of deep space spacecraft. Spacecraft Eng 16(6):82–86 (Chinese vision)
2. Gilmore DG (2002) Spacecraft thermal control handbook—volume I fundamental technologies. The Aerospace Corporation Press, California
3. Xiang Y, Chen J, Zhang B (2015) Thermal control for jade rabbit rover of Chang' E-3. J Astronaut 36(10):1203–1209 (Chinese vision)
4. LyraJC, Novak KS (1997) The mars pathfinder system level solar thermal vacuum test. In: AIAA thermophysics conference, Atlanta
5. Novak KS, Philips CJ, Sunada ET et al (2005) Mars exploration rover surface mission flight thermal performance. In: The 35th ICES, Rome
6. Novak KS (2003) Development of a thermal architecture for the mars exploration rovers. Space technology and applications international forum—STAIF, Albuquerque
7. Bhandari P, Birur G et al (2005) Mars science laboratory thermal control architecture. In: The 35th ICES, Rome
8. Pantano DR, Dottore F, Tobery W (2005) Utilizing radioisotope power system waste heat for spacecraft thermal management. AIAA, San Francisco
9. Robson A, Seward E (2005) Thermal design considerations for a mars rover. In: The 35th ICES, Rome
10. Gardini B, Berner C, Van Casteren J (1999) The technical and programmatic challenge of the international rosetta mission. In: The 50th international astronautical congress, Amsterdam.
11. Min G, Guo S (1998) Thermal control of spacecraft (Second edition). Science Press, Beijing (Chinese vision)
12. Hou Z, Hu J (2007) Principles and applications of spacecraft thermal control technology (First edition). China Science and Technology Press, Beijing (Chinese vision)
13. Xiang Y, Chen J, Zhang B (2014) Thermal balance test of Chang'e-3 "Jade Rabbit" inspector. In: Proceedings of the 2014 annual conference of the Chinese astronautical society, Beijing (Chinese vision)
14. Vickers JMF, Garipay RR (1968) Thermal design evolution and performance of the surveyor spacecraft. In: AIAA, 5th annual meeting and technical display, Philadelphia

Chapter 10
Propulsion Technology

10.1 Introduction

The targets for deep space exploration are far away from the Earth. In order to reach the target and complete the exploration mission, the probe needs long-distance flight and complex orbit maneuver. Therefore, the propulsion system of the probe is particularly important. Deep space exploration requires high thrust, high specific impulse, long life, high-precision variable thrust and adaptability to harsh environment. As the probe is required to provide sufficient propulsion capability to meet the delta-V required for orbital maneuver, the traditional chemical propulsion systems are increasingly difficult to meet the actual needs of deep space exploration, and the application requirements of new technologies such as electric propulsion and special propulsion have become increasingly strong.

This chapter first systematically introduces the classification, principle and characteristics of the propulsion system and then analyzes the application of the corresponding propulsion technologies in the field of deep space exploration. Besides, combined with the characteristics of deep space exploration missions, the mission requirements of the propulsion system are analyzed, and the design constraints of the deep space probe propulsion system are abstracted. In the design of propulsion system, combined with the project implementation, the design factors to be considered in the project are explained and corresponding design methods are put forward.

10.2 Propulsion System Classification

Generally, propulsion systems can be divided into active control and passive control. At present, most propulsion systems applied in spacecraft are active control, including cold gas propulsion, chemical propulsion, electric propulsion and new concept propulsion. The passive-controlled propulsion system is still in the research stage, relies on gravity or gravitational propulsion and has not yet been put into engineering application [1, 2].

10.2.1 Cold Gas Propulsion

The cold gas propulsion system is the simplest and earliest propulsion system which utilizes the high-pressure gas in the gas tank that is released, expanded and then accelerated through the nozzle to generate thrust. General gases can be applied as the working fluid of the cold gas propulsion system. At present, nitrogen or helium is relatively often used. The greatest features of the system include simple structure, reliable performance, low cost, non-toxic and non-polluting, etc. And the biggest disadvantage is poor performance. Under normal circumstances, the vacuum specific impulse is only 50–70 s.

Figure 10.1 shows a typical cold gas propulsion system in which the feed and vent valves are applied for nitrogen charging and the gas tank is separated from the downstream part by a high-pressure isolation valve. Once the isolation valve is actuated to the open position, high-pressure nitrogen will flow to the pressure regulating assembly. The latter then adjusts the high-pressure nitrogen pressure to a nominal pressure that is much lower than the tank pressure. The thruster valve is opened as commanded and nitrogen flows through the supersonic nozzle to generate thrust.

Because of its poor performance, the cold gas propulsion system is generally not the main propulsion system of deep space probes. However, due to its simple structure, no pollution and convenient maintenance, the system is still occasionally applied in the design of some small probes. In the Rosetta comet exploration mission, its small lander, Philae, carried a small cold gas propulsion system as an active descent control system. This cold gas propulsion system is equipped with only one thruster and used nitrogen as the propellant.

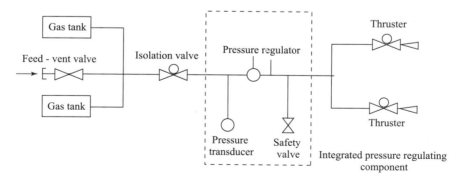

Fig. 10.1 Typical cold gas propulsion system

10.2.2 Chemical Propulsion

The chemically propelled energy system and the propellant supply system are fully integrated. The liquid or solid propellant combustion or propellant catalytic decomposition is used to convert the chemical energy of the propellant into the internal energy and pressure potential energy of the working medium and then release to generate thrust. The greatest advantages of chemical propulsion are large thrust, wide thrust range and high reliability, especially for spacecraft with high requirements of total impulse and maneuverability.

According to the physical state of the propellant working fluid at room temperature, chemical propulsion can be further divided into liquid rocket engine (LRE), solid rocket engine (SRE), solid–liquid hybrid propulsion and ramjet propulsion, as shown in Fig. 10.2. Among them, the solid–liquid hybrid propulsion and ramjet propulsion are seldom used in spacecrafts, which will not be described herein in detail.

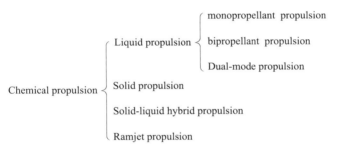

Fig. 10.2 Classification of chemical propulsion

10.2.2.1 Liquid Propulsion Technology [3–5]

Liquid propulsion technology is the most widely used technology in space. At present, most of spacecraft propulsion systems are equipped with liquid propulsion systems. The liquid propulsion system can be further divided into the bipropellant propulsion system, monopropellant propulsion system, and a dual-mode propulsion system. For probes with a thrust demand of a few Newtons to a hundred Newtons, a monopropellant liquid propulsion system is generally used. For probes with a thrust demand of tens of Newtons to tens of thousands of Newtons, a bipropellant propulsion system is generally applied; for probes with relatively large thrust different in attitude control thruster and orbital control thruster, the dual-mode propulsion system can be applied. In fact, the selection of propulsion system is not only based on the thrust size principle mentioned above but also various factors such as minimum thrust equivalent, comprehensive attitude control and orbit control demands and system dry weight constraints.

1) Monopropellant propulsion technology

The advantage of the monopropellant propulsion system lies in that the system is simple, the crucible (as the most commonly used propellant) is quite stable under normal storage conditions, and the decomposition products are clean. Therefore, the system can be used as the main propulsion system or form an electrochemical hybrid propulsion system with electric propulsion in deep space probes. The disadvantage of monopropellant propulsion is that the performance is still relatively backward. Although there are significant improvements compared to cold gas propulsion, the performance is still low compared to bipropellant propulsion.

Monopropellant propulsion engines generally consist of injectors, catalytic beds and nozzles. The monopropellant refers to the single chemical or a mixture of several chemicals (generally in a liquid state), which is then injected into a catalytic bed by an injector and decomposed into gas under certain conditions (high temperature, high pressure or catalyst action). The gas passes the nozzle and then is jetted at high speed to generate thrust. Commonly used monopropellant propellants include hydrazine, hydroxylamine nitrate, hydrogen peroxide, mixed amine, etc., of which hydrazine is most widely used in monopropellant propulsion engines while other propellants and their mixtures have potential advantages in improving performance and reducing toxicity.

During the descent stage of the Mars Science Laboratory, helium pressured propulsion system with the propellant of hydrazine is used. Eight 250 N Reaction Control System thrusters provide attitude control, and eight throttled 3300 N thrust engines were used for the deceleration and attitude control during the powered descent. Figure 10.3 shows the schematic of the descending stage propulsion system.

The throttled engine developed by Aerojet is based on the improved design of the Viking lander engine, with the thrust range of approximately 31–3603 N, specific impulse of 204–223 s, and burn time of 350 s. the engine inlet pressure is 4.14–5.24 MPa, the engine room pressure is 3.17 MPa (with the propellant flow rate of 1.4 kg/s), and the propellant flow rate is 0.02–1.633 kg/s.

2) Bipropellant propulsion technology

With the continuous enhancement of functions, mass and service life of satellite, the total impulse and the corresponding propellant consumption become more and more large, which makes the monopropellant propulsion system difficult to satisfy the demands. Therefore, for the main engines and high-performance auxiliary engines with long life, bipropellant propulsion engines are widely used.

The bipropellant propulsion system, as the name implies, requires two propellants (oxidant and fuel) for engine operation. The two propellants are stored separately and do not mix outside the combustion chamber. A typical bipropellant propulsion engine consists of the valve and the thrust chamber, of which the thrust chamber consists of the injector, the combustion chamber and the nozzle extension. Both propellants are injected into the combustion chamber through the injector, resulting in intense combustion to generate gas with high-temperature and high-pressure gas. Then, the

10.2 Propulsion System Classification

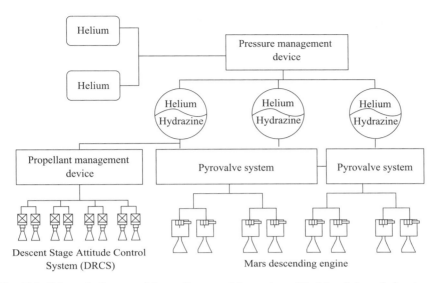

Fig. 10.3 Schematic diagram of descending propulsion system of the Mars Science Laboratory

high-temperature gas is accelerated by the nozzle to form a high-speed airflow to generate thrust.

The bipropellant used in spacecraft is mostly the combination of nitroxide and terpene fuel. Among them, propellant combinations such as N_2O_4/monomethyl hydrazine, N_2O_4/metamethyl hydrazine, green nitrous oxide (MON)/hydrazine (N_2H_4) have relatively high specific impulse, excellent vacuum ignition performance, material compatibility, storage stability and good combustion performance. This combination is the most widely used. Due to features such as limited specific impulse, high toxicity and corrosion, the development and use cost of the above propellant combination is high. Therefore, countries all over the world are intensifying the efforts to develop high-performance, low-pollution hydrogen–oxygen propellant, fluorine-based dual-component propellant, hydrogen peroxide-based bipropellant, etc.

The propulsion subsystem of the Chang'e 3 lander adopts the MMH/MON-1 bipropellant extrusion-type attitude and orbit control unified propulsion system, as shown in Fig. 10.4. The pressurized helium gas is stored in two high-pressure gas cylinders; four equal-volume metal diaphragm tanks are symmetrically arranged to store fuel and oxidant, respectively. Maximally, 2600 kg propellant can be carried in total. One 7500 N variable-thrust divert thruster is located in the center of the lander to provide the probe with the orbit control power for braking and suspending in the processes of Earth–Moon orbit transfer, braking at perilune and landing descent. Sixteen 150 N thrusters and twelve 10 N thrusters are divided into 2 branches based on same functions and installed around the lander to provide powers for attitude

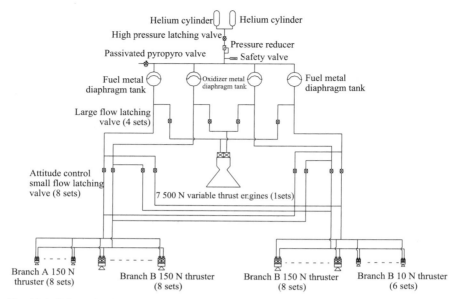

Fig. 10.4 Schematic diagram of propulsion system of the Chang'e 3 lander

control, midcourse correction, lunar orbit transfer, translation, etc., in the processes of on-orbit flight and landing descent of the probe [6].

The propulsion system consists of four parts: pressurized helium gas delivery, propellant delivery, propellant utilization and drive control circuit. The pressurized helium gas delivery part consists of the gas cylinder, pyro valve, pressure reducing valve, gas path latching valve, safety valve, pipeline, etc. and provides pressurized gas for the tank and driving gas for the pneumatic valve. The propellant delivery part consists of the tank, liquid phase latching valve, diaphragm valve, pipeline, etc. and is used to store and supply the propellant required for the engine. The propellant utilization part consists of the 7500 N throttled engine, 150 N thruster, 10 N thruster, etc. and provides the entire power for the attitude control and orbit control in the flight of the probe. And the drive control circuit consists of the propulsion power distribution box, propulsion line box, electromagnetic valves of various types, stepping motor, pressure transducer, etc. and is used to receive control commands, enable equipment power supply control and drive control of valves, motors, as well as telemetry acquisition of pressure transducers, contact switches and voltage.

3) Dual-mode propulsion technology

The dual-mode propulsion system is a new propulsion system developed since the 1980s. Generally, it adopts unitary hydrazine as fuel and nitrogen tetroxide as oxidant. The divert thruster is in bipropellant propulsion design, and unitary engine is applied for attitude control. The unitary hydrazine is used as one component of the binary engine as well as the propellant for the attitude control engine. Therefore, only

10.2 Propulsion System Classification

one storage tank is required, which greatly reduces the system mass and cost. The unitary engine applied in attitude control is developed and simple. Though the specific impulse is slightly lower than that of the binary design, the total impulse required for attitude control is relatively small, which will not have a great impact on the propellant consumption. It can also avoid the shortcomings of great plume pollution of the binary propulsion system. The dual-mode propulsion system is suitable for the spacecraft with a large difference in thrust between the divert thruster and the attitude control thruster, relatively small attitude control thruster thrust and total impulse as well as relatively long service life requirement.

The dual-mode propulsion system is an important improvement to the spacecraft propulsion system. It enables optimal design and use single monopropellant and bipropellant engine engines and organically integrates the superiorities of unitary propulsion system such as high reliability, low thrust and excellent pulsation with the advantages of binary propulsion system such as high specific impulse so that the system can be better integrated.

Since the 1990s, the dual-mode propulsion system has been widely applied. The dual-mode propulsion systems by Lockheed Martin (USA), LEROS (UK) and the Japan have been applied to the Earth orbit aircraft. Figure 10.5 shows the schematic diagram of a dual-mode propulsion system.

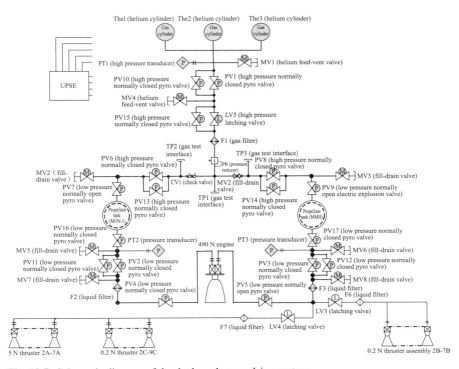

Fig. 10.5 Schematic diagram of the dual-mode propulsion system

4) Variable thrust technology [6]

With the deepening of space exploration, especially the demands for the entry exploration of extraterrestrial celestial bodies, the fixed thrust engine can no longer satisfy the mission requirements. Therefore, another advantage of the liquid engine—the thrust adjustment—is highlighted, which enables the liquid engine to adjust the thrust. In other words, the thrust can be freely changed over a wide range. This greatly improves the flexibility of the engine, making it easier to accomplish missions such as maneuvering, space docking, and orbit braking and landing onto extraterrestrial celestial body.

The 7500 N throttled engine of the Chang'e 3 lander adopts the scheme of flow positioning double regulation open-loop control system [12]. The schematic diagram of the throttled engine system is shown in Fig. 10.6. The engine adopts the extrusion-type propellant conveying mode, and the engine flow is controlled by the flow regulator. The injector adopts the flow regulating pintle injector that can independently adjust the injection area. The 7500 N throttled engine mainly consists of the oxidant shutoff valve, fuel shutoff valve, flow regulator, thrust chamber (including the pintle injector, combustion chamber and nozzle), pipeline, etc.

For the characteristics that the engine requires a wide range of thrust, the design of the pintle injector is applied, as shown in Fig. 10.7. In the process of the engine thrust change, pintle injector can automatically adjust the nozzle area of the injector according to the upstream engine flow, so that the injection pressure drop is maintained basically constant under different working conditions, thus ensuring the high performance of the engine under different thrust conditions.

To fully utilize the overall space and improve the engine nozzle area ratio, the overall layout structure of the engine has been optimized. The 7500 N throttled engine shutoff valve, flow regulator, etc., are circumferentially arranged along the injector head of the periphery of the thrust chamber to reduce the axial size of the engine within the range of the axial size of the injector. The engine is connected to the landing probe through the butt joint of the injector head and transmits the thrust, as shown in Fig. 10.8. This layout effectively shortens the axial length of the engine, reduces the spatial size, decreases the structural mass and enables the organic unification of the space size, mass and inlet pressure constraints.

10.2.2.2 Solid Propulsion Technology

The solid propulsion system is characterized by simple structure, reliable operation, small size, convenient use, instant ignition, long-term storage and excellent acceleration. However, it has relatively poor performance with the specific impulse of generally 280–300 s. The working time is quite short, and the thrust adjustment and repeated starting are relatively difficult. For mission such as landing thrust reverse deceleration, celestial surface ascent, the superiorities of the solid propulsion system that can provide large thrust are apparent as large thrust is required to reduce gravity losses.

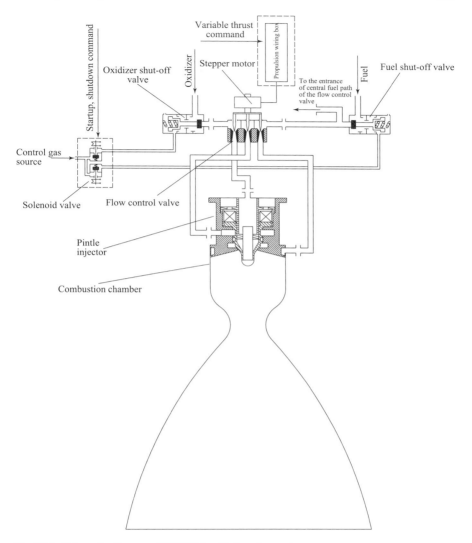

Fig. 10.6 Schematic diagram of 7500 N throttled engine system

The solid propellant is the energy and working fluid of the solid engine, which is composed of three basic components: combustion chamber, nozzle and igniter. The combustion chamber stores the solid propellant and completes the combustion of the solid propellant during operation. In the combustion process, chemical energy of the solid propellant is converted into high-temperature and high-pressure gas heat energy, and the casing can also form part of the spacecraft structure. The nozzle is generally the Laval-type nozzle consisting of the convergent section, throat and extension section. It accelerates the high-temperature and high-pressure gas in the

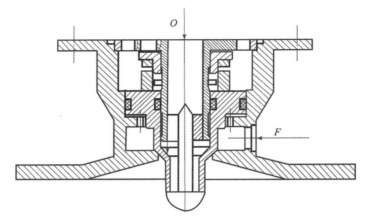

Fig. 10.7 Structure principle of the pintle injector

Fig. 10.8 Overall layout of 7,500 N throttled engine

combustion chamber through continuous expansion, converts the heat energy of the fuel gas into kinetic energy and discharges it from the nozzle at a high speed, generating thrust. The ignition device ensures the safe and reliable ignition of the propellant in the combustion chamber. Figure 10.9 shows a schematic of a typical solid rocket motor.

In the future Mars sampling return mission plan, the US Jet Propulsion Laboratory (JPL) proposed a two-stage solid-propelled Mars ascender propulsion system solution, in which the first-stage engine is Star-17 with thrust vector control and the second-stage engine is Star-13 with cold gas attitude control.

10.2 Propulsion System Classification

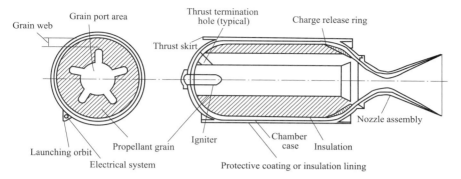

Fig. 10.9 Schematic diagram of typical solid propellant rocket engine

10.2.3 Electric Propulsion

Human space activities have greatly promoted the development of chemical propulsion technology research and application. However, as humans are exploring deeper and further, the dissatisfaction with chemical propulsion technology for its low specific impulse is also increasing. Therefore, the electric propulsion technology has drawn the attention of the humans.

Electric propulsion systems apply electrical energy generated by solar arrays or nuclear reactors to power propellants, so the propellant can produce jet velocity that is much higher than that produced by ordinary chemical propulsion. Electric propulsion system generally consists of the thruster, power processing unit (PPU), propellant feed subsystem (PFS) as well as data and control interface unit (DCIU).

The research on electric propulsion technology can be traced back to 70 years ago. So far, many forms of electric propulsion have been developed, and various electric propulsion systems have been applied to various space probes. Even now, there are still new types of electric propulsion concepts emerging. In general, electric propulsion can be broadly divided into three categories (Table 10.1).

Electrothermal type: It is similar to chemical propulsion system but applies electric energy to heat propellant and increase its enthalpy to achieve higher specific impulse. The electrothermal type mainly includes resistance thrusters (Resistojets), arc thrusters (Arcjets), microwaves electrothermal thruster (MET), etc. Its specific impulse is generally between 500 and 1200 s.

Electromagnetic type: It applies electric breakdown of the propellant to generate the plasma, and the plasma accelerates under the combined action of electric field force and magnetic field force. It mainly consists of stationary plasma thruster (SPT), magnetoplasma dynamic thruster (MPD), pulsed plasma thruster (PPT), pulse induction thrusters (PIT), etc. Generally, the specific impulse is higher than that of the electric thruster and lower than that of the electrostatic thruster and is within the range of 1000–7000 s.

Table 10.1 Electric propulsion and main technical indicators

Electric propulsion type			Power range	Specific range/s
Electric heating	Resistance thrusters (Resistojets)	Hydrazine	Several kW	500–600
		Hydrogen	Several tens of kW	900–1200
	Arc thrusters (Arcjets)	Ammonia	kW to several tens of kW	600–800
	Microwave electric thruster (MET)		100 W–10 kW	300–500
Electrostatic propulsion	Ion thruster (Ion Engines)		Several W to 100 kW	2000–10,000
	Field launch thruster (FEEP)		kW	6000
	Colloid thruster (Colloid Thruster)		kW	1000–1500
Electromagnetic propulsion	Hall thruster (HET)	Stationary plasma thruster (SPT)	Hundreds of W to tens of kW	1000–2500
		Electric thruster with anode layer (TAL)	Hundreds of W to tens of kW	1000–4000
	Magnetic plasma power thruster (MPD)	Steady state	kW	1000–4000
		Pulse	Hundreds of kW to MW	3000–7000
	Pulsed plasma thruster (PPT)		Tens of W to 100 W	1000–1500
	Pulse induction thruster (PIT)		Tens of W	3000–5000
	Electron cyclotron thruster (ECR Thruster)		kW to tens of kW	2000–4000

Electrostatic type: The propellant with lower ionization potential is selected to accelerate in the electrostatic field after ionization, and the specific impulse is the highest among the electric propulsion systems. Electrostatic type includes ion thruster (ion engines), field emission thruster (FEEP) and colloidal thruster (colloid thruster). The ion thruster has a relatively high specific impulse within the range of 1000–10,000 s.

The electric propulsion technology is being increasingly widely applied, and the Hall thruster and ion thruster have gradually become the prevailing trend. The Hall thruster is mainly represented by stationary plasma thruster (SPT). Such two types of electric propulsion technology will be discussed in a detailed manner below.

10.2.3.1 Hall Electric Propulsion

The working principle of Hall electric propulsion system: The gas (usually helium) is ionized by the hollow cathode through the anode injection into the circular passage;

10.2 Propulsion System Classification

the ions are accelerated by the negative cathode electrostatic field and only slightly deflected under the action of the radial magnetic field. However, the electrons are strongly magnetized and forced to deflect deflection, while axial diffusion is quite slow (Fig. 10.10). In the acceleration channel, the movement speed of ions and plasma is much smaller than that of the electrons. Under the action of electromagnetic field, the deflection movement of electrons in the radial direction is much larger than the diffusion movement along the axial direction. And the radial current density is much larger than that of the axial current. Therefore, the motion of the electrons constitutes an approximate loop, which is equivalent to an accelerator that accelerates the ions; that is, the ions are accelerated by the electric field in the plasma. The acceleration of ions in the acceleration channel is quasi-electrostatic. However, the discharge of the plasma is semi-sinusoidal. Therefore, this kind of engine is also called stationary plasma engine (SPT). The radial current density of the electrons is proportional to that of the axial current.

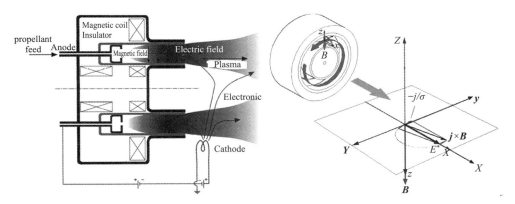

Fig. 10.10 Working principle of SPT

If the accelerating voltage is one order of magnitude smaller than that of the conventional ion engine, the ion current density of SPT is two orders of magnitude higher. In order to ensure the non-composite ion flow, the gas density is required to be relatively low, which is just the reason that Hall thruster is larger than the equivalent power arc thruster. However, compared with the ion thruster, its structure is much more compact.

The plume of SPT thruster is highly ionized, without cracking loss. The loss of ionization process is more or less offset by relatively high operating voltage. The gas flow of the hollow cathode causes a specific loss of about 5%, and the electrode loss is relatively small. The efficiency of the SPT thruster is 45%–55%,which increases slowly with the increase of the working voltage. Because the fluid in the discharge chamber mainly consists of ions, the flow rate of the control gas can control the discharge current. However, the flow rate is too small (generally only a few milligrams per second), which makes the gas flow control very difficult. It is usually realized by thermally changing the viscosity of the gas supplied to the capillary or by high-precision electric valve. Compared with the arc thruster, the PPU of the Hall thruster is more complex and heavier.

Main factors affecting SPT performance:

(1) Ionization efficiency of working fluid. The ionization efficiency is correlated to the flow density of the propellant and the magnetic flux of the magnetic field. By controlling these two parameters, it is possible to maximize the ionization efficiency.
(2) Acceleration direction of ions. The structure of the improved magnetic field can control the intensity distribution of the accelerating electric field, thereby minimizing the radial component of the accelerating electric field and improving the efficiency of the SPT.

The main factor affecting the service life of SPT is the wall corrosion of the discharge chamber caused by ion splash. Plume pollution of SPT is a major disadvantage of SPT in the system. Therefore, importance must be attached to the layout of SPT system to prevent the negative impacts on the normal operation of the probe. Besides, the operation of the thruster can bring deep current fluctuation, which will lead to all kinds of high-frequency plasma oscillations and then affect radio communications.

The main propulsion of the European SMART-1 probe is Hall electric propulsion system. The system consists of the Xe supply system, electric energy supply system and thruster, digital interface and communication system with a dry mass which is 29 kg.

The Xe supply system consists of the Xe tank, pressure regulator unit (BPRU) and Xe flow controller (XFC). The Xe tank is a cylindrical aluminum liner composite gas cylinder with a volume of 49L and can store 82.5 kg Xe propellant under the pressure of 15 MPa and density of 1.7 g/cm^3. The maximum working temperature is 50 °C.

The power supply system consists of power processing unit (PPU/TSU), filtering unit (FU), etc. The PPU/TSU has a power range of 462–1190 W and a total of 117 power adjustment gears.

The digital interface and communication system is mainly the pressure regulating circuit board (PRE board) and is responsible for processing and transmitting commands and telemetry signals of regulating Xe pressure, as well as sending the command of the thruster, PPU/TSU and telemetry signals to the digital interface of the PPU/TSU for further processing by the microprocessor of the PPU/TSU.

The working process is that the Xe rated gas pressure is reduced to 0.2 MPa by BPRU, and then Xe flows into the Xe flow controller; BPRU's two series-connected pressure regulating solenoid valves control the on/off state of the Xe gas circuit in the form of a switch to adjust the pressure. The valve is controlled by the PRE board with a simple and reliable closed-loop control program; the Xe flow controller applies the thruster discharge current to achieve closed-loop control under the control of the PPU/TSU and maintain stable Xe propellant flow into the thruster, thereby enabling stable system operation.

10.2.3.2 Ion Electric Propulsion

At present, ion thrusters are basically of Kaufman type, which mainly consists of the hollow cathode, discharge chamber surrounded by the magnet, coaxial anode, accelerating grid, neutralizer and helium supply system and has a complicated structure and relatively great number of parts.

The working principle of the ion thruster is shown in Fig. 10.11. The propellant gas (usually helium) is delivered to the cylindrical discharge chamber through the axial hollow cathode and the bypass distributor. The DC discharges between the cathode and the concentric cylindrical anode ionize the gas and increase the ionization efficiency by the deflected and slightly expanded magnetic field in the ionization chamber. The highly ionized plasma drifts toward a set of perforated grids at the bottom end of the discharge chamber, and the positive ions are extracted by the electric field between the grids and then accelerated. The grid spacing is quite small (typically 0.5–1 mm). However, the voltage between them is quite high (over 1100 V), and the ions are accelerated to a quite high speed within a short distance (generally 30–50 km/s). The positive ions from the ion beam are neutralized by electrons released by the external neutralizer.

The magnetic field of the ion thruster is not as critical as that of the Hall thruster and serves to improve the ionization effect. It is generally the permanent magnet and can be replaced by the coil. Acceleration grid generally has two layers and sometimes three layers. The outermost layer is the negative potential as a protection to prevent

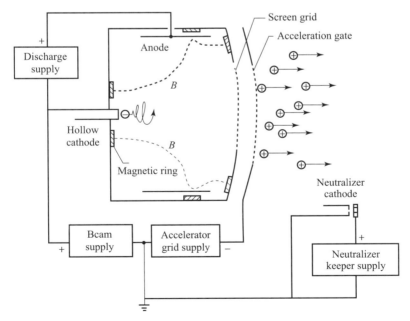

Fig. 10.11 Schematic diagram of working principle of ion thruster

electron reflow and reduce the splash loss caused by the ion explosion generated during charge exchange on the accelerator surface.

Due to the limitation of space charge, the size of ion thruster is larger than that of the Hall thruster with the same thrust. The advantages of the ion thruster include long life (the current verified is up to 10,000 h), high efficiency, large specific impulse (the specific impulse of the ion thruster can easily reach 3000 s and the efficiency is up to 65%).

Therefore, the ion thruster is particularly suitable for applications with high specific punching requirements and long life requirements, especially for north–south position holding missions and deep space exploration missions of geosynchronous orbiting spacecraft.

Disadvantages of the ion thruster: The power system of the ion engine is complicated by the simultaneous supply of the grid, field coil, hollow cathode, neutralizer and control system; the supply of helium is generally to the ionization chamber, the cathode and the neutralizer in a separate manner. Therefore, each path can be precisely controlled. However, the disadvantage is the increased complexity of the propellant storage system.

According to the structure and working principle of the ion thruster, the main factors affecting the performance of the ion thruster mainly include thrust feature size, magnetic field distribution and strength, thruster thermal stability, power supply and processing system performance.

The Dawn ion electric propulsion system consists of two digital control and interface units (DCIUs), two power processing units (PPUs), one xenon supply system (XFS), three thrust vector adjustment mechanisms (TGAs) as well as three 30 cm ion thrusters [7, 8]. To match the output power of the solar array under different solar distance conditions, the input power of the ion electric propulsion system also varies within a certain range, and the operating power range is designed to be 470–2500 W.

10.2.4 New Concept Propulsions

10.2.4.1 Nuclear Energy Propulsion [9]

In the absence of solar energy and other external energy which is not sensitive to space radiation, nuclear propulsion can provide energy for a long time. Therefore, nuclear propulsion technology has an irreplaceable advantage for deep space exploration missions. Nuclear propulsion technology is the general term of propulsion technology that converts nuclear energy into kinetic energy. There are two main forms: nuclear power propulsion and nuclear thermal propulsion.

The core of nuclear power propulsion system is the organic combination of the nuclear energy system and the electric propulsion system; that is, the system that uses nuclear energy to drive electric propulsion work. Obviously, this drive is indirect because of the need to convert nuclear energy and electrical energy. Figure 10.12 shows the basic block diagram of the nuclear power propulsion system. Contents

10.2 Propulsion System Classification

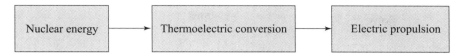

Fig. 10.12 Basic components of a nuclear power propulsion system

related to nuclear power conversion are discussed in a detailed manner in Chap. 11. The content related to electric propulsion is discussed in the previous section of this chapter.

Nuclear thermal propulsion technology uses fission energy to heat the working fluid and then the heated high-temperature and high-pressure working fluid which is injected through the engine nozzle to generate thrust (Fig. 10.13). It features large thrust, long life, and high specific impulse and is not easy to be affected by the external environment. The specific impulse of the nuclear heat engine with hydrogen working fluid can reach 1000 s.

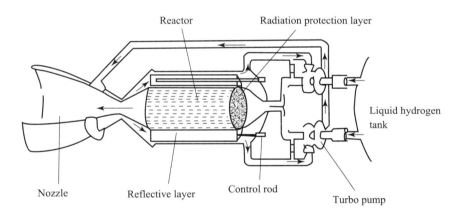

Fig. 10.13 Nuclear thermal propulsion system

10.2.4.2 Space Beam Flow Energy Propulsion [10]

The concept of space beam flow energy propulsion comes from the model of small ball collision, where many small balls are accelerated by one source before acting on the probe. The probe will then get the momentum of the balls. The concepts of laser beam current and microwave beam current belong to the space beam propulsion category. The laser beam current is taken as an example to introduce the principle of such propulsion technology.

The principle of laser beam propulsion is the same as that of chemical propulsion, which is achieved by recoil. Chemical propulsion relies on the combustion of chemical fuels, and the laser beam current can be propelled by the supersonic jet of plasma generated by the laser beam irradiating the surface of the target. Specific process: focus the strong laser beam on the target of the aircraft, and the laser energy is absorbed by the target. When the target is heated to a temperature of several thousand or even tens of thousands of degrees, the target begins to melt or even vaporize, the gasified target material and the gas on the surface of the target further absorb the laser energy, thereby being ionized to form the ion region. The high-temperature, high-pressure plasma expands to form a laser-supported detonation wave, thereby generating an impulse effect on the target surface, as shown in Fig. 10.14. For these propulsion modes, corresponding spatial acceleration sources should be established.

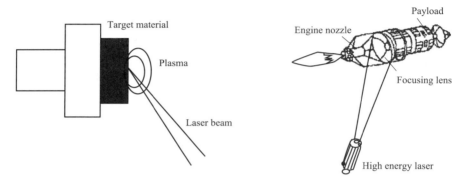

Fig. 10.14 Schematic diagram of "ablative mode" laser beam current propulsion

10.2.4.3 Solar Sail Propulsion

Solar sail is a device that uses a large number of photons generated by the Sun to apply light pressure to the light and thin reflective layer of the sail, so as to generate thrust. It belongs to the propellant-free propulsion technology. The thrust of the solar sail is proportional to the surface area of the sail. Therefore, in order to obtain the ideal acceleration, the solar sail must have a large enough area and a small enough mass. Therefore, the application of solar sail propulsion, breakthroughs must be made on key technologies such as large-scale ultra-thin sail manufacturing, lightweight high-strength support structure construction, small-volume packaging of large sails, lightweight deployment structure and sail/probe control.

10.2.4.4 Magnetic Field Sail and Plasma Sail Propulsion

The magnetic field sail is a propulsion device that applies a magnetic field around the probe to obtain thrust. Plasma sail is a device that uses the plasma of the solar wind to obtain thrust. As the magnetic field around the probe or the plasma field is uncertain, both propulsion modes should be provided with a set of propulsion systems as auxiliary.

10.2.4.5 Solar Thermal Propulsion [11]

Solar thermal propulsion (STP) is the process of converting the focused solar energy into thermal energy, heating the propellant with thermal energy, collecting solar radiation from the concentrator and then sending the collected solar energy to the optical waveguide transmission line. It sends high-intensity solar radiation to the heat collector, effectively generating high-performance thrust. The core technologies include the solar energy collection system and light energy delivery technology.

10.3 Design and Verification of Deep Space Exploration Propulsion System

As the actuator for the orbit and attitude control system of the probe, the propulsion system has a significant impact on the overall performance of the mission in terms of its performance, reliability and service life. The structural hardware of the propulsion system, together with the propellant it carries, occupies a considerable part of the mass of the probe. The propulsion system of the deep space probe is designed with greater total impulse requirement. Therefore, the proportion of the propulsion system will further increase. Therefore, the proportion of the probe payload can be significantly increased by selecting the appropriate propulsion system, improving the performance of the propulsion system, reducing the dry weight of the propulsion system and reducing the carrying capacity of the propellant.

This chapter introduces the design and verification of the propulsion system in a detailed manner. Generally, the design process of the propulsion system starting with the mission analysis of the deep space exploration mission, obtains related key parameters of the design, initially selects possible propulsion methods and then promotes the selected propulsion. The method is to carry out multi-program design and comprehensive comparison, determine the selection of propulsion system, carry out further refinement design and finally carry out systematic test verification of the propulsion subsystem [12, 13].

10.3.1 Mission Analysis

Generally, propulsion system mission analysis includes mission function requirement analysis, design constraint analysis, mechanical environment analysis and space environment analysis. Through mission analysis, the key parameters and system design boundaries that affect system design can be determined to provide input for the selection of subsequent propulsion system.

10.3.1.1 Mission Function Requirements Analysis

Deep space exploration started from the moon and was then gradually expanded to the Mars, the Venus, the Sun and the interplanetary space. The exploration method also evolved from the initial flyby, orbiting exploration to hard landing exploration, soft-landing in-position exploration, unmanned sampling return and manned landing, etc. Different exploration methods have placed different requirements for the propulsion systems, and such requirements can be summarized into the following categories:

(1) The long-distance over-the-fly and surround exploration missions, which require the propulsion system with high performance. On the basis of ensuring the probe speed increment requirements, the dry weight and propellant mass of the system should be minimized to realize the goals such as reducing the overall mass of the probe, improving the proportion of payload and reducing the cost of launching.
(2) Soft-landing exploration missions, especially soft landings with gravitational stars, generally apply the combination of powered descending or "pneumatic deceleration + power descending". The reduced power requires the propulsion system to provide greater thrust to satisfy the need to quickly reduce probe speed. In addition, the propulsion system should also be provided with variable thrust adjustment capability to accommodate changes in probe mass due to propellant consumption during the entry process.
(3) Unmanned sampling and returning technology, especially the sampling and returning mission of gravity celestial bodies, in which the propulsion system is required to provide the speed increase required for the powered descending and ascending from the surface of the star body. Therefore, the propulsion system should be provided with features such as large thrust, high specific impulse, relatively small system mass and relatively small system size.
(4) Manned landing exploration missions, in which the size of the probe must be larger than that of the unmanned soft-landing system due to the need to carry the life and environmental protection system required by the astronauts. Therefore, compared with the unmanned soft-landing missions, a significant characteristic lies in that the thrust and total punch of the propulsion system have been greatly improved, raising higher requirements on the performance of the propulsion system. In addition, the reliability requirements of the system have also been significantly improved for manned missions.

Through the analysis of the function demands of the mission, a clear understanding of the scale of the propulsion system can be established, and key parameters such as speed increment demand and engine thrust demand can be initially determined, so as to provide the necessary basis for the preliminary screening of the subsequent propulsion system.

10.3 Design and Verification of Deep Space Exploration Propulsion System

10.3.1.2 Design Constraint Analysis

As an important part of the probe, the propulsion system has related interfaces and design boundaries with other subsystems on the probe. To comb such constraints in a systematic manner is the key to further design. The design constraints of the propulsion system are mainly reflected in the following aspects.

1) Structural mass constraints

The launching capacity of the carrier rocket directly limits the total mass of the spacecraft, and the mass of the propulsion system generally accounts for about 60% of the total mass of the probe. Therefore, reducing the mass of the propulsion system is an effective approach to reduce the total probe. Therefore, in the overall design, there are strict constraints on the mass of the propulsion system. In this respect, on one hand, the propulsion system should increase the specific impulse of the engine to reduce the carrying capacity of the propellant; on the other hand, the propulsion system should also reduce the structural mass of the propulsion system by adopting new technologies, new materials and system optimization.

2) Configuration layout constraints

The tank, cylinder and divert thruster of the propulsion system are all large-sized components with special shapes (spherical, spherical-column, bell-shaped, etc.). And most of the space of the probe is occupied by such components of the propulsion system, which bears relatively great influence on the space layout efficiency of the probe. Therefore, it is necessary to strictly restrict the configuration of the propulsion system. There are also some special missions (such as the atmosphere entry mission) in which the propulsion system is wrapped inside the heat protection device. Therefore, the propulsion system configuration is limited by the heat protection structure. All these conditions require comprehensive consideration on the layout design of the propulsion system, and if necessary, the design of the probe should be integrated.

3) Energy condition constraints

Generally, probes apply solar arrays to provide energy. The area of the solar arrays is limited, and the power generation amount of solar cells is closely related to the distance of the probe. Therefore, the energy available for the probe is quite limited. Even though some probes apply nuclear sources, the power generation is also quite limited. Therefore, in the design of the propulsion system, it is necessary to detect the constraints of the mission energy and configure a reasonable propulsion approach.

10.3.1.3 Flight Environment Analysis

Taking the Mars landing exploration as an example, from launch to mission completion, the system is expected to experience the mechanical environment of the rocket launching stage, the space environment of the on-orbit flight stage, the martian atmospheric environment, the mechanical environment of the landing descending stage, the landing impact environment, the surface survival environment of the Mars after landing, etc. Such environments can be further divided into mechanical environment, spatial particle environment, atmospheric environment, temperature environment, etc. The complex environment directly affects the design of the system. For example, the mechanical environment affects the strength design of the system components, the space particle environment affects the propulsion system circuit design and material selection, the atmospheric environment of the exploration target directly affects the performance of the engine, and the ambient temperature environment affects the design of pressure vessels such as tanks and gas cylinders. These environmental factors should be divided into mission sections according to the detailed exploration missions, and the working state of the propulsion system, the environmental conditions and the impacts thereof in each mission section should be analyzed in detail to guide the propulsion system design.

Take the Mars landing exploration mission as an example. Table 10.2 gives the mission analysis table for a Mars landing rover propulsion system. Through mission analysis, the following design parameters can be basically determined:

(1) Determine the propulsion system functions required by the probe, such as orbital maneuver, orbit maintenance, attitude control and power deceleration;
(2) Determine the budget of Δv and the constraints on the thrust level, such as orbital maneuver, orbital maintenance and powered descending;
(3) Determine the total impulse of the attitude control, the required thrust level, duty cycle, total number of cycles and mission life required for the control function;
(4) Determine the design constraints of the propulsion system;
(5) Determine the flight environment conditions.

10.3.2 Propulsion System Selection

Section 10.2 describes different types of propulsion systems with different advantages and disadvantages. According to the main contents and outputs of the mission analysis, the types of propulsion systems can be initially screened to determine several feasible propulsion systems. If the speed increment demand is great, the possibility of cold gas propulsion should be basically eliminated, and selection should be made among chemical bipropellant, dual-mode propulsion and electric propulsion with relatively high specific impulses; if the speed increment is small, cold gas, monopropellant propulsion, etc., can be selected. The main work items at this stage are as follows:

(1) Whether the orbit control and attitude control system is unified;
(2) Selection of large thrust or small thrust;
(3) Selection of solid, liquid or electric propulsion mode.

10.3 Design and Verification of Deep Space Exploration Propulsion System

Table 10.2 Mission analysis of the mars landing rover advance system

No.	Flight stage	Working status of the propulsion subsystem	Flight environment condition	System function requirement
1	Launch section	The propulsion system is powered up, monitor system status	Carry rocket launch environmental conditions	Power-up of the sensors of the landing rover propulsion system at stages of launch, Earth–Mars transfer, Mars capture, Mars parking, the monitoring of the system states; during the Earth–Mars transfer flight, after the landing rover is powered up and the test state is completed, the propulsion system should accomplish the state test work and judge the health status of the propulsion system
2	Earth–Mars transfer		Earth space environmental conditions	
3	Mars capture		Mars space environmental conditions	
4	Mars parking		Mars space environmental conditions	
5	Off-orbit landing	Complete the working state setting, complete propulsion system pressurization, propellant filling, etc.	Mars space environmental conditions	The propellant is stored in the tank for a long period from the launch stage of the probe to the separation of the landing rover. The propulsion system should meet the long-term storage functions of the propellant and effective and reliable isolation of the propellant should be guaranteed. Check the health status of the propulsion system before separation, complete the working state setting of the propulsion system, and complete the actions of the pipeline exhaust, tank pressurization and pipeline filling

(continued)

Table 10.2 (continued)

No.	Flight stage	Working status of the propulsion subsystem	Flight environment condition	System function requirement
		Complete the separation attitude correction, complete the attitude control during the Mars entry process, complete the Mars braking, hovering translation	Mars atmospheric conditions Mars atmospheric entry aerodynamic heating conditions; inflow conditions of the Mars atmosphere Mars dust conditions	After the landing rover is separated from the orbiter, the propulsion system should provide power support for the spin-off and de-rotation of the landing rover. The attitude control torque is required for the landing rover to ensure that the angle of attack meets the requirements After the landing rover enters the martian atmosphere, it adopts pneumatic deceleration and brake parachute to decelerate. The landing rover throws out the heat-resistant outsole, and the propulsion system is exposed to the martian atmosphere, and it is necessary to adapt to such high-speed flow environment The parachute patrol and the landing rovers tarts to control autonomously. The system should provide reliable ignition under the inflow condition so as to enable orbit control and attitude control thrust for the powered descending of the landing rover When the landing rover is relatively close to the surface of Mars, hovering obstacle avoidance is required. The propulsion system should provide orbit control thrust for the landing rover to hover and provide horizontal orbital thrust for horizontal maneuvering
	Mars landing		Landing impact conditions Mars surface temperature conditions	The system should guarantee the integrity of the system structure under impact conditions The working medium such as system propellant should adapt to the temperature condition of Mars or have the ability of passivation. In the passivation process, the attitude of the probe should be ensured and no effects should be made on the surrounding environment of the probe

(continued)

10.3 Design and Verification of Deep Space Exploration Propulsion System

Table 10.2 (continued)

No.	Flight stage	Working status of the propulsion subsystem	Flight environment condition	System function requirement
6	Scientific exploration	No demand		

A targeted system design for the preselected feasible modes of propulsion (the content of the propulsion system design will be described in a detailed Sect. 10.3.3), and a comprehensive comparison will be made on various propulsion system solutions in terms of quality, reliability, cost and other factors. The main work items at this stage are as follows:

(1) Estimate the key parameters of each solution:
   a. Attitude and divert thruster configuration;
   b. Effective specific impulse of orbit and attitude control;
   c. Mass of propellant;
   d. Propellant tank volume and volume of extruded gas;
(2) Determine the system configuration and list the equipment.
(3) Design a circuit management scheme for the propulsion system.
(4) Calculate the dry weight of the system based on the propulsion system.
(5) Carry out comprehensive comparison of multiple solutions.

In the design of the probe, it is necessary to consider factors such as weight, performance, volume, restrictions on factors such as operation, reliability, cost, risk, schedule or technical feasibility, interface relationships. For a particular exploration mission, multiple rounds of iterations between various features of the propulsion system and the user requirements should be conducted to optimize the propulsion plan.

10.3.3 Scheme Design

Due to different types of propulsion systems, there will be some differences in the design of propulsion systems, but the overall design process is basically the same. This chapter takes the chemical propulsion system as an example to introduce the design of the propulsion system and gives design considerations and design methods.

10.3.3.1 Propellant Selection

Propellant selection is a comprehensive issue that is generally determined by prior demonstration. Various factors should be considered such as the physical properties of the propellant, mixing ratio, vacuum specific impulse performance, density specific impulse performance, cooling performance, vacuum ignition performance, storability, toxicity, safety and technical maturity. Key considerations include:

1) Engine specific impulse performance

The selection of propellant with high vacuum specific impulse and high density specific impulse is an important factor to reduce propellant carrying capacity, reduce

10.3 Design and Verification of Deep Space Exploration Propulsion System

tank volume and structural weight of the system. For the lunar lander propulsion system, the general total impulse is relatively large, and the structure is relatively light. The bipropellant propulsion system is often applied. The bipropellant N_2O_4 (or MON-1, MON-3)/MMH spontaneous combustion propellant combination is the main component. However, in recent years, with the increasing demand for high specific impulse and non-toxicity, propellant technologies such as LO_x/LH_2 and LO_x/HC (LCH_4, LC_3H_8, RP-1) have been rapidly developed.

2) Long-term space storability

The long-term space storage of propellant is also one of the important factors affecting system design. Based on the comprehensive analysis of physical properties such as propellant freezing point, boiling point, density, saturated vapor pressure and engine technology maturity, the selection of long-term storable propellant in conventional space can simplify the thermal control facilities of the propulsion system, reduce the structural mass of the system and reduce the difficulty of development technology. Taking the ascent mission of Mars as an example, the ascender is required to stay on the surface of Mars for a long period. Therefore, the propellant should adapt to the low-temperature conditions of Mars to reduce the demand for thermal control design and energy.

3) Cooling performance of the engine

For the dual-component propellant radiant cooling engine, as the combustion temperature is up to 3000 °C, the existing metal material cannot stand such high temperature, and the combustion chamber wall should also be cooled. Generally, a small amount of the fuel is injected from the injector to the inner wall of the combustion chamber for wall cooling. This method of cooling is called edge zone liquid film cooling. The liquid film cooling working medium generally adopts fuel components to reduce the oxidizing atmosphere near the wall surface of the combustion chamber, and to avoid oxidation of metal materials and coatings at high temperatures. The selection of the fuel liquid film cooling working medium should be based on the principle of wide range of liquid, large latent heat of vaporization, large specific heat capacity, small self-decomposition at high temperature, low thermal conductivity is and high density.

4) Uniformity of propellant density ratio and mixing ratio

In the design of radiant-cooled engine, the central zone mixing ratio is generally close to the theoretical optimum mixing ratio of the propellant, so as to obtain high specific impulse performance, while the side zone mixing ratio is relatively low to ensure effective cooling of the edge zone. Therefore, the specific impulse performance and cooling performance should be balanced to determine the operating mix ratio of the engine. In the selection of propellant, it is necessary to unify the ratio of the oxidant with the density of the fuel and the mixing ratio, so that the oxidant and the fuel tank can be designed in equal volume, and the tank type can be unified to easy assembly and reduce the development cost.

10.3.3.2 Engine Configuration Design

The engine configuration of the probe largely depends on the demands of the specific missions. The engines on the probe are generally classified into divert thruster and attitude control thruster according to the functions, of which the orbital divert thruster is mainly used for the orbital maneuver of the probe. Therefore, it must be stable for long-term operation. The attitude control thruster is mainly used for probe attitude control. It usually works in pulse mode, the shortest pulse can be just several milliseconds, and the pulse life is high. The two engines differ not only in function and thrust size, but also in propulsion and types, such as dual-mode propulsion system.

The selection of the divert thruster for deep space probes is different from that of the general Earth orbit spacecraft. It should not only satisfy the thrust demands of the orbital maneuvering, the demand for the number of times, the life expectancy, etc., but also consider the requirements of the divert thruster. For example, demands of landing mission of the gravitational star on maximum thrust and variable thrust of the engine, as well as the requirements for the working back pressure of the engine by the atmospheric celestial body entry mission.

The thrust of the divert thruster is generally large. Because the thrust vector does not have a large interference torque caused by the centroid of the probe, in the selection of the divert thruster, it is necessary to determine the thrust vector adjustment of the divert thruster. General probes can adopt the ground heat standard + fine adjustment mode. For some large engines, the universal frame adjustment can be applied, such as the Apollo descending stage divert thruster.

The selection of attitude control engine is similar to that of the near-Earth spacecraft. Considering factors such as matching with the divert thruster, the requirements of attitude control accuracy, the attitude stability and probe mass characteristics, the thrust and pulse equivalent of the engine are determined. The requirements of mission on orbit accuracy are analyzed to determine whether torque control or force coupled control should be applied for the attitude control system. The life requirements of the attitude control system are analyzed to determine the parameters such as the number of engine pulses and service life. Finally, comprehensive consideration is made on factors such as layout constraints and plume effects to determine the layout of the thrusters.

10.3.3.3 Design of Propellant Storage and Transportation Scheme

1) Propellant budget

Upon the completion of the initial orbit design of the mission, the speed increment Δv required for the mission is output. After the propellant selection and engine configuration design of the propulsion system are completed, the specific impulse of the engine is also determined accordingly. In this case, the propellant mass budget should be carried out, which is a key mission in propulsion system design.

10.3 Design and Verification of Deep Space Exploration Propulsion System

The relationship between the given speed increment and propellant consumption can be expressed by Eq. (10.1):

$$\Delta v = I_{sp} \ln\left[\frac{m_0}{(m_0 - m_1)}\right] = I_{sp} \ln\left(\frac{m_0}{m_f}\right)$$
$$m_p = m_f\left[\exp\left(\frac{\Delta v}{I_{sp}}\right) - 1\right] = m_0\left[1 - \exp\left(-\frac{\Delta v}{I_{sp}}\right)\right] \quad (10.1)$$

where m_p is the propellant mass-consumed to achieve a given speed increment Δv; m_0 is the initial mass of the probe; m_f is the final mass of the probe; I_{sp} is the specific impulse of the engine.

The budgeting of the propellant mainly includes the following items:

(1) Propellant mass consumed by orbit control;
(2) Propulsion mass consumed by attitude control (including attitude control, spin-off, attitude maneuver, etc.);
(3) Reserved propellant dose: generally about 10% of nominal propulsion mass (orbital consumption + attitude control consumption);
(4) Residual propellant dose: The propellant mass that cannot be used in the propulsion system, including tank residue, pipeline residue, etc., generally 1%–2% of the total propellant.

2) Tank scheme selection

The function of the tank is to store and supply the propellant that satisfies the demands of the liquid propulsion system. When the system is in operation, the propellant supply cannot carry the pressurized gas and should satisfy the flow requirements of the propulsion system.

In microgravity and non-gravity environment, liquid gas and pressurized gas exist in the same tank, which determines that it is quite different from ordinary pressure vessels. In the microgravity environment, the gas–liquid interface is no longer horizontal, and the distribution of the liquid can be discontinuous. In case of external disturbance, the position of the liquid is not stable, and phenomena such as migration, collision and gas–liquid mixing can occur. If special measures are not taken, there is no guarantee that the liquid will always cover the liquid outlet, and it is impossible to discharge all the liquid without entraining the gas. To satisfy the above requirements, a propellant management device is necessary.

The main forms of common propellant tanks include centrifugal tanks, bellows tanks, diaphragm tanks and surface tension tanks.

Centrifugal tank is generally applied in spin-type spacecraft and applies rotation to sink the liquid to achieve gas–liquid separation and propellant emissions.

The bellow-type tank separates the liquid from the pressurized gas by a retractable bellow, and the piston is moved to extrusion and discharge the liquid. The tank is mainly applied for some refillable propulsion systems, as shown in Fig. 10.15.

The diaphragm tank divides the tank into a liquid portion and a gas portion by means of the diaphragm. The diaphragm moves under pressure, and the propellant

Fig. 10.15 Schematic diagram of the bellow tank

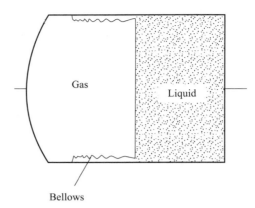

is extruded into the tank until the tank is emptied. As the gas and liquid are isolated, the tank can operate in any acceleration environment at any flow rate. At present, there are two types of diaphragms applied: elastic rubber diaphragms and aluminum metal diaphragms. The schematic diagram is shown in Fig. 10.16.

The surface tension tank separates the gas and liquid with the propellant management device (PMD) and delivers the untrapped propellant to the outlet. The working principle of PMD is the liquid flow principle driven by surface tension or the principle of surface tension liquid storage of capillary web. Surface tension tanks are the most widely applied propellant tanks at present. The schematic diagram is shown in Fig. 10.17.

In the design of a specific propulsion system, the type of tank can be screened according to the characteristics of the mission. The following factors should be considered in the design:

(1) Total mass of propellant: The total mass of propellant determines the size of the tank. If the tank size is excessively large, the bellow type tank and the diaphragm tank (especially the metal diaphragm tank) are relatively difficult to manufacture, and a surface tension tank should be selected.

Fig. 10.16 Schematic diagram of the diaphragm tank

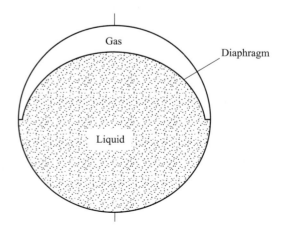

10.3 Design and Verification of Deep Space Exploration Propulsion System

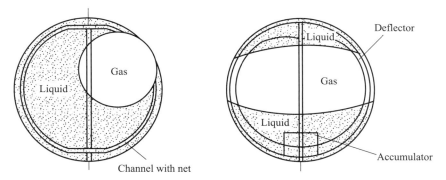

Fig. 10.17 Schematic diagram of surface tension tank

(2) Propellant sloshing effect: In the bellow type tank and the diaphragm tank, the controlling device is closely attached to the propellant without free pages. Therefore, there is little effect of liquid sloshing. It is suitable for probes with relatively high attitude control requirements.

(3) Compatibility requirements: The tank and propellant should be compatible for a long period. Therefore, the type of propellant directly determines the type of material used in the tank, and the tank of the rubber diaphragm is limited.

After selecting the tank type, the detailed design of the tank is carried out, in which the tank volume design mainly depends on the propellant consumption, tank discharge efficiency, volume of the tank controlling device structure, propellant temperature, filling volume of the pipeline, mixing ratio deviation. In addition, a safe air cushion volume is required.

The working pressure of the tank mainly depends on the inlet control pressure of the divert thruster/attitude control thruster, the flow resistance of the propellant conveying pipeline (including valves, filters, flow resistance adjusting components, etc.) and the flow resistance of the tank. The characteristic of the rise of the static pressure of the gas path reducing valve should also be considered. For example, the rated working pressure of the Chang'e 3 lander propulsion system tank is about 1.90 MPa. Considering the static pressure increase characteristic of the pressure reducing valve, the design working pressure of the tank is selected according to 2.0 MPa.

3) Propellant delivery mode selection

Generally, two methods are applied for system controlling: one is to pump the propellant to the combustion chamber, and the pump can be driven by the turbine, the gas booster or directly by the electric motor. This method is often applied in rocket propulsion systems, in some probes with quite large engine thrust or high total impulse, pump pressure controlling approach can also be applied, as in the case of the descending engines of the Soviet Moon series of probes. There is also an approach of squeezing a propellant from a storage container to an engine combustion chamber with high-pressure gas, which is known as extrusion propellant supply and is commonly applied in propulsion systems for probes. This method is simple, and the mass of the entire propulsion system is also much smaller. In addition, the system is simple and can improve the reliability of the system.

There are two kinds of common extrusion systems: constant pressure extrusion and pressure drop extrusion. A typical system schematic is shown in Fig. 10.18. The constant pressure extrusion system provides a stable pressure to pressurize the tank through the high-pressure gas in the high-pressure gas cylinder bypassing the pressure reducing device. The output pressure of the system is stable, guaranteeing the stable performance of the engine. The pressure drop system pressurizes the tank through the pre-pressurized gas in the tank. With the consumption of propellant, the pressure of the system drops, and the inlet pressure of the engine decreases. The engine performance also decreases accordingly. This system requires a strong adaptability of the engine.

When the engine thrust is large, it is difficult for the extrusion system to meet the requirement of the continuous supply of the propellant. Therefore, the pump-type supply method is to be considered. In addition, in some systems with large total impulse demand and high engine chamber pressure, the size of the propellant tank is relatively large accordingly, and the large-sized tank can satisfy the requirements

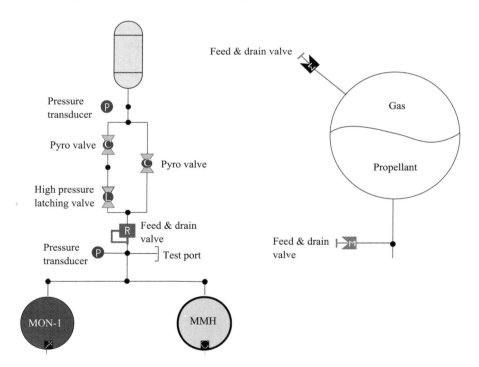

Fig. 10.18 Typical system schematic

of constant pressure, and the mass of the tank structure will increase. In this case, the pump system should be considered to reduce the pressure demands of large-size tanks, thereby reducing the mass of the tank and effectively reducing the mass of the system.

The selection of delivery system depends on factors such as mission, size, propulsion system mass, thrust size, length of work, engine combustion chamber pressure, etc. In addition, other factors such as reliability should also be considered. Therefore, when selecting the delivery mode of the propulsion system, repeated comparisons should be made to determine the reasonable supply plan.

4) Propellant management design

Propellant delivery and management designs mainly include divert thruster management, attitude control thruster management and propellant supply system status monitoring. To reduce the interaction between the divert thruster and the attitude control thruster, the divert thruster and the attitude control thruster are generally independently managed by the latching valve. The attitude control path latching valve directly adopts the electromagnetic latching valve, while the orbit control path latching valve can control the engine to push the power level according to the orbit. The electromagnetic latching valve or electromagnetic pneumatic latching valve can be applied. When the thrust level is relatively large (over 3000 N), the electromagnetic pneumatic latching valve is adopted (the electromagnetic latching valve structure is heavy). The attitude control thruster is usually configured with two branches for redundancy. The main and standby branches can be crossed, and single-branch isolation can also be enabled in case of failure.

10.3.3.4 Design of Pressurized Gas Path

1) Pressurized gas selection

The selection of pressurized gas mainly considers the factors of gas activity, relative molecular mass, compression characteristics and throttling temperature characteristics. Generally, the inert gas with relatively small molecular masses is selected as pressurized gases, and the gas with a relatively small molecular mass is required to provide a unit pressure. Small mass is beneficial to reduce the mass of the system. For the surface tension tank where gas and liquid coexist, it is also necessary to consider the compatibility and solubility of gas and propellant. However, there is no such problem for metal diaphragm tank featuring physical separation of gas and liquid. Commonly used pressurized gases include nitrogen and helium, of which helium is widely applied to reduce the mass of pressurized gas (under the same volume and pressure, the nitrogen mass is 7 times the mass of helium).

2) Working pressure and volume of gas cylinder

The working pressure and volume of the gas cylinder are mainly determined according to the comprehensive optimization of the storage tank volume, the working pressure of the tank, the flow rate of the tank, the type of pressurized gas, the mass and size of the gas cylinder structure, the difficulty in the development of the gas cylinder and the development cost. With the continuous development of carbon fiber material technology, winding technology and testing technology, it has become a common trend to apply high-performance, lightweight composite materials for the wrapping of gas cylinders. The working pressure of gas cylinders has reached 60 MPa or more, effectively reducing the structural mass of and size of the gas cylinder.

3) Supercharged gas pipeline transportation and controlling design

The probe propulsion system has relatively high requirements on the thrust accuracy of the divert thruster (generally not exceeding 3%). The pressure drop system cannot satisfy the relevant requirements. In general, the constant-pressure extrusion supercharging system regulated by the pressure reducing valve can be selected to make the thrust accuracy of the divert thruster reach the level of 1%. In order to control the mixing ratio precision, generally one-way pressure reducing valve is applied (considering the failure of the pressure reducing valve, the pressure reducing series or parallel connection approach can also be set) to simultaneously pressurize the two types of component tanks so as to ensure the same pressure of the oxidant and the fuel tank. For surface tension tanks where gas and liquid coexist, one-way valve management is required for each component tank pressurized air channel so as to ensure that the two-component propellant and steam do not mix in the pneumatic system. For diaphragm tanks that are physically gas-separated, no one-way valve management is required. In addition, this also avoids the problem of tank pressure deviation caused by the one-way valve flow resistance deviation and the possible gas flow system low-frequency oscillation due to check valve and gas flow.

Considering the fact that the "zero flow" of the pressure reducing valve would likely result in overpressure of the tank due to the rise of the locking pressure, the high-pressure latching valve is often installed upstream of the pressure reducing valve for self-controlling of overpressure failure, that is, automatic control in case of overpressure of the tank. The high-pressure latching valve is closed, the high-pressure gas supply is cut off, and after the tank pressure returns to normal, the high-pressure latching valve is automatically controlled to enable the high-pressure gas supply. To prevent the quick overpressure of the tank caused by serious failures such as failure of pressure relief valve seal and the valve core jam, the safety valve can also be installed behind the pressure reducing valve. In case of overpressure, the safety valve opens automatically to exhaust the gas and release the pressure, and the pressure then drops to the normal value. The safety valve is closed by itself to avoid permanent leakage of pressurized gas.

To guarantee cylinder filling/deflation, propulsion system leak exploration and condition monitoring, correspondingly, charge/discharge valves, test interfaces, pressure and temperature sensors are required.

In the celestial landing mission, the post-landing probes still need to survive for a long period. At this time, in order to reduce the energy consumption of the entire device, the propulsion system is generally not temperature-controlled, and the high-pressure vessels such as the cylinder storage tanks are exposed to long-term

10.3 Design and Verification of Deep Space Exploration Propulsion System 423

temperature alternating conditions, raising the risk of structural failure and the safety of the probe. In order to solve this problem, the special functional requirements of the post-landing propulsion system for the depressurization of the pressurized gas are proposed. Generally, the exhaust gas can be directly opened by the exhaust solenoid valve or the pyro valve provided downstream of the pressure reducing valve until that the pressure of the tank and cylinder drop to 0 MPa, so as to ensure the safety of the entire probe.

10.3.3.5 Propulsion System Assembly Plan Design

1) Assembly layout design

The structural size of the propulsion system generally accounts for the largest proportion of the entire probe, and the idea of propulsion and structural integration design can be adopted where necessary. In the layout of the propulsion system piping, the installation position of each component of the propulsion system and the direction of the pipeline system should be combined. The layout design should be carried out based on the principle of uniformity of symmetry, easy assembly, convenient testing, modularization and integration.

2) Propulsion system pipeline seal design

In the design of the assembly pipe seal for the propulsion system, there are two common connection methods: welding and tightening. In order to improve the sealing reliability of the pipeline system, the gas and liquid connecting pipelines are mostly connected by means of welding, especially the gas pipeline system in which gas (helium gas, nitrogen gas) leakage is highly probable. For the liquid path system, consideration should be made on particularity of engine/thruster orifice plate debugging, dissimilar material welding difficulty, pressure transducer calibration, welding operability, etc., and the parts that must be screwed should generally be connected by two or more seals. In order to ensure the overall leakage rate index of the propulsion system, the single-point leakage rate index of each joint shall be strictly controlled. Generally, the leakage rate of welded parts shall not be more than 1×10^{-7} Pa m³/s, and that of the threaded joint shall not be more than 5×10^{-7} Pa m³/s.

3) Divert thruster/attitude control thruster installation accuracy design

The installation accuracy design of the divert thruster/attitude control thruster for the propulsion system is one of the key links to guarantee the accuracy of orbit and attitude control. Divert thruster/attitude control thruster installation accuracy can be further divided into position accuracy and nozzle geometry axis angle, of which the installation reference of the divert thruster is the surface of the head butt flange, and the installation reference of the attitude control thruster is the docking flange face of each thruster head or the attitude control thruster mounting surface. The geometrical axis accuracy of the nozzle of the divert thruster/attitude control thruster in the single machine state is generally guaranteed by the single machine design or machining process, and the accuracy test can be performed if necessary. For the orbit-controlled engine, due to the large thrust, the actual thrust vector of the engine deviates from the geometric center of the installation and the geometrical axis of the nozzle. The influence on the attitude control is significant. Generally, the actual thrust vector data is obtained by the engine hot-code test, and the installation adjustment is made based on the measured results. The thrust point on the vertical axis of the lander and along the thrust vector direction should coincide with the vertical axis of the lander as possible.

The accuracy of the geometrical axis of the divert thruster/attitude control thruster nozzle relative to the head mounting surface is ensured by the combined machining process. The accuracy of the nozzle geometry axis is ensured by high-precision molds, and the combined machining accuracy is guaranteed by the tooling and equipment.

The geometric axis accuracy of the divert thruster/attitude control thruster is generally measured by optical method or CMM. For the whole probe, the optical method is generally adopted, and the reference surface of the object to be measured is set to the reference cube mirror, and the measurement data with respect to the reference cube mirror is coordinate-converted to obtain the measurement data with respect to the whole-probe reference. If the accuracy data does not satisfy the control requirements, the installation control link of single machine or probe can be adjusted to finally meet the accuracy control requirements.

10.3.3.6 Circuit Design of the Propulsion System

The circuit system of the propulsion system is the "pivot" between the overall control system (GNC, power supply and distribution, TT&C) and the propulsion system load. It is used to convert various types of control commands and drive the load to work, and meanwhile collect various state parameters of the propulsion system for the telemetry of the downlink, a propulsion control line box is generally configured to implement such functions. Given the importance of propulsion control wiring boxes, redundant designs are often employed to ensure reliable operation. The circuit system design of the propulsion system includes power supply/power-off control of various electrical equipment in the propulsion system, DC/DC secondary power conversion, valve (including solenoid valve, latching valve) on/off control, sensor acquisition and processing, autonomous control of the thermal control heating device, autonomous fault exploration and switching control of the propulsion system, internal state monitoring of the propulsion control line box as well as the command communication with the control system and the telemetry system. Generally, the power-on/off-off of the equipment adopts the relay control mode. The valve on/off and the heater on/off are widely used in the CMOS circuit control mode. The stepping motor adopts the H-bridge drive control mode based on the smart chip, and the command communication mode adopts the RS422 asynchronous serial interface or 1553B bus interface. The pressure, displacement, contact switch, etc., are directly under the output of voltage of 0–5 V, the temperature sensor is mainly the output of the voltage divider circuit.

10.3 Design and Verification of Deep Space Exploration Propulsion System

For special propulsion systems, the control of the variable thrust engine or the universal adjustment frame should also be considered. Therefore, it is necessary to configure the related circuit for the motor control.

Take Chang'e 3 lander propulsion system as an example. The A/B machine cold backup redundancy design with 80 C32 as the core is applied for propulsion wiring line box, and any control function can be completed by the normal operation of either device. The stepper motor adopts the chopped wave constant current closed-loop control method based on the H-bridge driver chip and is made commutation control through the RS422 asynchronous serial port commanding. In order to improve the resolution and smooth running of the stepping motor, the step angle is further subdivided.

10.3.3.7 Propulsion System Test Verification

Propulsion systems are generally mechanical, electromechanical or electronic products, with main features of pressure, load, movement, fluid, combustion, thermal control, etc. The system is complex, and the products are closely related to the production process and materials. In addition, many work processes are difficult to be accurately analyzed by theoretical simulation means, and past experience and experimental means are heavily relied on; for example, engine combustion performance is basically verified by a large number of test run. Therefore, a large number of ground test verifications at component level, subsystem level and probe level are required.

Generally, it can be divided into acceptance test, identification test, batch sampling test and special bottom test. The test items cover electrical performance, parameter adjustment, leak exploration, fine measurement, environmental test (centrifugation, vibration, shock, high and low temperature, damp heat, thermal vacuum, thermal cycle, comprehensive stress test), operating characteristics, engine/thruster test, pressure fatigue cycle, blasting strength, subsystem joint test, electromagnetic compatibility test, etc. The verification tests are widely applied to fully verify the function, performance, reliability, environmental adaptability, quality stability, etc., of the product. According to different categories of products, verification tests are required for several times or on several sets/batches. The acceptance test shall cover all relevant products, and the products participating in the flight test shall be subject to the acceptance test. The key points of batch sampling test are batch quality, performance stability of the product and the performance that cannot be directly tested by flight test.

In addition, special diagnostic tests can be conducted for different exploration missions. This type of test is a targeted verification test carried out in combination with the design of the propulsion system and the actual working characteristics. It is generally carried out for 1 or 2 times and is mainly to test the work adaptability and margin of the product.

The typical propulsion system test for the propulsion system is as follows:

1) Cold flow test of chemical propulsion system

The purpose of the test is to evaluate the performance of the coordinated work of the components of the gas path of the system, to obtain the characteristics of the system liquid pipeline, to obtain the characteristics of dynamic water hammer under various working conditions, to study the matching of system parameters under various working modes and to evaluate the coordination of components of the system.

2) Hot test run of the chemical propulsion system

The purpose of the test is to assess the reliability of the final assembly structure of the propulsion system, the matching of parameters, the coordination of the performance of each component, the performance of the system under different thrust conditions, the performance of the main components on the system and whether the parameters can meet the initial design specifications.

3) Plume test

The purpose of the test is to measure the key parameters characterizing the features of the plume and evaluate the influence of the thruster plume.

10.4 Outlook

With the continuous improvement of the level of aerospace science and technology, the pace of space exploration by the humans has been further accelerated, and the scope and depth of deep space exploration activities are continuously expanded. Therefore, higher requirements are put forward for the propulsion system. The future development trend of propulsion technology in deep space exploration mission includes:

1. **High-power electric propulsion technology**

High-power propulsion technology is the inevitable choice for large-scale deep space exploration missions in the future. In the missions of asteroid capture and manned Mars exploration demonstrated presently by the USA, high-power electric propulsion technology has exhibited excellent advantages, which can greatly shorten the exploration period.

2. **Working medium free propulsion technology**

In future interplanetary exploration programs, the speed increment requirement is huge. Traditional chemical propulsion and electric propulsion rely on the consumption of a large amount of propellant for thrust. Therefore, the mass of propellant to be carried is huge and is totally unfeasible. Therefore, the working medium free propulsion technology that does not consume propellant will be an important option for future deep space exploration and propulsion technology. At present, research is being

10.4 Outlook

carried out all over the world; for example, NASA has researched the microwave working medium free propulsion technology, and ground verification has also been carried out accordingly.

3. **Antimatter propulsion technology**

The energy produced by the annihilation of antimatter and positive matter is the largest energy so far has been recognized. Propulsion systems use charged particles produced by mixing matter with antimatter. This kind of particle has a high velocity and is directly released into space through the constraints of the magnetic field to generate thrust. The technical difficulty of antimatter propulsion lies in the control and guidance of huge energy in the process of antimatter action, which requires a great technological breakthrough.

References and Related Reading

1. Sutton GP, Biblarz O (2003) Rocket propulsion elements. Science Press, Beijing (Chinese vision)
2. Mao G, Tang J et al (2009) Spacecraft propulsion system and its application. Northwestern Polytechnical University Press, Xi'an
3. Liu G, Ren H, Zhu N, et al (1993) Principle of liquid rocket engine. Aerospace Education Press, Beijing
4. Zhu N, Liu G et al (1994) Design of liquid rocket engine. China Aerospace Publishing House, Beijing
5. Husel (2004) Modern engineering for design of liquid-propellant rocket engine. China Aerospace Publishing House, Beijing (Chinese vision)
6. Yu D et al (2016) Lunar soft landing detector technology. National Defense Industry Press, Beijing (Chinese vision)
7. Russell CT, Capaceioni F (2006) Dawn discovery mission to Vesta and Ceres: present status. Adv Space Res 38(9):2043–2048
8. Marc DR, Thomas CF et al (2005) Preparing for the DAWN mission to Vesta and Ceres. In: Japan: 56th international astronaufieal congress
9. Strobl WC, Hora R, Mildicl J (2011) Nuclear electric propulsion and human space exploration. AIAA, Long Beach
10. Chu G, Zhang H (2007) Lunar probe technology. China Science and Technology Press, Beijing (Chinese vision)
11. Gerrish H, Schmidt G, Rodgers S (2001) Advanced propulsion research at Marshall PROPULSION Research Center. AIAA, Salt Lake City
12. Wertz JR, Larson WJ (1992) Analysis and design of space missions. Aviation Industry Press, Beijing (Chinese vision)
13. Turner MJL (2008) Rocket and spacecraft propulsion: principles, practice and new developments. Springer, German

# Chapter 11
Power Supply Technology

11.1 Introduction

Nowadays, deep space probes have traveled much farther in more complicated environments in diversified missions, bringing about higher demands on power systems [1]. The characteristics of the power supply systems of the deep space probes are closely related to the location and mission type of the target celestial body. For inner planets (e.g., Mercury, Venus), the probes will be subject to high-temperature and high-illumination-intensity environment. Therefore, heating resistance and radiation resistance measures should be considered for power system design, especially solar arrays; while for outer planets (e.g., Mars, Jupiter, Saturn), the probes will be subject to low-temperature and low-illumination-intensity environment, so solar array should have high conversion rate, large area, high power and low mass, and MPPT-type power controllers should be applied as possible to maximize the utilization of solar energy for the power system. For landing exploration on the celestial bodies and sample return missions, the impacts of factors such as the atmosphere and dust of the celestial bodies on the solar spectrum should also be considered for the power system design [2], or nuclear energy should be applied to eliminate the effects of sunlight and other environmental conditions [3, 4].

11.2 Solar Cell Technology

For the landing exploration on the celestial bodies, sample return missions, as the solar spectrum on the surface of a specific celestial body may be affected by factors such as the atmosphere and dust. It is also necessary to consider the spectrum matching of the solar cell and introduce dust removal techniques in the power system design. This chapter takes the Mars landing exploration as an example to analyze the effects of the atmosphere and dust on the surface of the Mars on solar arrays and improvement measures.

11.2.1 Spectral Matching

The sunlight reaching the surface of Mars is affected by the atmosphere and dust. And the intensity of the sunlight is weaker than that at the orbit near Mars. The specific distribution varies with the latitude and longitude. Meanwhile, the spectrum on the Mars surface also varies compared with the AM0 spectrum on the Earth orbit. The blue light spectrum band of the Mars is relatively weak, while the red and infrared spectrum bands are relatively intensive. NASA has measured the solar spectrum of the Martian surface with a spectrometer carried by the Mars rovers Spirit and Opportunity in the Mars Exploration Rover (MER) mission. As can be seen from the results shown in Fig. 11.1 [5], after the solar spectrum passes through the dust, the spectral transmittance in the blue–violet band is lower than that in the red and near-infrared bands. The factors affecting the transmittance mainly include the absorption and reflection of the medium. The atmospheric components on the surface of Mars are mainly carbon dioxide and dust. The absorption spectrum of carbon dioxide is mainly in the infrared band, and there is almost no absorption in visible light. It can be speculated that on the surface of Mars, the modulation effects of dust on the spectrum is more significant. In addition, with the decrease of the solar elevation angle, the overall solar light transmittance decreases, because of the increase in the optical path of the sunlight passing through the surface of Mars and the reduction of light intensity by gas and dust. The reduction of the spectral transmittance of the blue–violet band will directly affect the structural design of the solar cell.

Therefore, the performance of the triple-junction GaAs solar cell developed for the AM0 spectrum will be changed under the influence of the surface illumination conditions on the surface of Mars, and the originally matched inter-junction current will be mismatched, affecting the output power of the solar cell. Assuming that the

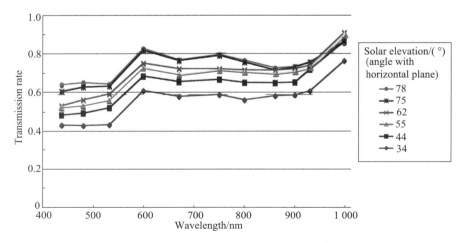

Fig. 11.1 Transmittance of sunlight in different wavelengths through the Mars atmosphere. *Note* The optical depth (the dimensionless unit of the amount of dust in the atmosphere) is about 0.93

11.2 Solar Cell Technology

photocurrent of each sub-cell of the multi-junction solar cell is balanced under AM0 illumination conditions, the top sub-cell in the multi-junction solar cell will have the minimum photocurrent because of the solar illumination on the surface of Mars, thereby limiting the current of the entire cell. And the loss coefficient has a certain relationship with the solar elevation. If the solar elevation angle is relatively low, the performance degradation will be more significant.

The triple-junction GaAs solar cell is formed by three sub-cells in series and by connecting three current sources in series in terms of volt-ampere characteristics, as shown in Fig. 11.2. Each sub-cell selectively absorbs and responds to different solar spectral bands. The overall cell voltage is the sum of the sub-cell voltages minus the tunneling junction voltage. The overall cell current satisfies the principle of continuity, and the current flowing through each sub-cell is equal and the output current is limited by the minimum photocurrent of the sub-cells. Therefore, in the triple-junction cell design, the photocurrent of each sub-cell should be equalized as possible to reduce current loss. The photocurrent of each sub-cell of the triple-junction GaAs cell is determined by the band gap and absorption coefficient of the material, of which the photocurrent generated by the bottom sub-cell is the largest (about twice that of the top or middle sub-cell) and has no limiting effect on the overall cell current. The overall output is determined by the top or middle sub-cell with relatively small photocurrents. Therefore, to improve the performance of triple-junction GaAs cells, the main effort is to optimize the top and middle sub-cells and the current matching thereof.

The triple-junction GaAs cells for conventional spaces application have basically achieved the current matching between the top sub-cell and the middle sub-cell under the AM0 illumination condition, but the top sub-cell becomes the bottleneck of the current output, limiting the current of the entire cell, because of the weakening of the violet spectrum of the Mars. Therefore, the optimization of solar cell for Mars

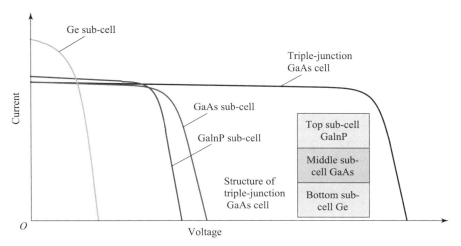

Fig. 11.2 Volt–ampere characteristics of a triple-junction GaAs solar cell

missions should mainly focus on the top sub-cell (violet light response) in the triple-junction GaAs cell, and the main optimization direction is the forbidden bandwidth and the thickness of the absorption layer. Possible measures include:

(1) Reduce the band gap of the top and middle sub-cells. The band gaps of the top and middle sub-cells of the triple-junction GaAs solar cell in current aerospace application have been specially optimized on better match to the AM0 spectrum. On the premise of satisfying the lattice matching of materials, current of the top sub-cell and the middle sub-cell can be basically balanced. However, on the surface of Mars, with the sunlight scattered and absorbed by the atmosphere and dust, the energy attenuation of the blue–violet band is stronger than that of the red band, directly resulting in smaller photocurrent of the top sub-cell compared with the middle sub-cell, and further increasing the current difference with the bottom sub-cell. On the premise of ensuring relative lattice matching, by appropriately reducing the material band gap of the sub-cells for the upper two junctions, especially the top sub-cell, the absorption edge is red-shifted, which can ensure the current balance of the sub-cells for the upper two junctions and is of help to improve the overall current output capability of the entire cell.
(2) Reduce current mismatch of the top and middle sub-cell. On the basis of expanding the spectral absorption range of the cell and improving the overall utilization of light energy, by adjusting and balancing the currents of the top and middle sub-cells, the current mismatch between the junctions can be reduced to maximize the current output of the cell.

11.2.2 Dustproof Techniques

The impacts of Martian dust on solar cells are mainly manifested in two aspects:

(1) The deposition and accumulation of Martian dust on the surface of the cell would increase the optical depth, reduce the transmittance of sunlight and change the spectrum of the working environment of the solar cell that further results in the decrease of the output power;
(2) If the Martian dust adheres to the surface of the solar cell, the thermal physical properties of the cell surface are altered, resulting in the increase in cell temperature and decrease in performance. Presently, the main dustproof techniques include active dust sweeping and passive dustproof. According to the working principle, it may be further subdivided into six types. The characteristics of various dust removal techniques are shown in Table 11.1 [6]. In practical applications, selection may be done according to mission requirements, mission constraints such as mass and volume.

11.2 Solar Cell Technology

Table 11.1 Analysis of dust removal techniques for solar array

Dust removal method		Principle	Main composition	Main features and application conditions
Active mode	Mechanical wiping	Wipe the surface of the solar cell wings with a "broom" or "feather duster"	Mechanical dust removal device, control circuit	It is easy to result in scratch on the surface of the cell, the mechanical structure reliability is limited, the frictional electric charge is increased, and the electrostatic adsorption effect is enhanced, so that the dust particles adhere closely to the surface of the cleaning tool and the material to be cleaned, and cannot be removed at all
	Blowing	Directly blow the dust on the working area with compressed gas	Injection line, gas cylinder, control circuit	Suitable for a wide range of dust sizes, basically covering the main feature sizes of Martian dust, being able to work well on multi-layer dust, but having more requirements for weight and volume resources
	Vibration	Apply piezoelectric ceramics or motor to create continuous oscillations and remove dust with the tilting of the solar array	Mechanical vibration device, control circuit	In the Martian environment, to remove the silicon dust particles attached to the surface of the alloy aluminum, an acceleration of more than 100 m/s^2 is required, which is difficult to be achieved in engineering. In addition, the solar array is required to be tilted at a certain angle

(continued)

Table 11.1 (continued)

Dust removal method		Principle	Main composition	Main features and application conditions
	Electric curtain dust removal	Form an electric field on the electric curtain surface to remove dust particles under the action of the electric field	Transparent electric curtain, secondary power change device	It is necessary to establish a high-voltage alternating electric field, and the surface of the solar array is encapsulated by a transparent electric curtain to generate electrostatic force acting on the fine dust. Electric field dustproof technology has been successfully applied in some industrial fields. However, for the atmospheric carbon dioxide low pressure environment of Mars, the minimum threshold of Paschen's discharge curve is only 100 V, making it difficult to directly apply electric field dustproof techniques on Mars
Passive mode	Natural dust removal	Remove dust by the wind on the surface of Mars	None	Based on the natural wind and gravity conditions of Mars, the adhesion of the Martian dust on the solar array is effectively reduced. The solar array substrate or the vehicle body is required to be tilted at a certain angle
	Superhydrophobic material	Prepare a "nano-needle bed" on the surface of the glass cover sheet to reduce the dust contact area and the dust adhesion	None	The adhesion of Martian dust on the surface of the cell is reduced, travel vibration or gentle wind may be exploited for dust removal purpose. The design of the glass cover plate should be improved

11.3 MPPT Technology

11.3.1 Basic Principles of MPPT

Solar array output power modulation may be divided into two categories: parallel modulation and series modulation. In case of parallel modulation system, the reference operating point of the solar array is generally designed according to the light intensity and temperature conditions at the end of life with main consideration of the effects of irradiation loss, temperature, etc. However, for deep space probes, the light intensity and temperature change in early and late life stages are quite different. For example, the average solar light intensity on the Mars orbit is only 0.43 of that on the Earth orbit, and within one revolution period, there is also a $\pm 19\%$ change in light intensity. Therefore, the adoption of a fixed reference operating point does not necessarily maximize the solar array output power. Under this condition, the maximum power point tracking (MPPT) mode is adopted to control the output power of the solar array according to the load power demand to maximize the potential of the solar array output power.

The principle of MPPT technology can be expressed as follows: at the maximum power point, the static and dynamic impedances of the solar array are equal, that is, the relationship between voltage (V) and current (I) and their variations dV, dI can be expressed as Eq. (11.1).

$$\frac{V}{I} = \frac{dV}{dI} \rightarrow \frac{dV}{V} = \frac{dI}{I} \tag{11.1}$$

In case of a certain disturbance to the solar array, the solar array is at the maximum power point (P_{max}) if the variation of the output voltage is equal to that of the output current. This process may be represented by Fig. 11.3.

11.3.2 MPPT Implementation

The solar array MPPT technology can be implemented through both software algorithms and hardware circuits. The implementation of MPPT software algorithm involves many methods, which commonly used include disturbance observation method, voltage feedback method, power feedback method, three-point phase method, intermittent scanning tracking method and optimal gradient method. With the application of large-scale integrated chips, there have also been many artificial intelligence-based algorithms such as fuzzy logic and neural network. Presently, although there are many algorithms for MPPT, as restricted by the special conditions and high reliability requirements of space applications, there are few algorithms available for actual on-orbit applications, which mainly include disturbance observation method, voltage feedback method and direct measurement method. With the

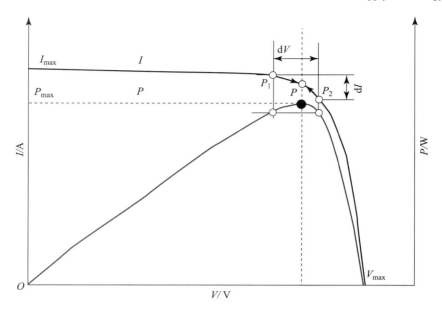

Fig. 11.3 Working principle of MPPT

enhancement of the autonomous management capabilities of the spacecraft power system, some more complicated algorithms have been applied. The basic flow of the algorithm is shown in Fig. 11.4.

The MPPT algorithm may also be implemented by an analog circuit, and its circuit diagram is shown in Fig. 11.5, which includes the sample retainer, comparator, RS flip-flop and integrating circuit. The MPPT circuit samples the present input current and voltage values of the present solar cell circuit corresponding to the two-way sample hold circuit and one-way flip-flop. The present current and voltage values are compared with current and voltage values of the previous state, and the output state of the trigger is changed to move the operating point of the solar array to the left and right (refer to the I–V curve in Fig. 11.3) and finally converges at the maximum power point.

11.3.3 MPPT Topology

The common power system topology applying MPPT technology mainly includes three types, as shown in Fig. 11.6 [7].

11.3 MPPT Technology

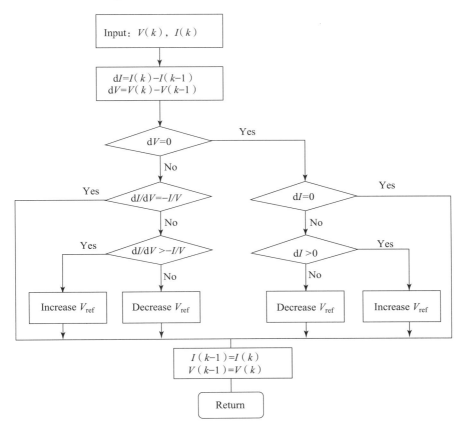

Fig. 11.4 MPPT software algorithm implementation. *Note* "k" is the present time; V_{ref} is the reference voltage

11.3.3.1 Regulated Bus MPPT Topology

In this topology design, DC/DC converter, battery charge regulators (BCR) and battery discharge regulators (BDR) are introduced to the solar array power regulator (APR), so that the bus can output stable bus voltages during both illumination and shadow periods. The main disadvantages lie in the increased power loss on the main power path, the reduced power utilization of the solar array, and the increased mass due to the presence of charge regulators and discharge regulators. This topology is mainly applicable to systems with high requirements for bus voltage during the illumination and shadow periods as well as for missions with large changes in light intensity, temperature and solar elevation. Spacecraft applying such topology include: Mars Express, Venus Express, Rosetta comet probe, GAIA satellite, the Laser Interference Space Antenna for the pathfinder mission (LISA Pathfinder) for gravitational wave detection, etc. [8–11].

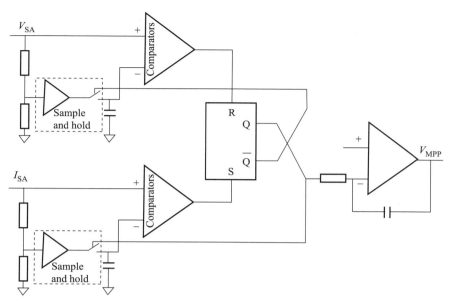

Fig. 11.5 MPPT hardware circuit implementation. V_{SA}—Solar array output voltage; I_{SA}—Solar array output current; V_{MPP}—Solar array maximum power point voltage

11.3.3.2 Unregulated Bus MPPT Topology

This type of topology is with simple structure, but the solar array power is output strictly according to the load demand so as to maximize the utilization of the solar array output power. In this topology, charge regulator or discharge regulator is unnecessary, so the mass of the solar array is smaller, thereby its power loss between the cell and the bus is reduced. The main disadvantages lie in that the bus voltage is unstable, and the bus voltage is subject to the clamping by the battery voltage during charging, so there is a relatively high requirement on the load power converter. Such topology is mainly applied in low-altitude spacecraft with short illumination periods and also missions with large changes in light intensity, temperature and solar elevation. There were quite a few early spacecraft using this MPPT topology, such as the low-orbit communication spacecraft Globalstar-2 [12], low-orbit Earth magnetic field exploration spacecraft SWARM [13], Messenger [14] Mercury probe launched by NASA in 2004, etc.

11.3.3.3 Sequential Switching Shunt Maximum Power Regulation (S³MPR) Topology

Sequential switching shunt maximum power regulation (S³MPR) is an improvement over the traditional direct energy transfer (DET) approach, which was first proposed

11.3 MPPT Technology

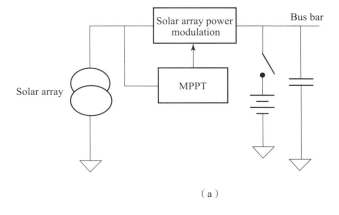

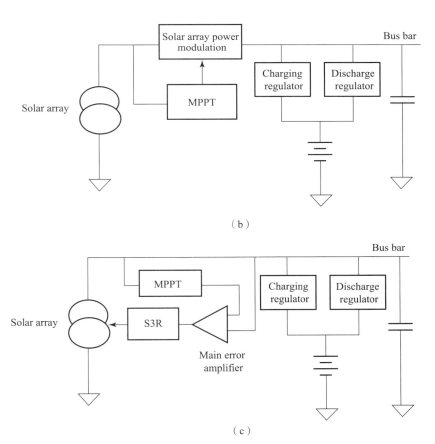

Fig. 11.6 Three common MPPT topologies. **a** No adjustment; **b** full adjustment; **c** sequential switch shunt maximum power modulation

by Weinberg in 2002. The main features lie in that the main error amplification signal (MEA) in the traditional sequential switching shunt regulator (S^3R) or series sequential switching shunt regulator (S^4R) topology is improved from a fixed-voltage reference point to real-time variation value given by the MPPT circuit, which indirectly achieves maximum power tracking. Compared with the fully regulated bus MPPT topology, this topology reduces power converter loss in the power transmission path and maximizes the use of solar array output power. The main disadvantage lies in that the bus voltage during the illumination period will be affected by the load change, and the bus voltage cannot be fully adjusted. Currently, spacecraft applying such topology include ESA's BepiColombo [15] probe for Mercury exploration.

11.4 Lithium-Ion Battery Technology

11.4.1 Overview of Lithium-Ion Batteries

The basic concept of lithium-ion cells may be dated back to the rocking chair-type batteries proposed by Armand et al. in the 1970s. Common lithium-ion cells are mainly composed of the positive electrode, negative electrode, diaphragm, electrolyte, a casing as well as various insulating and safety devices. The positive electrode is generally the ion intercalation compound, and the commonly used materials include $LiCoO_2$, $LiMn_2O_4$, $LiNiO_2$, $LiFePO_4$, $LiNi_{1/3}Mn_{1/3}Co_{1/3}O_2$, etc. The negative electrode is generally made of material enabling reversible delithiation and lithium intercalation with redox potential as low as possible, and the commonly used materials involve graphite, MCMB, silicon-based negative electrode materials, tin-based negative electrode materials, titanyl compound negative electrode materials, composite negative electrode material, etc.

Based on the application and development of positive and negative electrode materials, the research and development of lithium-ion cells may be roughly divided into three generations, as shown in Table 11.2. With the development of the third-generation lithium-ion cells, the upper limit of the cell charging voltage gradually increased from 4.25 V. For different positive electrode materials, the charging voltage is increased from 4.35 to 4.9 V. For positive and negative electrode materials, electrolytes, diaphragms, binders, conductive additives and current collectors working within the voltage range of 4.9–5 V, further developments are required.

Series lithium-ion battery products have become available in China and have been successfully used in on-orbit applications. According to the cell capacity, different specifications such as 10 Ah, 15 Ah, 20 Ah, 25 Ah, 30 Ah, 45 Ah and 50 Ah have been covered to satisfy different power requirements (Fig. 11.7).

11.4 Lithium-Ion Battery Technology

Table 11.2 Generation division of lithium-ion batteries based on positive and negative electrode materials

Generation	Positive electrode	Negative electrode	Time
First generation	$LiCoO_2$	Needle coke	1991–
Second generation	$LiMn_2O_4$	Natural graphite	1994–
	$LiNi_{1/3}Co_{1/3}Mn_{1/3}O_2$	Artificial graphite	
	$LiFePO_4$	Lithium titanate	
Third generation	High-voltage $LiCoO_2$	Soft carbon	2005–
	$LiNi_x \geqslant_{0.5} Co_y Mn_z O_2$	Hard carbon	
	$LiNi_{0.8}Co_{0.15}Al_{0.05}O_2$	SnCoC	
	$LiFe_{1-x}Mn_xPO_4$	SiO_2	
	$xLi_2MnO_3\text{-}Li(NiCoMn)O_2$	Nano-Si/C	
	$LiNi_{0.5}Mn_{1.5}O_4$	Si-M alloy	

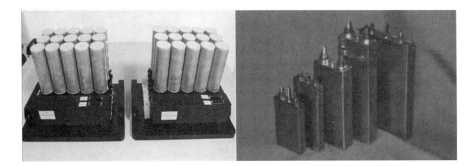

Fig. 11.7 Photographs of lithium-ion cells

11.4.2 Low-Temperature Resistance Technology for Lithium-Ion Cells

For the probes exploring extraterrestrial planets or the probes undergoing planet nights after landing on the surface of planets, they will face the low-temperature working environment. Therefore, the low-temperature resistance techniques of lithium-ion batteries are special requirements for deep space exploration missions. When the lithium-ion battery is discharged at a low temperature, the output voltage is lowered, and the usable energy is reduced. When the current is large at a low temperature, lithium deposition may occur on the surface of the negative electrode, which may affect the cycle life of the cell.

11.4.2.1 Effect of Low Temperature on Output Performance of Lithium-Ion Battery

The lithium-ion cell can be simplified as a model with a resistor r connected in series with an ideal cell, as shown in Fig. 11.8. The voltage across the ideal cell is greatly affected by the state of charges, and impacts of other factors such as temperature are relatively small. In the figure, r represents various impedances of lithium-ion batteries during charge and discharge (such as concentration polarization, ohmic polarization and electrochemical polarization), which is changed with charge/discharge, state of charge, current, temperature, etc.

The low temperature will result in the increase of the cell discharge resistance r. When the temperature is low, the voltage over the resistor r is large, so the output voltage of the lithium-ion cell is lower than that at normal temperature. Under the same discharge cutoff voltage, the low-temperature discharge capacity, the available power and the power supply time are negatively affected. Figure 11.9 is a schematic diagram showing the difference in low-temperature discharge performance of a lithium-ion cell. As can be seen in the figure, for the same operating voltage range, the low-temperature discharge capacity is significantly reduced, and the energy output of the cell is more significantly reduced after the influence of the output voltage drop is superimposed. In practical applications, when the cell exhibits a significant increase in voltage drop rate in a low-temperature environment, an over-discharge protection voltage may be triggered quickly.

At low temperature, the concentration polarization in the lithium-ion cell increases; compared with the same current, r increases. Therefore, in the charging process, the lithium-ion cell reaches the constant voltage charging point in a shorter time, and the current gradually decreases after entering constant voltage charging. The voltage across the ideal cell gradually approaches the constant voltage point. As

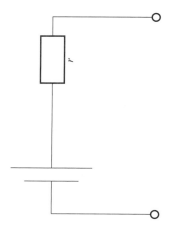

Fig. 11.8 Battery circuit model

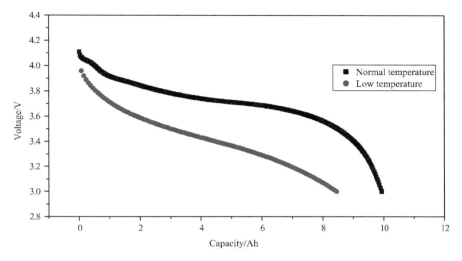

Fig. 11.9 Schematic diagram of the difference in low-temperature discharge performance of lithium-ion cell

a result, the charging voltage rises rapidly at low temperature, the charging time is long with low efficiency, and the actual state of charge is low. Figure 11.10 shows the difference in low-temperature charging performance of a lithium-ion cell. As can be seen from the figure, charging at a low temperature in the same working voltage range, the cell voltage quickly reaches a constant voltage point, and then enters constant voltage charging, but the charging current is reduced. In addition,

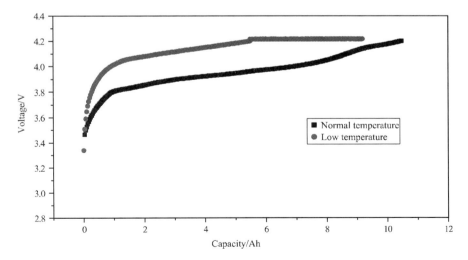

Fig. 11.10 Schematic diagram of the difference in low-temperature charging performance of the lithium-ion cell

in case of charging at a low temperature, if the current is excessively large and the penetration rate of lithium ions in the SEI film is not satisfactory, lithium ions will react to precipitate metallic lithium (i.e., the lithium dendrites) on the surface of the negative electrode.

11.4.2.2 Analysis of the Change Mechanism of Lithium-Ion Cell Output Performance

From the perspective of electrode process dynamics, factors affecting the low-temperature charge and discharge performance of lithium-ion cells are mainly the combination of the following factors:

(1) The migration rate of lithium ions in the positive electrode active material;
(2) The electrochemical reaction rate of the surface of the positive electrode;
(3) The migration rate of lithium ions in the electrolyte;
(4) The penetration rate of lithium ions in the SEI film on the surface of the negative electrode;
(5) The electrochemical reaction rate of the surface of the negative electrode;
(6) The migration rate of lithium ions in the negative electrode active material.

The above factors are mainly determined by the type of active material of the positive and negative electrodes, the grade and the electrolyte formulation. In addition, the process parameters such as electrode pad thickness, electrode porosity, diaphragm porosity and diaphragm thickness all affect the low-temperature performance of lithium-ion batteries.

When the lithium-ion cell is discharging, lithium ions are separated from the negative graphite layer, penetrate the SEI film on the surface of the negative electrode, then migrate onto the surface of the positive electrode in the electrolyte to be embedded in the active material of the positive electrode (e.g., $LiCoO_2$). It is generally believed that, in a low-temperature environment, the diffusion of lithium ions within the negative electrode sheet is a major control step. When the discharge current is relatively large, a large amount of lithium ions would migrate from the graphite layer of the negative electrode to the surface of the negative electrode. Due to the slow diffusion rate of lithium ions inside the negative electrode sheet, the polarization voltage drop occurs, resulting in a discharge voltage of the cell and the discharge capacity lower than those at normal temperature, which will return to normal after restoring to normal temperature.

When the lithium-ion cell is being charged, the lithium ions are removed from the positive electrode, migrate onto the surface of the negative electrode in the electrolyte and then penetrate the SEI film to the surface of the negative electrode and intercalate between the graphite layers of the negative electrode. It is generally believed that the diffusion of lithium ions within the negative electrode sheet is a major control step at low temperature. If the charging current is relatively large, a large amount of lithium ions migrate to the surface of the negative electrode. As the diffusion rate of lithium ions within the negative electrode sheet is slow, and the ion permeability of the SEI film is affected at a low temperature, part of the lithium ions fail to penetrate the SEI film and diffuse within the negative electrode. Instead, the lithium ions are deposited directly on the surface of the negative electrode to produce metallic lithium with the very active chemical property. On one hand, it is easy to react with the electrolyte to produce a gas; on the other hand, it is easy to form lithium dendrites, which can penetrate the diaphragm and result in short circuit inside the cell, forming an important source of cell capacity loss. Therefore, efforts should be made to avoid charging the cell in a low-temperature environment as possible.

11.4.2.3 Low-Temperature Performance Evaluation Method for Lithium-Ion Cell

To objectively evaluate the low-temperature performance of lithium-ion cells, importance should be attached to the following aspects:

1) Discharge cutoff voltage

From the previous analysis of the circuit model of lithium-ion battery, it can be seen that there is a relatively large difference between the true voltage (zero current) and the output voltage of the lithium-ion cell with the internal resistance of the cell increases at a low temperature. If the electrical load can accept a lower voltage, the discharge performance of the cell will be significantly improved.

2) Operating current

It can also be seen from the lithium-ion cell circuit model analysis mentioned above that the working current magnitude will affect the voltage over the internal resistance, resulting in a large difference between the real voltage and the output voltage. In addition, with the increase of the operating current, the concentration polarization may also increase accordingly which would superimpose with the low-temperature effect and the internal resistance increases. Therefore, high current discharge performance is more sensitive to temperature effects.

3) Energy and capacity

At low temperature, with the decrease of discharge voltage of the lithium-ion cell, the capacity (Ah) also decreases for the same cutoff voltage, the drop of energy (Wh) available for supply is even greater. In actual power supply, the power supply is generally at a constant power, and the energy will be more suitable than the capacity for evaluating the low-temperature performance.

4) Low-temperature charging

From the preceding lithium-ion cell circuit model analysis, it can be seen that the charging efficiency will decrease when charging at low temperature, and the actual state of charge after stopping charging will be quite different, also the gained energy is quite different from that when charged at normal temperature.

11.4.2.4 Electrochemical System Selection of Lithium-Ion Cell for Low-Temperature Usage

Lithium-ion cell systems are generally classified by the positive electrode materials. Reasonable selection of positive electrode materials is the key to improving the low-temperature performance of lithium-ion cells. Compared the low-temperature discharge performance of lithium-ion cells in $LiCoO_2$ system with that in $Li(Ni_xCo_yAl_{1-x-y})O_2$ system for space applications, it can be found that the lithium-ion cell of $LiCoO_2$ system has better low-temperature performance than $Li(Ni_xCo_yAl_{1-x-y})O_2$ lithium-ion cell. The ICP40 cell ($LiCoO_2$) and INP45 cell ($Li(Ni_xCo_yAl_{1-x-y})O_2$) are used for the $-20\ °C \pm 0.2\ °C$ low-temperature performance test. The results are shown in Table 11.3. It can be seen from the table that when the discharge cutoff voltage at $-20\ °C$ is 3.3 V, the $Li(Ni_xCo_yAl_{1-x-y})O_2$ system cell can only discharge 41.7% of the normal temperature capacity, while the $LiCoO_2$ system cell can discharge 62.4% of the normal temperature capacity. At the voltage of 3 V, the $Li(Ni_xCo_yAl_{1-x-y})O_2$ system cell can only discharge 59.8% of the normal temperature capacity, while the $LiCoO_2$ system cell can release 82.2% of the normal temperature capacity. Meanwhile, the average discharge voltage of the $LiCoO_2$ system cell is also higher than that of the $Li(Ni_xCo_yAl_{1-x-y})O_2$ system cell.

11.5 Space Nuclear Power

11.5.1 Overview of Space Nuclear Power

Space nuclear power refers to the power system that converts thermal energy generated by nuclear decay or nuclear fission into electrical energy for space probes. It mainly consists of the heat source and thermoelectric converter, as shown in Fig. 11.11.

According to the types of heat sources, space nuclear power may be divided into radioisotope thermoelectric generator (RTG) and fission reactor. The radioisotope power supply can provide medium and low power (<1 kW) with simple structure, low radiation and high reliability. In contrast, the power supply of nuclear reactor is large (>1 kW); however, the research and development of space nuclear reactor is difficulty, for example, the decay of uranium-235 produces a large number of fast neutrons and gamma rays, which may seriously affect the function of the electronic devices on the probes and require additional radiation protection measures.

From the perspective of the thermoelectric conversion methods, the space nuclear power supply may be divided into two types: static conversion and dynamic conversion. The static conversion systems without rotating parts inside, including thermoelectric power generation, thermal ion power generation and thermal photovoltaic power generation, generate direct current (DC); while dynamic conversion systems, including Rankine cycle, Brayton cycle, Stirling cycle, etc., generate alternating

11.5 Space Nuclear Power

Table 11.3 Comparison table of discharge performance of $LiCoO_2$ and $Li(Ni_xCo_yAl_{1-x-y})O_2$ cells at room temperature and $-20\ °C \pm 0.2\ °C$

Discharge cutoff voltage /V	Temperature /°C	INP45				ICP40			
		Capacity /Ah	Percentage of capacity /%	Energy /Wh	Average discharge voltage /V	Capacity /Ah	Percentage of capacity /%	Energy /Wh	Average discharge voltage /V
3.3	20	48.042	100	176.80	3.680	47.488	100	180.01	3.791
	−20	20.022	41.7	70.100	3.501	29.653	62.4	104.96	3.540
3	20	51.108	100	186.57	3.651	48.195	100	182.25	3.782
	−20	30.546	59.8	103.44	3.386	39.600	82.2	136.58	3.449

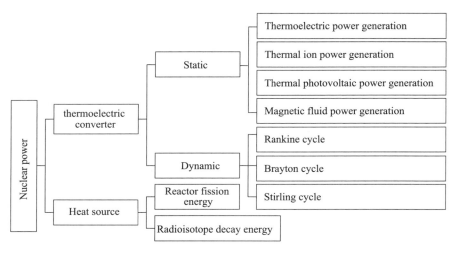

Fig. 11.11 Thermoelectric conversion method of space nuclear power supply

current (AC). The conversion efficiency of the dynamic conversion method is higher than that of the static method. Up to present, all nuclear power supply systems having been officially adopted in space applications belong to the static conversion system, which mainly use a temperature difference discharge method. The Stirling cycle in the dynamic conversion mode is a suitable method for space applications with relatively mature techniques.

The characteristics of the space nuclear power supply are as follows:

(1) It can work continuously without being affected by the distance from the Sun. For deep space explorations, it is always an essential technical challenge for sufficient electrical energy. The intensity of sunlight near the planets in the solar system is significantly different from that in the near-Earth space. The sunlight intensity near Mars is about 43% of that near the Earth. And the sunlight intensity near the Jupiter is only 3.7% of that near the Earth. With the increase of distance from the Sun, the solar intensity continues to decline. Therefore, for the outer planets exploration missions, the performance of solar cell is becoming increasingly low due to the decrease of the sunlight intensity. In contrast, the nuclear power is generated by the thermal energy created from nuclear decay or nuclear fission without depending on sunlight, making it an optimal choice at present.

(2) It can provide thermal and electric energy on occasions where the solar power density is small or the solar energy is discontinuous. It becomes difficult due to energy generation and storage problems in the shadow environment of celestial bodies and polar region missions. For example, the night time on the Moon equals to 14 Earth days. If solar cells are used, the lunar probe will be out of

11.5 Space Nuclear Power

power and heat supply during such 14 Earth days, and the scientific equipment will be frozen. Nuclear power may be utilized to remove such constraints or restrictions.

(3) It can provide power from tens of milliwatts to hundreds of kilowatts with prolonged service life. The radioisotope power systems on Voyager 1 and Voyager 2 have a working life of more than 40 years.

(4) It can work in harsh environments (radiation, climate, magnetic). In the shadow and strong natural radiation environment, the solar power system is not suitable or unavailable, and the radiation damage (charge and lattice change) of the solar cells caused by charged ions and neutrons would seriously degrade the current and voltage characteristics. In contrast, the space nuclear power systems are better adapted to such harsh environments.

11.5.2 RTG Technology

11.5.2.1 Overview

The radioisotope thermoelectric generator (RTG) refers to a process of static thermoelectric conversion and is a radioisotope power system (RPS) that utilizes the thermal energy generated by the decay of radioactive isotopes acting on materials to directly convert thermal energy to electrical energy with Seebeck effect. RTG is a dense energy source with high viability featuring compact structure, high reliability and excellent radiation resistance and can operate in the harshest environment. It has the highest mass specific energy (105 Wh/kg) and the longest life compared to other chemical and physical power sources known to date (design life of 5–10 years, actual operation for over 20 years). It requires no maintenance and is not affected by the environments.

1) Isotope material

The commonly used heat source (also known as the "fuel") for isotope thermoelectric generators is radioisotopes, referring to Table 11.4 for a brief description of **commonly used radioisotopes**.

Among isotope power supplies, the first isotope applied was ^{210}Po, which has a **calorific value** of 141 W/g and a half-life period of 4.5 months. The amount of heat

Table 11.4 Comparison of commonly used radioisotopes

Isotope name	Calorific value/(W·g^{-1})	Half-life	Radiation hazard	Price
^{210}Po	141	4.5 months	Yes	/
^{238}Pu	0.55	88 years	No	High
^{90}Sr	0.93	18 years	Relatively serious	Relatively low
^{242}Cm	2.8	18 years	No	/

released thereafter will be 1/2 of the level at the beginning. The half-life of ^{210}Po is excessively short, resulting in short life of the generator, so it is not suitable for deep space mission applications since a deep space mission lasts a long period of time.

The ^{90}Sr has a half-life cycle of 18 years and a calorific value of 0.93 W/g, and is relatively inexpensive. However, the radiation hazard of ^{90}Sr is larger than others, and therefore, it is generally not used. So ^{238}Pu is selected by engineers for space nuclear power supply. The fission half-life of ^{238}Pu is 88 years, and the generated thermal energy declines at a rate of 0.8% per year. The decay products are ^{234}U, alpha particles and quite few gamma rays.

2) Thermoelectric converter

Thermoelectric converters are generally based on the Seebeck effect, which utilizes the temperature difference between two different materials to create an electromotive force at the junction. The main components of the converter are the thermocouples, which consist of semiconductor elements such as lead–bismuth, tantalum–silver quaternary alloy or germanium–silicon alloy (P-type and N-type). Multiple pairs of thermocouples form the galvanic pile by means of connection in series and in parallel, the hot end is close to the heat source box with a temperature of generally 350–650 °C, and the cold end is connected to the heat sink with a temperature of generally 100–250 °C. The heat sink is made of metal with excellent heat conducting performance (aluminum alloy). The cold end temperature is maintained as low as possible to enhance the temperature difference and improve the thermoelectric conversion efficiency. The total thermoelectric conversion efficiency of the power generation system is 4.2%–6.6%, and the specific power is 1.3–4.2 W/kg. Early RTGs applied ruthenium-based material in the thermoelectric principle. Silicon–germanium alloys have been applied for all of the RTGs for the space missions since Lincoln test satellites 8 and 9 (1976), referring to Table 11.5.

Table 11.5 Various thermoelectric materials being applied or developed

Material name	Conversion efficiency/%
Silicon germanium (SiGe)	6–7
Lead telluride (PbTe)-TAGS	5–7
Bismuth telluride (BiTe)	2.5–4.5
Segmented PbTe–TAGS/BiTe	9
Skutterudites	>9

11.5.2.2 Development of RTG Technology

The US has successfully applied 41 RTG system in 23 spacecraft since the first RTG SNAP-3B7 was launched to space onboard Transit-4A satellite by America in 1961. The representative products in the US RTGs developmental history include SNAP-19, MHW-RTG, GPHS-RTG and MMRTG that is in research and development. The

Chinese Chang'e 4 RTG applies ^{238}Pu, the thermal power is 115-127.5 W, the output electric power is greater than 3 W at 1.3 V, and the annual decay rate of thermoelectric performance is less than 5.3% per year.

11.5.3 Nuclear Reactor Power Supply

11.5.3.1 Nuclear Reactor Technology Systems and Solutions

The function of the nuclear reactor power supply is to convert nuclear energy into a power source, and store, regulate and transform the power source as needed, then supply power to the probe subsystems.

The nuclear reactor power supply shown in Fig. 11.12 consists of six parts: reactor, power generation unit, thermal control, circuit, control and nuclear safety mechanism.

The reactor section mainly includes the core, control drum, safety mechanism and radiation shielding device. The reactor is the energy source for the entire nuclear power system, which is responsible for safe operation of the space reactor and radiation shielding. The core is responsible for providing nuclear energy. Generally, fuel components such as UO_2, UC and UN are applied. The mass is converted into heat energy by the chain reaction method, and the working medium is heated to a high-temperature plasma state. When the control drum is closed, the incoming neutrons can be shielded. After the nuclear power probe is on-orbit, the control drum is opened and the reactor is excited by the neutrons from the natural environment. The safety mechanism can quickly absorb neutrons to enable the reactor shutdown. The radiation shielding device isolates the radiation generated by the nuclear reaction product from the probe body as well as the radiation sensitive device of the nuclear power system itself.

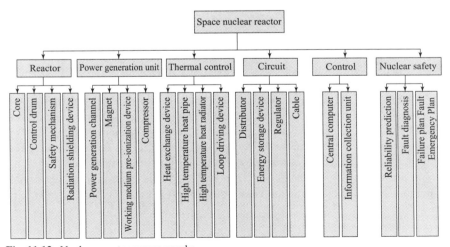

Fig. 11.12 Nuclear reactor power supply

The power generation device converts the thermal energy generated by the nuclear energy into electrical energy, including the power generation channel, magnet, working medium pre-ionization device and compressor, which are based on a magnetohydrodynamics (MHD) nuclear power conversion mode, that is, the thermal energy generated by the nuclear reactor is converted into the kinetic energy of the working medium, the plasma working medium passes through the magnetic field of the power generation channel at a high speed, and generates power by creating Hall current with the action of the magnetic field on the plasma. The working fluid pre-ionization device increases the ionization degree of the working medium, and the compressor closes the cycle of the entire process.

The thermal control mechanism discharges the residual heat generated by the space nuclear power source into the space outside the probe or collects it to provide thermal control for the probe. Meanwhile, the space nuclear power source is maintained within an appropriate operating temperature range mainly by heat exchange devices, high-temperature heat pipes, high-temperature heat radiators, loop driving device, etc. The heat exchange device enables heat recycling. In the high-temperature heat pipe, the high-power heat energy is transported, and the heat energy is directly discharged into the space through the high-temperature heat radiator by means of heat radiation. The loop driving device ensures stable operation of the entire closed system.

The circuit is responsible for the regulation, transfer and distribution of electrical energy generated by power generation, including the distributor, energy storage device, regulator and cables. The distributor implements the quality adjustment of the bus power supply and allocates the destinations of the power supply. The regulator adjusts the power, current and voltage of the input power based on load variations. Energy is transferred by the cables among different devices. The energy storage device is responsible for the startup of the entire system, the power supply function of the nuclear power system during the emergency control, the restart after the shutdown as well as the power supply for the subsequent system mission after the shutdown.

The control system is responsible for the safe operation, mode switching and technical state control of the entire space nuclear power supply, including the central computer and information acquisition unit. The main task of the control system includes three parts: state monitoring of the entire life cycle, power output control of the nuclear reactor as well as power output control of the power supply. The working modes of the power control include the monitoring mode, startup mode, low power maintenance mode, full load operation mode and safety mode. The central computer realizes comprehensive information processing and autonomous operation management, and the information collection unit is responsible for collecting initial signal and transmitting the relevant physical quantities to the central computer.

The nuclear safety mechanism is responsible for the protection of personnel, equipment and the environment under nuclear conditions, including reliability prediction, fault diagnosis and failure emergency planning. The nuclear safety protection cycle covers all elements from nuclear power reactor material preparation to ground testing, assembly, transportation, launch, on-orbit operation and disposal. During the design life cycle of a space nuclear power source, the nuclear safety mechanism is operated under the full lifecycle state monitoring.

The block diagram of the nuclear reactor based on magnetic fluid power generation is shown in Fig. 11.13.

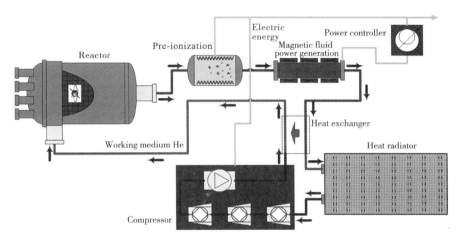

Fig. 11.13 Schematic diagram of the nuclear reactor power supply

In the working process of the system, the inert gas working medium He is heated by the reactor into a high-temperature plasma. The pre-ionization process further improves the ionization degree of the working medium, and the working medium enters the power generation channel, which generates induced current under the Lorentz force induced by the strong magnetic field. Then, the working medium enters the heat exchanger, transfers heat to the low-temperature working medium and discharges the waste heat through the heat radiator. The working medium enters the next cycle under the action of the compressor and is preheated in the heat exchanger. The electric energy so generated is managed and regulated by the power controller and sent to the probe main body for application, and also sent to support the work of pre-ionization and compressor.

11.5.3.2 Development of Nuclear Reactor Technology

Since the 1960s, the USA has invested intensively on research of space nuclear reactor energy systems that can be applied on the Moon or Mars bases and has proposed a variety of representative research schemes. In terms of different thermoelectric conversion methods, there are three typical cases:

(1) SNAP-8 series adopt Rankine cycle thermoelectric conversion system or thermocouple conversion system. It can provide 20–35 kW of electric power for a lunar base of 6–12 people. SNAP-8 is a zirconium hydride moderated, liquid NaK metal cooled, drum-controlled thermal neutron reactor, and the reactor core applies $U\text{-}ZrH_x$ fuel elements.

(2) The SP-100 Brayton energy system applies the Brayton cycle thermoelectric conversion system. The SP-100 reactor is a high-temperature liquid metal cooled fast reactor, in which the UN fuel rod is applied. The cladding material is refractory niobium alloy PWC-11, and the inner surface is made of metal crucible. The peak operating temperature of the fuel rod is 1400 K at the beginning of life and 1450 K at the end of life. The modular design of the SP-100 reactor enables multiple levels of energy output by varying the number of fuel assemblies, providing electrical power ranging from 8 to 15,000 kW.

(3) HOMER (heat pipe cooled Mars/Lunar surface nuclear reactor power system) applies Stirling cycle thermoelectric conversion system. It can supply 50–250 kW of energy for life support, propellant production and scientific experiments. It also provides high-intensity lighting for crop growth in Mars or lunar missions.

11.6 Deep Space Power System Design

11.6.1 Mission Analysis

Due to the obvious and unique characteristics of the deep space environment, the probe will be affected by multiple factors in the deep space during the deep space flight. This section lists the inputs and constraints closely related to the design of the power system, based on such constraints, gradually develops the system scheme design of the power supply system and key device design. The mission analysis and scheme design block diagram are shown in Fig. 11.14 [16].

11.6.1.1 System Scheme

Presently, the system schemes of the power system applied in deep space exploration mainly include the solar array + battery pack scheme and the nuclear power source + battery pack scheme [17, 18]. In the mission analysis, the power generation scheme to be selected should be determined first. The power generation scheme should be selected under the consideration of two constraints: the position of the exploration target relative to the Sun and the longest shadow duration experienced by the probe (Fig. 11.15). Jupiter is the planet farthest to the Sun that can be supported by solar energy under the current technical conditions. If the target planet is farther away from the Sun than Jupiter, nuclear power should be considered. In addition, for a

11.6 Deep Space Power System Design

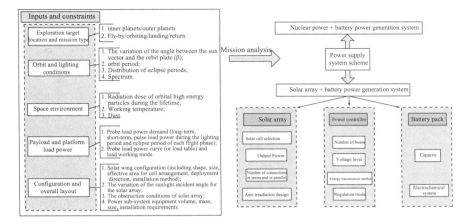

Fig. 11.14 Mission analysis and scheme design block diagram

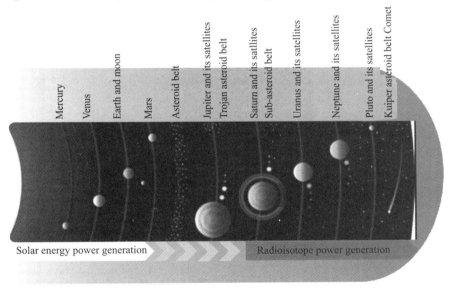

Fig. 11.15 Application of solar/nuclear energy in deep space exploration

celestial body (such as the Moon) with relatively long rotation period, if the probe is required to work during the shadow period (e.g., lunar night) without entering the sleep mode, it is necessary to consider the introduction of nuclear power for joint power generation [19, 20].

For the power system using the power generation scheme of solar array + battery pack, the scheme design of the solar array, power controller and battery pack are discussed next.

11.6.1.2 Solar Array Scheme

The design of solar arrays mainly involves the selection of solar cells, output power, and number of connections in series and in parallel and anti-irradiation design.

Solar cell types applied in aerospace applications include silicon cells and GaAs cells. Because of the higher conversion efficiency of GaAs cells relative to silicon cells, the current mass production efficiency of triple-junction GaAs solar cells is nearly 30%. Therefore, the silicon cells are being gradually replaced by GaAs cells.

According to the load power demands and variation curve of the entire probe, the output power is determined by the energy balance analysis. The output power value is affected by lighting conditions (including light intensity and spectrum), solar array operating temperature, particle irradiation dose, etc., and it should be guaranteed in the design that the minimum output power can satisfy the load power demands. The output power of the solar array is the product of voltage and current at the operating point. The output voltage is greatly affected by the operating temperature of the solar array, and specifically, the higher the temperature, the lower the output voltage is. The output current is greatly affected by the light intensity, and specifically, the higher the light intensity, the higher the output current is. Note that the light intensity decreases with the increase of the distance from the Sun. When designing the number of cells in series connection of the cell array, it is necessary to ensure that the bus voltage can be established at the highest temperature of the cell array. After the number of cells in series connection is determined, the number of cells in parallel connection can be determined according to the output power and the bus voltage. It should be noted that, due to the limitation of the solar panel area, the increase of the number of cells in series connection will inevitably lead to a decrease in the number of cells in parallel connection, which will further result in a decrease in the total output power. Therefore, when conducting the series connection design, it is not appropriate to leave excessive voltage margin. In addition, it is also necessary to consider the constraints and influences of the configuration and layout and occlusion condition of the entire probe when designing the solar array configuration and cell array layout.

The deep space radiation environment (as discussed in detail in Chap. 2 of this book) is a major factor affecting deep space exploration missions. Especially, during the Jupiter exploration process, the probe will be operating in an environment of high-energy radiation belt particles due to the radiation belt within the Jupiter magnetosphere. The space radiation environment is more severe than the near-Earth environment, and the charged particles have higher energy, and the particle flux is large, so the solar cells and the structural components made of carbon fiber composite materials will face extremely harsh space environment. Therefore, the targeted anti-irradiation measures should be taken for the glass cover sheet, the back sheet glue and the cover sheet glue according to the irradiation conditions of the exploration mission.

11.6.1.3 Power Controller Scheme

The power controller scheme design mainly includes factors such as the number of buses, voltage level, energy transfer mode and regulation mode.

The number of buses is determined based on the characteristics of the electrical equipment. There are mainly three kinds of bus system: single-bus power supply system, dual-bus power supply system and hybrid-bus power supply system. For the single-bus power supply system, the platform load and impulsive load share the same power supply bus. It has the advantages of the power system with a small number of devices and a small volume, which can fully utilize energy with relatively low cost. The disadvantage is that the impulsive load brings relatively large noise interference in the frequency domain and the time domain for the power supply bus, and the anti-interference ability of the electrical equipment on the bus and the filtering technology of the power supply system are highly required. In the dual-bus power supply system, two sets of solar arrays and batteries are equipped for platform load and impulsive load, respectively. The two buses are grounded at the same ground point of the spacecraft. The advantage lies in that it can effectively avoid the interference of the impulsive load against the stable load. The disadvantage lies in that the power system has a large number of devices and relatively large volume and mass. As each bus is independent in energy utilization, a certain safety margin is required, which leads to the decrease in bus energy utilization efficiency. For the hybrid bus system, a unified solar array and battery design is adopted, and an unregulated bus is separately extracted from the battery to supply the impulsive load. The advantage of the power supply system is that the solar array and the battery are uniformly arranged, the power system has a small number of devices and small volume and mass. Therefore, the energy can be fully and reasonably utilized. In addition, the platform adopts a fully regulated bus with high product inheritance.

The bus voltage level is determined according to the level of the load power. The primary bus voltage should be generally selected according to the following principles: (1) 28 V bus for power below 2.5 kW; (2) 42 V bus for power below 8 kW; and (3) 100 V or higher voltage bus for higher power.

The energy transfer mode can be divided into two major types, namely the series bus transfer mode and the direct energy transfer (DET). The typical mode of series bus transfer is non-dissipative peak power tracking (PPT), which outputs power according to the demands of the spacecraft with the maximum output up to the peak power of the solar array. However, the PPT method requires a series power conversion circuit to stabilize the output bus voltage. DET is a dissipative regulation mode because it consumes the inexhaustible power from the load. In the DET mode, a shunt is used to adjust the output power of the solar array, so that the bus voltage is maintained within the preset range, and the bus voltage is directly output, so that the transfer efficiency is higher. In the DET mode, to ensure the maximum utilization of the solar array output power at the end of life, the limit voltage value of the shunt regulator is generally set at the optimal operating voltage of the solar array at the end of the life, and as a result, the solar array output power at the early stage of life cannot be fully utilized. However, for spacecraft with little difference in power consumption between the early life and the end of life, the DET mode is an ideal choice as the transfer efficiency is the highest. The DET mode can be further subdivided into three modes: the unregulated bus voltage, the fully regulated bus voltage and the fully regulated bus voltage in the lighting area (also called the partially regulated bus voltage).

The advantage of the fully regulated bus is that the voltage of the power supply bus is always stable within the specified range, and the voltage stabilization accuracy is high. The voltage can be used directly for the appropriate electrical equipment. Even if a secondary transformation is required, the design difficulty of the secondary power supply can be reduced, and the conversion efficiency of the secondary power supply can be improved. The disadvantage lies in that the power controller has a large volume, mass and heat consumption, and the output impedance of the power system is also large. In case of short-term loading, unloading and impulsive loading, the bus voltage fluctuations and interference are relatively large.

The advantage of the partially regulated bus system lies in that the volume, mass and heat consumption of the power controller are relatively small. During the illumination period, when power supply is provided by the solar array alone, the bus voltage is stable, the solar array can work near the maximum operating point, and the solar cell utilization rate is same as the fully regulated bus. During the shadow period, the battery pack directly supplies power to the bus through the isolation diodes, the battery pack has high discharge efficiency, and the output impedance of the power supply system is quite small when battery pack is supplying power. The disadvantage is that the bus voltage range is large, secondary transformation is needed for the electrical equipments that have high requirement on the input voltage, and the design of the secondary power supply is relatively difficult. If joint power supply is required in the lighting period, the solar cell cannot operate at the maximum high-power point, and the solar cell utilization rate is lowered.

The advantage of unregulated bus system is that the output bus has a quite small output impedance and extremely fast response speed, fully satisfying the power supply demands for short-term peak load and impulsive load. In addition, the control of the power supply system is simple, and the power supply controller has a small volume, small mass, low heat consumption and high reliability. The disadvantage lies in that the bus voltage has a relatively large variation range, secondary transformation is needed for the electrical equipments that have high requirement on the input voltage, and the design of the secondary power supply is relatively difficult. For the unregulated bus system of the DET type, the output voltage of the solar array is always clamped by the battery pack voltage and cannot work at the maximum operating point, which reduces the utilization efficiency of the solar cells.

11.6.1.4 Battery Pack Scheme

Battery pack scheme design mainly includes electrochemical system, numbers of connections in series or in parallel, capacity design, etc. As lithium-ion battery pack has significant advantages over cadmium-nickel cells and hydrogen–nickel cells in terms of specific energy, specific power and self-discharge rate, the analysis of the electrochemical system in this section mainly focuses on lithium-ion battery pack. When determining the electrochemical system, the constraints, such as specific energy, operating temperature and cycle/storage life of the battery pack, should be considered. Presently, lithium-ion battery pack for space applications mainly includes $LiCoO_2$ system and $Li(Ni_xCo_yAl_{1-x-y})O_2$ system. Under the existing technical conditions, the $Li(Ni_xCo_yAl_{1-x-y})O_2$ system is superior to the $LiCoO_2$ system in terms of specific energy and cycle/storage life, but is inferior to the $LiCoO_2$ system in terms of high- and low-temperature performance. Deep space exploration has rigorous requirements on the low-temperature discharge performance of lithium cells due to environmental constraints. When selecting an electrochemical system, the influence of the above constraints must be comprehensively considered. For the design of the numbers of cells in series or parallel connections, it should be calculated and determined according to the bus voltage level and topology of the probe power system. The capacity of the battery pack should be determined according to the power demand of the probe during the shadow period, the number of cells in series connection in the battery pack and the average discharge voltage of the battery cells.

11.6.2 Solar Array Design

The solar array design mainly includes the design of the solar array output power and the series/parallel design.

11.6.2.1 Output Power Calculation

The solar array power output includes power supply and charging power. The output power of the solar array at the beginning of life is calculated as:

$$P_{BOL} = S_0 X X_S X_e A_C N F_j \eta F_C (\beta_P \Delta T + 1) \cos\theta \tag{11.2}$$

where

P_{BOL} is the maximum output power of the solar array at the beginning of life, the unit is watt (W);
S_0 is the solar constant: 1353 W/m²;
θ is the angle between the sunlight and the normal direction of the solar array, the unit is degree (°);
X is the correction factor when the sunlight is obliquely illuminating the solar array, and is generally 0.95–1.00;
X_S is the seasonal variation factor of solar intensity;

X_e is the gain factor of the Earth reflection to the output power of the solar array and is assumed to be 1.0 for the deep space probe;
A_C is the nominal area of a single solar cell, in cm^2;
N is the total number of all solar cells in the solar array;
η is the photoelectric conversion efficiency of the solar cell;
F_C is the solar array combination loss factor;
β_p is the solar array temperature loss factor;
ΔT is the difference between the solar array temperature and the standard test temperature (25 °C), in °C.

The output power of the solar array at the end of life is calculated as:

$$P_{EOL} = P_{BOL} \times F_{RAD} \times F_{UV} \times F \quad (11.3)$$

where

P_{EOL} is the maximum output power of the solar array at the end of life, in W;
F_{RAD} is the particle radiation attenuation factor of the solar array;
F_{UV} is the ultraviolet radiation attenuation factor of the solar array;
F is the other attenuation factor of the solar array.

The output power obtained by the Eqs. (11.2) and (11.3) is the output power at a distance of 1 AU from the Sun. As the sunlight intensity is inversely proportional to the square of the distance, for deep space probes, the actual solar array output power should be calculated according to the distance from the Sun. In addition, for the probes in the landing exploration missions of celestial bodies with atmosphere, when calculating the actual solar array output power, the attenuation effect of the atmosphere on the sunlight intensity should be also considered.

11.6.2.2 Series and Parallel Design

Whether it is a charging array or a power supply array, the solar cell circuit relies on a number of solar cells in series connection to achieve the required voltage. The factors determining the number of solar cells in series connection are: (1) bus voltage or maximum charging voltage of battery pack; (2) line voltage drop; (3) the output voltage of individual solar cell at the end of life under maximum solar incident angle, and the loss factors for the worst scenario and radiation dose margin should be also taken into account. The number of parallel solar cell strings is calculated according to the maximum load current of the solar array and the optimum operating point current of the single solar cell at the end of life under the worst scenario.

The solar cell circuit depends on a number of individual solar cells connected in series to achieve the required voltage. The number of solar cells in series connection is determined by factors such as the bus voltage or the highest charging voltage of the battery pack, the line voltage drop as well as the electric performance during

11.6 Deep Space Power System Design

operation of the single solar cell at the end of life (with the radiation margin and loss factors considered).

1) Number of solar cells in series of a sloar array.

The calculation of the number of solar cells in series connection is expressed in Eq. (11.4):

$$N_S = \frac{V_B + V_D}{V_{mp}} \qquad (11.4)$$

where

N_S is the number of solar cells in series connection in solar cell circuit;
V_B is the bus voltage or battery pack charging voltage;
V_D is the line voltage drop, including the voltage drop across the isolation diode, wire and connector.
V_{mp} is the end-of-life electric performance of the solar cell.

The calculation of V_{mp} is expressed in Eq. (11.5):

$$V_{mp} = V_{mp0} \times \phi_v + \beta_v \times (T - 25) \qquad (11.5)$$

where

V_{mp0} is the optimum operating point voltage of the solar cell at the beginning of life under standard test conditions;
Φ_v is the product of the loss factors (particle irradiation attenuation factor, measurement error factor and other loss factors) at the optimum operating point voltage;
β_v is the temperature coefficient of the optimal operating point voltage at the end of life;
T is the solar cell operating temperature at the end of life.

2) Number of solar cells in parallel of a solar array.

The total number of solar cells in parallel strings in a solar array is expressed in Eq. (11.6):

$$N_P = \frac{I_B}{I_{mp}} \qquad (11.6)$$

where

N_P is the total number of parallel solar cell strings of solar arrays;
I_B is the maximum current of the bus or the maximum charging current of the battery;
I_{mp} is the optimum operating point current of the single cell at the end of life.

The calculation of I_{mp} can be found in Eq. (11.7):

$$I_{mp} = I_{mp0} \times \varphi_i + \beta_i \times (T - 25) \quad (11.7)$$

where

I_{mp0} is the optimum operating point current of the solar cell at the beginning of life under standard test conditions;
Φ_i is the product of various loss factors (particle irradiation attenuation factor, ultraviolet radiation loss factor, combined loss factor, measurement error factor, and other loss factors) of the optimum operating point current;
β_i is the temperature coefficient at the end of life of optimum working point current;
T is the solar cell operating temperature at the end of life.

After determining the total number of parallel connections of the solar array, the number of parallel connections of each solar cell circuit is determined according to factors such as the number of power sliding rings of the driving mechanism, the number of shunting channels of the shunt, the shunting capacity at each level, the junction capacitance of the solar cells and the charging control mode of the battery pack. For the linear shunt regulation system in which the shunt regulator is placed outside the probe, the number of parallel connections per solar cell circuit depends mainly on the maximum adjusted current value of shunt adjusting pipe at each level. For a switching shunt regulation system in which the shunt regulator is located within the probe, the number of parallel connections per solar cell circuit depends mainly on the ability of the power sliding ring of the driving mechanism to convey current. The number of parallel connections of single solar cell circuit is determined by Eq. (11.8):

$$N_{PL} = \frac{I_L}{I_{mp0}} \quad (11.8)$$

where

N_{PL} is the number of parallel connections of each solar cell circuit, taking the largest integer less than the value to the right of the equation;
I_L is the maximum current of each shunted level or the rated current delivered by a single power sliding ring.

11.6.3 Battery Pack Design

The battery pack design includes battery output power calculation, battery discharge depth calculation and rated capacity calculation.

11.6.3.1 Battery Pack Output Power Calculation

The formula for calculating the output power of the battery pack is:

11.6 Deep Space Power System Design

$$P_D = \frac{P_L}{\eta K_L} \quad (11.9)$$

where

P_D is the output power of the battery pack, in W;
P_L is the load power, in W;
η is the discharge regulator efficiency;
K_L is the discharge loop loss factor.

11.6.3.2 Battery Pack Discharge Depth Calculation

The calculation formula of battery discharge depth is as follows:

$$D = \sum \frac{t_d I_d}{C} \times 100\% \quad (11.10)$$

where

D is the depth of discharge, %;
I_d is the discharge current of the battery pack in one or several orbital periods, in A;
t_d is the discharge time of the battery pack in one or several orbital periods, in h;
C is the rated capacity of the battery pack, in Ah.

The charge/discharge cycle life of the battery pack is closely related to the depth of discharge, and the relationship is approximately exponential. The general allowable values of the discharge depth are shown in Table 11.6.

Table 11.6 Allowable values of battery discharge depth

Battery type	Probe with frequent charge and discharge cycles /%	Probe with infrequent charge and discharge cycles /%
Nickel–cadmium battery	15–30	50–60
Hydrogen–nickel battery	30–40	70–80
Lithium-ion battery	15–30	70–80

11.6.3.3 Calculation of Rated Capacity of Battery Pack

According to the power requirements of the probe in shadow period, the number of single cells in series connections and the average discharge voltage of the single cell, the capacity of the battery pack is calculated according to the following formula:

$$\begin{cases} C = \dfrac{P_1 \tau_1 + P_2 \tau_2}{\eta U_{bd} D} \\ U_{bd} = (N - n)U_d - U_{dd} - U_{cd} \end{cases} \quad (11.11)$$

where

P_1 is the power requirement during the shadow period, in W;
P_2 is the average power requirement of supplementary discharge of the battery pack in the lighting zone, in W;
τ_1 is the battery discharge time during the shadow, in h;
τ_1 is the battery discharge time during lighting time, in h;
U_{bd} is the average discharge voltage of the battery pack, in V;
D is the discharge depth of the battery pack;
U_d is the average discharge voltage of the battery cell at the end of life, in V.

For lithium-ion battery pack, the combination mode, in which multiple single cells are connected in parallel first and then connected in series, may be applied, and the capacity of the battery pack is the sum of the capacities of the parallel single cells.

11.6.4 Power Controller Design

The function of the power control device is to adjust the power balance among the solar array, the battery pack and the load, so as to stabilize the bus voltage and realize the management functions of charging and discharging the battery. The power control device mainly consists of the main error amplifier circuit, solar array shunt and voltage-stabilizing regulation circuit, battery pack charge regulation circuit, battery pack discharge regulation circuit, etc. In the design of the power controller, the following aspects should be included:

(1) Shunt regulator design. It should be determined by the series connection-type regulation technique or the parallel connection-type regulation technique adopted; if the parallel connection regulation technique is adopted, the sequential partial linear shunt regulation technique, S^3R type or S^4R type, should be selected.
(2) Charge control design. It includes charging system selection, end of charge voltage control methods, etc.
(3) Discharge control design. The battery pack discharge mode includes direct discharge and indirect discharge through the discharge regulator; the discharge regulator mode may be further subdivided into buck-type discharge regulation and boost-type discharge regulation according to the relationship between the battery pack voltage and the bus voltage.
(4) MEA design. The main error amplifier (MEA) module controls the operating states of the shunt regulation module, the charging control module and the discharge regulation module, by which the power balance among the solar array, the battery pack and the load is also adjusted to ensure the stability of the bus output.

11.6 Deep Space Power System Design

The specific design method should be comprehensively considered based on the bus voltage, load characteristics as well as the charge and discharge characteristics of the battery.

11.6.5 Example of Power System Design

Chang'e 3 probe consists of a lander and a rover, both of which adopt the triple-junction GaAs solar array + lithium-ion battery power generation system. The lander can provide no less than 1600 W power supply capacity during the Earth-to-Moon transfer, Moon orbiting and descent phases. As for the power supply capacity of the rover, its primary power supply provides an output power of no less than 220 W during the lunar daytime. As the lander and rover adopt similar design methods for solar array, battery pack and power controller, this section only takes the lander as an example.

11.6.5.1 Lander Power System Design

The primary power subsystem of the lander adopts the fully regulated bus system, and the bus voltage is kept within the range of 29 ± 1 V, providing the load power of the lander in the Earth-to-Moon transfer, Moon orbiting, lighting period of lunar surface exploration. The lander primary power subsystem consists of the solar array (circuit part), the lithium-ion battery pack, power controller and wake-up load. The component block diagram of the primary power subsystem is shown in Fig. 11.16, and the principle block diagram is shown in Fig. 11.17.

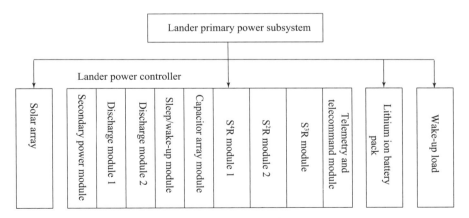

Fig. 11.16 Block diagram of the lander primary power subsystem

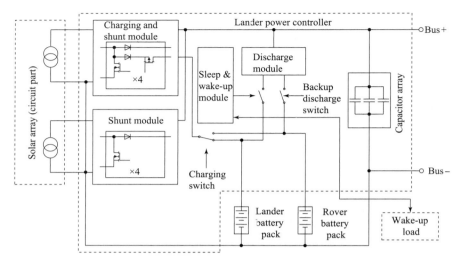

Fig. 11.17 Principle block diagram of the lander primary power subsystem

11.6.5.2 Lander Solar Array Design

1) Solar cell

As the basic power generation unit of the solar array, triple-junction GaAs solar cells (GaInP$_2$/InGaAs/Ge) with an average efficiency of better than 28.3% are adopted. The main design parameters are shown in Table 11.7.

2) Solar cell loss factors

Table 11.7 Design parameters of three-junction GaAs solar cell

Parameter	Value
Solar cell structure	N/P-type triple-junction GaInP$_2$/InGaAs/Ge solar cell
Solar cell size /mm	60.5 × 40.0 (lack of two corners), cell area 24 cm^2
	40.0 × 30.3 (lack of a corner), cell area 12 cm^2
Absorption rate α_s	≤0.92
Hemispherical radiance ε_H	0.85 ± 0.02
Open-circuit voltage V_{oc} /mV	2650
Short-circuit current density J_{sc}/(mA·cm^{-2})	17.1
Working voltage V_{mp} /mV	2280
Working current density J_{mp}/(mA·cm^{-2})	16.8
Average photoelectric conversion efficiency (AM0, 135.3 mW/cm^2, 25°C)/%	28.3

11.6 Deep Space Power System Design

The solar cell loss factors are shown in Table 11.8:

Table 11.8 Solar cell loss factors

Loss factor	K_I	K_V	K_P
Combined loss	0.98	1	0.98
Ultraviolet loss	0.98	1	0.98
Temperature alternating loss	0.99	1	0.99
Temperature coefficient	+0.007 mA/cm^2·°C	−6.50 mV/°C	−0.24%/°C
Irradiation loss	0.98	0.952	0.93

3) Solar cell circuit layout and arrangement design

The lander solar array contains a total of four solar panels, namely ±Y panel, +X panel and +X panel of −Y module, of which the latter two are body mounting panels. Under the comprehensive consideration of the carbon fiber aluminum honeycomb substrate area, as well as the solar cell arrangement direction and coefficient, the layout is designed based on the theoretical calculation results of the solar cell series/parallel connection design. The solar cell arrangement is shown in Fig. 11.18.

4) Solar array output power

When the solar array is perpendicular to the sunlight direction, the output power is shown in Table 11.9 for the unfolded panel at 125 °C and for the body-mounted panel at 138 °C.

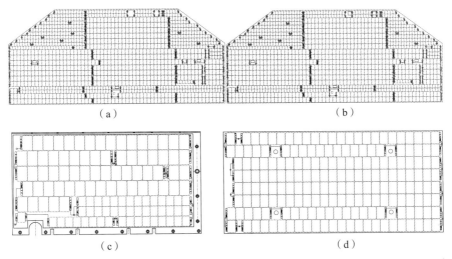

Fig. 11.18 Solar array circuit cell arrangement. **a** +Y panel; **b** −Y panel; **c** −Y Cabin +X panel; **d** +X panel

Table 11.9 Output power of solar arrays perpendicular to sunlight

Name	Beginning of life			End of life		
	Voltage /V	Current /A	Power /W	Voltage /V	Current /A	Power /W
+Y panel	29	16.25	471.25	29	15.44	447.76
−Y panel	29	16.25	471.25	29	15.44	447.76
+X panel	29	3.30	95.7	29	3.14	91.06
+ X panel of − Y module	29	1.65	47.85	29	1.57	45.53
Total	29	37.45	1086	29	35.59	1032

11.6.5.3 Lander Lithium-Ion Battery Pack Design

As restricted by the mass of the probe, lithium-ion battery equalization control is not applied. To ensure that the performance of each unit of the battery is not becoming significantly non-uniform during the 1-year lifetime of orbital operation, the individual cells are screened to ensure the uniformity of these cells. The cell formation efficiency, charge and discharge efficiency, internal resistance, self-discharge efficiency and monomer capacity index are compared, and the cells with the uniform indexes are selected to form a battery pack.

The lithium-ion battery pack adopts a drawbar-type structure to achieve tight assembly between the individual cells, which can prevent the single cells from being deformed due to internal pressure, and also facilitate heat dissipation. The single cell is separated from the structural member by an insulating polyimide pressure-sensitive adhesive tape to ensure excellent insulation between the single cell and the structural member. Lithium-ion cells are electrically connected by a guide strip, and the cell and the wires are connected via terminals. All current lines and voltage lines converge on different electrical connectors according to different functions, respectively, so that signal transmission is isolated from power transmission (Fig. 11.19).

With the increase of the number of charge and discharge cycles, due to the accumulation of the differences in charge-accommodating capability and self-discharge rate of the individual cells in the lithium-ion battery pack, the voltage difference between the cells in the battery pack is getting increasingly large with a diverging trend. To ensure the safety of the lithium-ion battery pack, each single cell in the battery pack must be subjected to voltage-limiting equalization charging control during the charging process of the battery pack.

The lithium-ion battery pack consists of a group of 60 Ah lithium-ion cells and is formed by fourteen 30 Ah lithium-ion single cells arranged in 2 × 7 parallel/series connection.

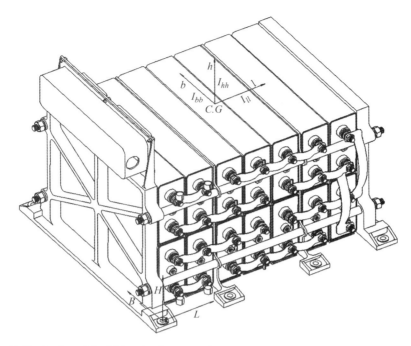

Fig. 11.19 Outline of the lithium-ion battery pack

11.6.5.4 Power Controller Design

The power controller consists of four shunt regulation circuits, four charge shunt regulation circuits and six discharge regulation circuits. The bus output voltage of the shunt regulation module is in the range of 29.65–29.80 V, the bus output voltage of the charging shunt regulation module is in the range of 29.45–29.60 V, and the bus output voltage of the discharge regulation module is in the range of 28.2–29.2 V. The working principle is as follows.

(1) If the load power is small, the shunt regulation module can provide the load power, and the charging shunt module can provide the charging power of the battery pack, and the excess power is shunted to the ground.
(2) If the load power increases and the shunt regulation module is with insufficient output power, the charging shunt regulation module reduces the charging power of the battery pack to supplement the load power. If the charging shunt regulation module still fails and does not satisfy the load power demand, the battery pack would discharge through the discharge regulation module and supply power jointly with the solar array to the load.
(3) During the shadow period, the battery pack would supply power to the load via the discharge regulation module.

11.7 Prospects

Among the power generation technologies, high-efficiency solar cells and nuclear power technologies are regarded as the key development directions for deep space exploration power generation technology.

Presently, triple-junction GaAs has become the preferred choice of solar arrays in deep space exploration. Continued improvement of solar cell conversion efficiency is the focus of future research. On one hand, the efficiency of the triple-junction GaAs cell may be improved with improving the existing techniques and processes; on the other hand, the research on the multi-junction GaAs solar cell is carried out through the research results of the new material, and the lattice mismatching problem of multi-junction GaAs cells is solved by the method of reverse growth and double-sided growth so as to reduce the influence of mismatching and misalignment on the overall cell performance. In addition, the spectral matching must be so designed that the conversion efficiency of the cell should be further improved based on the lighting conditions of the target planet. For inner planet exploration, high-temperature-resistant cells, low absorption rate glass cover sheets, high-temperature-resistant and strong-radiation-resistant back sheet glues and cover sheet glue are also required for solar arrays, and appropriate thermal control and space protection measures should be taken. For outer planet exploration, ultra-flexible solar arrays may be applied to expand the cell array area, or concentrator solar array may be used to increase the light intensity and the generated power in low-light conditions.

The nuclear power technology has a number of advantages such as being able to work continuously without being affected by the distance from the Sun and shadow period, being able to work in a harsh radiation, climate and magnetic field environment, and having a compact structure and a small volume, which make it a useful supplement to the solar cell power generation technology. Presently, it mainly includes two types, namely RTG and nuclear reactor, in which RTG technology is the most developed and widely applied, while the nuclear reactor has the superiorities of high output power and conversion efficiency.

In terms of the energy storage technologies, high-density lithium-ion battery pack still occupies a prevailing position and has gradually replaced cadmium–nickel batteries and hydrogen–nickel batteries. Presently, the lithium-ion battery pack with a specific energy of 170 Wh/kg has been developed by SAFT company and was flight-proven, and the $LiCoO_2$ lithium-ion battery pack developed in laboratories has a specific energy of 250 Wh/kg. The research directions for key technologies of lithium-ion battery in future space applications include advanced materials manufacturing technology, safety research and new energy storage power management technologies. In addition, the flywheel energy storage offers superiorities of quick release of energy and low self-discharge rate. If its applicability became comparable with the storage cell, it would be expected to be combined with the battery cell for application in the exploration missions in the far deep space.

Among the power control technologies, MPPT has a strong advantage and has been widely applied in spacecraft power supplies. In the future, it is expected to be combined with other power control technologies to complete the power regulation of solar arrays. With a newly developed digital power supply control technology, power supply, loop protection and communication interface can be controlled via programming. Through programming and algorithm improvement, the performance of the power supply can be improved flexibly.

To sum up, the development of new technologies in power generation, storage, transmission, transformation and utilization of space power systems should be promoted to adapt to the particularity and diversity of future deep space exploration missions. With the application of new technologies, in the next 20 years, the total mass of power supply systems is expected to be reduced by 10%, and the energy utilization efficiency is expected to be improved by 5%–10%, and the power supply technology for deep space exploration will be developing into new stage. Keywords such as high efficiency, high power, long life, high reliability, light weight and intelligence will become the development directions and main features of the new generation power systems for deep space exploration.

References and Related Reading

1. Li G (2008) Introduction to spacecraft power system technology. China Aerospace Publishing House, Beijing (Chinese vision)
2. Zhuang J, Wang X, Feng J (2011) Overlapping effect of lunar dust sediment on solar cell. Spacecraft Environ Eng 27(4):409–411 (Chinese vision)
3. Yu D (2015) Lunar soft Landing detector technology. National Defence Industry Press, Beijing (Chinese vision)
4. Sun Z, Zhang H, Wu X et al (2010) Mission analysis of a lunar soft lander. Spacecraft Eng 19(5):12–16 (Chinese vision)
5. Geoffrey AL, Dan H (2006) The solar spectrum on the Martian surface and its effect on photovoltaic performance. In Proceedings of the IEEE 4th world conference on photovoltaic energy conversion. IEEE, New York, pp 151–154
6. Liu Z, Wang F, Chen Y, Huang S et al (2016) Impact analysis and solution of solar array design in Martian surface environment. Spacecraft Eng 25(2):39–45 (Chinese vision)
7. Liu Z, Cai X, Chen Q et al (2011) Overview of space power system design using MPPT for deep space spacecraft. Spacecraft Eng 20(5):105–110 (Chinese vision)
8. Jensen H, Laursen J (2002) Power conditioning unit for Rosetta/Mars Express. ESA, Paris
9. Loche D (2008) Mars Express and Venus Express power subsystem in-flight behavior. ESA, Paris
10. Fiebrich H, Haines J, Tonicello F (2004) Power system design of the Rosetta spacecraft. In: Proceedings of the 2nd IECEC. AIAA, Washington, 1–7; Croci L, Caccivio M (2008) Electrical power system for GAIA. In: Proceedings of the 8th European space power conference. ESA, Paris, pp 386–392
11. Luca AD, Gray R, Otero J (2008) The LISA Pathfinder power system. In: Proceedings of the 8th European space power conference. ESA, Paris, pp 296–301
12. Lempereur V, Jauquet D, Liegeois B (2008) Power conditioning and distribution unit of Globalstar-2 constellation. In: Proceedings of the 8th European space power conference. ESA, Paris, pp 119–125
13. Breier N, Kiewe B, Mourra O (2008) The power control and distribution unit for the SWARM satellites. In: Proceedings of the 8th European space power conference. ESA, Paris, pp 289–295

14. Dakermanji G, Person C, Jenkins J (2005) The MESSENGER spacecraft power system design and early mission performance. In: Proceedings of the 7th European space power conference. ESA, Paris, pp 1–8
15. Maset E, Sanchis-Kilders E, Weinberg AH (2008) Ion drive propulsion MPP power conditioning system without battery. In Proceedings of the 8th European space power conference. ESA, Paris, pp 239–247
16. Liu Z, Cai X, Du H (2013) Analysis on power System Design for Mars orbiting probe. Spacecraft Eng 22(1):60–64 (Chinese vision)
17. Li C, Zhu L, Wang B, Guan Z (2011) Design and implement of power control unit for a lunar orbiter. Chin J Power Sources 35(10):1255–1258 (Chinese vision)
18. Cai X, Liu Z, Zhang M, Lin W (2013) Design and simulation of a mixed MPPT electrical power system topology. Spacecraft Eng 22(3):77–82 (Chinese vision)
19. Lei Y, Zhang M, Jin B, Ma L (2014) Research on autonomous sleep-reboot of lunar probe. Spacecraft Eng 23(6):13–16 (Chinese vision)
20. Lei Y, Zhang M, Jing Y, Jin B (2014) A method of power sharing for deep space probe. Spacecraft Eng 23(1):58–62 (Chinese vision)
21. Ma S (2001) Satellite power technology. China Aerospace Publishing House, Beijing (Chinese vision)
22. Peng C (2011) The system design of spacecraft. China Science and Technology Press, Beijing (Chinese vision)
23. Patel MR (2010) Spacecraft power system. China Aerospace Publishing House, Beijing (Chinese vision)
24. Ouyang Z, Li C, Zou Y et al (2002) Progress of deep space exploration and chinese deep space exploration stratagem. Chinese Society of Astronautics, Beijing (Chinese vision)
25. Ye P, Peng J (2006) Deep space exploration and its prospect in China. Eng Sci 8(10):13–18 (Chinese vision)
26. Yang M, Jia Y, Chen J (2008) Research on system design of lunar rover. Aerosp Control Appl 34(3):12–16 (Chinese vision)
27. Peng J, Liu Z, Zhang H (2008) Conceptual design of a lunar lander. Spacecraft Eng 17(1):18–23 (Chinese vision)
28. Wu W, Liu W, Jiang Y (2011) The revelation from the development of foreign deep-space exploration beyond the moon. China Aerosp 7:9–12 (Chinese vision)
29. Carr GA, Jones L, Moreno V (2012) Mars Science Laboratory (MSL) power system architecture. AIAA, Atlanta
30. Huang C (2004) Research status and prospects of space power. Electron Sci Technol Rev 5:1–6 (Chinese vision)
31. Zhang J, Xu J, Jia W, Qiu B, Xiao J (2016) Application of solar array and analysis of key technologies in deep space exploration. J Deep Space Explor 3(1):3–9 (Chinese vision)
32. Liu Y, Wang C (2015) Research progress of solar cells for space aircraft. Power Technol 39(10):2325–2337 (Chinese vision)
33. Xi Z (2013) Design of new deep space explore power conditioning unit. Tianjin University, Tianjin (Chinese vision)
34. Bergsrud C, Straub J, Noghanian S (2013) Space solar power satellite systems as a service provider of electrical power for lunar industries. AIAA, USA
35. Butler (2013) The Van Allen probes power system launch and early mission performance. AIAA, USA
36. Sun D, Huang C (2005) Space electrical power subsystem with lithium-ion battery. Power Supply Technol 10:687–690 (Chinese vision)
37. Wang C, Li K, Ren J, Pan Y, Li G (2006) Research ideas on energy storage technology for lithium ion batteries for deep space exploration. In: Xi'an: The 3rd academic conference of the deep space exploration technology committee of the Chinese Society of Astronautics (Chinese vision)
38. Zhu A, Zou D, Tang Y, Hao X, Zhao S (2012) Application of space nuclear power in deep space exploration. Chinese Society of Astronautics, Beijing (Chinese vision)

39. Zhang M, Cai X, Du Q, Lei Y (2013) Research on nuclear reactor in space application. Spacecraft Eng 22(6):119–126 (Chinese vision)
40. Kang H (2011) Review of isotopic power system. Chin J Power Sources 35(8):1031–1033 (Chinese vision)
41. Wang T (2015) Progress of radioisotope thermoelectric generator for deep space exploration. Chin J Power Sources 39(7):1576–1579 (Chinese vision)
42. Hou X, Wang L (2007) Introduction of US space radioisotope power systems. Spacecraft Eng 16(2):41–49 (Chinese vision)
43. Zhang J, Wang Z, Ren B, Tang Y, Cui P (2007) Strategic study on radioisotope thermoelectric generators in lunar exploration program and deep space exploration program. J China Acad Electron Inf Technol 2(3):319–324 (Chinese vision)
44. Xu H (2009) Design and simulation of deep space detector power system scheme. Harbin Institute of Technology, Harbin (Chinese vision)
45. Guo W, Liu Z, Cai X, Chen Q (2014) Review of electrical power system scheme for mars probe during EDL. Spacecraft Eng 23(5):111–115 (Chinese vision)
46. Stella PM, Mardesich N, Edmondson K (2008) Mars optimized solar cell technology (MOST). In: Proceedings of the IEEE 5th world conference on photovoltaic energy conversion. IEEE, New York, pp 84–89
47. Crisp D, Pathare AV, Ewell RC (2003) The performance of GaAs/germanium solar cells at the Martian surface. ActaAstronautica 54(2):83–101
48. Edmondson K (2005) Simulation of the Mars surface solar spectrum for optimized performance of triple-junction solar cells. In: Proceedings of the 19th space photovoltaic research and technology conference. NASA, Washington, D.C., pp 67–78
49. Croci L, Caccivio M (2008) Electrical power system for GAIA. In: Proceedings of the 8th European space power conference

Chapter 12
Autonomous Management and Tele-operation Technology

12.1 Introduction

With the continuous development of space technology, space competition is becoming increasingly fierce. And it has become a highlight and tough task to improve the on-orbit survivability of spacecraft and performance of the spacecraft. In deep space exploration missions, it has become the key research content in spacecraft operation and control to improve the efficiency of completing deep space exploration missions. The technical approaches to solve this problem include autonomous management technology and tele-operation technology, focusing on improving the autonomy of the probe and the level of ground operation and control technology, respectively.

With the development of computer technology, especially the artificial intelligence technology, it has become technologically feasible for deep space probes to complete mission planning, failure diagnosis and troubleshooting autonomously. Therefore, the spacecraft autonomous management technology is being increasingly highlighted.

For spacecraft autonomous management technology, an intelligent autonomous management software system is installed in the spacecraft, which autonomously performs planning and scheduling of engineering tasks and scientific tasks, command execution, monitoring of onboard conditions as well as system reconstruction from failure state, to enable unmanned long-time autonomous and safe operation of the probe. In the field of deep space exploration, the autonomous management technology has gradually become a key technology under rapid development.

Tele-operation technology solves the problems of remote interactions between humans and probes. It enables astronauts to carry out on-orbit operations by technical means, such as cameras and ground tele-operation by personnel on the Earth.

The roving exploration of extraterrestrial celestial bodies is an important form of deep space explorations. And the design of tele-operation and autonomous functions of the rovers on the surface of the planet is directly related to the efficiency of the completion of the rover mission. It is an important part of the design of the rover system. Based on the system design of Chinese lunar rover and Mars rover, this chapter briefly introduces the application of autonomous management technology and tele-operation technology in deep space exploration and looks forward to the future development direction of these technologies.

12.2 Autonomous Management Technology for Deep Space Probes

Autonomy refers to the attribute of a system, which does not need external support for a specified period of time according to the mission. In the deep space exploration mission, the target has a long distance, a long flight time and a dynamic and variable environment. Therefore, the operation and control of deep space probes vary greatly from those of the near-Earth satellites in terms of multiple factors such as the delay of uplink instructions, spatial occlusion of the stars, low data transmission rate and long-term reliable operation requirements. The traditional large loop operation and control mode of ground TT&C station-spacecraft limit the real-time control and long-term safe operation of deep space exploration missions. Therefore, the establishment of the on-orbit autonomous management system based on the onboard computer hardware and software systems is an important direction for the future technological development of deep space explorations.

The on-orbit autonomous health and autonomous management technology of the spacecraft involves many aspects. It applies technology such as artificial intelligence, intelligent control and image processing and proposes extremely high demands for spacecraft hardware such as onboard computers, various sensors, payloads and actuators.

With the extension of exploration missions to more distant celestial bodies, the requirement of spacecraft autonomous management technology is more and more urgent, which is mainly manifested in facts that the deep space probe is far away from the Earth, the communication delay increase and real-time control capabilities decrease. The time delay poses challenges to spacecraft control with high real-time requirements, and the mission execution efficiency is lowered. With the continuous increase of deep space missions, deep space stations on the Earth are becoming resource-intensive, and time-sharing tracking measurements may also result in decrease in exploration efficiency. Deep space probe payloads, especially for the imaging equipment, generate a large amount of scientific data during the exploration process. The contradiction between the transmission demands of massive data and the declination in data transmission rate due to long distance is becoming increasingly prominent. The environment in which deep space probes are operated is complicated, and emergency situations such as failures require immediate processing to ensure the safety of the probes.

Therefore, the designers of spacecraft are inspired to build a system-level autonomous system on the spacecraft, rationally allocate the tele-operation and autonomous functions of the spacecraft, so that the spacecraft can autonomously perform tasks such as mission planning, command execution, failure diagnosis and recovery. A planning sequence that satisfies flight constraints and resource constraints within a certain period of time may be established according to mission-level commands received from the Earth, and then, the mission planning sequence may be converted into commands executable by the hardware system of the probe at a lower level. In the process of execution, the execution status of the commands is monitored, and the health conditions of the probe are inferred based on the measured information to enable further troubleshooting, system reconstruction, etc. The probe may be set to safe mode provided that the probe encounters a failure that cannot be handled by the autonomous operating system. In this case, the relevant status information is transmitted to the Earth station to request the maintenance commands from the personnel on the Earth. This autonomous system for spacecraft operation may greatly save the labor of the control personnel, reduce the demands for the deep space communication network, increase the reliability and real-time capability of the mission and reduce the cost for mission operation.

The advantages of spacecraft autonomous management technology are mainly reflected in the following aspects: The daily management of spacecraft is mainly carried out by spacecraft to save manpower and material resources; the interaction between the probe and the Earth is reduced by improving the autonomy level, making the spacecraft less susceptible to interference and attack and improving the timeliness for the spacecraft to handle failures and emergencies. The tasks of autonomous control include autonomous attitude control, navigation, thermal control, spacecraft health monitoring, system reconstruction, planning and scheduling, intelligent execution, autonomous onboard data processing, etc.

Based on the design of the autonomous management system of Mars rover, the implementation concepts and technical development directions for the deep space probe autonomous management are introduced. With major scientific issues of human interests such as the origin of life, Mars has continuously become a highlight for deep space exploration for a long period. Polar regions, soil thermostatic layers and Mars satellites are still important targets for explorations. In the landing and roving missions of Mars, the impact of dust storms, low air pressure and low-temperature environment are problems that the probes must encounter. The distance between the Earth and Mars is so distant that the information time delay may be up to several tens of minutes. Considering other factors such as the influence of the occlusion of the stars, the limited communication window, the efficiency of the tele-operation of Mars rover is greatly reduced. In the process of movement on Mars, the rover may encounter obstacles, slips and other difficulties, which will seriously affect safety. Stated thus, the ability of autonomous mission planning and behavior planning of the rover should be improved.

12.2.1 Development of Autonomy Capabilities

Autonomous operation of spacecraft refers to the modern control technology based on artificial intelligence and other modern control technologies. It establishes a remote agent on the spacecraft, which enables the spacecraft to self-manage and complete the mission. The goal is to achieve information injection and control independently from the outside world or to rely on external control as little as possible to accurately sense its own state and external environment and make appropriate decisions based on such information and the tasks assigned by the user and control spacecraft independently to accomplish various missions. The tasks of autonomous operation include autonomous survival and autonomous mission execution. The gradual improvement of the autonomy of Mars rovers reflects the development of the autonomous management technology for spacecraft.

At present, in the past 50 years, four Mars rovers have been successfully launched onto the surface of Mars for roving missions, all of which are developed by the USA. The autonomy technologies applied by such rovers are as follows.

1. **Sojourner**

The Sojourner rover for Mars Pathfinder project was launched in December 1996 with a tele-command mode. The time delay between the Earth and Mars made it impossible for tele-operation engineers to control Sojourner in real-time. The rover communicated periodically with the Earth through the lander, and the rover only performed simple command sequences. The communication between the lander and the Earth was carried out twice a day for 2 h. The rover telemetry analysis was done by the rover control station on the Earth. At the end of each mission day, the scientists on the Earth analyzed the images captured by the rover and lander to determine and designate the target position and path for the next movement of the rover. The limited autonomy of Sojourner was mainly reflected in three aspects: autonomous terrain crossing, autonomous emergency handling and autonomous resource management.

2. **Spirit and Opportunity**

Mars Exploration Project (MER) adopted Mars rovers Spirit (launched in June 2003) and Opportunity (launched in July 2003). The main mode of operation was autonomous navigation and tele-operation. Specifically, the mission execution mode of Mars rover in MER was carried out on a daily cycle of Mars. The sequence of commands sent in the morning was based on the images and data returned the previous day and was completed by the team of tele-operation scientists on the Earth for a specific Martian day. After such activities, the images and data obtained by the rover were transmitted back to the Earth. The work time of each Martian day was about 4 h, which was generally around the noon. The autonomy technology applied by MER solved the problems of unmanned monitoring, motion control over complicated terrain, etc.

3. **Curiosity**

Mars rover Curiosity (launched in November 2011) used in Mars Science Laboratory Project (MSL) had the main working mode of long-distance autonomous navigation plus tele-operation, and the mode might switch from a flight state to a Mars operation state just by maintenance of the on-orbit software. The efficient path planning decisions made by curiosity during Mars roving as well as the precise control of the mechanism in the process of sampling were all autonomously performed by the probe.

The autonomous safety management design of curiosity is similar to that of MER. Most of autonomous safety management contains failure handling strategies. If a failure is detected, the system only stops the related activities, and the other activities could continue without being interfered. However, in case of some serious failures, such as unexpected processor restarting, low voltage and failure to receive the Earth commands for a period, the autonomous safety management will respond. This response would abort the current activities, switch to critical device if necessary and then automatically place the rover in a safe state until the Earth personnel figure out the cause of the failure and switch the rover to a normal state.

In the future Mars exploration mission, using Mars rover to explore the surface of Mars can overcome the impact of landing accuracy, expand the scope of fine exploration of the surface of Mars and select the target more accurately, which becomes an indispensable means of exploration. From the perspective of the technical development of the rovers, with the continuous improvement of the carrying capacity, the size of the rover has developed from small (10 kg level) to medium (100 kg level) and large (1000 kg level). Technology related to Mars will be further developed aiming at improving exploration capability, moving speed, terrain adaptability, autonomy, etc. Especially for autonomous management, related technology is developing very rapidly. In the future, the intelligence level of Mars rovers will be greatly improved, making it adaptable to the complicated environment of Mars surface as well as the complicated mission planning of Mars rover itself.

12.2.2 Mobile Intelligent Agent

Intelligent agent refers to a computing unit that obtains information of surrounding environment by means of perception and performs actions or behavior sequences through actuators to implement mission-level goals, including mobile intelligent agent, fixed agent, software agent, etc. Among then, mobile intelligent agent technology is gradually developed with the development of autonomous control technology of AGV, ground robot, unmanned combat platform and extraterrestrial celestial rover.

Extraterrestrial celestial rover is a kind of movable space-based robot. With the development of robot technology and artificial intelligence technology, the rover has gradually evolved from a remote control type to a tele-operation type and an intelligent agent type. As a special spacecraft for conducting roving explorations on the surface of Mars, Mars rover is characterized by mobility, adaptability, autonomy and functionality. Due to the limitation of communication delay and other factors, the rover must be provided with certain autonomy abilities to improve work efficiency and ensure its own safety. From the completion of simple tasks to more complicated scientific exploration missions, the autonomy technology of rovers has been continuously improved, upgrading rovers from the remote-controlled rovers to the tele-operation-type rovers and further to the mobile intelligent rovers. Figure 12.1 shows the development course of the autonomous technology for successfully launched lunar and Mars rovers. Figure 12.2 shows that the intelligent rover is the result of integrated development of roving technology, robot technology and intelligent agent technology.

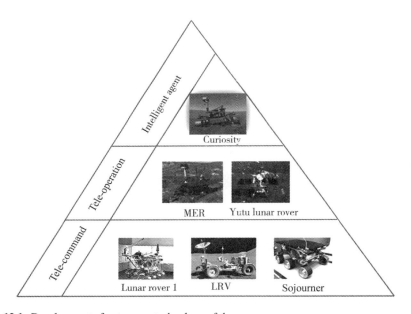

Fig. 12.1 Development of autonomy technology of the rover

12.2.3 Mars Mobile Intelligent Agent

From the perspective of spacecraft health and safety, the large communication delays may make the spacecraft in a dangerous state that the Earth operators cannot be

12.2 Autonomous Management Technology for Deep Space Probes

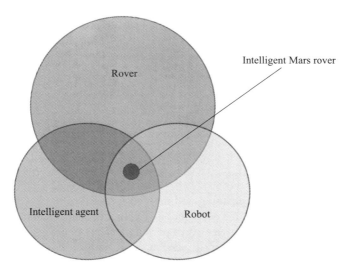

Fig. 12.2 Intelligent Mars rover

informed and the spacecraft may be out of control before receiving a disposal command sent from the Earth. Therefore, countermeasures should be embedded in the onboard system and updated with the progress of the missions. This also means that the work efficiency will be lowered by relying on control and decision-making from the Earth. In order to upgrade the probe into an intelligent agent, it is necessary to improve the autonomy and place the control loop on the probe as much as possible. The Mars rover based on the autonomy of operations on the surface of Mars is just an example of Mars mobile intelligent agent, and its technical function requirements are analyzed as follows.

12.2.3.1 Mission Planning

Spacecraft mission planning is a complicated process that combines scientific objectives as well as management requirements and constraints. Therefore, a scientific plan is generally developed into a mission plan with requirements on ensuring the health and safety of the spacecraft. The mission planning requirements of Mars rover are based on the Mars environment, ephemeris, topographic map, working capability and constraints of the rover. It gives the basic driving routes and working modes along such routes to the rover and works on the planning of solar orientation of the solar wing, the orientation of the mast to the Earth, the orientation to the orbiter and tracking of the orbiter. To sum up, it guides the whole roving exploration process of the rover.

The principle of mission planning is to achieve mission objectives efficiently and safely on the premise of satisfying constraints. The core of mission planning is the mathematical model and its behavior sequence search algorithm thereof.

12.2.3.2 Autonomous Energy Management

The autonomous energy management functions of Mars mobile intelligent agent mainly include

1) Energy balance estimation

According to the mission planning and the current state of the system (discharge depth of battery pack, etc.), the Mars mobile intelligent agent estimates the balance of power consumption of the rover over a period (usually about 24 h) in combination with the ephemeris. If the energy is safe in the overall process, no measures are required. Otherwise, it is necessary to reduce or delay the working time of high-power equipments, cut off the consumption level or adopt the means of increasing the power input by adjusting the solar wing orientation.

2) Adjustment of Solar wing Orientation to the Sun

This function can adjust the orientation of the solar wing without adjusting the heading of the Mars rover, maximize the power generation as much as possible and maintain the orientation adjustment state. The effect of this adjustment is related to the heading of the rover. In case of the North–South heading, one-dimensional orientation adjustment provides the greatest benefit in a Martian day. In case of the East–West heading, the energy that the rover can obtain is maximized at noon. However, in terms of the comprehensive effect of a Martian day, the gain is relatively small.

The adjustment method is mainly based on the ephemeris, the latitude and longitude of the position of the rover and the attitude of the rover body. When the attitude parameters provided by GNC subsystem are not reliable, the solar scanning method can also be used to rotate the solar wing to the allowable range, so as to deal with the inverse relationship between the current generated by the solar wing and the rotation angle and finally achieve the orientation with the maximized power.

As the Sun moves, the optimal orientation changes accordingly. In this case, the approach of solar wings power generation current combined with the qualitative control may be applied to maintain the tracking state. Such calculations may also be conducted by the ground or orbiter, and a sequence of the relationship between the angle of rotation and time is transmitted to the rover.

In special cases, the lunar rover body scans the output current of the solar wing in two directions, in which the heading is approximately orthogonal, so the heading angle can be roughly determined.

12.2 Autonomous Management Technology for Deep Space Probes 483

3) Low power mode of the system

To save energy, if the battery discharge depth is greater than a certain threshold (60%, for example), the system enters the minimum working mode, in which only a small number of devices such as the central computer work to receive commands from the Earth, and other devices will be shut down.

4) System sleep

In extremely special cases, for example, due to long-term dust storms, the energy state of the rover is further deteriorated (e.g., the discharge depth is greater than 80%, and it is expected that the state cannot be alleviated within a short time), the entire system will be powered off and switched to the sleep mode. The sleep mode continues until the power can maintain the minimum operating mode of the system, and the battery pack capacity is in a safe state when the system automatically restarts and enters the charging mode.

5) Mathematical model self-learning of power system

The power output of the solar wings changes with the deposition of Mars dust. The intelligent agent may establish an energy prediction model for onboard self-learning according to the output attenuation of the solar wings to ensure the accuracy of the energy balance estimation.

12.2.3.3 Independent Thermal Management

The control of the heating circuit is the general function of all spacecraft. Generally, two temperature thresholds (high and low) are set. When the temperature measured by the sensor is lower than the low temperature threshold, the heating circuit is turned on; when the temperature is higher than the high threshold, the heating circuit is turned off; if the temperature is between the two thresholds, the current state is maintained. From the perspective of reliability, the heating circuit is generally designed with a backup, which is available for use when a failure of the main loop is confirmed. Moreover, by setting an appropriate temperature threshold, if the main control fails to provide the desired control effect, the backup may participate in the operation spontaneously.

Thermal management functions required by Mars mobile intelligent agent include

1) Heading adjustment for thermal control

At noon on the surface of Mars, if the temperature of the rover equipment is excessively high, the heading of the rover body may be adjusted, and the solar wings may be lifted to form a parasol, which is conductive to the heat dissipation of the device. As the temperature of Mars soil is only 30 °C at noon, the high temperature problem of the equipment will only occur if the high-power equipment is working for a long time at noon. The strategy is to adjust the direction of the heading of the rover to the east–west direction at noon and then independently control the corresponding solar wings to offer a parasol effect, which can reduce the heat entering the heat collector on the top surface.

2) Roll angle adjustment for thermal control

If the active suspension is adopted, the thermal control ability can be improved by adjusting the roll angle of the rover body when the general thermal control method cannot meet the requirements. When the heading and the solar azimuth are approximately perpendicular, in order to increase temperature, the roll angle may be adjusted toward the Sun, and vice versa, which is conductive to cool down, and the roll angle may be adjusted within $\pm 10°$.

In more extreme cases, local terrain can also be built by adjusting the roll angle and pitch angle through wheel digging. The angle adjustment range is generally within $\pm 3°$.

3) Working mode adjustment for thermal control

The system adjusts the upcoming working mode and behavior sequence by means of task planning to achieve the device temperature adjustment, which is an advanced adjustment method. When the temperature of the equipment is low, earlier execution of communication, sensing and detection modes is conductive to improve the temperature level of the platform; otherwise, the startup time of the high-power working mode may be postponed so that the heat dissipation burden can be alleviated.

12.2.3.4 Autonomous Communication Management

Generally, the rover is designed with digital transmission links for UHF band and X-band. Although the X-band digital transmission link can be directly transmitted to the Earth, the code rate is too low and is only of values in emergency. The commonly applied digital transmission link is the UHF link when the orbiter is in the periareon TT&C arc. However, UHF antenna is generally fixed, and there is no requirement for autonomous orientation control. When the orbiter is in the periareon arc and apoareaon arc, the X-band link may realize digital transmission, which requires the orientation control. In addition, due to the rapid changes of the azimuth and elevation angles, higher requirements are put forward for the orbiter orientation control.

1) Code rate self-adaptation

When UHF band data transmission is adopted, due to the changes of communication distance in a large range, the corresponding applicable code rates vary greatly. In order to transmit as much data as possible in a limited time, it is necessary to adjust the bit rate adaptively according to the link margin.

12.2 Autonomous Management Technology for Deep Space Probes 485

2) Directional antenna orientation control toward the Earth

The difficulty of pointing to the ground lies in achieving the Earth direction tracking with poor positioning accuracy. The scheme is spiral scan within the corresponding space area, and the scanning strategy should guarantee the search range coverage in consideration of the trajectory of the Earth motion.

3) Orientation control of the directional antenna to the orbiter

When the orbiter is in the apoareon TT&C arc, the angular velocity of the visual motion is relatively small, and the strategy is the same as the ground orientation control. In the periareon arc, the apparent motion velocity is large, and the search process may adopt the orthogonal search strategy to autonomously realize the sawtooth tracking of the orbiter trajectory.

12.2.3.5 Autonomous Navigation Control

When Mars rover is moving autonomously, there are several optional implementation modes available.

1) Blind walking mode

In the relatively flat area, the ground station determines the target direction, and the rover controls the mobile device to walk autonomously and continuously detects changes in parameters such as pitch angle, roll angle, speed, current and temperature of the moving mechanism. When the detection result exceeds the limit, the rover stops moving, otherwise, it will continue to move until the target distance is reached.

2) Path tracking

The traveling path is planned by the ground station according to the obstacle locations on Mars surface, which is generally composed of straight line segments and circular arcs described by curvature and arc length. When the rover moves along the path, it is necessary to estimate the slip rate in real-time, and the path tracking error is corrected in the motion, until the rover finally reaches the target point.

3) Autonomous obstacle avoidance

The ground engineer specifies the target point, and the Mars rover autonomously perceives the surrounding information, plans a safe motion path independently, avoids obstacles that cannot pass and finally reaches the target point.

In summary, the navigation control functions to be implemented by the intelligent agent mainly include

(1) Environmental perception and three-dimensional recovery (imaging parameter determination, sequence binocular 3D vision image processing);
(2) Analysis of terrain suitability (impassable regional analysis, analysis of safety levels of passable areas, global hazardous area marking, etc.);

(3) Position and attitude determination (including slip rate, movement mileage estimation, etc.);
(4) Motion control (decomposed to the rotational speed and rotation angle of the wheel, controlled in real time with special cases including rover crab motion, creeping, wheel lifting, etc.);
(5) Safety monitoring during the movement (checking whether the parameters such as pitch and yaw angle, temperature and current are beyond the limit), if necessary, takes emergency action to enter safe mode.

12.2.3.6 Autonomous Scientific Exploration

The scientific instruments carried by Mars rover are various in types with different functions, so the demands for independent scientific exploration are quite different, including:

1) Autonomous calibration

Due to the complexity of space environment and Mars surface environment, many detection instruments are required to be calibrated before performing the detection tasks. Some instruments may be self-calibrated and have relatively low requirements for system autonomy. However, some instruments require the probe to adjust the detector head orientation autonomously, and the calibration can be accomplished only after the sample is measured.

2) Intelligent detection

It is a typical process for scientific instruments to guide probes to carry out intelligent detection by manipulating and placing instruments to achieve scientific detection of specific targets. The rover autonomously travels to the vicinity of the detection target specified by the scientists, uses the camera to perceive the surrounding environment of the target, establishes a local working scene, plans the motion path of the robot arm considering the end attitude, and then, the robot arm carries the instrument to reach the detection target to complete the detection task.

12.2.3.7 Autonomous Data Management

There are many functional requirements for the autonomous data management of the rover, including at least:

1) Intelligent telemetry

In order to reduce the amount of telemetry information data characterizing the state of the probe, it is possible to adopt the data group information extreme value transmission method under normal conditions or to determine the three-level thresholds so that only the over-limit data information can be transmitted. In different working

12.2 Autonomous Management Technology for Deep Space Probes

modes, the focus of telemetry is different, therefore, variable frame rate and variable priority transmission modes can be designed to focus on the key information of the current state and adapt to the very low code rate telemetry transmission mode [1].

2) Data organization in data transmission

The downlink data is transmitted autonomously according to the order in which the transmission contents are adjusted according to the priority. The relationship between the data tag and the storage address may be established with abilities such as accurate autonomous data file query, address selection playback and breakpoint retransmission. After the transmission is completed, the storage space is released according to the data deletion mechanism rules. It is also necessary to design an image compression scheme compatible with lossless and loss compression based on the characteristics of low channel code rate and high bit error rate of Mars exploration missions, including wavelet domain block protection coding strategy, progressive transmission scheme, region of interest transmission scheme, etc., which has the characteristics of high performance and low complexity.

3) Autonomous management at night and Sun transit

Mars rover is controlled autonomously into the night working mode and transmits data at night according to the demands; at the end of the night, Mars rover is set to the normal working mode. With the authorization of the ground station, it can operate autonomously, complete the autonomous orientation toward the Sun and adjustment of the working mode and ensure the safety of the probe during the Sun transit.

4) System software maintenance

The system requires storage and recovery functions for important data, including important data such as automatic temperature control parameters, and time delay commands (group) and time code to ensure normal system operation after system reset and primary/standby switchover. The periodic working template is established for the daily tasks of Mars rover, and the equipment switch in regular activities is managed by Mars rover without intervention from the ground.

5) Safety check and failure diagnosis

The attitude of Mars rover is monitored in real time to keep it within a safe range to avoid tipping over. In case of an emergency during the rover's motion, an emergency response command may be issued. According to the motion logic of the mechanism, the safety control lock of the mechanism is formed, namely, when the target position of the mechanism under control interferes with the current position of other mechanisms or devices, the control command may not be executed.

No matter what happens during mission execution (such as hardware/backup hardware failure and flight software anomalies), the safety mode acts as a final contingency to ensure that the probe enters a safe mode without permanent damage.

In the event of a failure, reconfiguring system is acceptable, including the use of heterogeneous cameras to achieve environmental awareness, in-situ steering through differential speed, etc. [2].

12.2.4 Autonomous Management Implementation Framework of Mars Rovers

The realization of the above-mentioned Mars intelligent agent functions is technically complicated. Figure 12.3 shows the framework of the implementation scheme, which is mainly composed of three parts: mechanism-driven control layer, navigation control layer and intelligent planning layer.

The input of the autonomous mission planning of the probe includes initial state, mission objectives and related knowledge. The output is mainly a rational plan, that is, a series of actions from the initial state to meet various constraints of the

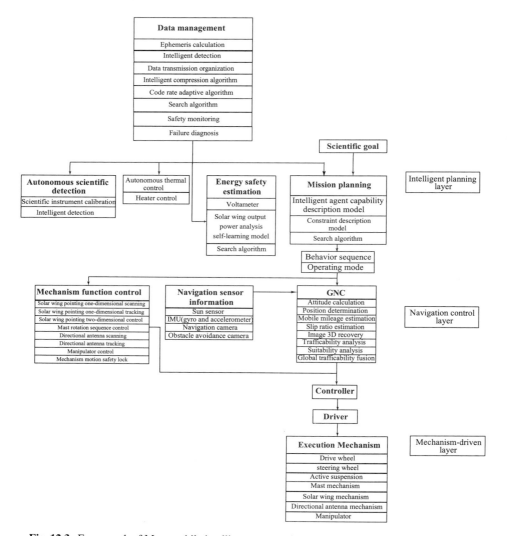

Fig. 12.3 Framework of Mars mobile intelligent agent scheme

12.2 Autonomous Management Technology for Deep Space Probes

probe, and after successful implementation, the state of the probe will be transferred to the target state. Through the processing of the existing knowledge of the probe and the surrounding environment obtained by the existing knowledge and various sensors on the probe, a rational sequence of activities that can reach the target state is obtained [3].

The on-orbit mission planning technology not only reduces the operational cost for collecting information, processing and forming command sequences in deep exploration missions but also improves the ability to cope with deep space variable environments of the mission. However, some complicate situations in the field of deep space explorations have posed new challenges to traditional independent mission planning technology. These special cases include complicated resource constraints, time constraints and concurrency constraints between or among different activities, uncertainty in the exploration environment as well as limited resources on the planet. Therefore, compared with the traditional planning and scheduling technology, many key techniques are required to be mastered in the mission planning of deep space exploration, including numerical/logical hybrid planning knowledge modeling, planning space fast search, resources optimization and processing, time-constrained processing, etc.

Autonomous failure processing technology is applied to monitor the execution of mission planning results, identify failure modes, perform failure recovery and ensure the safe operation of the spacecraft. It mainly includes two parts: failure pattern recognition and failure recovery. The failure pattern recognition tracks the execution of the planned activity sequence and identifies whether each activity in the planning sequence is executed correctly; while failure recovery determines the current state of relevant components/parts on the probe through the current state information of the system. In case of failure, the failure recovery module is responsible for proposing a reconfiguration or recovery strategy.

Deep space exploration mainly serves the purpose of scientific exploration. The data obtained by the probes will be transmitted to the Earth and analyzed by scientists for new understandings of the space. However, due to the relatively long distance for deep space exploration, the data transmission rate is relatively low, and there is a large time delay in the transmission of deep space exploration data, resulting in the missing of some important scientific exploration opportunities for the deep space probes. In order to discover the current scientific phenomena and to generate viable scientific observation task sequences, the rover should be designed with certain scientific data analysis and independent reaction capabilities. With the accumulation of scientific data, a large amount of original exploration data cannot be all transmitted back to the Earth via deep space network for further processing. Therefore, the rover must have the ability to perform preliminary screening and data compression on the collected scientific data, which may be of great significance so that useful or sensitive data can be efficiently transmitted back to the Earth for analysis by scientists.

In instrument operation planning, not only the operations, data acquisition, analysis, processing and transmission of instruments are considered and coordinated from a global perspective, but also all the instrument operation process are planned locally to eliminate the conflicts in resources and time when using different instruments.

In conclusion, the deep space probe autonomous management technology has become the key technology in the field of space explorations. The algorithm implementation of the mission planning is described below in conjunction with the UHF band communication.

12.2.5 *Mars Rover Autonomous Mission Planning*

Space mission planning is a typical knowledge processing, involving complicated logical reasoning and numerous constraints. A mission planning starts from the initial condition, according to the pre-established inference rules, infers the operating conditions required for each step until the end of the mission, based on the state of each step in the mission process and the resource constraints it is subject to. The result is to form an action time series of operations on the entire process of the mission. Finally, it sorts the set of events and eliminates conflicts among such events, resulting in an executable sequence of events, that is, and mission planning.

Mission planning is a multi-constrained and complicated decision-making task involving optimization of multiple objectives. In the spacecraft mission planning process, the constraints have the characteristics of complicated constraints, strict time requirements, strong resource constraints and strict action sequence constraints.

Due to strict communication constraints (limited communication window, long time delay, limited bandwidth, etc.), unstructured unpredictable environment, complicated missions and systems, etc., Mars rover exploration missions only permit limited ground intervention, thereby requiring a certain degree of autonomy of Mars rover. The autonomy capability is mainly reflected in ground mission-level instruction decomposition, autonomous data transmission to the ground (UHF inter-probe communication, Mars night X-band communication to orbiter), detection mode and detection at Martian night, etc. The following is a description of UHF inter-probe communication for Mars roving exploration, as a representative example of autonomous mission planning of Mars rover.

In the UHF inter-probe communication mode (periareon TT&C arc), the starting time of the transmission is calculated, and the plan is executed. The plan can be further divided into two parts: autonomous planning part and execution part. The autonomous planning calculates the transmission arc, plans the transmission start and end time as well as the execution conditions based on time. The execution performs UHF inter-probe communication according to the plan. The algorithm procedure for autonomous planning is shown in Fig. 12.4.

The main algorithms for inter-probe communication are described as follows:

1) Distance calculation between the orbiter and the rover

Calculate the distance between the orbiter and the rover according to the position of the rover (longitude and latitude) and the orbital equation of the orbiter, and output the distance variation between the orbiter and the rover in the future communication arc.

12.2 Autonomous Management Technology for Deep Space Probes

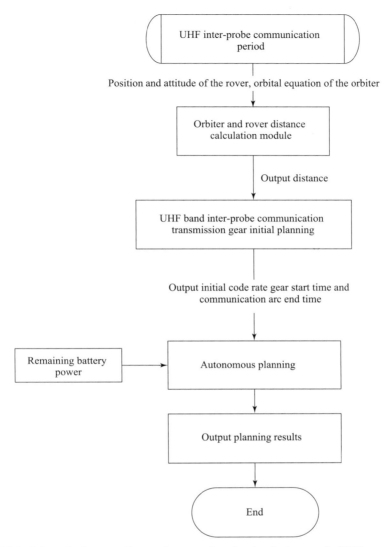

Fig. 12.4 Schematic diagram of instruction execution planning for inter-probe UHF communication

2) Initial planning for UHF band inter-probe communication transmission

According to the distance between the orbiter and the rover, the initial code rate range, the start time and the end time of the communication arc of the UHF band inter-probe communication in the communication arc are preliminarily planned.

3) Communication effective duration planning

Through the analysis of the rover power balance, the power balance estimation results of the next Martian day are obtained and whether the lowest point of the discharge depth is lower than the maximum allowable discharge depth is judged. The input parameters include the position of the rover on the surface of Mars (latitude and longitude), attitude (yaw, pitch and roll), the deployment angle of the $\pm Y$ solar wing A as well as the work plan of the rover, and the output parameters include the expected power generation of the rover on a Martian day, the expected power consumption and the change in battery power to give the estimated minimum battery level.

If the estimated minimum value of the remaining battery capacity is not less than the allowable discharge depth of the battery (e.g., the allowable discharge depth is set at 70% for a certain safety margin), the preliminary planning results are performed. Otherwise, the communication duration is adjusted according to the allowable discharge depth of the new planning results.

12.3 Tele-operation Technology of Rover

It is well known that there are many environment elements on the Earth and outside the Earth that are not suitable for humans to work in. Therefore, various concepts such as tele-operation have been proposed from different perspectives so that people can retreat from such harsh working environment and stay in a distant, safe and comfortable environment and operate and manipulate the machines in the harsh environment. With the development of space shuttles, manned spacecraft, orbital space stations and planetary surface roving exploration technology, the tele-operation technology has been more and more widely used.

The operation of deep space probes, such as lunar rovers, is a process of interaction between the Earth and the probe with continuous support from the ground. It is obviously different from on-orbit TT&C mode of the traditional near-Earth spacecraft. Instead, it is technically difficult with features such as long distances, complicated missions (multiple modes, many constraints), unstructured exploration environment, frequent ground support with long duration as well as requirement on accompanying flight simulation verification. Therefore, designing an on-orbit operational mode that matches the rover mission is a problem that must be addressed.

The research and development of the ground mission support and tele-operation system for the rover, also known as the research of tele-operation technology, include key technologies such as absolute orientation, relative orientation, image fusion, overall mission and exploration cycle planning, mechanism (mast, solar wing, manipulator, mobile device) planning, travelled mileage mapping and comprehensive comparison, path planning and mathematical and physical simulation verification.

12.3.1 Tele-operation in Space Environment

The tele-operation in space environment is different from the traditional tele-operation technology on the ground. The difference between the ground environment and space environment brings unique characteristics to the tele-operation in space environment, but there are problems of long distance and delay. Specifically, the characteristics of tele-operation in space environment are as follows:

(1) The spatial position of the tele-operation object is far away from the operator. Therefore, the tele-operation in space environment is limited by the time delay and the limited data transmission bandwidth in the communication between the ground and the space;
(2) The mission environment model is incomplete. The working environment of the tele-operation object is rarely known or just partially understood, making it necessary to set two working modes: tele-operation and automatic control.
(3) The tele-operation mission in space environment is non-repetitive and unpredictable, increasing the difficulty of tele-operation technology;
(4) Tele-operation technology requires the far-end control equipments have sufficient flexibility and reliability, but the logistic service capabilities for the tele-operation equipment are limited, requiring the tele-operation equipment to be able to automatically recover from an emergency.

Tele-operation of operating objects in a space environment may be divided into two types based on the operating locations: space tele-operation and ground tele-operation, of which space tele-operation refers to the control by the astronauts on the orbiter to the operating objects in space environment, while the ground tele-operation refers to the control of the spacecraft at the control station on the Earth.

In the space tele-operation mode, if the human operator works under the condition that the surrounding environment is basically known, the delay of the space tele-operation loop is very small and basically has no impact on the operation. In ground tele-operation, the human operator may work in a scenario with unknown environment and uncertain trajectory. Due to the time delay, the ground tele-operation generally adopts the "Operation-Wait" or monitored control mode, allowing intervention of the space operations at any time by the operators. This would result in poor system dynamic performance and make it difficult for continuous operations and refined operations.

The tele-operation of the surface roving on a celestial body is a kind of tele-operation in space environment, which satisfies the basic characteristics of tele-operation in space environment.

12.3.2 Planetary Surface Roving Tele-operation

For the rover working on the surface of the planet, the tele-operation technology is to generate the remote scene on the ground according to the state of the probe, the environment of the area to be explored and the exploration target. Scientists and engineers work together to make corresponding planning and carry out test verifications, propose control strategies, carry out the control according to certain rules and complete online demonstration, failure simulation and disposal methods on the ground.

The tele-operation of the planet surface roving is different from the tele-command technology for the spacecraft. The unstructured and unpredictable nature of the surface exploration environment of the planet, the complexity and non-repetition of missions and the differences in lighting conditions caused by the changes of day and night on the planet surface will all affect the tele-operation of the planet surface roving.

According to the level of the rover autonomy, the rover can be controlled in four ways: space tele-operation, ground tele-operation, ground tele-operation with semi-autonomous mode and ground tele-operation with autonomous mode.

12.3.2.1 Space Tele-operation Mode

Space tele-operation mode refers to the tele-operation mode implemented by the manned rover during the missions such as manned lunar landing. The space tele-operation control commands are sent directly or indirectly to the rover by the astronaut. In this control mode, the rover has no autonomy at all, and the control is completely carried out by the astronaut. The unique advantage of this operation mode is that the tele-operator (i.e., the astronaut) can clearly understand the surrounding environment.

12.3.2.2 Ground Tele-operation Mode

The ground tele-operation mode refers to the mode, in which the human operator controls the motion of the rover according to the detailed control strategies based on the images returned by the rover and the state information of the rover; the rover is only responsible for converting the speed and orientation commands from the tele-operator into the speed and rotation angle of the wheels using forward and backward kinematics or converting the speed and rotation angle of the wheels into the speed and orientation information that is sent back to the tele-operator for the preparation of the next step. In the ground tele-operation mode, the human operator should send detailed work instructions, posing higher requirements on the communication bandwidth. The tele-operation quality is determined by the human operator. It not only requires a long time to train the human operator, but also leads to intensive labor and fatigue-prone work for human operators, and it is difficult to realize high-precision operation. However, in this way, the rover has only limited safety autonomy, reducing the requirements for its autonomy as well as the burden on the computer system equipped by the rover itself.

12.3.2.3 Ground Tele-operation with Semi-autonomous Mode

The ground tele-operation in semi-autonomous mode means that the rover can be controlled by the human operator according to the instruction of the graphical simulation interface, and also, local autonomous decision can be made by the computer equipped by the rover itself. For example, in the initial stage of the experiment or application and for complicated areas, ground tele-operation is conducted, while in the later stages or for simple terrain conditions, tele-operation is mainly done by the rover autonomously. This control mode can improve the reliability, robustness and the speed of dealing with uncertain problems, reduce the work intensity of the tele-operator and greatly improve the work efficiency. However, the rover is required to be provided with relatively high autonomy.

12.3.2.4 Ground Tele-operation with Autonomous Mode

The autonomous mode refers to the mode that the rover can make its own decision by its own computer during the rover exploration process, and the human operator only issues phase operation instructions and monitors the operation as needed. Although this mode requires high intelligence level of the rover, it excludes the human operator from the remote working loop, avoiding the influence of time delay compared with the semi-autonomous mode. In addition, the human operator can take controls over the rover at any time. It not only greatly improves the working efficiency of the rover, but also increases the flexibility of the tele-operation system.

To sum up, although the rover has a certain degree of autonomy, the important steps of the rover in scientific exploration still require ground participation and control. In view of the fact that the lunar communication delay is not very long, the focus of the design of the ground tele-operation system is to make full use of the high processing capability of the ground support system and the experience of the ground operators as well as the advantages of man–machine integration, thus improving the detection efficiency of the rover and ensuring the safety of the rover.

12.3.3 Key Technology for Rover Tele-operation System

In the design process of the rover mission support and tele-operation system, the core technologies to be mastered mainly include the following aspects:

(1) Absolute positioning technology;
(2) Relative positioning technology;
(3) Image fusion technology;
(4) Mission planning technology;
(5) Planning technology for mechanisms (mast, solar wing, manipulator, mobile device);

(6) Path planning technology.

12.3.3.1 Absolute Positioning and Relative Positioning Technology

The landing images play a quite important role in the deep space exploration missions. For example, the US Courage and Opportunity rovers successfully landed on Mars in 2004, relying on the landing image motion estimation system to achieve the horizontal velocity estimation of the landing process and position determination.

In the soft-landing mission of the Chang'e 3 rover, the landing camera carried by the lander acquired the sequence of landing images with clear and continuous quality, providing guarantee for accurate positioning of the landing site. In the initial positioning of the lander in the tele-operation system, the landing image was utilized, and the SIFT feature matching, the affine transformation based on the ground relative level hypothesis and the image mosaic method were applied to achieve the accurate matching between the landing image sequence and the lunar topographic map acquired by the Chang'e 2 probe, which resulted in accurate positional coordinate information for the lander. The flowchart of the positioning process is shown in Fig. 12.5.

To visualize the initial positioning effect based on the landing images, a set of mosaics is selected, as shown in Fig. 12.6, in which seven landing images at typical heights are selected, and the inlaid image corresponds to the registration position in the DOM image of the Strip 0236 from Chang'e 2 mission. The red marked point in the figure is the position on the map of the final landing point of the lander.

Presently, there are many position determination methods used by deep space rovers, and each has its own advantages and disadvantages. Among them, the image bundle adjustment technology is based on the image information during the movement of the rover as well as the orbiter image and the landing image for joint positioning, under the consideration of the characteristics of the rover.

Taking the rover Yutu as an example, the odometer and the inertial navigation system work together during the motion of the rover, and the position and attitude information of the camera at the imaging moment can be obtained. The accuracy of the telemetry position information of the rover is reduced with the accumulation of systematic errors in the odometer and inertial navigation system, as well as the influence from some factors such as the slip of the rover wheels. Therefore, a navigation and positioning algorithm based on the 3D image of the rover is proposed. The schematic diagram of the stereo image positioning process of the rover is shown in Fig. 12.7. Compared with the traditional visual measurement method (the stereo vision positioning method of the rover used in the MER mission), which only

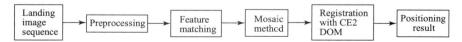

Fig. 12.5 Initial positioning flowchart based on landing images

12.3 Tele-operation Technology of Rover

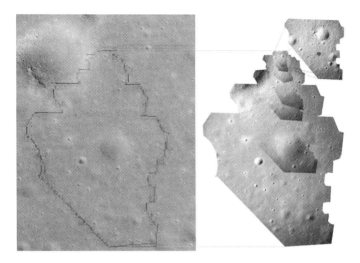

Fig. 12.6 Landing image mosaic results (right) and registration with CE2-DOM (left)

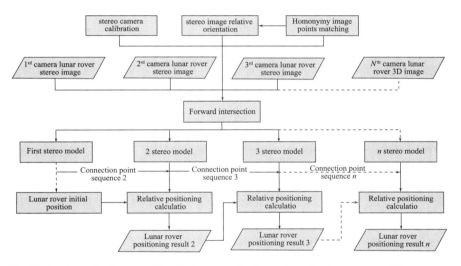

Fig. 12.7 Schematic diagram of the 3D image navigation and positioning process of the rover

performs positioning calculation between the front and rear camera images, the new algorithm has superiorities as follows: The prior knowledge of the telemetry data of the position and attitude of the rover is not needed, the construction and iterative calculation of the bundle adjustment method equation in the visual measurement method are avoided, and the calculation is simple and fast; the relative orientation elements of the stereo camera are recalculated to avoid the decreasing of positioning accuracy from still applying the ground checking value, with the variation caused by the influence of environment and landing impacts; in addition, it has strong adaptability and stability for the geometric calculation of the large angle 3D model of the rover camera.

12.3.3.2 Image Information Fusion Technology

The 3D terrain reconstruction based on 3D camera images generally adopts binocular 3D vision method. Based on the dense homonymy points from matching, the 2D to 3D space transformation is carried out. However, due to the existence of factors such as low lunar illumination and poor texture information, this method cannot satisfy the current demands.

To solve the technical difficulties of high-precision three-dimensional topographic reconstruction, such as high-accuracy automatic matching of the connection point between the rover navigation/panoramic camera lunar stereo image pair and the splicing of the inter-site images under large difference shooting angle condition, an iterative dynamic programming matching algorithm proposed has improved the dynamic programming matching algorithm and the accuracy of homonymy points matching. With reference to the positioning method of Mars rovers, a multi-site image data geometric correction and splicing method based on bundle adjustment processing is proposed, which enables efficient and high-precision 3D reconstruction of lunar surface terrain and multi-site ortho-image mosaic. The specific technical route is shown in Fig. 12.8.

The fusion result of the on-orbit navigation of the rover and the lunar stereo image pair of the panoramic camera for the single station and the multi-station is shown in Fig. 12.9, of which the width is 72 m (left and right directions) × 77 m (up and down directions), and the upper direction is north.

The application of mileage calculation and display software based on the Geographic Information System (GIS) framework in the field of planetary exploration are proved by developed precedents. The Cartography and Geographic Information Systems Laboratory of Ohio State University has developed Mars WebGIS system for the release and management of the positioning path information and topographic mapping products of Mars rovers. Moreover, Mars Space Foundation of Arizona State University has developed JMARS, a WebGIS site for storing and managing large amount of Mars exploration data.

The mileage calculation and path display management module is used for data management and visualization of the rover exploration paths, enabling data browsing, processing, support for decision analysis as well as the production and distribution of topographic map products. The module adopted a cross-over data management model based on data attributes. The classification basis includes time attribute classification, grid and vector classification, real-time and non-real-time classification, point-line-surface classification, received and self-generated classification, etc., and the data attributes of different categories are cross-correlated. The moving mileage of the rover Yutu in the first lunar day is shown in Fig. 12.10.

12.3 Tele-operation Technology of Rover

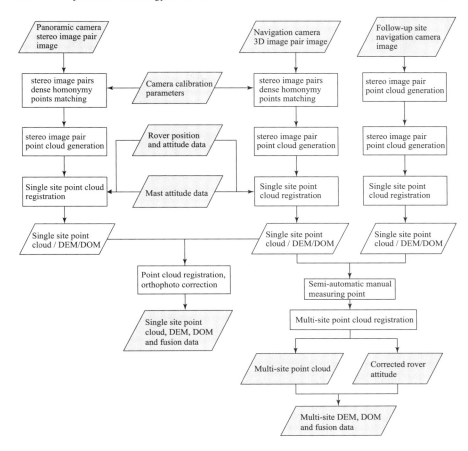

Fig. 12.8 Lunar rover image processing technology roadmap

12.3.3.3 Mission Planning

The mission decision system is the core system of ground mission support and tele-operation system. It mainly completes the position determination, image fusion, integrated planning and mobile mileage drawing of the rover and creates the execution strategies such as the overall rover mission planning, the exploration cycle planning, the navigation unit planning, mast-to-Earth orientation planning, solar wing-to-Sun orientation planning as well as manipulator in-situ detection planning.

The mission planning of the rover Yutu is based on the heuristic planning method based on the "Mosaic" obstacle map, as shown in Fig. 12.11, to adapt to the heading constraints and plan the moving path of the rover.

The performance constraints of the rover mechanism include +Y wing control motion constraint, mast motion constraint, omnidirectional communication antenna constraint (as shown in Fig. 12.12). The TT&C and data transmission, energy supply

Fig. 12.9 Lunar rover image fusion map

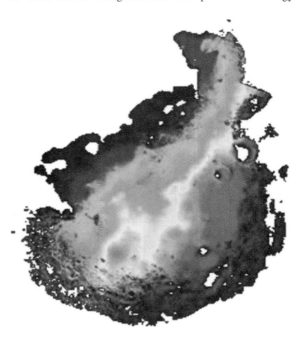

of the rover are determined by the ephemeris, the Earth station deployment, and the position and attitude of the rover. The rover mission planning function matrix is shown in Fig. 12.13. Factors such as temperature control constraint relationship, communication visibility constraint, solar visibility constraint, rover working mode and action relationship constraint, +Y wing motion angle limitation, occlusion of solar array by mast, sleep condition and time consumption of ground operation must all be considered during the mission planning process.

12.3.3.4 Path Planning

On the basis of obtaining telemetry and data transmission information, the navigation and control system first completes the global environment perception, extracts the digital elevation model of the surrounding terrain and then performs global path planning and motion control according to the mission behavior sequence generated by the mission decision system so as to further generate sequence of control commands. The overall process should be completed within a certain period of time, and the work contents of several units included in the system are generally executed in sequence and should not be completed in parallel.

12.3 Tele-operation Technology of Rover

Fig. 12.10 Mileage chart in the first lunar day by the lunar rover

12.3.3.5 Manipulator Planning

The rover manipulator in-situ detection planning is a 9-DOF (degree of freedom) trajectory planning problem for the manipulator arm, in which the rover body attitude has six DOFs and the manipulator has three DOFs. The in-situ detection planning of the manipulator on the rover is a difficult part in the manipulator trajectory planning.

For the in-situ detection planning of the Yutu rover manipulator, the rover position is estimated on the basis of the safety of the rover when parking, such as whether there are obstacles under the wheels, whether there are obstacles under the rover body, whether the rover is in the point cloud map. According to the position where the rover is parked, the position of the rover wheels is calculated to obtain the point cloud data under the wheels. With the point cloud data under the wheels, the pitch angle and the roll angle when the rover is parked can be estimated. The rover in-situ exploration planning function matrix is shown in Fig. 12.14. The planning results

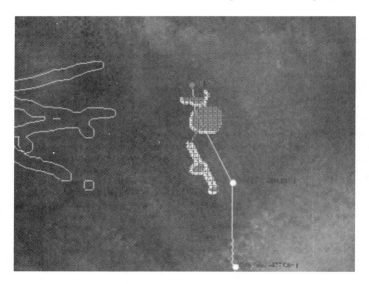

Fig. 12.11 Heuristic planning principle

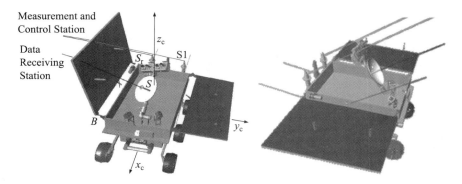

Fig. 12.12 Schematic diagram of omnidirectional antenna occlusion analysis

include rover parking position, parking attitude and manipulator deployment action sequence.

The manipulator obstacle detection is the problem of whether the virtual sensor line segment intersects the obstacle triangle surface. In collision detection, the virtual sensors are attached on the manipulator members, as shown in Fig. 12.15. Each sensor acts as a laser range finder, measuring the distance from the emitting point to the receiving point. If the laser line is blocked by the obstacle, it indicates that the corresponding member collides with the obstacle.

12.3 Tele-operation Technology of Rover

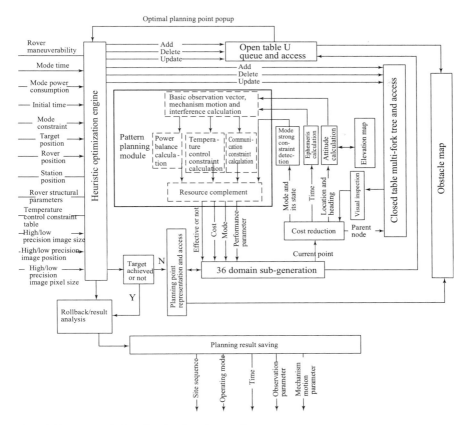

Fig. 12.13 Rover mission planning function matrix

12.3.4 Tele-operation System

The ground mission support and tele-operation system consists of the main control system, mission decision system, navigation and control system, control command generation system as well as simulation verification system. The main functions are as follows.

(1) Receiving engineering telemetry data, engineering image data and scientific exploration data from the rover performs operations such as data processing, storage, query, distribution and management.
(2) Acquiring image data of different cameras and completing image processing, generating lunar topographic map and realizing the positioning of the rover in the roving area.
(3) Completing mission planning in different parts of the roving area, determining the basic working mode at different positions of the rover and generating the sequence of mission behavior.

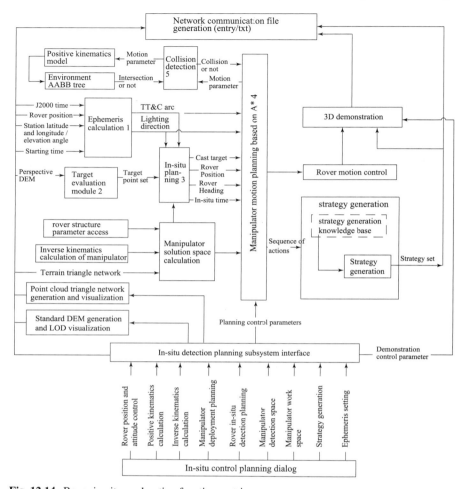

Fig. 12.14 Rover in-situ exploration function matrix

(4) Completing the global environment perception, global path planning and motion control of the rover and generating the sequence of control commands under the guidance of the mission planning to complete the tele-operation and control of the rover.

(5) Completing the sequencing and confirmation of the control commands and generating the control command packet.

(6) Supporting the requirement of on-orbit concurrent verification of the rover with the simulation system; providing simulation analysis and verification means and supporting the simulation verification demands of the rover control strategy under normal conditions (mainly including mission behavior sequence and control instruction sequence) and supporting the simulation verification of the rover failure resolution strategy under abnormal conditions.

12.3 Tele-operation Technology of Rover

Fig. 12.15 Manipulator collision detection

(7) Providing indoor test site to support the in-field test requirements of rover ground mission support and tele-operation system; supporting the in-field comprehensive test in the rover ground comprehensive test; realizing the abnormality and failure reproduction in the rover on-orbit operation, conducting the in-field test verification of the failure strategy; and assisting in the formulation of a new control strategy.

(8) Forming a self-closed loop and supporting the simulation verification of the system.

(9) Providing a platform for simulation trainings to ground operators.

The main control system is mainly responsible for unified management of operating environment and software such as data processing, image processing and mission planning of all tele-operation systems of the rover, completing data management such as automated execution, scheduling and management of planning processes, imaging, planning and analysis of tele-operation system missions, also realizing the functions such as automatic generation of the mission planning report [4].

The interface diagram of the main control system is shown in Fig. 12.16.

The framework of the rover mission support and tele-operation system is shown in Fig. 12.17. The system framework represents the basic relationship and basic workflow among the various components of the system and is applicable to both the engineering development phase and the on-orbit flight phase.

In the orbital operation phase, the system first acquires telemetry and data transmission data via the ground TT&C center and the ground application center. The telemetry data and data transmission data, after being received by the main control system, are transmitted to the mission decision system, and the processing results of the global environment perception unit in the navigation and control system are obtained. After the mission behavior sequence is obtained, it is sent to the navigation and control system, and the control instruction sequence of the rover is obtained. The control instruction requiring no verification is directly sent to the control instruction

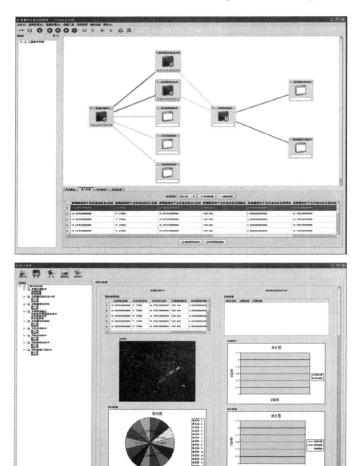

Fig. 12.16 Schematic diagram of the main control system interface

generation system, and the control command to be verified is verified by the simulation verification system and then sent to the control instruction generation system. The image and data products obtained by the mission decision system and the navigation control system are sent to the simulation verification system for the purpose of modeling. Finally, after the control instruction sequence is sequenced and confirmed by the control instruction generation system, the final control command sequence is obtained, and the tele-command instruction or data is input into the uplink through the ground TT&C center and is then received and executed by the rover.

In the simulation verification phase by the system itself, as the simulation verification system contains the rover simulation and physical model, it can simulate sending telemetry and data transmission data, simulate receiving tele-command instructions

12.3 Tele-operation Technology of Rover

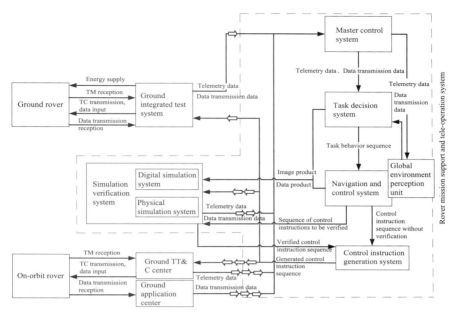

Fig. 12.17 Framework of rover mission support and tele-operation system

and perform the verification of the system itself for the entire operation process of the rover mission support and tele-operation system, using digital or physical model.

In addition, it should be noted that the mission decision system needs scientists to provide scientific exploration demands, including the scientific exploration targets and behavioral sequence.

12.4 Prospects for Technology Development

With the diversification and complexity of deep space exploration missions, system-level autonomous requirements of spacecraft are becoming increasingly urgent, and long-term system-level autonomy in the absence of manual intervention is an inevitable direction for the development of autonomous management technology. Autonomous management technology is developing from the engineering-oriented system autonomy to the science-oriented system autonomy. More importance should be attached to the degree of realization of the core goal of the deep space exploration mission—the scientific objective. Therefore, the autonomous system will focus on the science-oriented autonomous functions such as processing of scientific data, the on-orbit discovery of scientific phenomena and the autonomous tracking and observation of sudden scientific events. The architecture of autonomous system of deep space probes will be developing from a centralized autonomous operation mode to a distributed autonomous operation mode, further improving exploration efficiency and reducing software complexity. In addition, with the increase of the complexity of space missions, spacecraft is often required to cooperate to complete complicated mission objectives, such as interference imaging and stereo imaging. Therefore, the cooperative work of multiple spacecraft also poses a new challenge of autonomous control.

In the sample return missions, the ground tele-operation system support is also required for the soil and rock collection in the extraterrestrial celestial environment. The technical problems to be solved are the mission implementation is greatly affected by the environment of the sampling target area, and the 3D environment recovery of the sampling area and the sampling target is indispensable; the mission implementation process is complicated, and mission planning and simulation verification are required on the ground; there are many unknown factors in the mission implementation process, and the physical and mechanical properties of the soil and rock distribution on the surface of the planet are rarely known; moreover, the requirements for environmental recovery and mission planning timeliness are high.

References and Related Reading

1. Dai SW, Sun HX (2002) Autonomous control for spacecraft. Chin J Space Sci 22(2):147–153 (Chinese vision)
2. Gong J, Yang H, Zhao W et al (2011) Knowledge inference based self-failure diagnosis method for spacecraft. Aerosp Control Appl 37(4):19–23 (Chinese vision)
3. Li ZB, Wu HX, Xie YC et al (2001) Experimental platform for spacecraft intelligent control. Acta Autom Sin 27(5):695–699 (Chinese vision)
4. Roland DP (2006) Space mission operations DBMS. Space mission challenges for information technology. In: Second IEEE international conference, SMC-IT 2006, ISBN: 0-7695-2644-6
5. Sun Z, Jia Y, Zhang H (2013) Technological advancements promotion roles of Chang'e-3 lunar probe mission. Sci China Technol Sci 56(11):2702–2708

Chapter 13
Mechanism Technology

13.1 Introduction

There are many definitions of mechanism. The traditional definition is: Mechanism is a physical combination with a certain motion, which has a component system with a frame, which is connected by moving pairs to transfer motion and power. Spacecraft mechanism is a mechanical component that enables spacecraft and its components or accessories to complete specified actions or movements. It is generally composed of moving components and power supply. The research scope of spacecraft mechanism technology covers the transmission of mechanism motion, the force between various components, as well as the functional performances to be realized, mechanism manufacturing technology, test technology, transmission technology and reliability technology.

As a special spacecraft, deep space probes usually have complex space environment, objectives of scientific missions and many kinds of scientific payloads. In order to accomplish a specific scientific exploration mission, specific mechanisms are needed to perform corresponding actions and realize specific functions, such as landing buffering of the probe on the surface of the target celestial body, release and movement of the rover, surface sampling and sample transfer. The specific mechanisms for performing these activities include landing gear system, rover transfer and release system rover mobility system, sampling collection and transfer and release system. Whether the functional performance of these mechanisms is reliable and often directly determines the success or failure of deep space exploration missions.

It is worth noting that there are many specific mechanisms in the field of deep space exploration. This chapter mainly introduces the design methods and test methods of some typical mechanisms, which provides reference for the design of similar mechanisms.

13.2 Landing Gear System

The landing gear system is one of the important mechanisms to realize the soft-landing exploration on the surface of celestial bodies other than the Earth. It reduces the landing impact load to the required range through energy transformation or dissipation to ensure the lander to land stably, maintain the required landing attitude, ensure the safety of personnel and equipment and provide long-term reliable support for the lander [1].

Generally, a plurality of landing gear system gears are uniformly distributed on the lander, and the landing impact energy is absorbed through the deformation of the single buffer or a plurality of buffers of the gear system. The landing gear system has the advantages of easy control of landing attitude, no rebound after landing, long-term stable and reliable support. It is suitable for soft landing of lander with small landing speed, large landing mass and volume, and certain requirements for the attitude after landing. In addition, with the support of the landing gear system, the lander can make a launching platform, which is suitable for the exploration mission that needs to return to the Earth again after the landing exploration is completed. Figure 13.1 shows a typical landing gear system.

In practical engineering applications, multiple landing gear systems are usually adopted and installed at the bottom of the lander structure in a symmetrical or uniformly distributed manner to form a landing buffer system that satisfies the buffer function. During the launch phase, the landing gear system is in the state of collapse and compression, as shown in Fig. 13.1a; after reaching the predetermined orbit, the landing gear system is released, deployed and locked, as shown in Fig. 13.1b.

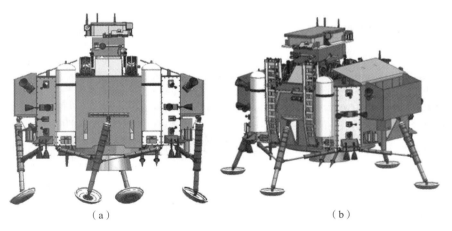

Fig. 13.1 Schematic diagram of landing gear system, **a** in a collapsed and compressed state; **b** deployed and locked state

13.2.1 Functions and Composition Characteristics of Landing Gear System

In general, the landing gear system shall have the following functions[1]:

(1) It can be collapsed, compressed, deployed and locked. In order to meet the envelope size requirements of the launch vehicle and withstand the mechanical loads in spacecraft launching, orbit transfer, orbit landing and other phases, the landing gear system is in a state of collapse and compression during launching. Before landing, the landing gear system shall be locked according to the flight procedure to obtain larger support area and better guarantee the stability during landing.
(2) Effective buffer of landing impact during landing. This is the core function of the landing gear system. Before landing, the lander usually has kinetic energy of tens of thousands of joules, which needs to be absorbed at the moment when the landing gear system collides with the landing surface, and the maximum impact load on the lander shall be controlled within the required range to ensure the safety of personnel and equipment on the lander machine.
(3) Able to withstand the corresponding space environment. In the process of flying to the target celestial body, as well as during and after the landing, the lander usually experiences the temperature cycle, high vacuum, aerodynamic heat of the surface atmosphere (if any) of the target celestial body or thermal environment caused by engine operation. It is required that the landing gear system can withstand the above environment with the assistance of corresponding thermal control measures.
(4) Ensure stability during landing. In general, the lander lands with horizontal velocity in addition to vertical downward acceleration. In addition, the topography of the landing area is also very complex, including the slope, the law of size distribution of pits and bulges, and the characteristics of soil. Under the above complex conditions, the landing gear system shall ensure the stability of the landing process, prevent the lander from tipping over and ensure that all kinds of detection work are carried out after landing.
(5) Provide long-term and effective support. After landing, the landing gear system can provide long-term and effective support to meet the working needs of the payload on the lander. Sometimes the landing gear system is also used as a launch support for sampling return or manned return to ensure the relevant attitude requirements for return and takeoff.
(6) Other functional requirements. In some cases, there are other functional requirements for landing gear system. For example, trigger signal can be given at the moment of landing for accurate judgment of landing safety.

Landing gear system generally includes main buffer, auxiliary buffer, footpad, compression release assembly, hold-down and locking assembly, as shown in Figs. 13.2 and 13.3. The main components are as follows:

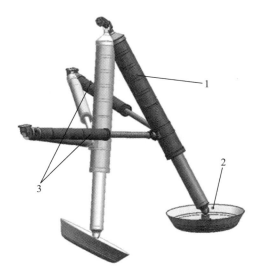

Fig. 13.2 Schematic diagram of landing gear system. 1-main buffer; 2-footpad; 3-auxiliary buffer (the light color shows the collapsed state and dark color deployed state)

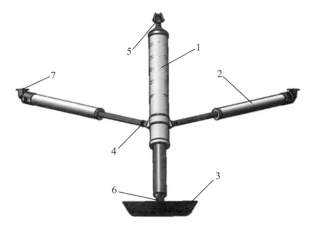

Fig. 13.3 Connections between components. 1-main buffer; 2-auxiliary buffer; 3-footpad; 4-ball hinge II; 5-universal joint I; 6-ball hinge I; 7-universal joint II

(1) Main bumper: It is the main energy absorbing and supporting component, which is mainly used for withstanding the longitudinal load during landing buffer, and bearing the landing gravity on the celestial body surface to keep the lander stable. The upper end of the main bumper is connected with the main structure of the lander through the universal joint, the lower end is connected with the footpad through the ball hinge, and the middle part is connected with both ends of the two auxiliary buffers through the ball hinge.

13.2 Landing Gear System

(2) Auxiliary bumper: It is an auxiliary energy absorbing and supporting component, which is mainly used to buffer the lateral impact load during landing and assist the main buffer to bear the lander gravity on the celestial body surface. Generally, two sets of auxiliary buffers are designed, one end of which is connected with the main buffer through a ball joint, and the other end is connected with the main structure through a universal joint.
(3) Footpad: It is the main supporting component to prevent the lander from sinking excessively due to the impact load. When the horizontal velocity is high and the ground topography of the landing area allows, the lander slip can be effectively ensured. When a rigid obstacle is encountered during sliding, the impact energy can be absorbed by the plastic deformation of the footpad.
(4) Hold-down and release assembly: It is used to realize compression and release of landing gear system. During launching, it shall be ensured that the appearance of the landing gear system meets the requirements of the fairing envelope, and that it can be released after reaching the predetermined orbit, creating conditions for deployment and locking.
(5) Deployment and locking component: It is used to realize deployment and locking after release, ensure the lander to land with a larger supporting area, improve the landing stability and create conditions for full play of the buffer capacity.

13.2.2 *Design and Verification of Landing Gear System*

13.2.2.1 Design of Landing Gear System

1) Configuration and key parameters design of landing buffer system

(1) Configuration analysis of landing buffer system

Usually, multiple landing gears are uniformly arranged at the bottom and side of the lander. Because the connection point between the gear system and the structural body is subject to a large concentrated load, it is required that the connection point of the structural body has better rigidity and strength. In general, the installation layout of the landing gear system needs to be designed integrally with the configuration of the lander structure body.

① Number of landing gears. The regular polygon with the center of the footpads of multiple landing gears as the vertex is called the stable polygon, as shown in Fig. 13.4. During the landing process, if the projection of the landing center along the gravity direction of the land surface always falls in the stable polygon, the landing process is stable; that is, the lander will not tip over. Under the same other conditions, the more the number of landing gears, the larger the area of the stable polygon, the better the landing stability, but it will lead to complicated installation interfaces and increased mass. Therefore, when determining the layout and

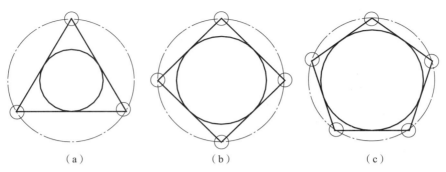

Fig. 13.4 Schematic diagram of stable polygons for different numbers of landing gears. **a** Three sets of landing gear systems; **b** four sets of landing gear systems; **c** five sets of landing gear systems

quantity of landing gears, the mass of the system and the installation interface of the landing gears shall also be considered.

② Configuration of landing gear system. There are mainly "cantilever beam type" (as shown in Fig. 13.1) and "inverted tripod type" (as shown in Fig. 13.5). "Cantilever beam type" has better landing stability and higher buffering reliability than "inverted tripod type." The composition of the "cantilever beam" landing gear system is more complex, the stress state is also more complex, and the processing and assembly technologies are more difficult. Usually, small landers are designed in an "inverted tripod" configuration, while large ones are designed in a "cantilever beam" configuration.

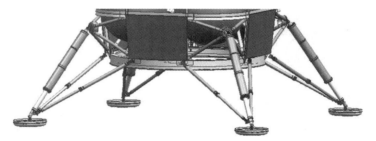

Fig. 13.5 "Inverted tripod type" landing gear system

(2) Analysis of key geometric parameters of landing gear system

① Key geometric parameters. The core of landing gear system design is to ensure stable and reliable landing. The main geometric parameters thus determined include L_h and L_v, as shown in Fig. 13.6. L_h is the initial height from the lowest point of the lander structure to the bottom surface of the footpad, and L_v is the shortest distance between the connecting line between the centers of two adjacent pads and the center axis of lander structure.

13.2 Landing Gear System

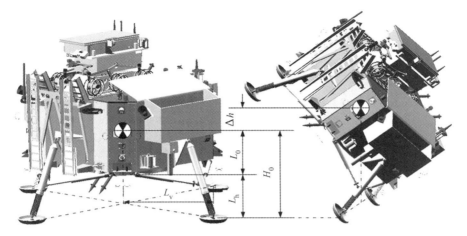

Fig. 13.6 Changes in potential energy during landing vehicle overturning

② Determination of parameter value. Considering the reduction of system mass and volume, L_h and L_v values should be as small as possible on the premise of meeting the landing stability requirements. The minimum value L_{hmin} for L_h can be estimated by the following equation:

$$L_{hmin} \approx S + S_0 + T + \Delta H \qquad (13.1)$$

where

S is the displacement of the main buffer in the vertical direction;
S_0 is the safe reserved distance between the lowest point of the structure and the landing surface after the lander lands;
T is the height of the footpad;
ΔH is the subsidence depth of the lander relative to the land surface after landing.

In general, the distance L_0 from the lander center of masscenter of mass to the lowest point of the structure is known. Therefore, once the L_{hmin} value is determined, the center of masscenter of mass height H_0 of the lander can be determined according to the following equation:

$$H_0 = L_{hmin} + L_0 \qquad (13.2)$$

The precise value of L_v needs to be determined through landing stability analysis. In the initial stage of design, the energy conversion principle can be used to preliminarily estimate the value of L_v. In order to ensure that the lander will not tip over during landing, the following conditions must be met:

$$W_D \leqslant W_H = m_{\max} g' \Delta h \tag{13.3}$$

where

- W_D is the kinetic energy of the lander at the beginning of turning over.
- W_H is the increase of the relative landing instantaneous potential energy of the lander when the lander center of masscenter of mass moves to the vertical plane of the connecting line of the ball hinge of the adjacent footpad.
- $m_{\max}$ is the maximum possible landing mass of the lander.
- g' is the gravitational acceleration on the landing surface.
- Δh is the increase of the center of masscenter of mass height during overturning.

It can be seen from Eq. (13.3) that as long as the kinetic energy of the lander during landing is not greater than the increase of the potential energy of the lander during the overturning process.

(3) Analysis and determination of buffer capacity

The total energy to be absorbed by the landing gear system mainly includes: kinetic energy of the lander at the moment of landing; reduced potential energy of the lander due to bumper operation and footpad subsidence.

① Buffer energy hypothesis. In the initial stage of design, in order to simplify the calculation, it can be assumed that all the potential energy of footpad subsidence is absorbed by the ground, and the landing gear system is considered to land on the rigid ground.

During landing, the total energy W absorbed by the main buffer is:

$$W = \frac{1}{2} m_{\max} v_v^2 + m_{\max} g' \Delta H \tag{13.4}$$

where v_v is the maximum vertical landing speed of the lander relative to the landing surface.

② Extreme conditions of landing. When landing on a slope with pits or bulges at a certain horizontal speed, the energy absorption of a certain landing gear system is significantly greater than that of other landing gear systems. At the beginning of the design, it is assumed that the maximum energy absorption capacity $A_{\max}$ of each main buffer is:

$$A_{\max} = a_0 W \tag{13.5}$$

where a_0 is a correction coefficient not greater than 1, and the greater the uncertainty of landing conditions, the greater the value.

For the auxiliary buffer, it is mainly used to offset the impact load in the horizontal direction. It can be considered that single auxiliary buffer absorbs all kinetic energy in the horizontal direction of the whole lander; that is, the maximum energy absorption capacity of single auxiliary buffer can be estimated as follows:

13.2 Landing Gear System

$$B_{\max} = \frac{1}{2} m_{\max} v_h^2 \qquad (13.6)$$

where

$B_{\max}$ is the maximum energy absorption capacity of single auxiliary buffer;
v_h is the maximum horizontal landing speed.

For different configuration of landing gear system, the energy absorbed by auxiliary buffer during stretching and compression can vary. The former is generally larger than the latter and can be larger than the value of calculated as per Eq. (13.6). The larger the angle between the main bumper axis and the lander axis at the moment of landing, the more likely the auxiliary buffer will be stretched and deformed, and the more energy will be absorbed during landing.

On the basis of numerical analysis of key geometric dimensions L_h, L_v and damping force, the requirement of maximum impact overload should also be considered, namely

$$a' \geqslant F/m_{\min} \qquad (13.7)$$

$$F \leqslant /m_{\min} a' \qquad (13.8)$$

where

a' is the allowable maximum impact acceleration response;
F is the maximum damping force allowed during landing.
$m_{\min}$ is the minimum possible landing mass of the lander.

According to Eqs. (13.5) and (13.7), the minimum stroke $S_{\min}$ of the main buffer can be further estimated:

$$S_{\min} = A_{\max}/(F/n) \qquad (13.9)$$

where n is the number of landing gears on the lander.

2) Main buffer design

The design of the main buffer includes two core contents: buffer capacity design and strength design. The design of buffering capacity includes selecting reasonable buffering method, appropriate buffering force, buffering stroke, buffering material, etc. Strength design is to ensure the strength margin requirements through structural design.

(1) Design of buffer method. Generally, cellular material deformation methods include foam aluminum deformation and aluminum honeycomb deformation. Thin-walled metal tube deformation method includes axial crushing deformation method, diameter expansion deformation method and overturning deformation method; metal cutting method; tensile deformation method of metal rod;

composite material crushing method; other buffering methods, such as gas–liquid damping method, electromagnetic damping method, magnetorheological fluid method, friction damping method, local structural deformation method. For choosing the buffer method, it is necessary to take into account many factors such as environment, landing conditions and impact response, surface characteristics of target celestial body, combination of multiple methods. Judging from the international successful application, the cellular material deformation method is the most common used solution.

(2) Design of buffering force and stroke. Generally, buffering is realized through the combination of low and high buffering forces, as shown in Fig. 13.7. The damping force F_1 of the lower level is realized by a buffering material with weak bearing capacity and is used for absorbing most impact energy when multiple sets of landing gears land simultaneously under ideal working conditions. On the one hand, the acceleration growth rate at the beginning of buffering can be reduced, which is beneficial to ensuring the safety of personnel and equipment; on the other hand, the height of the center of masscenter of mass of the lander can be rapidly reduced, the landing stability of the lander increased, and the impact response under ideal landing conditions reduced. The damping force F_2 of the second level is realized by using a buffering material with strong bearing capacity, which is used for coping with various extreme landing conditions. It can absorb more impact energy and will lead to a significant increase in impact response. It is required to ensure that the area enclosed by the damping force and buffer stroke meets the requirements of Eq. (13.5) and that the maximum damping force meets Eq. (13.8).

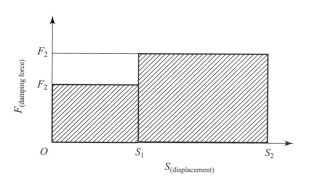

Fig. 13.7 Energy absorption diagram of main buffer

(3) Structure design of main buffer. The main buffer structure is piston cylinder structure, mainly composed of outer cylinder, inner cylinder, buffer material, guide ring, universal joint and ball hinge, as shown in Fig. 13.8. The ball hinge is connected with the footpad. During landing, the footpad first contacts the land surface, and the impact load generated therefrom is transmitted to the inner cylinder through the ball hinge. When the impact load exceeds the crushing load of the buffering material, the buffering materials with different bearing capacities undergo crushing deformation in turn from weak to strong, and the impact

13.2 Landing Gear System

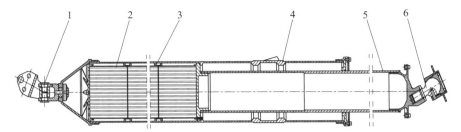

Fig. 13.8 Schematic diagram of main buffer. 1-universal joint; 2-aluminum honeycomb; 3-guide ring; 4-outer cylinder; 5-inner cylinder; 6-ball hinge

energy is also converted into the plastic deformation energy of the buffering material, thus realizing buffering of landing impact.

A corresponding limiting device is arranged on the ball hinge and between the inner and the outer cylinders to limit the relative rotation between the ball hinge and the inner and the outer cylinders in the launching phase, so as to keep the footpad at a desired position and to prevent the footpad from contacting and colliding with the carrier rocket. In order to prevent the plastic deformation of aluminum honeycomb buffer material from causing the axial clearance between the inner and outer cylinders, and then increase the amplitude of the footpad under the vibration load during the launch phase, an elastic link is arranged between the inner and outer cylinder to compensate the corresponding axial deformation.

Usually, aluminum honeycomb is used as buffer material, and when it in use, a plurality sections of aluminum honeycombs with different strengths are connected in series. When landing, the aluminum honeycomb with lower strength will crush and deform, and the aluminum honeycomb with higher strength will be deformed only after those with lower strength are completely deformed. In order to eliminate the peak load at the initial stage of aluminum honeycomb deformation, all aluminum honeycombs are generally pre-crushed before use. In order to prevent local instability of aluminum honeycomb and ensure the stability of load during deformation, guide rings are installed between aluminum honeycomb.

The outer cylinder and the inner cylinder are generally made of the same aluminum alloy material, and the universal joint and ball hinge are generally made of the same titanium alloy material, so as to ensure that the thermal expansion coefficient of each moving joint material is consistent, thus avoiding joint sticking, and ensure that the bearing capacity of all materials meets the requirements. In order to ensure that the position of the footpad does not change in the launch phase, a limit device is set in the ball hinge and between the inner and the outer cylinders to limit the relative movement between the ball hinge and the inner and outer cylinders in the launching phase. In order to prevent the plastic deformation of aluminum honeycomb buffering material from causing axial clearance between the inner and outer cylinders and then increase of the amplitude of the footpad under the vibration load in the launch phase, an elastic link can be arranged between the inner and outer cylinders to compensate for the corresponding axial deformation. In light of the characteristics that both ends of the inner cylinder bear less bending moment and the middle part bears more bending moment during landing, the inner cylinder adopts variable wall thickness design to minimize the mass of the landing gear system on the premise of meeting the strength requirements.

3) Design of auxiliary buffer

Considering that the auxiliary buffer may bear tensile load and compression load during landing, the auxiliary buffer shall have the ability of bidirectional buffer for stretching and compressing. In addition, the included angle between the auxiliary buffer and the main buffer is constantly changed due to the deformation of the auxiliary buffer during landing, thus the requirements for stretching and compressing buffering capability are often different. Under the influence of configuration changes, the auxiliary buffer bears tensile load in most cases, and its tensile buffering capacity is slightly larger than that calculated by the equation [2, 3].

The multi-functional auxiliary buffer not only has the bidirectional buffer capabilities of stretching and compressing, but also has the functions of compression release, deployment locking and locking indication. So its composition is relatively complicated. The multi-functional auxiliary buffer is of double piston structure, including universal joint, compression release device, outer cylinder, aluminum honeycomb, middle cylinder, steel ball lock, inner cylinder, deployment spring, ball hinge, pull rod, microswitch and other components, as shown in Fig. 13.9. The inner and the outer cylinders are locked together by a compression release device to form a stable structure, and the whole landing gear system is hold-down to meet the envelope requirement of the carrier rocket. When the landing gear system needs to be deployed on the orbit, the hold-down release device ignites to release the connection between the outer cylinder and the inner cylinder. Under the action of the deployment spring, the inner cylinder extends outward relative to the outer cylinder. After reaching the predetermined position, the inner cylinder and the outer cylinder are locked relative to each other, and a microswitch is triggered at the same time to give a deployment locking signal. The deployment spring is a compression spring, so even if the compression spring breaks, the driving force for deployment will not disappear completely as the tension spring, and reliable deployment can still be ensured.

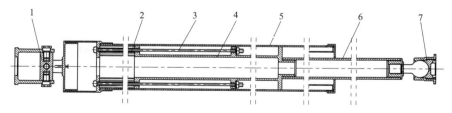

Fig. 13.9 Schematic diagram of auxiliary buffer structure. 1-universal joint; 2-aluminum honeycomb; 3-pull rod; 4-middle cylinder; 5-outer cylinder; 6-inner cylinder; 7-ball hinge

4) Footpad design

Footpads are mainly used for supporting, sliding and assisting buffering during landing, and giving landing signals at the moment of landing. It mainly consists of flange, trigger switch and basin body, as shown in Fig. 13.10.

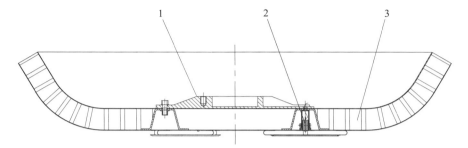

Fig. 13.10 Schematic diagram of footpad structure. 1-flange; 2-trigger switch; 3-basin body

The reliable connection between the ball hinge of the main buffer and the footpad with larger area is realized through the flange, thus ensuring the stable support for the lander. The basin is used to realize load transfer, smooth sliding and local buffering functions. It is an aluminum skin honeycomb sandwich structure, wherein the inner panel and the outer panel are, respectively, formed by spinning thin aluminum plates [2, 3]. The cap-shaped beam is embedded in the part where the basin body and the mounting flange are screwed together with stronger aluminum honeycomb material filled in the middle of the cap-shaped beam for good support, and a weaker aluminum honeycomb material is filled in the periphery for buffering the local deformation. The side surface of the basin body inclines outward at a certain angle with a certain height, so that dust on the land surface can be pushed away in the sliding process and a smooth landing can be achieved. In the sliding process, if the side of the footpad collides with a rigid object, the weaker aluminum honeycomb material will have local plastic deformation, which plays an auxiliary role in buffering.

In order to give a landing signal at the instant of landing so as to judge the landing situation or execute relevant instructions after landing, a trigger switch is also arranged at the bottom of each footpad. Considering the impact of the landing topography, the footpad may land at various deflection angles, and a plurality of switches are arranged on the sole of the footpad. After the footpad contacts the landing surface, the switch is closed (opened) and triggered. Any one of the switches that is triggered can give a landing signal. In practical application, in order to prevent giving wrong trigger signals, switch signals should be logically processed to ensure correct judgment.

The trigger switch can be a microswitch. In order to ensure that the footpad is reliably triggered when landing on the soft-landing surface, a leaf spring is arranged below the switch to increase the trigger force acting on the switch when landing. Besides, a certain distance is kept between the leaf spring and the switch to ensure that during the descent process before landing, the switch will not be mistakenly triggered due to the working pulsation of the engine, etc. In addition, the adoption of the leaf spring can also conveniently realize ground trigger detection for many times and ensure the consistency of trigger force in the ground detection process.

5) Dynamics analysis and verification of landing impact

Scheme analysis is an effective method to verify the rationality, effectiveness and correctness of schematic design. With the improvement of computer technology and model accuracy, the importance of analysis and verification becomes more and more prominent.

For landing gear system, the impact dynamics analysis of landing process is an important means to verify its core function. It is very difficult to establish an accurate simulation model of landing impact dynamics, which requires dynamic modeling of mechanical properties of celestial soil, landing gear system and landing impact process. The analysis and verification of landing impact dynamics are a systematic and complicated task, which plays an important role in supporting and guiding the research of the whole landing gear system.

As shown in Fig. 13.11, the analysis and verification work are generally carried out according to the following technical route [4]:

(1) Establish a landing impact dynamics simulation analysis model of a landing gear system and realize parameterized construction of a landing dynamics simulation model under any working condition and different terrains.
(2) The impact simulation analysis of single set of landing gear system is carried out. The correctness of the model is verified, and the simulation model is corrected by comparing and analyzing the simulation results and corresponding test results. Besides, on the basis of comparing the advantages and disadvantages of rigid body simulation model and flexible body simulation model, a better modeling method is determined.
(3) Establish typical extreme landing conditions, such as lander overturning extreme condition, main buffer compression extreme condition, auxiliary buffer compression extreme condition and auxiliary buffer stretching extreme condition and carry out landing impact analysis to obtain the safe landing boundary of the lander.
(4) Create a three-dimensional model of a typical celestial landing site (including slope gradient, location and depth of pits and bulges), as shown in Fig. 13.12. Then the initial landing conditions of the lander are randomly created, and the landing stability analysis is carried out on the surface of typical celestial bodies by Monte Carlo method.

Generally, the research of landing impact dynamics of lander mainly focuses on its three-degree-of-freedom and six-degree-of-freedom spatial models and landing impact models of single set of landing gears. In order to improve the efficiency of simulation analysis and calculation, it is usually assumed that the lander is a rigid model in the research, and the rigid assumption is made on the celestial body surface. However, the flexibility of the lander structure, the flexibility of the landing gear system and the elastic–plastic deformation of celestial soil have great influence on the

13.2 Landing Gear System

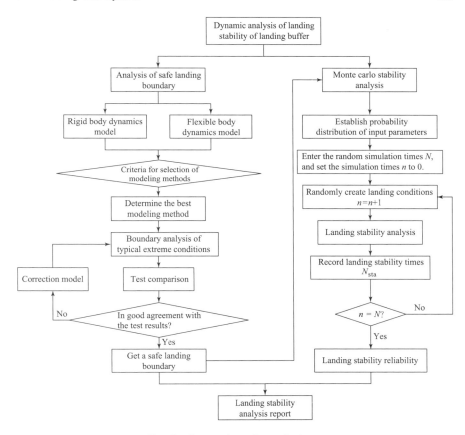

Fig. 13.11 Technical route of landing impact dynamics analysis

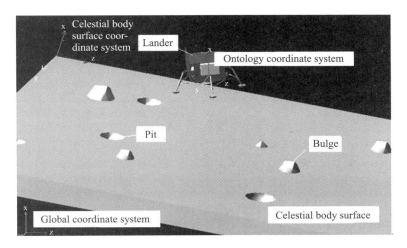

Fig. 13.12 Typical celestial landing field and lander model

landing impact performance of the lander. Therefore, in order to verify the rationality of the design scheme of the landing gear system of the lunar lander, the finite element model of the landing gear system can be established by using transient dynamics analysis method combined with mature commercial analysis software (such as MSC and Patran-Dytran), as shown in Fig. 13.13. According to the landing conditions in the landing area and the surface characteristics of celestial bodies, numerical simulations of various landing processes are carried out, which providing more accurate guidance for design improvement and ground impact test.

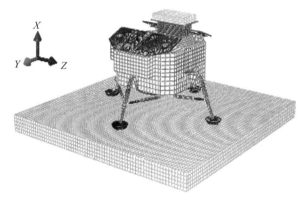

Fig. 13.13 Finite element model for landing impact dynamics analysis

13.2.2.2 Ground Verification of Landing Gear System

As the landing gear system is a multi-functional mechanism product which integrates the functions of collapsing and compressing, deploying and locking, landing buffer and supporting, it is quite different from a general mechanism, especially for certain particularity in ground testing and verification. In the development process, different levels of test and verification are usually carried out, such as material level, component level, mechanism level and integrate machine level. Typical ground verification tests include:

1) Performance verification of buffering materials

The performance verification of buffer material includes static test and impact test. In the initial stage of development, the bearing capacity and cushioning characteristics of buffering materials were obtained through static and impact tests, and the specifications of buffering materials meeting the requirements are determined. In the manufacturing stage of the flight model, the stability and consistency of the buffer performance of the buffer material in the production process are verified through static tests and impact tests.

13.2 Landing Gear System

2) Deployment and buffer friction test

The friction force between the moving pairs of the landing gear system will have a certain impact on the deployment and buffering performance of the mechanism itself. Therefore, it is of significance to obtain the friction force between the moving pairs for the design and performance evaluation of the mechanism.

The purpose of friction force test during deployment is to obtain the friction of each moving pairs related to deployment, so as to evaluate the deployment torque margin of the landing gear system. The purpose of friction test in buffer process is to test the value and change trend of friction under different bending loads and to provide basis for design and simulation analysis of landing gear system.

3) Performance verification of landing gear system

The purpose of the buffer performance test of single landing gear system is to verify whether the buffer and support capabilities meet the design requirements, and to reveal the possible problems in the design and manufacturing processes. The purpose of the combined buffer performance simulation test is to verify the combined buffer performance of the landing gear system. In this test, all landing gears are installed on the simulated structure of the lander to simulate the load conditions during landing and carry out landing buffer test.

13.3 Rover Transfer and Release System

The rover transfer and release system is a mechanism to transfer the rover from the lander to the surface of the celestial body to be explored. For different transfer modes of the rover, the composition and specific working mode of the transfer and release system may vary. Sometimes it is required to transfer the rover directly to the celestial body surface through the transfer and release mechanism, which is called direct transfer and release system [5]. Sometimes it is required to provide a ramp through the transfer and release mechanism, and then the rover will leave the lander by itself, which is called ramp transfer and release system. There is also a combination of the two; that is, the rover must be transferred to a specific position first and then leave by itself. This kind of transfer and release mechanism is called a hybrid transfer and release system. Judging from the current engineering application, ramp transfer and release system is widely used, as shown in Fig. 13.14.

The transfer and release system of China's Chang'e 3 rover adopts a hybrid transfer and release system, as shown in Fig. 13.15. The transfer and release system is first deployed on the lunar surface, the lunar rover moves to the parallel surface of the transfer and release system, then the transfer and release system carries the lunar rover to down to the lunar surface, and the rover finally leaves the ramp of the transfer and release system.

Fig. 13.14 Ramp transfer and release system

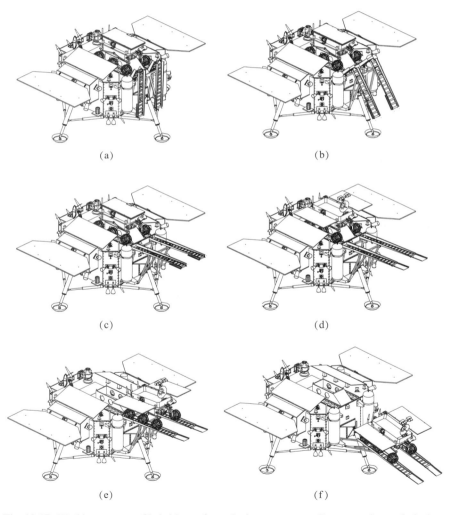

Fig. 13.15 Working process of hybrid transfer and release system. **a** Compressed state; **b** deployment process; **c** full deployment; **d** rover deployment; **e** transfer process 4; **f** touchdown

13.3 Rover Transfer and Release System

13.3.1 Functions and Composition Characteristics of Rover Transfer and Release System

In general, the rover transfer and release system has the following functions:

(1) It has the functions of compressing, unlocking and driving the rover away from the landing platform. Before reaching the celestial body surface, the transfer and release system needs to undergo the processes of spacecraft launching, on-orbit flight and orbit transfer, descent and landing, release and separation, etc. The transfer and release system is in a compressed state before landing and can be unlocked after landing.
(2) It has the ability to provide initial conditions for the rover under celestial gravity and complex environment to ensure that it can leave the lander. Given the complexity of the terrain on the celestial body surface, it is difficult to accurately predict and evaluate the departure conditions after the transfer and release system is deployed. In the design, various complex factors such as lander inclination angle, landing terrain inclination angle, soil bearing capacity and compactness in the landing area, rock density and size in the landing area should be comprehensively considered.
(3) It has the ability to adapt to all kinds of environment on the surface of celestial bodies. The transfer and release system is usually of large size, low rigidity and large area, with large span of structure installation, large length of slope after landing, and high requirement for traveling constraint error of the rover. The deformation caused by temperature and the influence of celestial body surface wind force on the deformation of the transfer and release system during deployment need to be analyzed and confirmed.
(4) Ensure the stability and controllability of the transfer process. For example, for the ramp transfer and release system, when the rover leaves the lander, it is affected by the angle difference between the ramps on both sides and the inclination angle of the lander. Under the gravity environment, there are not only downward pressure on the transfer and release system, but also lateral force, and tangential force in the traveling direction, so the stress state is complicated. In case of rigid slope, the stress state may be more complicated due to deformation. The transfer and release system needs to ensure stable structure, small deformation and no structural mutation under complex stress state. In addition, the ramps on both sides shall make constraints on the wheels of the rover to prevent slipping out of the transfer and release system.
(5) Other functional requirements. In general, the transfer and release system has some other requirements during its on-orbit operation, such as remote-sensing information feedback (unlocking information and deployment information). When necessary, it can also give detailed information such as ramp angle and non-planar angle to judge the working state of the transfer and release system.

The solution choice for rover leaving the lander is mainly related to the structural design of the lander system. For the lander with landing gear system, ramp transfer and release system is often used to facilitate the rover to leave from the lander, as shown in Fig. 13.16. The two ramps are, respectively, used as the traveling paths for wheels on both sides of the rover. The structural design of the two ramps is identical. A connecting piece is designed between the two ramps to ensure that the relative position relation remains unchanged.

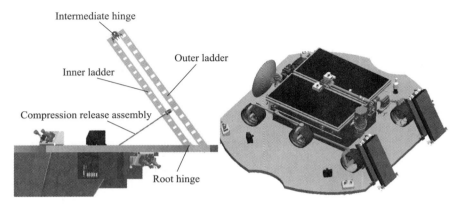

Fig. 13.16 Relationship between ramp and wheel position

Each set of ramp contains two ramp mechanisms. The collapsed state of single ramp mechanism is shown in Fig. 13.17. It is mainly composed of inner ladder, middle ladder, outer ladder, root hinge, middle hinge, angular displacement sensor, interlocked device and compression release device. In the collapsed state, the ramp forms an included angle with the landing platform to avoid interference with the rover, and the folded ramp is compressed together with the compression release assembly. The ramp is connected with the landing platform through the root hinge, and the hinge and the compression release assembly jointly maintains the configuration stability in a collapsed state. The driving torque of the deploying process is provided by the hinge spring at the folding position.

13.3.2 Design and Verification of Rover Transfer and Release System

13.3.2.1 Design of Rover Transfer and Release System

Taking the ramp transfer and release system as an example, the basic design process of the transfer and release system and the problems that should be paid attention to are explained as follows:

13.3 Rover Transfer and Release System

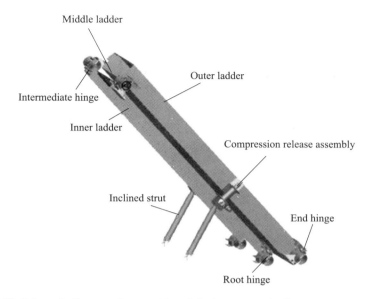

Fig. 13.17 Schematic diagram of composition of single ramp mechanism

1) System design

The following describes a symmetrically arranged foldable ramp type transfer and release system. Ramps on both sides are distributed at both ends of the lander. The layout is shown in Fig. 13.18.

After landing on the surface of the celestial body to be explored, the compression release assembly is unlocked, and the ramp mechanism is unfolded under the driving torque of the two hinges. The unfolding process is shown in Fig. 13.19. After the compression release device is unlocked, the folded ramp starts to unfold immediately, and the inclined stay rod used for supporting and fixing the ramp presses on the lander under the action of the root spring to prevent the interference to moving of the rover. Under the action of the interlocked device, all sections of ramps can be unfolded synchronously. When the ramp is unfolded horizontally, the hinges at the middle and

Fig. 13.18 Ramp layout. **a** Collapsed configuration; **b** deployed configuration

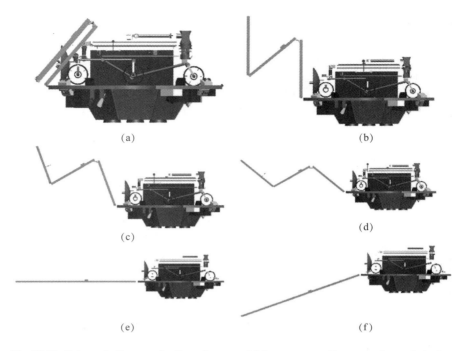

Fig. 13.19 Schematic diagram of unilateral ramp unfolding process. **a** Compressed state; **b** deployment position 1; **c** deployment position 2; **d** deployment position 3; **e** deployment position 4; **f** full deployment

the end part are locked, the interlocking between the whole ramp and the root hinge is released, and the ramp continues to rotate around the root hinge rotating shaft until the ramp is fully unfolded.

2) Ramp

The ramp consists of three parts, namely the inner ladder, the middle ladder and the outer ladder. The middle ladder is slightly shorter, and the inner ladder and the outer ladder are slightly longer. They all made of aluminum alloy plates. The overall configuration after deployment is shown in Fig. 13.20. In order to adapt the rover wheels to travel in the ramp with a certain amount of movement to prevent crushing and sticking, the ramp width is slightly larger than the wheel width, and the outer ladder is usually larger than the inner ladder and the middle ladder, so as to facilitate the rover to obtain greater freedom when driving to the road surface.

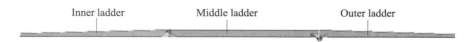

Fig. 13.20 Schematic diagram of single ramp deployment configuration

13.3 Rover Transfer and Release System

Flanged guardrails are designed on both sides of the ramp to prevent the wheels from falling off the ramp in case of tilting or driving deviation. The upper end of the guardrail is of circular arc configuration. On the one hand, it is used to improve the longitudinal rigidity of the ramp; on the other hand, it can also provide dotted line contact with the rover wheels to reduce friction, as shown in Fig. 13.21. The flanging height of the inner ladder gradually increases along the length of the ramp. When the rover travels to the middle position of the ramp, the ramp bears the maximum force, and the middle ladder bears the maximum load and the maximum strength is designed accordingly.

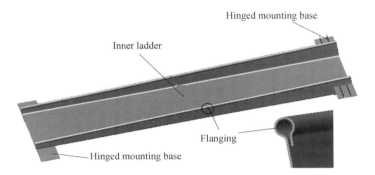

Fig. 13.21 Schematic diagram of inner ladder

There are two guide rails on the ramp, which can adopt a high-friction surface or a crushable sinking structure, or a baffle structure. The wheel of the rover pressed on the guide rail cannot only prevent the wheel from sliding, but also strengthen the longitudinal rigidity of the ramp (Fig. 13.22).

3) Unfold the hinge

The root hinge, middle hinge and end hinge can be designed in different forms according to different design factors such as deployment position, deployment force, bearing capacity, rigidity.

Fig. 13.22 Ramps and guide rails

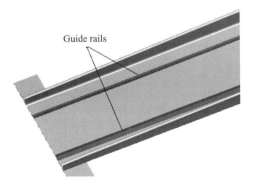

The unfolding hinge is mainly composed of locking components (male hinge, female hinge and latch hook), plane scroll spring, etc. One or two coil springs can be designed for each hinge according to the required unfolding force, and the unfolding force can be adjusted by changing the rigidity of the coil springs and the number of revolutions. The position synchronization is maintained between hinges through interlocked rope and interlocked wheel (Fig. 13.23).

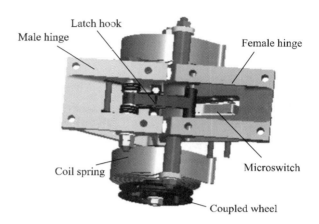

Fig. 13.23 Schematic diagram of middle deployment hinge

The outer ring of the plane scroll spring is connected with the female hinge, and the inner ring is fixed on the male hinge through a spring collar. When the ramp is in a folded state, the plane scroll spring of the unfolding hinge is in a pre-tightening state, and the magnitude of the pre-tightening moment can be adjusted by a plurality of uniformly distributed round holes and fixing pins distributed on the spring fixing bracket along the circumferential direction. When unfolded, the plane scroll spring drives the male hinge and the female hinge and drives the ramp to move. During the movement, the latch hook installed on the male hinge will contact with the guide surface of the latch hook on the female hinge, where self-lubricating material is installed. In the unfolding process, the latch hook on the locking hinge slides stably on the guide surface with low friction coefficient under the compression force of the torsion spring.

After deployment, the latch hook slides into the latch hook groove which is well-matched with the latch hook along the guide surface, and the locking state is maintained by the pressing force provided by the torsion spring. By reasonably designing the clearance, locking depth and torsion spring pressing force of the locking mechanism, it is ensured that the hinge has adequate locking rigidity after locking. After deployment in place, once the male hinge and the female hinge are locked, the microswitch installed on the deployment hinge will be triggered to provide the indication signal of deployment in place.

It can be seen from the unfolding process that in the initial folded state, gravity hinders the deployment of the torque of root hinge and acts as a resistance moment. After the inner ladder and the outer ladder are extended to a vertical state, the gravity generally performs positive work, which is beneficial to the deployment of the ramp.

13.3 Rover Transfer and Release System

In the process of unfolding the middle ladder, gravity must be overcome, but gravity does positive work to the whole ramp. According to the above situation and the simulation results, the coil spring torque at each rotating shaft of the ramp mechanism is designed with different values.

4) Interlocked device

The function of the deployment interlocked device is to prevent each ramp from interfering with each other during deployment and realize synchronous deployment. The interlocked device comprises two parts: one is the interlock between the landing platform and the middle hinge, the other is the interlock between the middle hinge and the end hinge.

For the interlocked rope shown in Fig. 13.24, one end is fixedly connected to the landing platform, and the other end is fixedly connected to the interlocked wheel of the intermediate hinge and controls the root hinge to deploy synchronously with the intermediate hinge. By adjusting the position of the O point and the radius of the interlocked wheel, the increase of the length OA of the interlocked rope is equal to the length of the interlocked rope released by the rotation of the interlocked wheel of the intermediate hinge in the unfolding process. When the ramp extended to the folding angle state, the length of the interlocked rope reaches the maximum value, i.e., (OO'' + $O''A'$). Among them, point O and point O'' are the fixed point of the interlocked rope and the center of the root hinge, respectively, point O' is the intersection point of the connecting line of point O and O'' and the edge of the landing platform, and point A is the center of the rotating shaft of the intermediate hinge.

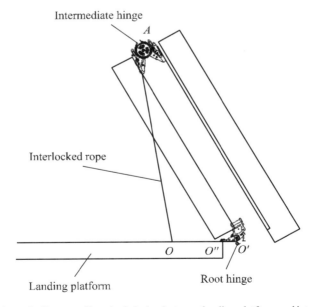

Fig. 13.24 Schematic diagram of interlock design between landing platform and intermediate hinge

As the ramp continues to unfold, the interlocked rope is centered at the edge O' of the landing platform, and the rotation radius $O'A$ gradually decreases. At this time, the actual length of the interlocked rope between the O' point and the hinge center point A is $O'A'$. According to the geometric relationship shown in Fig. 13.25, it is established that $O'A' > O'A'' > O'A'''$; that is, the actual length of the interlocked rope is larger than the rotation radius dimension, and the rope is loose, so that the interlock between the landing platform and the intermediate hinge can be released and the whole ramp can rotate around the center of the root hinge.

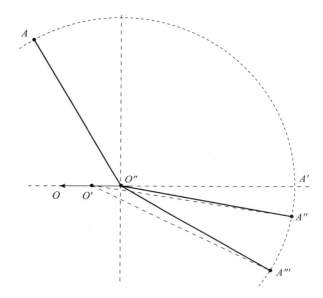

Fig. 13.25 Interlock analysis

13.3.2.2 Ground Verification of Rover Transfer and Release System

The transfer and release system is a mechanism that integrates the functions of collapsing and pressing, deploying and locking, supporting and protecting and is an important supporting mechanism for the rover to leave the lander reliably. In the development process of transfer and release system, it is usually necessary to carry out tests and verifications at different levels, such as material level, component level, mechanism level, whole device level. The key materials mainly include the materials for ramp support surface and wheel guardrail. Friction force, surface hardness and strength are the key points subject to verification, and the verification method is relatively conventional.

The departure test of the transfer and release system and the rover is a comprehensive test to fully verify that the ramp meets the functional performance, which is introduced in this section. The departure test is to load the landing gear system on the simulated structure of the lander. The simulation structure can perform functions such as roll, pitch and height adjustment to meet different landing attitudes of the

13.3 Rover Transfer and Release System

lander on orbit. After the ramp is deployed and contacted with the ground, the ground needs to simulate various working conditions on the ground such as rocks and pits. Firstly, simulated rover with the same moving function is used to carry out the test verification, and finally real rover is used to carry out the test verification.

The departure test is generally carried out under the condition of gravity field. The simulated rover can simulate the celestial body gravity through the counterweight. A real rover needs to use the gravity compensation device to carry out the test equivalent to the celestial body gravity. Figure 13.26 shows the schematic diagram of the transfer test of the rover and the photographs of the test site. The test system consists of test pieces, lander simulator, simulated rover, test equipment and simulated terrain. The lander simulator is used to adjust different attitude and terrain to simulate different pits and bulges, and different transfer and release system configurations are constructed to test the performance of the rover when it leaves the ramp.

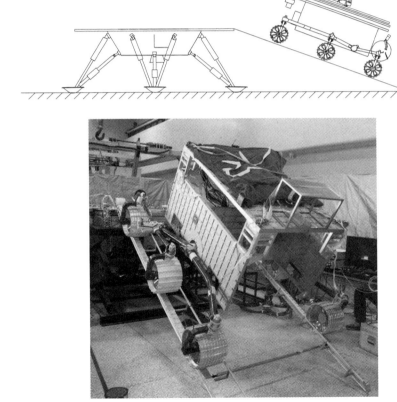

Fig. 13.26 Schematic diagram of rover transfer and photograph of test site

13.4 Rover Mobility System Mobility System

The rover mobility system is a kind of mechanism that can realize the movement and roaming of the rover on the surface of the celestial body to be explored, as well as the transportation of materials (including collected samples) or personnel. It is in direct contact with the soil and rocks on the surface of the celestial body, doing work and generating power to move the rover.

When designing the mobility system of the rover, it is first necessary to make a comprehensive, systematic, in-depth and detailed analysis of its functional requirements, so as to accurately grasp the functional requirements and avoid omission. Besides, the key technical indicators are analyzed, decomposed and refined, and the requirements of core performance indexes need to be met in the initial stage of design.

13.4.1 Functions and Composition Characteristics of Rover Mobility System

In general, the rover mobility system has the following functions:

(1) It has the ability to move under celestial gravity and complex environment. This is the core function of the mobility system, which can carry the rover under the celestial gravity environment and bear the impact of bumping and impact. Due to the lack of accurate data of soil bearing capacity, compactness, topography, obstacles, rock distribution, rock sharpness and hardness of celestial bodies, the mobility system shall have strong terrain adaptability.

(2) It has the functions of compressing, unlocking and leaving the landing platform. Before reaching the celestial body surface, the mobility system needs to undergo the processes of carrying and launching, on-orbit flight and orbit transfer, descent and landing, release and separation, etc. The mobility system is in a compressed state before the landing and roving and can be unlocked after landing [6].

(3) The mobility system should be able to adapt to the climatic characteristics of the celestial body surface, such as temperature, atmosphere, dust and other environmental characteristics. The temperature and atmospheric characteristics of different celestial bodies vary greatly, so temperature adaptive design shall be carried out for the mobility system in a targeted manner.

(4) Satisfying the functions of long service life and long-distance driving. In order to carry out roving and detection, the mobility system is expected to be able to move in a large range to obtain more detection information, so it shall have a travel distance as far as possible. The travel distance of the mobility system is divided into single continuous travel distance and accumulated travel distance. Constraints such as energy consumption and temperature rise in the operation

of the mechanism should be taken into account in single travel distance. Accumulative travel distance is mainly the total life obtained in mechanism design, which is related to load, temperature, lubrication mode, etc.

(5) Other functional requirements. Usually, some other requirements is applicable to the mobility system during the working process, such as remote-sensing information feedback (driving speed and steering angle) and judging the working state of the mobility system.

Six-wheel mobility system is one of the most widely used mobility systems in the world. Examples include the Mars rover in the USA, the Moon rover "Yutu" in China and the Mars rover in Europe. This section takes a six-wheel mobility system with active suspension function as an example to introduce the design, test and verification of the mobility system.

The configuration and hardware components of the active suspension mobility system are shown in Fig. 13.27. It is mainly composed of a drive steering module, a suspension adjustment module, a differential support module and a compression release mechanism. The functions of each part are as follows:

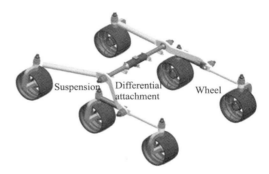

Fig. 13.27 Configuration of mobility system

(1) Drive steering module: It is mainly composed of wheels, drive mechanism, steering mechanism, steering arm and other parts. It is the core product of the rover to implement the functions of walking, steering, climbing, obstacle surmounting, static holding. It is a product that directly contacts the surface of celestial bodies for a long time. The tread of the wheel structure also needs to meet the requirement of resisting rock penetration.

(2) Suspension adjustment module: It is mainly composed of angle adjustment mechanism, clutch, front end of main rocker arm, rear end of main rocker arm and auxiliary rocker arm. The function of the suspension adjustment module is to maintain the suspension configuration, connect and drive the steering module, provide support for the carriage during normal driving, and use the clutch of the main rocker arm and the auxiliary rocker arm (open state) to realize free bumping of the four wheels on the auxiliary rocker arm, thus adapting to terrain changes. The suspension adjustment module can also realize the functions of vehicle body lifting, center of mass position adjustment, wheel lifting and so on, which is suitable for different terrains of celestial body surface.

(3) Differential support module: It is mainly composed of differential attachment, main shaft connecting shaft and angle adjustment structure adapter tube and is the main connection point between the rover body and the suspension, with overturning support function in the event of deviation of the vehicle body center of mass. When the wheels on both sides are running on uneven ground with different rotating speeds, differential balance is provided to prevent the vehicle body from rotating or the wheels from dragging on the ground.
(4) hold-down and release assembly: It is used to lock the mobility system reliably in spacecraft launching, so that the mobility system can bear larger load without damage and meet the rigidity requirement of the pressing state of the mobility system. After landing, the constraints on the mobility system are released smoothly to enable it to move normally.

13.4.2 Design and Verification of Rover Mobility System

13.4.2.1 Design of Mobility System of the Rover

1) Working principle design of active suspension

The working principle of the six-wheel active suspension mobility system is shown in Fig. 13.28. Wheel drive mechanism, steering mechanism and differential attachment are the basic components of the conventional fixed suspension mobility system. Angle adjusting mechanism is a mechanism product to realize the function of active suspension. Through the rotation of the angle adjusting mechanism, the vehicle body, the front main rocker arm and the rear main rocker arm change in a certain proportion, so as to realize the horizontal lifting and lowering of the vehicle body. The brake is matched with the angle adjusting mechanism to realize the lifting of each wheel [7].

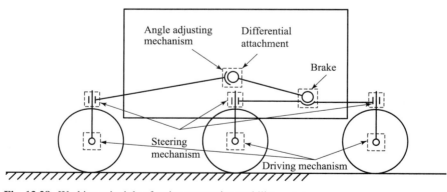

Fig. 13.28 Working principle of active suspension mobility system

13.4 Rover Mobility System Mobility System

Angle adjustment mechanism is the key element of active suspension adjustment function design. If the vehicle body is fixed with a rocker arm, only the angle relationship between the two main rocker arms can be adjusted. Although the vehicle body can be lifted, the vehicle body will tilt, as shown in Fig. 13.29 [8].

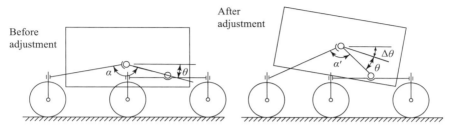

Fig. 13.29 Principle of vehicle body lifting and tilting

During the adjustment of the active suspension, if the horizontal position of the vehicle body is kept unchanged, the relationship among the three shall be established, and the angular correlation is shown in Fig. 13.30. Single-input double-output mechanism is needed to realize the proportional change, which can usually be realized by planetary gears. The shell of the planetary gear mechanism is fixed with the vehicle body, and the sun gear and the internal gear are respectively connected with two main rocker arms. The reduction ratio of the planetary gear shall conform to the following equation.

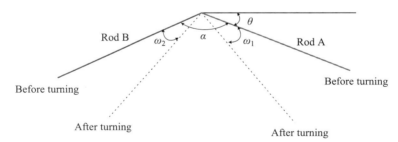

Fig. 13.30 Angle relationship of active suspension

$$k = \frac{\omega_2 - \omega_1}{\omega_1} = \frac{\Delta\alpha}{\Delta\theta} \qquad (13.10)$$

The specific design of reduction ratio and its variation characteristics is related to the structural dimensions of suspension. The geometric correspondence is shown in Fig. 13.31, where

l_1 is the length of the front section of the main rocker arm;
l_2 is the length of the rear section of the main rocker arm;
l_3 is the length of the auxiliary rocker arm;

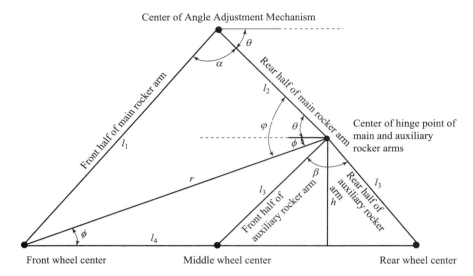

Fig. 13.31 Angle relationship of active suspension

l_4 is the distance between the center of the front wheel and the center of the middle wheel;
r is the connecting line length between the front wheel center and the hinge point of the main rocker arm and the auxiliary rocker arm;
h is the vertical distance between the hinge point of the main rocker arm and the axle;
α is the included angle of the main rocker arm;
θ is the included angle between the rear section of the main rocker arm and the vehicle body;
ϕ is, in most cases, the included angle between the line connecting the center of the front wheel and the center of the hinge point of the main rocker arm and the horizontal plane of the vehicle body;
β is the included angle between the auxiliary rocker arms.

According to the above geometric configuration relationship, θ is expressed as a function of α, as shown in Eq. (13.11), which has obvious nonlinear characteristics. Therefore, the adjustment process of the included angle adjustment mechanism cannot ensure the theoretical level of the vehicle body. The following two requirements are expected to be satisfied in the design optimization:

(1) The vehicle body is horizontal under the state of compression and normal driving;
(2) Within the adjustable height range, the included angle change shall be as small as possible.

13.4 Rover Mobility System Mobility System

$$\theta = \arccos\left[\frac{l_2 - l_1 \cos\alpha}{\sqrt{l_1^2 + l_2^2 - 2l_1 l_2 \cos\alpha}}\right] - \arcsin\left(\frac{h}{\sqrt{l_1^2 + l_2^2 - 2l_1 l_2 \cos\alpha}}\right) \quad (13.11)$$

In actual engineering design, the following issues are to be considered:

(1) Within the adjustable range of the vehicle body, there is the problem that the axis of the angle adjustment mechanism is at the same level as the wheel center, which requires the angle adjustment mechanism to provide a large torque and the wheel drive cannot provide assistance to lifting;
(2) Within the lifting range of the vehicle body, the wheel spacing will increase first and then decrease.
(3) After the wheel is lifted by the active suspension, the load of the rocker arm and the wheel changes greatly and needs to be checked.
(4) During the lifting process of the vehicle body, the center of masscenter of mass will be interlocked in both horizontal and vertical directions. In order to ensure the smooth lifting of the wheels, the position of the center of mass of the vehicle body should be reasonably configured.

2) Design of driving steering mechanism

The driving steering mechanism is mainly composed of steering mechanism, driving mechanism and wheels. The design principle and configuration of a drive steering mechanism are shown in Fig. 13.32.

The driving mechanism provides power for the mobility system to move forward and is the core of the moving ability of the mobility system. The driving mechanism design focuses on the traveling distance and driving ability of the mobility system, and whether the driving force of a small number of wheels or even single wheel meets the traveling function requirements of the mobility system should be considered.

Different materials and different structural forms shall be adopted for wheels according to the difference of terrain characteristics. For the surfaces of celestial bodies mainly composed of soil, the mobility system generally adopts an elastic screen wheel structure, the frame and pawls are made of aluminum alloy, and the screen is made of titanium alloy wire. For the surfaces of celestial bodies with more rocks, the mobility system usually adopts all-metal structure, usually aluminum alloy. With the technological innovation and the application of new materials, aluminum-based silicon carbide and other new materials have high hardness and high strength characteristics and have also been applied. The tread of the wheel has different shapes such as cylindrical wheel and drum wheel. Generally, drum wheel is helpful to reduce steering torque.

Steering mechanism usually has two modes of orthogonal steering and oblique steering. From the design analysis and test, when walking on sandy soil ground, the steering torque of oblique steering is large, the wheel subsidence during steering is large, and it turns to be a nonlinear growth trend with the increase of oblique angle. However, for oblique steering, the steering mechanism is installed near the inner cavity of the wheel, making the structural layout more compact and convenient for collapsing and folding. If the orthogonal steering is designed with eccentricity, the

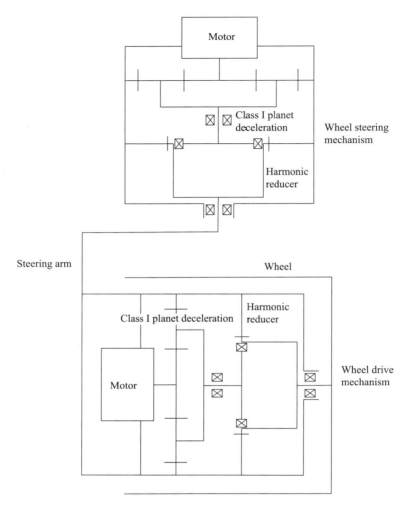

Fig. 13.32 Principle of driving steering mechanism

steering torque will increase significantly with the increase of eccentricity, even larger than the oblique steering. Therefore, the orthogonal steering is usually designed in centering.

3) Suspension adjustment mechanism design

Active suspension adjustment can realize the lifting of the vehicle body and change the position of the center of mass of the vehicle body. If a K-joint is added between the main rocker arm and the auxiliary rocker arm, the lifting of the wheels can be realized through the opening and closing of the K-joint [9].

13.4 Rover Mobility System Mobility System

The principle of active suspension adjustment is to coordinate the relative positional relationship between the front rocker arm, the rear rocker arm and the vehicle body through the mechanism and maintain the good coordination among them according to the specified proportion. The following describes single-input double-output mechanism composed of planetary gearing mechanism to realize the coordinated movement of the three. Its working principle is shown in Fig. 13.33.

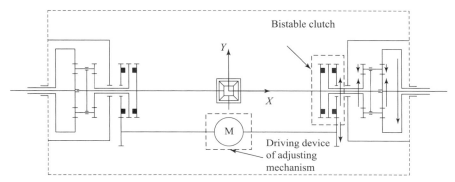

Fig. 13.33 Working principle of planetary gearing mechanism with active suspension

The vehicle body is connected with the shell of the differential attachment, and a pair of bevel gears of the differential attachment is connected with the inner shell of the angle adjusting mechanism, so that the vehicle body establishes a connection relationship with the shell of the angle adjusting mechanism through the shell of the differential attachment. The internal gear of the I-stage planetary gear is connected with the front main rocker arm through rod A, and the sun gear is connected with the rear main rocker arm through rod B, thus establishing a transmission ratio relation among the front, rear main rocker arms and the vehicle body. Under the condition of keeping the vehicle body horizontal and motionless, rod A and rod B move per a fixed reduction ratio.

The reduction ratio of rod A and rod B is related to the wheel tread, the height from main shaft to wheel center, whether the main shaft and middle wheel axes are coplanar, and the height of connecting point of main and auxiliary rocker arms, among other parameters. It is suggested to pay attention to the following problems in the design of suspension adjustment mechanism:

(1) The reduction ratio of planetary gearing is difficult to adjust continuously under the constraints of structure and load design. The reduction ratio determined by the relationship between suspension rod systems should be matched with the reduction ratio of planetary gearing as much as possible.
(2) Even if the reduction ratio of the planetary gearing is completely consistent with the calculated reduction ratio of the rod system, the vehicle body will still roll at a small angle within the adjustable range due to the nonlinear influence of the change process;

(3) By reasonably optimizing the structure of the rod system, the vehicle body can be kept horizontal at two points and the roll angle can be reduced as much as possible within the range of motion;
(4) The angle adjustment mechanism will bear a large load for a long time as it remains stationary or moving, and it must have self-locking function.

4) Lubrication design

The lubrication design of the mechanism is the basis of ensuring it can work for a long time, reduce resistance and power consumption. The choice of lubrication scheme needs to take into account the design constraints in the environment of use and supporting conditions. For rovers on different celestial body surfaces, the lubrication design may vary greatly. Lubrication methods for different types of moving parts are also different. For example, moving parts such as bearings and lead screws are usually sputtered with molybdenum disulfide. Harmonic reducers may take grease lubrication, solid lubrication, mixed lubrication and other methods. Planetary gears are usually lubricated with bonded molybdenum disulfide or grease.

Generally, grease lubrication can obtain a longer service life than solid lubrication, but it is greatly influenced by the environment. The following factors are usually considered:

(1) The resisting moment of grease lubrication increases under low-temperature environment; once it exceeds the freezing point, it will seriously affect the lubrication effect.
(2) In vacuum environment, grease lubrication has volatilization characteristics. It is necessary to design a sealed circuit, especially in high-temperature environment, which will accelerate volatilization. Lubrication performance is not only related to working life, but also to storage life.
(3) For moving parts lubricated with grease, it is recommended to coat anti-creep layer to prevent lubricating oil from seeping along the mating surface.

Solid lubrication is less affected by the environment and can adapt to a larger range of working temperature. Its lubrication performance will not change with the change of temperature. The service life of solid lubricated moving parts is shorter than that of grease lubrication. It is recommended to consider the following factors in design and use:

(1) Solid lubrication has two states: solid lubrication film and solid lubrication coating. The thickness is different with different process methods. The thickness of the former is generally no more than 5 μm, and that of the latter is generally no less than 5 μm.
(2) For moving parts with small clearance of moving pairs, such as bearings, lead screws and harmonic reducers, solid lubricating films are usually adopted, and for moving parts with large clearance, both methods can be adopted.
(3) Different solid lubrication methods have different requirements on the roughness of the friction surface. When the roughness Ra > 1.0 μm, it is not recommended to use solid lubrication film, and when Ra < 0.4 μm, it can be preferred.

(4) The average Hertz contact stress on the surface of the moving pairs has a great influence on the life of solid lubrication. Measures should be taken to reduce the average Hertz contact stress of the moving pairs in the design, and different lubrication technologies should be selected according to the stress.
(5) For closely matched moving pairs, such as harmonic reducers, bearings and lead screws, the influence of temperature changes on the gap changes of moving pairs and the resulting changes in average Hertz contact stress need to be considered.
(6) Some solid lubrication processes are sensitive to environmental characteristics such as space atomic oxygen and need special attention. Effective protective measures should be taken in design.

Taking Mars exploration as an example, the surface temperature of Mars is relatively low, ranging from -103 to $+27$ °C in the latitude of $0°$–$30°$. If temperature control measures are not taken, the mobility system will be in a low-temperature environment for most of the time, and the solid lubrication scheme is suitable.

If the travel life of mobility system with solid lubrication cannot meet the travel distance life requirement of the rover, grease lubrication is required, and active temperature control measures are required for the mobility system to ensure that the temperature meets the applicable range of grease lubrication. The relationship curve between lubricating performance and temperature of different lubricating greases is different, and targeted measures should be taken in design.

5) Thermal design

The purpose of thermal design is to make the mobility system work in a comfortable temperature range by analyzing and taking proper measures. Different celestial bodies have great differences in their temperature environment and heat dissipation conditions, and their mechanism design and measures to be taken are also different. If the lunar surface is in a vacuum environment and the temperature changes greatly, the mechanism should adapt to both the high-temperature and low-temperature environment. There is a pressure of 600–1000 Pa on the surface of Mars, and mostly in a low-temperature environment, and the high temperature generally does not exceed 55 °C. The adaptability of low-temperature environment is mainly considered in the mechanism design.

The influence of temperature environment on the working performance of the mechanism is mainly considered in the following aspects:

(1) In order to reduce weight, different materials are usually adopted for the shells, shafting, gears, bearings, etc., in the mechanism. Due to different thermal expansion coefficients and large deformation differences at high and low temperatures, the movement clearance and error range of the mechanism can be affected.
(2) The deformation of the mechanism caused by temperature change may lead to change in transmission accuracy, transmission efficiency, etc.
(3) The high-temperature and low-temperature environment will cause the change of lubricating characteristics. For example, low temperature will cause increase of the viscosity damping coefficient of lubricating grease, while high temperature will accelerate the volatilization of lubricating oil in lubricating grease.

(4) If temperature-sensitive electronic components, such as Hall devices, are used in the mobility system, the electronic components will be damaged at the extreme temperature.

The design and implementation of the thermal control system mainly focus on the key products and key parts of the mobility system, such as the driving mechanism, steering mechanism, differential attachment, compression release mechanism, igniters. Heat conduction path should also be considered in the mechanism design. Motor-related products will generate more heat in the work, resulting in rapid temperature rise, which causes not only temperature changes, but also the large temperature difference between the motor, reducer and structural members. The following issues are to be considered in thermal design and thermal implementation:

(1) The design of the working temperature range of the mobility system should be matched with the overall design of the rover. The wider the temperature range of the mobility system, the smaller the pressure on the energy of the rover, but the more difficult it is to design the mechanism itself.
(2) Comfortable temperature environment and temperature uniformity are favorable for ensuring the performance of the mobility system and improving the service life of the mechanism on the orbit.
(3) The temperature rises rapidly during the operation of the motor. The temperature characteristics of the initial working temperature and working time should be considered simultaneously in thermal design and thermal implementation.
(4) For implementation of thermal control, the movement range of the mechanism and the risk of interference with products within the movement range of the mechanism under fluffy states such as thermal control multilayer shall be considered.
(5) Generally, the mechanism products will be equipped with a housing grounded to release the accumulation of electrostatic charges on the surface. As for materials in thermal control implementation, such as multilayer materials that are apt to generate static electricity, it is also necessary to set up an electrostatic discharge circuit.
(6) During the implementation of thermal control, attention should be paid to the accumulation of dust on the surface of celestial body. Prevent the formation of funnel-shaped or local pits where dust is apt to accumulate.

13.4.2.2 Ground Verification of Rover Mobility System

In order to ensure that the design of the mobility system is correct, the functional performance index and environmental adaptability meet the design requirements and ensure that the technique and control of the flight model production process are in place, and the product quality meets the requirements of launching into the orbit; it is necessary to carry out corresponding tests on the mobility system. In different stages, enough tests and verifications shall be carried out in a targeted manner. For example, in the schematic research stage, key technical research and verification of technical

13.4 Rover Mobility System Mobility System

principles shall be emphasized; in the initial sample stage, research and development tests, identification tests, reliability tests and other verifications shall be emphasized; in the flight model stage, acceptance tests shall be emphasized to verify the quality control results in the production and manufacturing process, eliminate early failures and ensure the stable operation of products.

In the mobility system, the stress state of the single-axis mechanism is relatively complex, and the forces and moments in six directions act on the mechanism simultaneously. The test should consider the high- and low-temperature loading test under combined working conditions. In addition, the wheel is an important component for the long-term contact between the mobility system and the celestial body surface, and it is also an important product that determines the moving performance. There are two performance requirements for wheels in use: Firstly, they can travel on the ground with high efficiency and can adapt to various terrains; secondly, it has sufficient mechanical strength and toughness to meet the environmental requirements of bearing, impact, rock spikes, etc.

This section focuses on the single-axle mechanism and wheel test items carried out according to the characteristics of the mobility system.

1) Thermal vacuum test of single-axis mechanism

The purpose of the thermal vacuum test of single-axis mechanism is to test the functional performance index of the driving mechanism, steering mechanism, angle adjustment mechanism, clutch and differential attachment in the mobility system under vacuum high- and low-temperature environment, so as to verify the survivability and performance index change on the orbit environment.

The stress condition of the single-axis mechanism of the mobility system is different from that of the mechanism of the common Earth orbit spacecraft. In addition to torsional load, it also bears combination of various loads, such as bending load, radial load and axial load. It is a technical challenge to fully carry out combined load test under vacuum high- and low-temperature environment, especially eliminate the influence of error in combined load. In addition, the mechanical characteristics of products such as angle adjustment mechanism, clutch and differential attachment in the moving device are relatively complex. For example, the angle adjustment mechanism is single-input double-output double-redundancy mechanism, which needs to apply two different torques and the bending loads at the same time.

The stress state of the angle adjustment mechanism is very complicated, including torsional moment and radial force generated by the gravity of the rover and bending moment generated by wheel offset. The angle adjusting mechanism also has two output shafts, which are coaxial structure and need to be loaded at the same time. The loading principle of the test is shown in Fig. 13.34.

In addition to the issues related to vacuum loading to which attention shall be paid like the driving and steering mechanisms, the following issues need to be paid attention to for the angle adjustment mechanism:

(1) The angle adjustment mechanism not only increases bending moment, but also needs to increase the radial force load, which can be loaded separately. The structural design needs to avoid load interference.

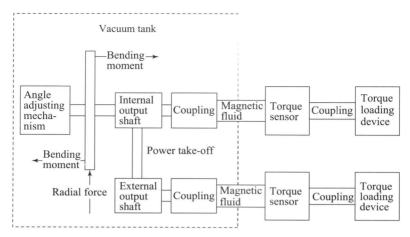

Fig. 13.34 Loading principle of test of angle adjustment mechanism

(2) One of the double-output shafts needs to be led out to the other loading system through the mechanism. The power takeoff can be achieved by different transmission forms such as gears, wire ropes and chains. Attention should be paid to the influence of various factors such as transmission efficiency, high- and low-temperature expansion on the loading torque.

(3) The torque of two sets of torque loading equipment shall meet the requirements of loading torque, respectively, and mutual interference shall be avoided in the movement process; otherwise, the final torque will be affected.

The following suggestions can be referred to in respect to the mechanism failure judgment and test success criterion in the thermal vacuum test process of single machine:

(1) The functional performance of the mechanism shall be tested before and after the test. Compare the current changes under the same load at normal temperature and high and low temperature.

(2) In case of excessively long test cable, attention should be paid to the influence of the test cable voltage drop on the performance of the mechanism. The larger the working current, the more obviously the voltage drops.

(3) Under the same load and temperature, if it is found that the current or rotating speed of the motor changes obviously, the test needs to be stopped to check the product status. If necessary, open the cover to check whether the reducer is seriously worn.

(4) The results of the mechanism life test can be comprehensively judged according to the actual working requirements and the change of product state. Performance indicators can be used as judgment criteria, such as whether the performance indicators such as accuracy, efficiency, current, starting torque are out of tolerance or whether the variation exceeds the specified range. The characteristics

of the product itself can also be judged by the wear of lubrication film, gear or bearing.

2) Single-wheel soil groove test

The wheel soil groove test mainly verifies the wheel driving subsidence, lateral driving subsidence and the influence of subsidence on steering torque. The wheel soil groove test can be conducted with a variety of soils to verify the influence of different soil characteristics on wheel performance. It is also possible to out tests with different wheels to verify the impact of size, shape, material, structural form and other aspects on wheel performance.

The wheel subsidence test mainly verifies the influence of wheel rotation speed, slip ratio and positive pressure on wheel subsidence and driving force. The test is carried out on the soil groove test platform, and the main working principle is shown in Fig. 13.35. Wheel drive provides power for wheel rotation, and the wheel speed can be changed by adjusting the speed. A horizontal moving power source is installed on the horizontal guide rail to realize the actual forward speed of the wheels by adjusting the speed. By adjusting the difference between wheel speed and forward speed, different slip ratios can be satisfied. The positive pressure of the wheel is applied to the wheel through the load on the bracket. After the wheel is started, the descending height till the sinking balance is the subsidence amount, and the driving torque measured by the torque sensor on the wheel drive is the wheel driving torque.

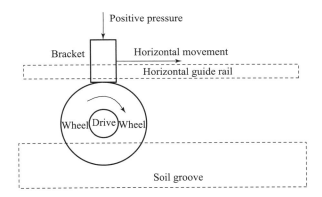

Fig. 13.35 Principle of wheel soil groove test

Through the test and verification, the wheel speed has little effect on the subsidence. The increase of slip ratio will cause the wheel subsidence to increase and cause the driving torque and traction to increase, but it is not a linear relationship. The pressure under the wheel load has a great effect on the subsidence, resulting in a great increase of driving torque and traction. When the lower pressure increases to a certain extent, the traction force is negative, which means that in the case the wheels cannot drive the mobility system forward.

The lateral subsidence test mainly assesses the subsidence and resistance increase caused by roll when the mobility system is running transversely on the ramp. The working principle is to change the wheel axis and the forward direction shown in

Fig. 13.35 from a vertical state to a non-vertical state. The test results suggest that the greater the wheel deflection angle, the greater the wheel subsidence and driving resistance, and the resulting lateral force will also increase.

The purpose of steering torque test is mainly to examine the different subsidence caused by different downward pressures and different steering speeds, resulting in different matching relationships of steering torques. According to the test results, the downward pressure has a greater influence on the steering torque, while the rotational speed has a smaller influence on the steering torque, but has a certain influence on the torque fluctuation.

3) Single-wheel mechanical property test

The purpose of single-wheel mechanical property test is mainly to verify the mechanical strength, bearing capacity, service performance after damage and other aspects of the wheel itself and is a comprehensive verification of the main structure, tread, pawls, spokes and other aspects of the wheel.

According to different verification purposes, wheel performance tests can be divided into three types: overall performance verification, pawl performance verification and damage performance verification.

The overall performance verification is mainly to verify the overall structural strength and stiffness of the wheel, such as radial and axial tensile and compressive stiffness, torsional stiffness, bending stiffness, wheel bearing capacity and spoke-free bearing capacity. Considering the unilateral stress of the wheel when driving on the stone-block pavement, a unilateral or local area bearing capacity test need to be carried out. When the wheel passes over the barrier, it is still be impacted by drop, so it is necessary to carry out single-wheel impact resistance test.

The performance verification is mainly to test the strength of pawls and to test the strength of pawls with different contact areas, at different contact positions and in different load conditions, so as to verify the pawls' ability to grasp the ground and withstand bumping on rock pavement.

The damage performance test is mainly to test and verify the wheel's resistance to sharp rock penetration and crack propagation characteristics after damage. On the surface of many celestial bodies, wheels will encounter hard and sharp rocks, which are easy to cause deformation or even puncture of the tread of metal wheels, as shown in Fig. 13.36. If the travel continues after the wheel is pierced, the damaged parts should have non-propagation characteristics.

Considering the quantification and standardization of experimental verification, it is difficult to take quantitative evaluation of the rock properties. Usually, a standard pressure head is used to simulate the puncture characteristics of rock to wheels by static loading. The hardness size and shape of the pressure head can be quantified. For example, alloy steel can be used. For the head, ball heads with different diameters are used to statically press the weak parts of the wheel, as shown in Fig. 13.37.

Prefabricated defects are adopted for the propagation characteristics of damaged wheels, and the shapes of the defects can be round, triangular, crack and other different shapes. Repeatedly apply reciprocating load to the damaged parts of the

13.4 Rover Mobility System Mobility System

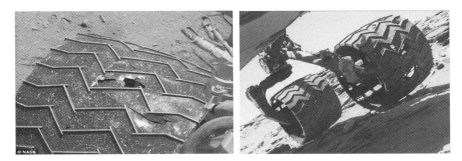

Fig. 13.36 Piercing effect of wheel on orbit

Fig. 13.37 Wheel piercing test

wheel to test the propagation of defects. The applied loads include different directions and forms of loads such as tension, compression, bending and torsion.

4) Terrain life test of mobility system

The terrain life test of the mobility system is mainly to test and verify the life of the wheel in the environment of complete machine. The focus of the test is to simulate the terrain accurately according to the characteristics of different celestial bodies and the terrain characteristics of the landing area, which can be properly toughened, such as increasing the load of the mobility system, extending the driving range, increasing the proportion of driving over harsh terrain. Terrain life test with rock as main feature is shown in Fig. 13.38.

Fig. 13.38 Rock pavement test

The characteristics of soil on celestial bodies have a great influence on the driving ability of the mobility system, so it is usually necessary to carry out targeted experiments on the influence of soil on the performance of the mobility system. Soil-related tests generally focus on the wheel sinking and relieving ability test of mollisol, test of climbing ability on mollisol and the evaluation of performance indexes such as moving speed and slip ratio of soft ground. For the mobility system with the active suspension function, it is also necessary to carry out special verification on the ability of active suspension to extricate itself from difficulties and wheel failure.

5) Drop impact test of mobility system

The drop impact test of mobility system is mainly to verify the impact resistance of wheels under various environments such as crossing over rocks, transferring from the landing rover, releasing from the rover for landing. Different methods such as single wheel release impact test and whole vehicle wheel release test can be considered. The drop impact test is carried out according to the actual possible impact conditions of the mobility system. Figure 13.39 shows single-wheel impact test and whole vehicle impact test, respectively.

13.5 Sampling Mechanism

The sampling mechanism is a mechanism of collecting samples on the surface of celestial bodies, including the soil and rock on the surface of celestial bodies, as well as the soil and rock below the surface. Other functions such as sample transfer and

13.5 Sampling Mechanism

Fig. 13.39 Impact test of mobility system

processing are integrated in the some of the sampling mechanisms. According to different requirements, the sampling mechanism can sample on the surface of large celestial bodies such as Mars and the Moon or on the surface of asteroids.

According to the different functions, sampling mechanisms are generally divided into two types: surface sampling and drilling sampling. In the design of the mechanism, it is necessary to conduct a comprehensive, systematic and in-depth analysis of its functional requirements and to carry out targeted design according to the characteristics of celestial bodies and sampling objects, so as to refine and decompose the key technical indicators and functional performance requirements.

13.5.1 Functions and Composition Characteristics of Sampling Mechanism

Generally, the sampling mechanism has the following functions:

(1) Collect celestial samples. The core function of the sample collection mechanism is to collect samples and carry out targeted design according to the characteristics of the samples that need to be collected and collection methods. To collect soil, rocks and other samples on the surface, sampling mechanisms such as shoveling, clamping can be adopted; if the core of the rock and the soil to be collected are located at a certain depth underground, the sampling mechanism for drilling samples shall be adopted, and the design shall be carried out according to the different hardness and sampling amount of the rock.

(2) It has the functions of compressing and unlocking. Before reaching the celestial body surface, the sampling mechanism needs to undergo the processes of carrying and launching, on-orbit flight, orbit transfer, descent and landing. The sampling mechanism is in a compressed state before landing and roving and can be unlocked and released after landing to carry out sampling work.

(3) The sampling mechanism shall be able to adapt to the climatic characteristics of the celestial body surface, such as temperature, atmosphere, dust and other environmental characteristics. The temperature and atmospheric characteristics of different celestial bodies vary greatly. In particular, during the drilling and sampling process, on the one hand, a large amount of dust will be generated, which will affect the work of the mechanism; on the other hand, a large amount of heat will be generated during the drilling process. In a vacuum or thin atmosphere environment, the heat dissipation conditions are very poor. Therefore, it is necessary to design and analyze the heat transfer path of the drilling mechanism reasonably and optimize the working time and mode.

(4) Other functional requirements. The sampling mechanism is generally installed on the lander or rover. If it is installed on the lander, the available sampling range and area are limited. The sampling mechanism should generally have wider adaptability to ensure that the landing area can obtain samples. If it is installed on the rover, the rover can reach a specific place to sample specific samples, and the design of the sampling mechanism can be more targeted.

The surface sampling mechanism is generally installed on a mechanical arm and can conveniently sample in a selected area on the celestial body surface. The sampling mechanism of shovel excavation is generally adopted, as shown in Fig. 13.40 [10].

The following two surface sampling mechanisms are available: open shoveling and closed shoveling. The open sampling mechanism is relatively simple and usually consists of a sampling shovel and a rotating motor. According to the characteristics of the sample, some auxiliary sampling structures can also be designed, such as sampling shovel blades, and auxiliary tools such as dynamic destruction and polishing of the

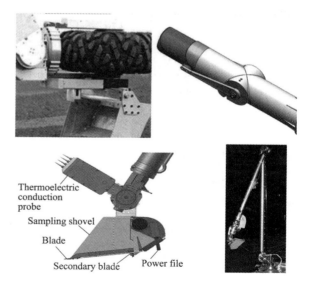

Fig. 13.40 Shovel-digging sampling mechanism

13.5 Sampling Mechanism

sample. On the basis of the open shovel mechanism. A seal-capping mechanism and an excess sample remobility system are added to the closed sampling mechanism.

Open and closed sampling mechanisms have different characteristics and are suitable for different application scenarios:

(1) The open sampling mechanism usually has a large bucket and takes more samples at one time. When larger rock samples are taken, there are less spatial constraints.
(2) The open sampling mechanism has simple structural design and high reliability and can only dump samples vertically;
(3) The enclosed sampling mechanism can seal the sample, the sample will not be scattered in the sample transfer process, the sample can be delivered voluntarily, and the range of applicable sample containers is wide;
(4) The closed sampling mechanism is generally suitable for sampling soil and fine rocks. Excessively large rocks will affect the seal-capping.

The drilling and sampling mechanism is a mechanism that drills down the surface of soil or rock to take samples. The drilling and sampling mechanism can directly sample on the surface of the celestial body, and can drill down to a certain depth to take samples below the surface. The drilling and sampling mechanism can also take, transfer and drop samples on the surface of asteroids with very low gravity.

In order to meet the requirements of drilling and sampling depth and the complicated composition of celestial bodies, drilling and sampling mechanisms have also been developed in different directions. For example, MARTE mechanism uses multisection drill pipes to assemble and extend the drill pipes on orbit, reducing the emission envelope size; the mechanism has the function of automatically adjusting feed rate and rotation speed according to rock characteristics. The drill pipe can control the drilling track, realize curve drilling, avoid hard rock and obtain different subsurface rock or soil samples, as shown in Fig. 13.41.

With the development of technology and more extensive requirements for deep space exploration missions, various celestial body exploration applications represented by robots are also one of the development directions in the future. Smart hands are usually configured as sampling mechanisms, which can complete various anthropomorphic actions at the same time, making the application range wider and flexibility better.

After sampling, two methods are generally adopted: one is to pack and transport the sample back to the Earth; the other is to place the sample in a fixed position and analyze the sample characteristics on the orbit. At present, on-orbit analysis is mostly conducted directly on samples, with relatively low technical difficulty and economic cost.

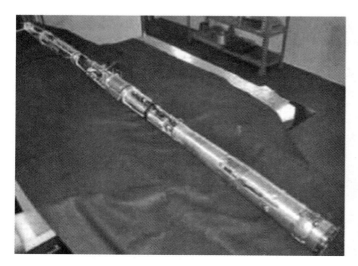 

Fig. 13.41 Flexible drilling and sampling mechanism

13.5.2 Design and Verification of Sampling Mechanism

13.5.2.1 Design of Sampling Mechanism

There are many kinds sampling mechanisms, which are specifically designed according to different aspects of sampling purposes, functional performance requirements, return or on-orbit testing, sampling methods, etc. This section takes the drilling and sampling mechanism with on-orbit analysis capability as an example to go deep into the design of the sampling mechanism.

1) System design

The main function of drilling and sampling mechanism is to sample soil or rock through drilling. Before the system design, the function and performance requirements of the mechanism are analyzed and confirmed, mainly including the following aspects:

(1) The sampling object is soil or rock. The applicable scope of soil and rock characteristics, rock hardness grade, etc., shall be determined.
(2) Whether there is sample transfer function, transfer position, whether the sample placing process depends only on gravity or requires pushing by external force, and whether the number of samples placed needs to be measured;
(3) Sampling depth, single sampling amount, service life, sampling time, sampling force and other performance indicators of the sampling mechanism;
(4) Basic design constraints such as volume, mass and power consumption of sampling mechanism;

13.5 Sampling Mechanism

(5) Environmental design constraints such as mechanics, heat, vacuum and radiation of sampling mechanism.

This section introduces a mechanism suitable for asteroid rock sampling, which analyzes the volatile gas information of rocks on the orbit to carry out on-orbit exploration research. The main composition is shown in Fig. 13.42.

The working process of the sampling mechanism is shown in Fig. 13.43, including sampling, sample placing, measuring, sealing test analysis, etc.

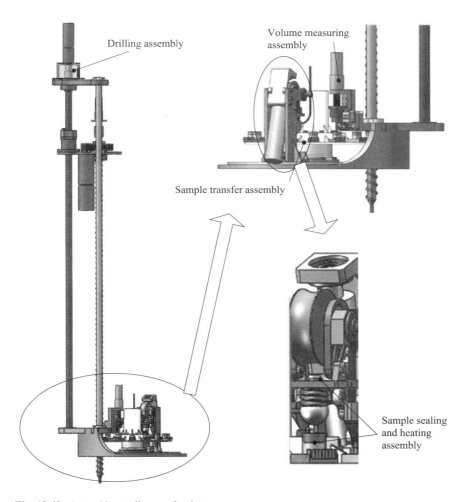

Fig. 13.42 Asteroid sampling mechanism

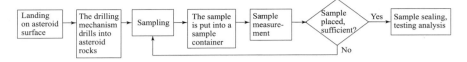

Fig. 13.43 Working flow of asteroid sampling mechanism

2) Design of drilling and sampling mechanism

Drilling and sampling mechanism is the core and the key mechanism to realize sampling. It is mainly composed of feed assembly, rotating assembly, sampling assembly. Its working principle is shown in Fig. 13.44.

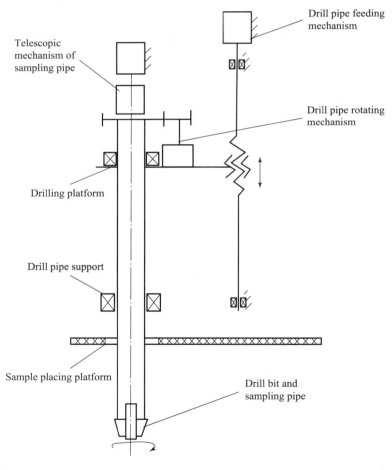

Fig. 13.44 Working principle of drilling and sampling mechanism

13.5 Sampling Mechanism

The main function of the feed assembly is to push the drill bit to extend downward to achieve a certain drilling depth. The feed assembly needs to adopt a linear motion mechanism. Usually, lead screw, gear and rack, wire rope and other mechanisms can be selected, or a linear motor can be directly adopted as the drive. In the process of drilling deeper and harder rocks, a larger thrust force is usually generated in the linear direction. In the process of rock drilling with non-uniform materials or complicated characteristics, vibration and impact in the feeding direction will also occur, which requires the feed assembly to bear large axial force. If necessary, vibration reduction or active vibration mechanism shall be designed to realize the vibratory drilling of rock. In the process of evaluating the drilling feed speed and feed force, it is necessary to configure a force sensor for accurate measurement to prevent damage to the bit from being damaged due to excessive feed force.

As the main mechanism of drilling, the rotating assembly drives the drill bit to rotate. The rotating assembly needs to provide high-speed rotation to meet the severe load changes caused by the difference of rock characteristics and the complex stress in the drilling process. The transmission mechanism of the rotating assembly should adapt to the vibration and impact load during drilling and carry out special tests and verifications on fatigue life.

A sampling assembly is an assembly that takes the sample from a borehole. According to different sampling requirements, there are different sample types such as rock cylinder sampling and rock powder sampling. The sampling assembly is usually installed in the drill pipe of the central controller. The sampling mechanism is pushed out of the drill pipe by the lead screw nut mechanism, and sample is taken by means of rotation, extrusion, etc., and then retracted to the inside of drill pipe. When the sample is placed, the sample pushing mechanism in the sampling tube works to push out the sample and place it into the sample container.

3) Design of sample transfer and sealing components

The drilling and sampling mechanism installed at the end of the mechanical arm can transfer the sample to the designated position through the movement of the mechanical arm. Because of the large depth of drilling, the large volume of drilling sampling mechanism and the large stress during drilling, it is difficult to realize sample transfer through mechanical arm. Generally, a movable and sealed component of the sample container is adopted to move to below the drill bit, and the sample is pushed into the sample transfer and sealing component through the sampling component. Then the sample is transported to the designated location of the analysis instrument to carry out sample detection and analysis.

The design principle of a relatively simple sample transfer assembly is shown in Fig. 13.45. When drilling, the turntable turns to a fixed position, so that the drill pipe can pass through the reserved through hole of the turntable to carry out drilling sampling. After sampling, the drill pipe is lifted, the turntable rotates, the sample container on one turntable is moved to below the drill pipe, and the sampling mechanism in the drill pipe completes the sample placing. The turntable rotates again to transfer the sample container to below the sealing mechanism, which seals the sample container under the action of the slider-crank mechanism.

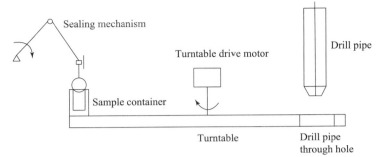

Fig. 13.45 Working principle of sample transfer and sealing assembly

4) Design of sample measuring assembly

Sample measuring assembly is a component for evaluating the quantity of sample after placing. This is to ensure that the sample analysis test is carried out under the design conditions. The sample measuring assembly mainly measures the volume of the sample. It consists of motor, lead screw, lead screw nut, detector, microswitch, etc., as shown in Fig. 13.46. The motor drives the lead screw to rotate, and the lead screw nut moves up and down along the lead screw nut support seat. In the initial state, the lead screw nut moves upward to the current limit and is set to the zero position. Then, the lead screw nut moves downward to the current limit, recording the number of turns of the motor and interpreting the sample volume. If the sample container is empty or the capacity does not meet the requirements, the lead screw nut will touch the microswitch in the descending process, the motor will stop, and the sample will be placed again.

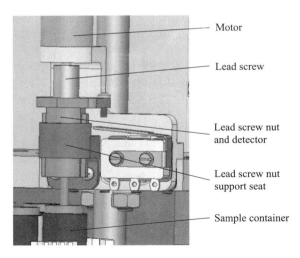

Fig. 13.46 Composition of sample volume measurement assembly

13.5 Sampling Mechanism

5) Drill pipe and drill bit design

Drill pipe and drill bit are the core components of drilling and sampling mechanism. They are designed according to the hardness, drilling size, rotating speed, feeding speed, drilling depth and other characteristics of the drilling samples. As long as the installation layout requirements are satisfied, solution of single drill pipe is preferred. If the drilling depth is very large, it is necessary to use an assembled drill pipe, as shown in Fig. 13.47. The total drilling length of the sampling mechanism is 10 m, consisting of 10 drill pipes in total, and each drill pipe has a drilling length of 1 m.

Fig. 13.47 Space drilling and sampling mechanism

According to the requirements of rock drilling with a rock hardness of grade 6, a drilling sampling mechanism is designed, which combines insert welding PDC composite sheet and low-temperature electroforming impregnated diamond. It is suitable for fast drilling with low hardness and grinding–drilling with high hardness. One-way layer and two-way layer can be considered in the design of bit tooth arrangement, and the number of cutter wings can be designed according to bit size and requirements. The design scheme of the bit is shown in Fig. 13.48.

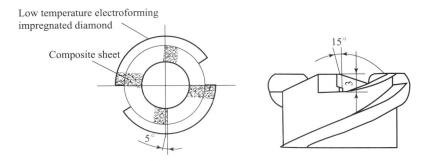

Fig. 13.48 Bit design

In the design of drill pipe and drill bit, it is recommended to consider the following design factors:

(1) The rotation speed and dynamic balance at high speed shall be considered in the design of drill pipe. The length of drill pipe has a great influence on the stability of dynamic balance.
(2) For drill pipe junk slots, the following factors shall be considered, such as drilling speed and characteristics of cuttings to ensure smooth discharge of cuttings.
(3) The diameter of the drill pipe shall be matched with the drill bit to avoid friction with the rock side wall during the rotation of the drill pipe.
(4) Rock hardness and vibration and impact load during drilling shall be considered for design of strength and hardness of drill pipe and drill bit.
(5) Single continuous drilling time, drilling heating and heat dissipation path shall be considered for the design of drill bit and drill pipe.
(6) If the drilling depth is too large, it may be considered to extend the multistage drill pipe connected on track to ensure the connection accuracy and strength.

13.5.2.2 Ground Verification of Sampling Mechanism

In order to ensure that the design of the sampling mechanism is correct, the functional performance indexes and environmental adaptability meet the design requirements, and the technique and control of the flight model production process are in place; it is necessary to carry out corresponding tests on the sampling mechanism. At different stages, enough tests and verifications shall be carried out in a targeted manner. For example, in the schematic research stage, the key technological research and verification of technical principles shall be emphasized; in the initial sample stage, research and development tests, identification tests, reliability tests and other verifications shall be emphasized; in the flight model stage, acceptance tests shall be emphasized to verify the quality control results in the production and manufacturing process, eliminate early failures, and ensure the stable operation of products.

This section focuses on the test items according to the characteristics of the mobility system.

1) Sampling test

Sampling test is to evaluate the function and performance index of the sampling mechanism. The evaluation of the characteristic change and performance of the mechanism during sampling is the key verification test that determines the success or failure of sampling mechanism design. Taking the drilling and sampling mechanism as an example, the core functions and performance tests of drilling and sampling are mainly carried out.

In terms of drilling sampling performance, different drilling parameters are designed according to different rock hardness, such as drilling rotation speed, feeding speed, so as to test the working characteristics of drilling sampling mechanism, such as discharge of cuttings, drilling resistance moment, drill pipe vibration, temperature rise of drill bit and drill pipe, drilling reaction force, drilling depth, pulling resistance,

13.5 Sampling Mechanism

residue of cuttings pulling. The basic principle of the drilling process test is shown in Fig. 13.49. In order to obtain more detailed test information, it can be considered to improve the drill pipe properly, or omit the sampling mechanism to facilitate the placement of sensors.

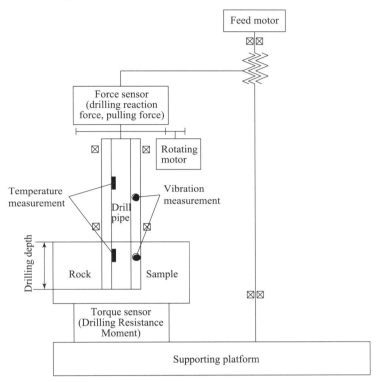

Fig. 13.49 Drilling test principle

The performance test of the sampling function includes sampling and placing. Different sampling mechanisms have different sampling methods and processes. For direct drilling to obtain rock pillar samples, the drilling and sampling can be the same process. The rock core samples obtained are shown in Fig. 13.50.

Fig. 13.50 Rock core samples. **a** Frozen soil/pebble/rock layer/basalt mixture core; **b** basalt core

The sampling mechanism designed by the integration of sampling pipe and drill pipe uses the expansion and contraction of sampling pipe and the rotation of drill pipe to realize the sampling and placing process, as shown in Fig. 13.51. In the sampling process, firstly, the drill pipe is lifted up to a certain height to make room for the extension of the sampling pipe; then, driven by the drill pipe, the sampling pipe extends out and rotates to bring the sample into the sampling pipe. Finally, the drill pipe is lifted up to drive the sampling pipe to pull out together to complete one sampling action. After sampling, the sampling pipe is not sealed and is stabilized in the sampler by the cohesion of the sample itself. This is suitable for sampling in microgravity environment such as asteroids. If the gravity of celestial body is large and the cohesion of rocks is small, it shall be considered to seal the sampling pipe after sampling.

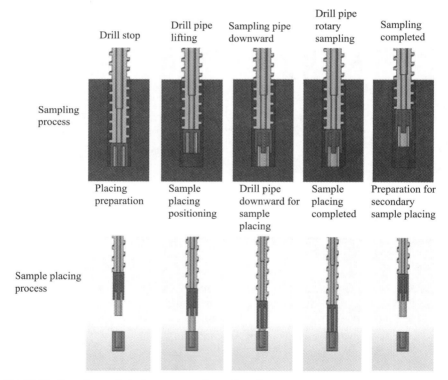

Fig. 13.51 Sampling and placing process of sampling pipe

The placing process is exactly the opposite. First, place the sample container under the sampling pipe, move the drill pipe down, align the sampling pipe with the sampler, and in the process of further moving the drill pipe down, the sampling pipe is lifted up, and the sample is pushed out by the pushing force of the mandrel and

13.5 Sampling Mechanism

placed into the sample container. During the sample placement, the impact of the sample falling near the container shall be considered to ensure the center.

2) Sampling test of space environment simulation samples

The influence of the space environment on the sampling process is generally considered from the following aspects:

(1) Gravity environment: different gravity environment and particle characteristics of different rocks after drilling have influence on cuttings removal during drilling and sample retention by sampling pipe, thus affecting the authenticity of sampling.
(2) Atmospheric environment: the atmospheric environment is mainly affected by atmospheric pressure and air humidity. During the drilling process, the high-speed rotation and temperature rise of the drill pipe have an impact on the change of atmospheric pressure in the cavity and have an impact on the characteristics of cuttings removal, friction, etc. Changes in humidity may affect the cohesion of the sample and the binding force between the sample and the sampling pipe.
(3) Vacuum and atmospheric environment have great influence on bit temperature rise and heat dissipation and also have great influence on drilling feed speed, bit rotation speed, drilling time, temperature balance time, etc.
(4) With high- and low-temperature changes, the structure of the sampling mechanism will be greatly deformed, which will affect the accuracy of drilling direction, motion stability, dynamic balance and other aspects, requiring overall consideration.

Given the influence of vacuum environment and temperature, the test is generally carried out in a vacuum tank. The test process and method are the same as that in normal temperature and pressure. The technical difficulties to be solved in the test mainly include the following aspects:

(1) The vacuum tank shall be able to withstand dust generated during rock drilling. The high- and low-temperature range of the vacuum tank shall meet the index requirements.
(2) The sensors and testing equipment required for measurement shall be able to meet the requirements of vacuum, high- and low-temperature environment. When necessary, heating measures shall be taken for adaptation to low temperature.
(3) The deformation of the test tool itself at high and low temperatures shall not affect the test results of the sampling mechanism.
(4) In order to ensure that a vacuum high- and low-temperature test covers as many working conditions as possible, it is recommended to design a rock movement and replacement device. One test can realize the drilling test at different positions of the same rock, allowing the drilling test of different rocks, so as to save the test cost.
(5) During the vibration measurement during drilling, the microvibration effect of the vacuum tank itself during operation shall be considered, and vibration isolation measures shall be taken when necessary.

There are two environmental impact of gravity: gravity greater than the Earth's gravity and gravity less than the Earth's gravity. For gravity simulation of gravity environment larger than the Earth's gravity, acceleration centrifuge can be used. Centrifuges are usually horizontal centrifugal forces that form a composite gravity direction with a certain inclination angle related to the Earth's gravity. The tooling is designed so that the drilling direction coincides with the gravity direction, which can simulate the test in an environment larger than the Earth's gravity.

For the test methods smaller than the Earth's gravity environment, the method of inclining at a certain angle can be considered to simulate the low gravity environment, as shown in Fig. 13.52. In this simulated environment, attention should be paid to the influence of the Earth gravity on the motion characteristics of the drill pipe and the transmission mechanism, especially the cantilever beam structure formed by the drill pipe extending out, which is apt to produce tail swing phenomenon. The weaker the rigidity of the drill pipe, the greater the influence.

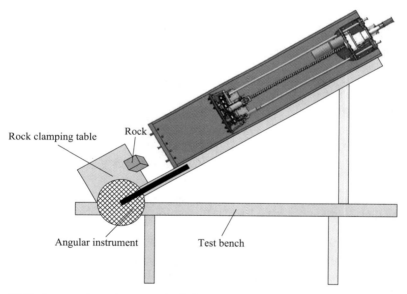

Fig. 13.52 Low gravity environment simulation

13.6 Outlook

The environmental uncertainty and diversity of targets in deep space exploration determine the need for different new probe mechanism technologies, especially for roving and sampling on the surface of celestial bodies. The implementation of complex missions such as Mars and small celestial bodies has led to the development of the mechanism technologies. On the one hand, it will promote the application of new materials and processes, and on the other hand, higher requirements are put forward for the integrated design of the mechanism. Future deep space exploration missions not only require environmental adaptability and miniaturization of the mechanism itself, but also, more importantly, functional integration for mission requirements. For example, future asteroid exploration missions need to integrate functions such as buffering, fixing, sampling and sample transferring, and use a set of mechanisms to achieve the above tasks.

References and Related Reading

1. Yang J, Zeng F, Zhang H (2005) Overview of lunar probe soft landing mechanisms. In: The 1st academic conference of China Aerospace Society's Deep Space Exploration Technology Professional Committee, Harbin (Chinese vision)
2. Chen L (2005) Spacecraft structures and mechanisms. China Science and Technology Press, Beijing (Chinese vision)
3. Yuan J (2004) Satellite structure design and analysis. China Aerospace Publishing House, Beijing (Chinese vision)
4. Yang J (2015) Landing buffer mechanism of spacecraft. China Aerospace Press, Beijing (Chinese vision)
5. Zhang D (2008) Elevator-like landing ladder design and analysis. Harbin Institute of Technology, Harbin (Chinese vision)
6. Tao Z, Chen B (2016) The inching locomotion of a martian rover on loose soil. Spacecraft Environ Eng 33(3):262–268 (Chinese vision)
7. Zou D, Yang L, Qu G (2008) Research on the attitude control of active suspension rover. Aerosp Control Appl 34(3):12–16 (Chinese vision)
8. Tao Z, Chen B, Jia Y (2016) Optimization of geometric parameters for martian rover active suspension. Spacecraft Eng 25(6):48–54 (Chinese vision)
9. LtBo (2011) Design and simulation analysis for folded deployed suspension of single drive six-wheeled rocker-bogie explore rover. Harbin Institute of Technology, Harbin
10. Chen Z, Qing Q, Wen G, Hou P (2012) Mechanical property analysis of a swing-arm transport mechanism. China Mech Eng 23(21):2562–2567 (Chinese vision)
11. Bonitz R, Shiraishi L, Robinson M et al (2009) The phoenix mars lander robotic arm. In: 2009 IEEE aerospace conference. IEEE, pp 1–12
12. Jiang WY, Liu AM, Howard D (2004) Optimization of legged robot locomotion by control of foot-force distribution. Trans Inst Measur Control 26(4):311–323

Chapter 14
Ground Test Verification Technology

14.1 Introduction

There are many similarities between deep space probe and Earth orbit spacecraft in terms of environmental adaptability and environmental testing, but there are also differences between them. Especially, to adapt to the special space environment, entry environment, landing environment and the surface environment of celestial bodies in the orbiting and landing exploration missions, a series of specific technical challenges to the simulation test of the ground environment need to be solved. At the same time, it is necessary to build and improve the relevant environmental testing facilities.

The environments encountered by deep space probes can be classified into natural environment and induced environment. The former is an objective existence, while the latter is generated by the task environment and transportation process of the probe. The influence of the environment on the probe involves various professional and technical fields such as machinery, electricity, thermal, magnetism, optics, radiation. The environmental effect on the products can be embodied in various assembly levels such as material level, element level, subassembly level, subsystem level, system level. The verification of environmental tests runs through the whole development process of the probe model. Different environmental tests need to be carried out in different development stages to achieve different test purposes.

For the environments experienced by the probe during the mission, it is necessary to verify each environment on the ground as much as possible considering the cost efficiency, scheduling and feasibility to verify the correctness and rationality of the product design scheme and process scheme, expose the defects in manufacture, materials and process, eliminate early failure and improve the service reliability during the working life.

This chapter only systematically introduces the test items, test purposes, test methods, test system design and application examples of characteristic system-level ground verification tests that simulate the entry, descent and landing (EDL), sampling and takeoff processes and need to be carried out during the development process of deep space probes. This chapter does not introduce the conventional system-level ground verification tests (such as acoustic test, sinusoidal vibration test, spacecraft rocket docking and separation tests, thermal test, electrical integrated test, EMC test), for which the readers may refer to the relevant monographs on spacecraft environmental tests.

14.2 Technological Development Status

14.2.1 Aerodynamic Deceleration Test Technology

14.2.1.1 Aerodynamic Shape Deceleration and Ablation Test

The aerodynamic shape and ablation performance of the entry capsules must be verified through wind tunnel tests.

During the US Mars exploration missions, systematic and comprehensive wind tunnel tests were conducted on the ground for "Viking." These tests were mainly for air medium, including the tests for CO_2 or CF_4 medium under some hypersonic conditions. Since then, limited by various factors such as development funds and progress, no ground wind tunnel tests have been carried out for later probes from "Mars Pathfinder (MPF)" to "Phoenix," until the Mars Science Laboratory (MSL) had significant changes in design status.

During Venus exploration and Saturn exploration missions, a series of ground aerodynamic tests were carried out on the static aerodynamic force, dynamic stability and thermal environment of the entry vehicles.

14.2.1.2 Parachute Test

The test and verification items of parachute system generally include function/performance/interface test, design verification test, environmental adaptability test, offset test, process verification test, etc.

Taking the Mars parachute system as an example, several typical test verification methods are introduced below.

1) Ground test of parachute deployment system

The ground test of parachute deployment system is a mortar function performance test conducted on the ground. It is used to verify the ejection performance of the mortar, to check whether the ejection speed setting is reasonable and whether the parachute system straightening process is normal and to obtain key performance data such as ejection speed and ejection recoil force.

2) Parachute strength test

Parachute strength is an important property of parachute, which must be verified through the tests such as wind tunnel test, airdrop test and rocket sled test.

The parachute-opening conditions of Mars parachutes are supersonic speed, low density and low dynamic pressure. Due to the existence of low density and low dynamic pressure, the parachute-opening process is a typical infinite mass deployment process; i.e., the maximum parachute-opening load occurs almost at the moment when the bottom edge of the canopy is fully stretched. See Table 14.1 for the differences between infinite mass deployment and finite mass deployment. Since the maximum stress of the canopy appears at different positions of the canopy during the infinite or finite mass deployment, the conditions for opening the parachute with infinite mass must be created to verify the strength of the parachute. This is the most significant characteristic of the Mars parachute strength test.

Wind tunnel test is the best method to test the strength of a parachute with an infinite mass. In the process of developing parachutes for "Mars Exploration Rover" (MER) and "Mars Science Laboratory" (MSL), the US developers used wind tunnel tests to assess the parachute strength. Compared with airdrop test, wind tunnel test has the advantages of short test period, low cost, accurate control of test conditions and good repeatability of test conditions. The US Mars parachute strength test is conducted in a NFAC 24 m × 36 m low-speed wind tunnel with the maximum wind speed of 50 m/s, the maximum section size of 24 m × 36 m and the length of 46 m, which can be used for a full-scale parachute strength test.

In the absence of super-large wind tunnel, the parachute strength test can be conducted by air drop test. During the development of the parachutes for "Viking," "MER" and "PHX," the US developers conducted the airdrop tests specifically aimed

Table 14.1 Infinite mass deployment and finite mass deployment

Item	Infinite mass deployment	Finite mass deployment
Parachute-opening load	Heavy load	Small load
	The maximum load usually occurs when the bottom edge of the canopy is fully stretched	The maximum load generally occurs in the middle of the canopy inflation process
	The parachute-opening load is borne by the whole canopy, and the maximum stress occurs at the top hole and near the bottom edge of the canopy	The parachute-opening load is mainly borne by the local canopy, and the maximum stress occurs at the top hole and the upper canopy
Front body velocity (dynamic pressure)	In the parachute-opening process, the resistance of the parachute has little influence on the movement of the front body, and the variation of the velocity of the front body is small	In the parachute-opening process, the resistance of the parachute has significant influence on the movement of the front body, and the variation of the velocity of the front body is significant

at parachute strength assessment. In order to simulate the conditions of infinite mass deployment, two methods of airdrop test can be adopted in general: One is to use the thin air at high altitude to implement the high-altitude airdrop, as done in the "Viking" parachute test; the other is to increase the mass of the recovered object to implement the heavy-load airdrop test. The main disadvantage of airdrop-type strength test is that the control precision and repeatability of test conditions are poor [1].

3) High-altitude parachute-opening test

The Martian atmosphere is quite different from the Earth's atmosphere, so the Martian parachute has the characteristics of low density, low dynamic pressure and high Mach number during parachute opening, inflation and steady descent and deceleration. The simulation of the parachute-opening conditions for a Mars probe on the Earth can only be realized at an altitude above 33 km by using the test method of high-altitude deployment. This test method can verify the function and performance of the parachute deceleration subsystem in the most realistic way under the most realistic deployment conditions and obtain the most real and effective test data. The high-altitude deployment test methods mainly include balloon platform test method and rocket platform test method.

Before and during the development of the "Viking" Mars probe, the NASA organized several high-altitude parachute-opening test items, including the Planetary Entry Parachute Program (PEPP), Supersonic Planetary Entry Decelerator (SPED), Supersonic High Altitude Parachute Experiment (SHAPE), Parachute Deployed Behind a Bluff Body (PDBABB) and Balloon Launched Decelerator Test (BLDT) for Viking.

14.2.2 Test Technology for Dynamic Deceleration

In the 1960s and 1970s, during the implementation of the Apollo lunar exploration program, the USA carried out systematic and comprehensive ground tests and verifications for the dynamic deceleration of the landing process, specifically including the verifications of Lunar Landing Research Facility (LLRF) and Lunar Landing Research Vehicle (LLRV) and others.

14.2.2.1 Lunar Landing Research Facility [2]

The main function of LLRF is to simulate the gravity field on the lunar surface in order to obtain the flight dynamics characteristics of the lander, verify the servo control system of the landing module in the low gravity field and its control characteristics, and provide long-term simulation training for astronauts to land on the lunar surface.

14.2 Technological Development Status

14.2.2.2 Lunar Landing Research Vehicle [3]

The LLRV is mainly used to simulate the flight path in the final phase of lunar soft landing and train astronauts to control the landing module during soft landing and takeoff. The LLRV can simulate the landing process when descending from the altitude of about 457.5 m, including hovering and horizontal movement and can also simulate the process of taking off from the lunar surface. It is used to train astronauts to control the lunar module to hover and descend and find the landing site.

14.2.3 Verification Technology of Soft-Landing Process

There are generally three ways for soft landing on the surface of extraterrestrial bodies: airbag buffer landing, landing-gear buffer landing and crane landing. The Mars Pathfinder (MPF) and Mars Exploration Rover (MER) in the Mars exploration mission used airbag for landing. The US lunar exploration missions ("Surveyor" series and "Apollo" missions), the Soviet Union's "Luna" missions and the US Mars probe missions "Viking" and "Phoenix" all used the landing gear for landing. The US Mars probe mission "Mars Science Laboratory" (MSL) used crane for landing. For all these different landing modes, targeted tests and verifications have been carried out on the ground. Typical tests include the landing stability tests and drop tests to verify the landing gear performance and the airbag landing tests to verify the buffering performance of airbags.

14.2.3.1 Landing Stability Test [4]

In order to verify the landing gear performance and landing stability of the Apollo lunar module, the USA conducted the 1/6 scale model tests under the Earth's gravity field and the full-scale model tests under the simulated lunar gravity field. The buffering stroke, centroid acceleration and attitude motion process obtained by the two test methods have good consistency. The full-scale model can perfectly reproduce the dynamic process under symmetrical landing conditions and can be used to study the stability during landing.

1) 1/6 scale model test

The 1/6 scale model is designed based on the principle of similar dynamic characteristics. During the 1/6 scale model test, the tester is installed on the release device at the bottom of the pendulum test device, and the release device can adjust the attitude of the tester when releasing. When the release device passes through the lowest point of the pendulum in the test, the release device releases the tester. Then, the tester performs parabolic motion in a certain attitude at the set horizontal initial speed and simulates the soft-landing process after contacting with the landing surface.

2) Gravity field simulation test with a full-scale model

In order to simulate the soft-landing process more truly and comprehensively, the lunar gravity simulation test device can be used to complete the landing stability test of the full-scale model. The lunar gravity simulation test device consists of an inclined landing surface, a one-dimensional servo crane and a supporting cable. The normal component of the static force applied on the landing surface by the lander supported by the cable is equal to the lander gravity on the moon.

14.2.3.2 Drop Test [5]

The purpose of drop test is to reproduce the impact dynamics process of soft landing and verify the adaptability of lander structure and various systems to extreme landing conditions.

In the drop test of Apollo program, LTA-3 tester (LM test article 3) was used to carry out 16 drop tests.

During the lunar exploration mission, the Soviet Union carried out a large number of drop tests for the landing process. In the impact test, a pointer-shaped steel frame structure system was adopted to drop the lander at different heights so as to obtain the required vertical landing speed. In order to simulate the horizontal movement of the lander when contacting the simulated lunar soil, the pointer-shaped steel frame structure can rotate relative to the vertical axis at the required speed. The drop testbed can simulate the landing process with the vertical landing speed of 5 m/s and the horizontal landing speed of 3 m/s.

14.2.3.3 Airbag Landing Performance Test

In order to improve the survivability in the uncertain landing environment, both Mars Pathfinder (MPF) and Mars Exploration Rover (MER) missions used airbag as the final deceleration mode to conduct a large number of simulated drop verification tests on the ground, in which the atmospheric pressure of Mars, landing velocity, landing attitude and landing surface characteristics are simulated. There are two main purposes in the tests: One is to find the weak link and failure mode of the airbag and use them as the basis for improving the design; the other is to obtain the load environment in the buffering process and obtain the failure modes ("stroke-out" failures) outside the buffering stroke of the airbag, such as direct stone impact on the lander located in the airbag, etc.

14.2.4 Validation Technology for Takeoff Process

During the sampling and return mission, due to the limited space between the lander and the ascender, the force and thermal effect of the engine plume borne by the ascender and the lander will be more significant, and the plume flow field will be more complex. They will have a direct impact on the takeoff stability of the ascender and the thermal protection of the landing and ascending assembly. Under this condition, new requirements are put forward for the normal operation of the engine. At present, the problem of engine plume is recognized as a global technical problem. Especially for the complex configuration of landing and ascending assembly, it is difficult to directly obtain effective plume field force and thermal data by relying entirely on numerical analysis. Therefore, it is necessary to verify through experiments the force and thermal effect of engine plume on the ascender and the lander as well as the compatibility between the engine and the flow guiding device during takeoff and ascending.

During the initial development of the Apollo lunar module, the force and thermal effect of engine plume and the engine performance during the takeoff and ascending from the lunar surface were specially studied. This effect was defined as "fire in the hole" (FITH); i.e., the engine was ignited at the ascent stage while the ascending and descending stages remained connected. In 1963, Grumman carried out the pioneering test of scale model and deduced the full-scale force and thermal effect through the similarity criterion, which was used as the basis for the preliminary design. Later in 1965, aiming at the development of the engine, a two-stage separation simulation test was carried out at the Arnold Engineering Development Center (AEDC) to verify the engine performance under different relative angles and distances of the ascent stage and flow guiding device. Then, a full-scale system-level two-stage separation test (PA-1 ground test) was carried out at WSMR test site in 1967 to verify the force and thermal effect of the ascent stage engine at the beginning of two-stage separation. Finally, in 1968, the "Apollo 5" flight tester was launched to carry out the test of after-ignition separation between the ascending and descending stages on the lunar module in the Earth orbit. In this test, the force and thermal effect data of the engine plume field under vacuum condition were obtained.

14.3 Demand Analysis

14.3.1 Principles of Test Planning

The following basic principles should be followed when planning the system-level ground verification tests:

(1) The test object can be fully verified.
(2) The boundary conditions of the test should not be lower than the real experience environment, in order to achieve the purpose of stricter examination.
(3) On the prerequisite of satisfying the test purpose, the test method shall be as simple, safe and reliable as possible.
(4) The test conditions shall be as few as possible under the premise of covering the on-orbit working conditions.

14.3.2 Test Requirement Verification

14.3.2.1 Analysis of Test Requirements for Aerodynamic Deceleration Process

1) Aerodynamic and ablation tests

When landing on the surface of a celestial body with atmosphere, aerodynamic shape deceleration is the main means to reduce the entry speed—possibly from about several kilometers per second to several hundred meters per second. In the process of aerodynamic deceleration, the constraints of aerodynamic heat, aerodynamic force, fall points distribution and the like need to be taken into account. In order to obtain the aerodynamic characteristic data and thermal environment data during the atmospheric entry of the entry capsule and verify the ablation characteristics of ablation materials, a large number of aerodynamic and ablative tests need to be carried out to ensure the correctness of the research methods and models. Specifically, the tests include wind tunnel force test, wind tunnel pressure test, wind tunnel heat test, free flight ballistic range test, wind tunnel ablation test, etc.

2) Deceleration parachute test

For the celestial bodies with low atmospheric density, the deceleration capability of their aerodynamic shape still cannot meet the task requirements, and a deceleration parachute is needed for further deceleration. Deceleration parachute can improve the stability of the entry vehicle in the transonic phase and ensure that the entry capsule can pass through the transonic phase with the required attitude. Deceleration parachute can generally reduce the entry speed to a level of tens of meters per second. In order to verify the performance parameters of key parachute processes such as ejection, stretching and inflation, it is necessary to simulate the initial conditions of parachute deployment, such as speed, atmospheric density, dynamic pressure, and the attitude of the entry vehicle, and to carry out the parachute deployment test at high altitude.

3) Dynamic load test during parachute opening

During the deceleration process of the deceleration parachute, a harsh dynamic load environment will be generated in the parachute ejection, deployment and capture.

14.3 Demand Analysis

The parachute ejection load is affected by the parameters such as explosive payload, mortar size, parachute package mass, exit velocity and temperature. The parachute deployment load is affected by the parachute parameters such as dynamic pressure, dynamic load coefficient and resistance area. The capture load is affected by the parameters such as capture package mass, capture speed, capture bridle parameters. The main structure design of traditional spacecraft generally adopts the design verification methods such as static load design, static load assessment and dynamic load recheck. The design criteria and dynamic load verification methods for composite structures under dynamic load environment are still in the research stage. Due to the aerodynamic heating in EDL process, the entry vehicle structure is in a high-temperature environment when subjected to the parachute-opening dynamic load, and the high-temperature environment has certain influence on the strength of the main structure. In order to verify the dynamic strength of the structure under the coupling effect of the dynamic load environment and the temperature environment and obtain the environmental conditions in key areas and equipment installation interfaces, it is necessary to accurately simulate the time history of dynamic load and the dynamic-load boundary conditions of the entry capsule on the ground, which are used to simulate the actual load-carrying process of the entry capsule on orbit to verify the dynamic strength of the structure and obtain the environmental conditions of the components.

4) Separation test of heat shield and backshell

In the process of entering a celestial body with atmosphere, the heat shield and the backshell need to be thrown away after aerodynamic deceleration so that the internal lander can decelerate and land. The separation process of the heat shield and the backshell needs to meet the requirements such as pyrotechnic devices unlocking environment controllability, no interference in short-term separation, no safety collision risk in long-term separation, etc. Because the heat shield and the backshell need to experience the powered-phase load environment, the on-orbit flight storage environment and the force and thermal environment during the entry process before separation and several pyrotechnic devices are detonated and unlocked at the same time during the separation of heat shield and backshell and creates a relatively severe shock environment for the component products on the entry vehicle, aerodynamic coupling effect will exist in the short-term separation process, and sufficient ballistic coefficient difference will be required in the long-term separation process. The separation process is interfered by aerodynamic force and atmospheric environment and affected by many influencing factors. It is subject to the constraints of the trajectory parameters of parachute-opening point, the aerodynamic drag to parachute/backshell/heat shield and landing platform, the attitude control, the mass and inertia, the separation ejection speed, the Mars environment (atmospheric density and wind field) and other elements. Therefore, in order to obtain the shock environment during the separation of heat shield and backshell, to verify the influence of asynchronous pyrotechnic devices unlocking on the separation process, and to verify the performance of the mechanism and the short-term separation, collision and interference after experiencing the mechanical environment of the powered phase, it is necessary to simulate the load conditions and boundary constraints during the separation of heat shield and backshell, so as to demonstrate the safety of the separation process.

14.3.2.2 Analysis of Test Requirements for Powered Deceleration, Soft Landing and Takeoff

1) Comprehensive verification test for landing and takeoff

Before soft landing, the entry vehicle usually decelerates in the powered descent phase by using an orbit-control engine. The powered descent process, involving complex processes such as ranging, velocity measurement, terrain recognition, autonomous navigation control of navigation sensors and the coordination of control and propulsion systems, needs to be tested and verified on the ground. However, under the existing technical conditions, it is almost impossible and unnecessary to verify the whole process of powered descent on the ground (the descending distance is about 15 km during lunar landing and about 1.5 km during the Mars landing). In the powered descent phase, the entry vehicle needs to complete key actions such as deceleration, terrain identification, obstacle avoidance, hovering and shutdown, which are also included in the hovering, obstacle avoidance and slow descent phases. Considering the limitation of outlays, development schedule and ground simulation method, it is necessary to verify the whole powered descent process through the ground tests for hovering, obstacle avoidance and slow descent phases.

2) Test for the influence of thrust pulsation of high-thrust engine on GNC sensors

A large liquid rocket engine will generate thrust pulsation when working, which is determined by its inherent combustion characteristics. The thrust pulsation value can be generally controlled within 3% of the average thrust. The entry capsule is equipped with a dual-component liquid engine for orbit change (for example, the 7500 N variable thrust engine on Chang'e 3), which is generally used for the deceleration during final braking at perilune and powered descent. In the process of powered descent, the mechanical environment caused by the pulsating load of the high-thrust engine will have certain influence on the measurement accuracy of the high-precision navigation sensor and landing sensor in GNC system, and the pulsation response of the structure will also have certain influence on camera imaging. Therefore, it is necessary to carry out tests and verifications on the ground to obtain the pulsation environment of the whole spacecraft and its influence on GNC sensors or other sensitive equipments.

3) Combination test of variable thrust engine and GNC computer

The real-time thrust of the entry capsule during the powered descent phase is controlled by GNC computer, which, according to the results of ranging and velocity measurement, autonomously develops the control strategy to enable real-time engine control. The high control frequency requires quick response of the engine, which is realized jointly by GNC system and the propulsion system. In order to evaluate their

14.3 Demand Analysis

joint performance, it is necessary to carry out special tests and verifications on the ground.

4) Drop test

After going through the powered descent phase, the entry capsule lands on the celestial body surface in a soft-landing mode. In the process of soft landing, the entry capsule will experience the landing impact load environment, and the onboard working equipments (payload and rover) must be exposed to the load environment. In the process of soft landing, the entry capsule is affected by the factors such as the mass characteristics, landing velocity and landing attitude of the vehicle, the surface topography and mechanical characteristics of landing surface and the buffering characteristics of landing gear. The force transmission path of the entry capsule during landing impact is that the landing impact load is transmitted to the main structure connection through the landing gear, then extended to the structural plate, and then is transmitted to each onboard device. The force transmission path and load characteristics are different from the load environment of the launching phase.

In order to assess the impact resistance of the main structure of the entry capsule during soft landing and to obtain the impact response at the key parts of the entry capsule and at the installation places of key equipments, it is necessary to test and assess the landing impact process of the entry capsule on the ground.

5) Landing stability test

The primary task of the soft-landing probe is to achieve soft landing on the celestial body surface. Whether the entry capsule can land on the celestial body surface stably with the required attitude and direction determines the success of the entire probe mission. The landing stability of the entry capsule is affected by the factors such as the mass characteristics, landing velocity and landing attitude of the vehicle, the topography and mechanical characteristics of landing surface, and the buffering characteristics of landing gear. In order to successfully realize soft landing on the celestial body surface, the landing stability must be verified by ground tests during the implementation of the project to assess the impact of the initial landing conditions of the entry capsule on its landing stability.

6) Airbag buffering performance test

Airbag buffering is generally used for the final stage of aerodynamic deceleration and has the advantages of small mass, small volume, strong terrain adaptability, high buffering efficiency, high buffering protection speed, reliable buffering performance and the like. It is suitable for the exploration tasks with small landing mass, complex terrain in the landing area and less strict requirements for initial landing conditions and post-landing attitude. The buffering performance of the airbag is affected by the material characteristics of the airbag itself, internal and external pressure difference, landing velocity and landing terrain. In order to obtain the buffering performance of the airbag (overload and the safety clearance between the airbag and the lander), the boundary conditions (such as pressure) on orbit need to be simulated and the airbag buffering performance test should be carried out.

7) Test for the influence of plume on thermal control

In the powered descent phase, a high-thrust orbit-controlled engine (such as the 7500 N variable thrust engines installed on Chang'e 3 and Mars landing probe) is generally used to realize soft landing and deceleration by reverse thrust. When the entry capsule is close to the landing surface, the plume of the high-thrust engine will interact with the landing surface and will be reflected to the bottom of the vehicle, thus creating severe thermal environment conditions for the thermal control material at the bottom of the vehicle, the equipments installed on the bottom plate and the landing gear. In addition, the force exerted by the plume on the entry capsule will also have certain influence on the landing attitude of the entry capsule. In order to determine the thermal effect of engine plume on the vehicle, it is necessary to test the plume effect on the entry capsule.

8) Test for the influence of plume on takeoff

After completing the probing and sampling tasks on the celestial body surface, it is necessary to use the engine to push the ascender into the celestial body orbit or the celestial-body-Earth transfer orbit. Due to the short distance between the ascender and the lander or the ascender launch platform, high pressure and heat flow will be generated at the top plate of the lander or the ascender launch platform and at the bottom of the ascender until the lander and the ascender are fully separated; i.e., the engine plume can diffuse freely without plume reflection from the lander or from the ascender launch platform. A takeoff plume impact test is required for this kind of thermal effect produced by the ignition of the engine in the limited area between the two stages.

14.4 Test Verification Technology

This section mainly introduces the test objectives, test constraints, test methods and test system design and cases of typical test verification technologies in the landing exploration missions.

14.4.1 Aerodynamic Deceleration Test Technology

This subsection mainly introduces the test items required for the acquisition and verification of key parameters in the process of aerodynamic deceleration, such as wind tunnel (aerodynamic force, aerodynamic heat and ablation) test, high-altitude parachute deployment test, dynamic load test during parachute ejection and deployment and heat shield and backshell separation test.

14.4 Test Verification Technology

14.4.1.1 Wind Tunnel Test

In order to determine a reasonable aerodynamic shape and verify the correctness of the aerodynamic environment, aerodynamic thermal environment simulation results and ablation model, it is generally necessary to carry out wind tunnel tests on the ground, mainly including wind tunnel force measurement tests (sub-trans-supersonic wind tunnel force measurement test, conventional hypersonic wind tunnel force measurement test, gun tunnel force measurement test and heat-shield-throwing separation wind tunnel force measurement test), wind tunnel pressure measurement tests (trans-supersonic wind tunnel pressure measurement test and hypersonic wind tunnel pressure measurement test), wind tunnel heat measurement test (shock waves tunnel heat measurement test) and wind tunnel dynamic derivative measurement tests (free flight ballistic range test and transonic shock wind tunnel test).

1) Test purpose

Through wind tunnel tests, the following objectives can be achieved:

(1) Determine the optimal aerodynamic shape.
(2) Determine static and dynamic aerodynamic data during the entry process.
(3) Determine the aerodynamic thermal environment data during the entry process.
(4) Complete the ablation mechanism research in the atmospheric environment, verify the correctness of the thermal protection material design and improve the ablation analysis model.

2) Design of test system

(1) Wind tunnel test of aerodynamic force and heat

① Select the appropriate wind tunnel according to the entry environment atmosphere, entry trajectory conditions and aerodynamic shape of the entry capsule.
② Design the scaled test piece considering the blockage ratio of the wind tunnel.
③ Select a special force measurement balance to measure aerodynamic force, a pressure measuring system to measure aerodynamic pressure and a thin-film uranium resistance to measure heat flux.

(2) Ablation test

① The engineering algorithm of one-dimensional isentropic flow hypothesis is adopted to estimate the wind tunnel operating parameters (power, pressure, enthalpy, Mach number, etc.) in accordance with aerodynamic heating parameters (heat flow, enthalpy, pressure, etc.).

② Taking the predicted parameters as boundary conditions, the high-temperature flow field and model bypass parameters are calculated by numerical methods to obtain the high-temperature flow field parameters, model surface heat flow distribution and other data under the predicted conditions.

③ The flow field of the nozzle is debugged and tested, and the heat flow distribution of the model is measured. Then, the measurements are compared with the theoretical calculation results. If the error is within the permissible range, a formal test will be conducted; otherwise, the adjustment will be continued until the state meets the design requirements.

④ The measurement parameters concerned in the test mainly include cold wall heat flux, stagnation pressure, total enthalpy of airflow, surface temperature and back surface temperature, mass and thickness, etc.

The stagnation pressure P_s can be measured by opening a pressure measuring hole on the test model and installing a pressure sensor.

The total enthalpy of airflow is calculated by measuring the stagnation point heat flux and stagnation point pressure of the hemispherical head model using the following formula:

$$q_s = K\sqrt{\frac{P_s}{R}}(H_0 - H_w) \tag{14.1}$$

where

H_0 is the total enthalpy of air flow, kJ/kg;
H_w is cold wall enthalpy, kJ/kg;
q_s is stagnation point heat flux, kW/m^2;
P_s is stagnation point pressure, atm;
R is the radius of the ball head, cm;
K is the coefficient, 1.113 for air, 1.210 for CO_2.

3) Test constraints and test methods

In order to obtain the aerodynamic force, thermal environment and ablation process data in the real entry process, the following constraints need to be considered when designing the wind tunnel tests:

(1) Atmospheric environment parameters (altitude-density, atmospheric composition, etc.).
(2) Entry trajectory parameters and entry corridor.
(3) The shape and size of the entry capsule. An appropriate scale model shall be selected considering the wind tunnel size and test capability constraints.
(4) Thermal boundary conditions of ablation materials.

4) Application examples

Figures 14.1, 14.2, 14.3 and 14.4 are, respectively, the photographs of the test processes of sub-trans-supersonic force measurement, hypersonic pressure measurement, gun tunnel heat measurement and arc-heated wind tunnel stagnation ablation for the Mars exploration mission.

14.4 Test Verification Technology

Fig. 14.1 Schlieren screenshot of sub-trans-supersonic force measurement test

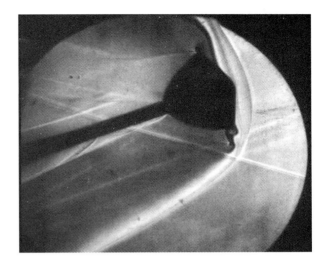

Fig. 14.2 Schlieren screenshot of hypersonic pressure measurement test

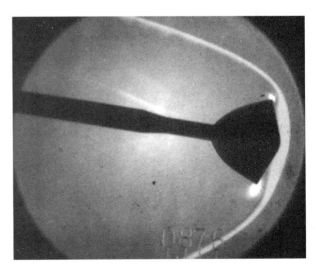

14.4.1.2 High-Altitude Deployment Test of Parachute System

1) Test purposes

(1) To simulate the parachute-opening conditions during the entry process to verify the working performance of the full-scale parachute system.
(2) To simulate the parachute-opening conditions during the entry process to obtain the key performance parameters of the full-scale parachute system during the parachute ejection, capture, straightening and inflation.
(3) To obtain the output load during the working process of the parachute.

Fig. 14.3 Thermal test in gun tunnel

Fig. 14.4 Stagnation point ablation test in arc-heated wind tunnel

2) Test constraints and test methods

In order to simulate the real parachute-opening process, the high-altitude parachute deployment test needs to simulate the real parachute-opening conditions (Mach number, atmospheric density, dynamic pressure, angle of attack, etc.), the mass characteristics during parachute opening, and the wake interference effect of the forebody.

Given the big difference between the atmosphere of the target celestial body and that of the Earth, the density, dynamic pressure and Mach number of the parachute in the atmosphere of the target celestial body are different from those in the Earth atmosphere in the process of parachute deployment, inflation, and steady descent and deceleration. In order to simulate the parachute-opening conditions on the Earth, the method of high-altitude parachute deployment test can only be used at a specific altitude. This test can verify the function and performance of the parachute system in the most realistic way under the most realistic deployment conditions and obtain the

14.4 Test Verification Technology

most real and effective test data. There are generally two test methods to simulate the parachute-opening process at high altitude on the Earth, namely, the balloon platform parachute-opening test method and the rocket parachute-opening test method, as described in Sect. 14.2.1.

Considering the test cost and the system complexity, the rocket parachute-opening test method is generally selected. The rocket is fixed on the launcher and adjusted to the launching angle. After being ignited and launched, the rocket flies along the predetermined trajectory and brings the mortar and parachute to a high altitude. After the rocket head is separated from the propelling part of the rocket body in the ascending phase, the parachute-opening channel is opened and the deployment device of rocket head begins to work. The maximum outer diameter of the rocket head is stretched to the required size. When the parachute-opening condition is satisfied, the mortar is controlled to eject the full-scale parachute. The parachute is inflated and fully expanded and brings the rocket head back to the target range. During the whole flight process of the rocket head, the measuring device on the rocket head measures and records various data in real time and transmits the data to the ground telemetry receiver. At the same time, the target range synchronously measures the external trajectory of the rocket head.

3) Design of test system

The test system is mainly composed of parachute, rocket system, ground equipment (test-launch-control system, launching device, telemetry and data processing system and ground tracking and measuring equipment), onboard measuring system, etc.

The parachute under test is a product in real state, and its onboard installation and working condition are simulated.

Rocket system is the carrier of the parachute under test. It provides mechanical and electrical interfaces for effective installation and release and creates the required conditions for parachute ejection and release during flight. The carrier system needs to have sufficient orbit/attitude control capabilities to ensure that the parachute-opening conditions meet the requirements.

The ground equipment is mainly used for the testing before the rocket is launched. It provides a launching platform, receives and processes the telemetry data during flight and optically measures key flight data.

The onboard measuring system is mainly used to measure the image and load data in the parachute-opening process and record the key parachute-opening steps such as parachute ejection, capture, straightening and inflation.

14.4.1.3 Dynamic Load Test for Parachute Ejection and Opening

The dynamic load test for parachute ejection and opening is used to verify the influence of the dynamic load generated during the parachute ejection and inflation on structural strength and onboard load environment.

1) Test purposes

(1) To verify the rationality of the structural scheme.
(2) To verify the rationality of the structural process plan.
(3) To assess the ability of the structure to bear the dynamic impact load under the working conditions of parachute ejection and opening.
(4) To obtain the responses of key structural parts to the dynamic impact load under the working conditions of parachute ejection and opening, so as to provide a basis for correcting the environmental conditions of the components.
(5) To provide a basis for revising the dynamic load analysis model under the working conditions of parachute ejection and opening.

2) Test constraints and test methods

The dynamic load test for parachute ejection and opening needs to simulate:

(1) Free boundary conditions.
(2) The time history, action point and direction of the load.
(3) Mass properties of the whole vehicle.

The dynamic load test for parachute ejection and opening generally adopts a special hanging concept to simulate free boundary conditions. The parachute ejection load is generally simulated by real pyrotechnic device, and the parachute-opening load is generally simulated by special loading device.

3) Design of test system

In the dynamic load test for parachute ejection, a real mortar is used. The entry capsule is hung by a hanging system to simulate the free boundary condition on orbit. The entry capsule is equipped with a real mortar. The recoil impact force generated when the ignited mortar throws the parachute will act directly on the entry capsule structure, thereby assessing the structure. The test concept is shown in Fig. 14.5.

The free falling body concept based on spring vibrator model can be used for the dynamic load test for parachute opening, as shown in Fig. 14.6. The connection between the parachute bridle and the structure is simulated. During the test, the load time course of the parachute-opening process can be simulated by adjusting the stiffness coefficient of force control spring and the dropping height. The mathematical model is as follows:

$$T = 2\pi/\omega = 2\pi (m/k)^{1/2} \qquad (14.2)$$

$$F = (2mgHk)^{1/2} \qquad (14.3)$$

where

m is the mass of the entry capsule;
k is the stiffness of the force control spring;
H is the dropping height.

14.4 Test Verification Technology

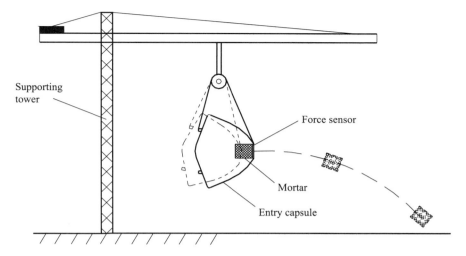

Fig. 14.5 Dynamic impact load test for parachute ejection

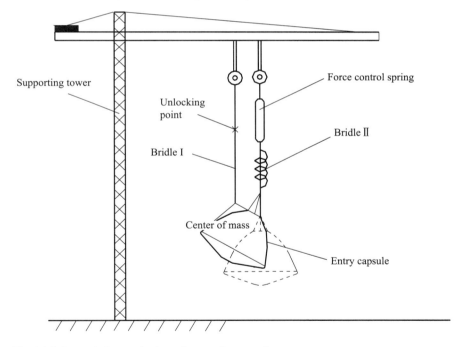

Fig. 14.6 Dynamic impact load test for parachute opening

14.4.1.4 Separation Test of Heat Shield and Backshell

1) Test purposes

(1) To obtain the shock response of the separation surface, structure and equipment installation locations when the pyrotechnic devices is detonated.
(2) To obtain the key parameters of the separation process of the backshell and the heat shield, including relative distance, speed, angular velocity and attitude.
(3) To verify the correctness and safety of the separation design of the heat shield and the backshell, including the harmony of the connection interfaces and no collision in the short-term separation phase.

2) Test constraints and test methods

As shown in Fig. 14.7, the stress on heat shield in the separation process is a free boundary condition. F_{ah} is the transient separation force generated by the operation of the pyrotechnic device, F_{pa} is the pulling force of the parachute on the entry capsule, A_a is the axial resistance of the atmosphere to the entry capsule, N_a is the normal resistance of the atmosphere to the entry capsule, G_a is the gravity of the entry capsule, M_a is the aerodynamic torque of the entry capsule, A_h is the axial resistance

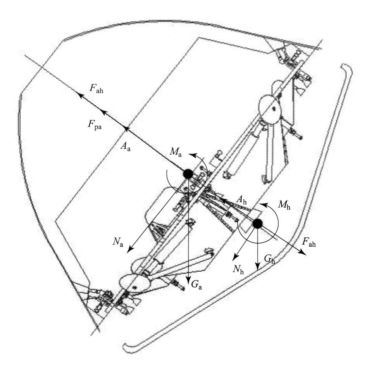

Fig. 14.7 Stress state of heat shield separation process

of the atmosphere to the heat shield, N_h is the normal resistance of the atmosphere to the heat shield, G_h is the gravity of the heat shield, and M_h is the aerodynamic torque of the heat shield.

As shown in Fig. 14.8, the stress on the backshell in the separation process is a free boundary condition. F_{ah} is the transient separation force generated by the operation of the pyrotechnic device, F_{pb} is the pulling force of the parachute on the backshell, A_b is the axial resistance of the atmosphere to the backshell, N_b is the normal resistance of the atmosphere to the backshell, G_b is the gravity of the backshell, M_b is the aerodynamic torque of the backshell, A_a is the axial resistance of the atmosphere to the landing platform, N_a is the normal resistance of the atmosphere to the landing platform, G_a is the gravity of the landing platform, M_a is the aerodynamic torque of the landing platform.

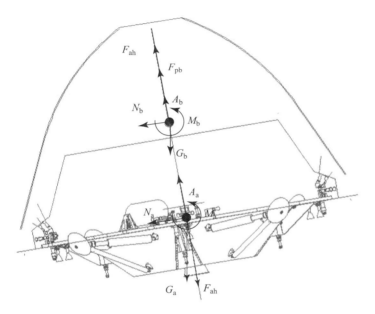

Fig. 14.8 Stress state of backshell separation process

The separation test of the heat shield and the backshell simplifies the whole separation process into the simulation of the relative motion of the heat shield and the backshell, namely the simulation of their relative acceleration (their relative velocity before separation is 0).

3) Design of test system

The heat-shield separation test adopts the test concept in which the backshell has a fixed support at the top, the heat shield has active load compensation, and they are separated in the reverse direction of gravity. The stress state in the ground test is shown in Fig. 14.9. By using axial, normal and compensation torques, the relative motion between the heat shield and the separation body is simulated.

The backshell separation test adopts the test concept in which the landing platform has a fixed support, the backshell has active load compensation, and they are separated in the reverse direction of gravity. The stress state in the ground test is shown in Fig. 14.10. By using axial compensation force, normal compensation force and torque compensation, the relative motion between the backshell and the separation body during the separation process is simulated.

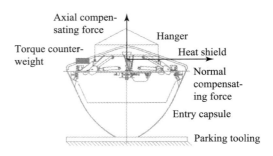

Fig. 14.9 Stress state of heat shield separation in ground test

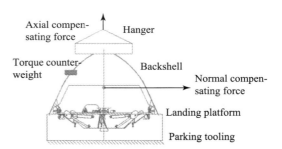

Fig. 14.10 Stress state of backshell separation in ground test

A unified test system is adopted for both the heat shield separation and the backshell separation. The test system is mainly composed of six parts: test platform, hanging system, shock measurement system, high-speed camera system, test auxiliary system and test product, as shown in Figs. 14.11 and 14.12. Among them:

(1) Test platform: provide the functions of axial force servo application, normal force application and rotation moment application in the separation process of the hanging capsule. The bottom of the platform is reliably fixed to the ground.
(2) Hanging system: lift the capsule under various separation conditions, connect with the capsule and adjust the hanging point to pass through the center of mass; and the hanging system is connected with the test platform through a lifting interface.
(3) Shock measurement system: measure the shock response of separation surface and other measurement points and process the measurement data.
(4) High-speed camera system: measure the separation speed, angular velocity and attitude angle in the separation process and record the separation process.

14.4 Test Verification Technology

Fig. 14.11 Diagram of heat-shield separation test system

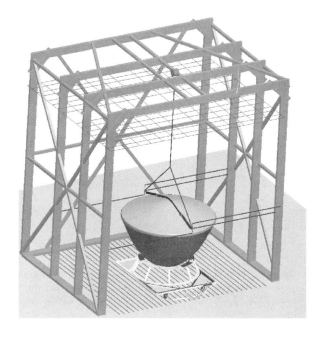

Fig. 14.12 Diagram of backshell separation test system

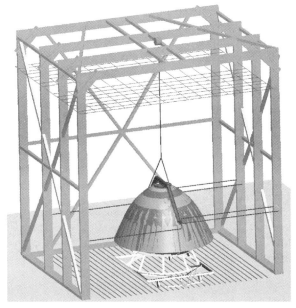

(5) Test auxiliary system: auxiliary tooling and equipment used in the test process.
(6) Test product: entry capsule.

14.4.2 Verification Technology for Powered Deceleration, Soft Landing and Takeoff

14.4.2.1 Comprehensive Verification Test for Hovering, Obstacle Avoidance and Slow Descent

1) Test purpose

The ground verification test for hovering, obstacle avoidance and slow descent is mainly aimed to simulate the real flight condition in the hovering, obstacle avoidance and slow descent phases and to comprehensively check the correctness of flight program design. It can also simulate the kinematic characteristics of landing process to verify the autonomous navigation control algorithm based on the navigation sensor or simulate the reflection characteristics of the landing surface and the motion of the lander to evaluate the dynamic working performance of the navigation sensor.

2) Test constraints and test methods

In the ground environment, the variable thrust engine of the lander is not enough to balance the gravity of the lander, so it is difficult to meet the test requirement of lander velocity and attitude control by GNC. How to construct the gravity environment of the target celestial body such as the moon or Mars is the key factor to be considered in the test plan. According to different solutions, there exist the following different schemes:

(1) Helicopter lifting scheme

The idea of this scheme is to use helicopter to lift the lander with a cable. During the test, the helicopter lifts the lander to the test height and then remains hovering. After the helicopter gets ready, the cable begins to balance part of the lander's gravity through the mechanical sensor and control system installed on the helicopter. At the same time, the engine on the lander is ignited and the test begins.

The disadvantages of this scheme are that the hovering stability of the helicopter does not meet the requirements, and the pilot's reaction time is much longer than the control cycle of the lander GNC. In addition, this scheme has potential safety hazards.

(2) Balloon lifting scheme

This scheme is similar to the helicopter lifting scheme except that the helicopter is replaced by a helium balloon. The disadvantage of this scheme is that the balloon volume will be more than 1000 m^3 and the maximum air resistance to the lander during descent will exceed 4000 N. Therefore, this scheme does not meet the test requirements.

14.4 Test Verification Technology

(3) Scale model scheme

The idea of this scheme is to develop a dynamic scaling model of the lander and design its mass and the corresponding engine according to the similarity principle in order to meet the test requirements. The disadvantages of this scheme are as follows:

① It is difficult to develop a scaling model based on similarity theory. For example, the thrust of attitude control thrusters needs to be scaled down.
② The technical condition of the scaling model is inconsistent with the real lander, which makes it difficult to verify the real lander.

(4) Alternatives for aero-engines

The idea of this scheme is to balance part of the gravity of the lander by using an aeroengine and push the lander to the test height by using a jet engine during the test, and then reduce part of its thrust. At the same time, the lander engine is ignited and the test begins. This scheme has the following three shortcomings:

① Limited by the lander structure, it is difficult to install both the aeroengine and the variable thrust engine, which will lead to the changes in the configuration of the lander.
② The aeroengine needs to be controlled after installation and needs to cooperate with the variable thrust engine, thus increasing the control difficulty.
③ There are potential safety hazards in the test; for example, the LLRV in the USA had an accident of out-of-control explosion.

(5) Test tower scheme

The idea of this scheme is to use a test tower to lift the lander. During the test, part of the lander gravity is balanced by the controlled cable to meet the test requirements. Meanwhile, the lifting cable follows the horizontal movement of the lander in real time to avoid the interference of cable angle deviation with the test, as shown in Fig. 14.13.

In order to simulate the actual hovering, obstacle avoidance and slow descent phases of the lander, and to assess the working performance of the lander, the lander-related equipments are required to work in the design condition and have normal working conditions in the ground test. Therefore, the following constraints need to be considered when designing the test system:

① Simulate the stress state of the lander in the process of slow descent, so as to realize the simulation of the motion parameters such as the motion speed and acceleration of the lander.
② The test site shall ensure the freedom of the lander's spatial motion and provide sufficient motion space, and the lander shall ensure the freedom of its own attitude motion.
③ The lander GNC system and propulsion system under test shall be real products. In the test, the GNC shall work in cooperation with the propulsion engine in light of the real control law and can control the velocity and attitude of the lander.

Fig. 14.13 Schematic diagram of tower test scheme [6]

④ During the test, the mass characteristics of the lander such as mass center and rotational inertia shall be consistent with the landing state of the lander.
⑤ In the hovering phase, the test site shall be able to simulate the topography and geomorphology of the landing surface so that the GNC system can complete the task of "obstacle identification and safe area selection." In the obstacle avoidance phase and the slow descent phase, the accurate measurement of the lander altitude and velocity is respectively completed by a laser ranging sensor and a microwave ranging and velocity sensor. The test site shall meet the working conditions of the sensors and simulate the reflection characteristics of laser and microwave of the landing surface.
⑥ The selected test facilities and test windows shall not interfere with the test system.

In order to address the above constraints, it is necessary to design a special servo tracking system for attitude DOF and a servo tracking system for translational DOF to simulate the on-orbit free flight state. A simulated landing site shall be designed to simulate the topography, surface reflection characteristics and sunshine characteristics of the landing surface. In addition, a test tower system shall be built to provide sufficient support and test space.

3) Design of test system

The hovering, obstacle avoidance and slow descent tests need to simulate the on-orbit stress state, the freedom of motion, and the terrain and reflection characteristics of landing surface, the synergy between GNC system and propulsion system and the autonomous navigation control algorithm. The whole system mainly consists of

14.4 Test Verification Technology

lander subsystem, landing test tower subsystem, simulated landing surface subsystem and test control subsystem.

(1) Simulation of stress state and translation process

The gravity environment on the ground is different from the surface of the target celestial body. During the ground test, the maximum output thrust of the engine is less than the gravity of the lander, so the lander cannot hover. Therefore, when designing the test system, the stress state simulation of the lander, which is one of the core factors of the whole test system, must be considered first.

The test tower is constructed, and the lander is suspended by a cable. In the test, the cable is controlled by a servo control system to maintain a constant tension so as to balance part of the gravity of the lander. In addition, the output thrust of the variable thrust engine is adjusted. As a result, the resultant force received by the lander in the test is consistent with that in the soft-landing process, thereby simulating the motion characteristics of the lander in the real process. The cable also has a horizontal servo function. During the test, the cable follows the horizontal movement of the lander in real time. The pulling direction of the cable acting on the lander shall be always vertical upward, thus avoiding the interference of the horizontal component of the cable's pulling force with the horizontal movement of the lander in the test. The schematic diagram of the test tower is shown in Fig. 14.14.

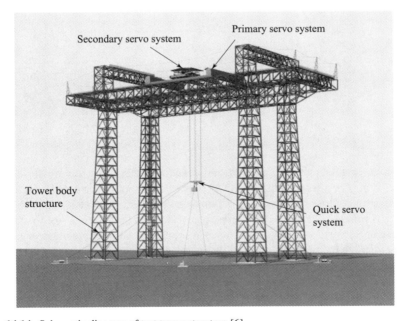

Fig. 14.14 Schematic diagram of test tower structure [6]

(2) Simulation of attitude motion DOF

In order to simulate the attitude freedom of the lander, it is necessary to design a universal hanger system. One end of the hanger is connected with the lander through two interfaces passing through the centroid of the lander, and the other end is connected with the cable of the test tower. In the test, the universal hanger system can provide the motion DOF of the lander in pitch, roll and yaw. The principle is shown in Fig. 14.15.

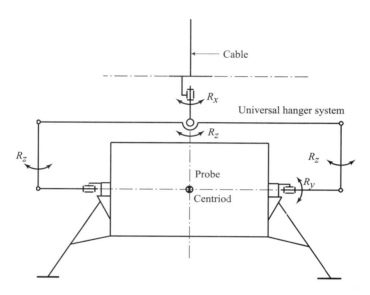

Fig. 14.15 Schematic diagram of universal hanger system [6]

(3) Simulation of terrain and reflection characteristics of landing surface

In order to simulate the topography and reflection characteristics of the landing surface, it is necessary to build a landing surface simulation system. The materials with reflectivity close to that of the landing surface can be selected and laid on the test site to simulate the laser and microwave reflection characteristics of the landing surface respectively, so as to meet the working conditions of GNC navigation sensors and to complete the tasks such as ranging and velocity measurement in the test.

(4) Test control

In the test, the lander and the servo system of test tower are controlled by their own control devices, which are subject to the scheduling of the general control system of the test site. In the test, the general control system collects the data information generated by the lander and the tower, and comprehensively analyzes the test results with the data from the external measuring equipment on the test site.

14.4 Test Verification Technology

4) Application examples

During the development of Chang'e 3 lander, the design of hovering, obstacle avoidance and slow descent tests and the development of test system were completed, and these tests were conducted. The test process is shown in Fig. 14.16.

Fig. 14.16 Hovering, obstacle avoidance and slow descent test process of Chang'e 3 lander [6]

14.4.2.2 Drop Test

1) Test purpose

The purpose of the drop test is mainly to evaluate the impact resistance of the main structure of the lander under the typical landing conditions during the soft landing and to obtain the impact response at the key parts of the lander and at the installation places of key equipments, thus providing references for establishing the equipment environment conditions.

2) Test constraints and test methods

In order to truly simulate the soft-landing impact process and obtain the impact response of the lander, the following constraints should be considered when designing the test system.

(1) The mass characteristics, load transfer characteristics and force transfer path of the lander during soft landing are the same as the actual ones.
(2) During soft landing, the buffering characteristics of the landing gear and the impact load environment of the whole lander are the same as the actual ones.
(3) The landing surface can simulate the landform characteristics and physical and mechanical properties of the real landing surface.

(4) The velocity and attitude of the lander when touching the ground are the same as the actual ones.
(5) The lander is in a free state during landing and there are no additional constraints on the lander.

When designing the test conditions, the simulation analysis of all conditions should be carried out, covering all landing modes, initial landing parameters and landing surface characteristics. Initial landing parameters shall cover the mass characteristics, landing speed parameters and landing attitude parameters of the lander. The landing surface characteristics shall cover the friction characteristics, load bearing characteristics and slope of the landing surface, from which the most severe typical landing condition shall be selected as the test condition.

3) Design of test system

The drop test needs to simulate the energy and motion DOF of the lander and the terrain and physical and mechanical characteristics of the landing surface during soft landing. The test system mainly consists of lander system, lifting and releasing separation device, simulated landing surface and ground measurement system, as shown in Fig. 14.17.

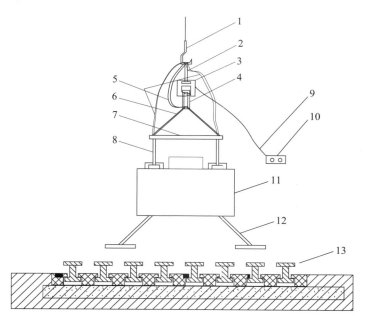

Fig. 14.17 Composition diagram of drop test system [6]. 1—Crane hook; 2—Sling of releasing separation device; 3—Releasing separator; 4—Lifting ring; 5—Limitation protection device; 6—Hanging beam sling; 7—Hanger beam; 8—Lander sling; 9—Releasing separation control cable; 10—Releasing separation control switch; 11—Lander; 12—Landing gear; 13—Simulated landing surface

(1) Lander system

In order to obtain the real landing impact response of the lander, it is necessary to simulate the real force transmission path and stress state during the soft-landing process of the lander. Therefore, the lander system shall have the real main structure and real landing gear, while other equipments can be simulated.

(2) Lifting and releasing the separating device

In order to determine the energy of the lander during soft landing, it is necessary to lift the lander to the set height by lifting and releasing the separation device. After the initial parameter setting is completed, the releasing separation device is started, and the lander is separated from the lifting and releasing separation device. Then, the lander contacts the landing surface through free falling.

(3) Simulating the landing surface

In order to simulate the contact process between the foot pad of the landing gear and the landing surface during soft landing, it is necessary to design a simulated landing surface that can simulate the typical landform characteristics and physical and mechanical characteristics of the landing surface. The parameters to be simulated include friction coefficient, load bearing strength, slope of the landing surface, etc.

(4) Ground measurement system.

The ground measurement system consists of a high-speed camera system and an acceleration/strain measurement system. During the test, the high-speed camera system measures the time histories of the parameters such as displacement, buffering stroke, speed, attitude angle, angular velocity and angular acceleration at the points of interest, while the acceleration/strain measurement system measures the time histories of acceleration and strain response.

4) Application examples

During the development of Chang'e 3 lander, the design of the drop test and the development of the test system were completed, and the drop test was conducted. The test process is shown in Fig. 14.18.

14.4.2.3 Landing Stability Test

1) Test purpose

The purpose of the landing stability test is mainly to examine the influence of landing attitude, landing velocity, landing surface topography and landing surface mechanical parameters on the landing stability of the lander, to verify the landing stability under extreme conditions and to modify the simulation model of landing stability test.

Fig. 14.18 Drop test process of Chang'e 3 [6]

2) Test constraints and test methods

When designing the landing stability test system, the following constraints need to be considered:

(1) The system can simulate the gravity environment during soft landing.
(2) The overturning direction of the lander during landing is in free state without additional constraints.
(3) The energy of the lander during soft landing is the same as that on orbit.
(4) The parameters such as the velocity and attitude of the lander during soft landing are the same as those on orbit.
(5) The test system can simulate the landing buffering characteristics during soft landing and the interface load between the landing gear and the main structure.
(6) The appearance and mass characteristics of the lander are the same as those of the real lander, and the buffering characteristics of the landing gear are the same as those of the real landing gear.
(7) The landing surface can simulate the landform characteristics and physical and mechanical characteristic parameters of the real landing surface.

In order to meet the above constraints, the simulated main structure and the real landing gear are usually used as the lander structure. A low-gravity simulation system is designed to simulate the gravity environment on the surface of the target celestial body. A follow-up system is designed to simulate the free movement of the lander. A tower is designed to support the lander and provide enough test space. A hanging and

14.4 Test Verification Technology

releasing system is designed to obtain the initial landing parameters. A simulated landing surface is designed to simulate the topography and mechanical properties of the lunar surface.

3) Design of test system

The landing stability test system mainly consists of tester system, low-gravity simulation system, test tower and one-dimensional servo system, hanging and releasing system, simulated landing surface and ground measurement system, as shown in Fig. 14.19.

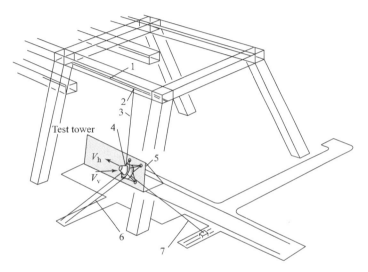

Fig. 14.19 Composition diagram of landing stability test system [6]. 1—Guide rail of one-dimensional servo system; 2—One-dimensional servo system; 3—Support cable; 4—Lander system; 5—Simulated landing surface; 6—Swinging cables; 7—Horizontal pulling cable

(1) Tester system

The tester system consists of a simulated main lander structure and a landing gear, which can simulate the mass characteristics (location of the center of mass and moment of inertia) of the lander during soft landing. The simulated main lander structure has the same physical dimensions as the real structure and can simulate the connection interface with the landing gear, which is a real product.

(2) Low-gravity simulation system

An inclined landing surface is used to simulate the gravity field of the target celestial body. When the angle between the normal line of the inclined landing surface and the horizontal plane is $\alpha = $ arsin (gravity acceleration on of surface of the target celestial body/gravity acceleration on the Earth surface), the gravity component of the lander in the normal direction of the inclined landing surface coincides with the tester system gravity during soft landing on the surface of the target celestial body. Therefore, the low gravity field can be simulated in the normal direction of the inclined landing surface. During the test, the downward gravity component of the tester system along the inclined landing surface is balanced by the support cable of the hanging and releasing device.

(3) Test tower and one-dimensional servo system

The test tower is used to provide support for one-dimensional servo system and can provide enough test space. Generally, the one-dimensional servo system is passive. To minimize the interference force applied by follow-up hysteresis on the tester system, the mass and follow-up friction factor of the servo system should be as small as possible. The follow-up capability of the one-dimensional servo system should be able to cover the maximum horizontal landing velocity of the lander with a certain margin.

(4) Hanging and releasing system

The hanging and releasing system is composed of a cable-hanging device, a lifting cable, a swinging cable, a swinging-cable release device, a swinging-cable length adjustment device, a horizontal pulling cable, a release device of horizontal pulling cable and a length adjustment device of horizontal pulling cable. The specific functions of each part are as follows:

① Hanging device: used to connect the lifting cable and enable one-dimensional cable movement along with the tester by means of the servo device.
② Lifting cable: used to connect the tester with the hanging device and balance the downward gravity component of the tester along the inclined landing surface during the test. In order to reduce as much as possible the change in the cable angle caused by the change in the normal displacement of the tester centroid on the landing surface due to the change in attitude and landing buffer stroke during the test, the length of the lifting cable needs to be specially designed.
③ Swinging cable: used to obtain the initial attitude and vertical landing speed of the lander.
④ Swinging-cable release device: to trigger the pyrotechnic device at the set position and release the swinging cable.
⑤ Length adjustment device of swinging cable: used to adjust the length of the swinging cable and obtain the initial attitude angle of the tester by adjusting the relative cable length.
⑥ Horizontal pulling cable: used to obtain the horizontal landing speed of the tester parallel to the guide rail direction.
⑦ Release device of horizontal pulling cable: to trigger the pyrotechnic device at the set position and release the horizontal pulling cable.
⑧ Length adjustment device of horizontal pulling cable: used to adjust the length of horizontal pulling cable to obtain different horizontal landing speeds.

14.4 Test Verification Technology

(5) Simulating the landing surface

The simulated landing surface can simulate the typical landform features and physical and mechanical characteristic parameters of the landing surface. The parameters to be simulated include friction coefficient, load bearing strength, slope of the landing surface, etc.

When designing the size of the simulated landing surface, it is necessary to consider the space motion of the tester and the slip during its contact with the landing surface, which is generally determined by theoretical calculation and simulation analysis. Stiffness and strength constraints should be considered comprehensively in the design of simulated landing surface. Its fundamental frequency should be kept away from the main modal frequency of the tester to avoid the resonance interference with the test.

(6) Ground measurement system

The ground measurement system consists of a high-speed camera system and a load measurement device. During the test, the high-speed camera system measures the time histories of displacement, buffering stroke, speed, attitude angle, angular velocity and angular acceleration at the points of interest. The load measurement device measures the time histories of acceleration and load.

4) Application examples

During the development of Chang'e 3 lander, the design of the landing stability test and the development of the test system were completed, and the landing stability test was conducted. The test process is shown in Fig. 14.20.

14.4.2.4 Test for the Influence of Thrust Pulsation of High-Thrust Engine on GNC Sensors

1) Test purpose

The purpose of this test is to determine the influence of the thrust pulsation of a large thrust engine on the GNC landing navigation sensor during on-orbit operation and to verify the adaptability of GNC inertial navigation device, microwave ranging and velocity sensor and other sensors to the thrust pulsation environment.

2) Test constraints and test methods

The following constraints should be considered when designing the test system to evaluate the impact of high-thrust engine pulsation on GNC sensors:

(1) It can simulate the free boundary of the lander when the engine is working on orbit.
(2) It can simulate the pulsating load condition when the engine is working on orbit.

Fig. 14.20 Landing stability test process of Chang'e 3 [6]

(3) It can simulate the interface and mode of action of pulsating load when the engine is working on orbit.
(4) It can simulate the load transfer path and characteristics of the lander under the pulsating load.
(5) It can simulate the mass characteristics of the lander.
(6) The tested lander has a real main structure and a real landing gear.
(7) The GNC landing navigation sensors of concern shall be real products and powered up during the test.

In order to satisfy the above constraints and to simulate the impact of the pulsating load of a large thrust engine on the on-orbit mechanical environment of the lander as realistically as possible, real main structure and landing gear are needed to simulate the force transmission path and load transmission characteristics. A spring gravity compensation system with near-zero stiffness shall be designed to simulate the free boundary of the lander. A pulsating load simulation method shall be used to simulate the pulsating load and simulate the pulsating load input through the exciter.

When designing the test conditions, it is necessary to first make statistical analysis on the time domain data of the pulsating load under various thrust conditions during the ground vacuum test of the high-thrust engine. Then, the time slice with the largest time domain load is selected from the rated thrust condition and all variable thrust conditions respectively as the input excitation of the test conditions, and the test is carried out in a closed-loop control mode.

14.4 Test Verification Technology

3) Design of test system

The test system for evaluating the impact of high-thrust engine pulsation on GNC sensors is mainly composed of lander system, spring gravity compensation system with near-zero stiffness, pulsation excitation simulation and input control system, measurement system and other parts, as shown in Fig. 14.21.

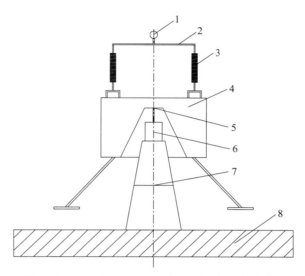

Fig. 14.21 Composition diagram of test system for evaluating the influence of pulsation of high-thrust engine on GNC sensors [6]. 1–Lifting ring; 2—Hanging beam; 3—Spring gravity compensation system with near-zero stiffness; 4—Lander system; 5—Pulsation load action interface; 6—Pulsation excitation simulation and input control system (exciter); 7—Exciter support; 8—Ground

(1) Lander system

The lander system uses a real main structure and a real landing gear and should be able to simulate the mass characteristics of the lander during soft landing. The landing navigation sensors that the GNC subsystem is concerned with are real products. The exciter is used to apply a load at the installation interface of the high-thrust engine.

(2) Spring gravity compensation system with near-zero stiffness

Metal spring is a common elastic element, which can be used as a gravity compensation unit to form a gravity compensation system for gravity compensation. The basic principle of gravity compensation with springs is to utilize the stiffness characteristics of the springs. Under the action of load gravity, the elastic restoring force generated by the elastic deformation of the springs counteracts the load gravity. At the same time, the load mass and the springs form a spring mass oscillator model with a natural frequency. The key to gravity compensation is that the spring meets the load-carrying capacity and has a sufficiently low natural frequency in the spring-based pulsation test.

The spring gravity compensation system with near-zero stiffness mainly simulates the free boundary condition of the lander that is excited by high-frequency low-amplitude pulsation load on orbit. As long as its support stiffness is low enough, the influence of its natural oscillation caused by elastic recovery on the test system during the test process can be ignored.

(3) Pulsation excitation simulation and input control system

In the dynamic response test and transfer path test of the lander under fluctuation excitation, the fluctuation source of the lander should be used as the excitation source. However, in order to obtain the response and transmission characteristics of the structure of the lander, it is not convenient to install a pulsation source in the structure and simulate it by real ignition of the engine, so an external excitation source is needed to simulate the pulsation excitation. In the ground test, the exciter is generally used for simulation. When selecting the exciter, the amplitude and frequency range of pulsating load should be considered.

(4) Measurement system

The measurement system mainly consists of acceleration measurement system, angular displacement measurement system and GNC-sensor ground test system. Acceleration measurement system and angular displacement measurement system are mainly used to measure the acceleration response and angular displacement response at the GNC sensor installation interface and the key parts of the whole lander. As the pulsating load features small amplitude and wide frequency range, the sensors with low range, high sensitivity and wide measurement bandwidth are generally selected. During the test, GNC system focuses on the power-up of landing navigation sensor, monitoring its performance and analyzing the impact of pulsating load on it.

4) Application examples

During the development of Chang'e 3 lander, the test design and test system for the influence of 7500 N pulsation on GNC sensor were completed, and the pulsation influence test was conducted. The test process is shown in Fig. 14.22.

14.4.2.5 Combination Test of Variable Thrust Engine and GNC Central Control Unit

1) Test purpose

The combination test of the variable thrust engine and the GNC system central control unit (CCU) is mainly to verify the control harmony and working compatibility of the CCU in controlling the variable thrust of the engine and to measure the average

14.4 Test Verification Technology

Fig. 14.22 Experiment of the influence of 7500 N pulsation of Chang'e 3 on GNC sensor [6]

specific impulse under frequently variable thrust conditions in the approaching phase and hovering translation phase.

2) Test constraints and test methods

The following constraints should be considered when designing the combination test system of variable thrust engine and CCU.

(1) It can simulate the vacuum environment when the engine works on orbit.
(2) It can measure the actual thrust of the variable thrust engine.
(3) It can simulate the variable thrust control system.
(4) It can cover the on-orbit ignition condition of variable thrust engine.

In order to meet the above constraints, a vacuum test bench is required to simulate the vacuum environment, a real engine is required to simulate its thrust characteristics, and a real CCU and a real variable thrust control system are required.

When selecting the test conditions, all working processes of the variable thrust engine on orbit should be comprehensively considered. In general, the variable thrust condition of the engine in the powered descent phase is the most complex, where the engine thrust change has a high frequency, a large range and a high response requirement. In addition, the propulsion system is also required to quickly respond to and correctly execute the variable thrust command of GNC system, so the powered descent phase is generally selected as the test condition for evaluation.

3) Design of test system

The combination test system of variable thrust engine and CCU is mainly composed of variable thrust engine, variable thrust control system, vacuum simulation system, test parameter measurement system, test control system and test bench pipeline, as shown in Fig. 14.23.

(1) Variable thrust engine

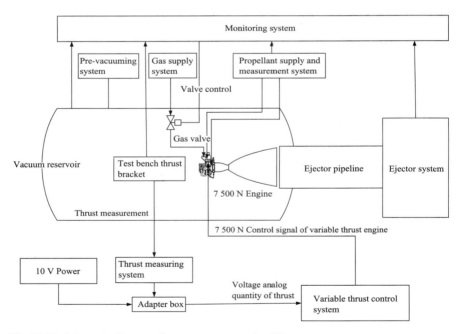

Fig. 14.23 Schematic diagram of test system connection [6]

14.4 Test Verification Technology

As the control target of the whole test system, a real variable thrust engine product should be selected during the test in order to truly simulate the output load characteristics of the variable thrust engine on orbit.

(2) Variable thrust control system

The variable thrust control system is mainly used for variable thrust control of the engine during the test. The variable thrust control system is mainly composed of real spacecraft products, the corresponding ground inspection equipment and various test cables.

(3) Vacuum simulation system

In order to truly simulate the on-orbit operating characteristics of the engine, a corresponding vacuum simulation environment must be created. The vacuum simulation system is mainly used to simulate the vacuum environment during the test run and consists of a vacuum reservoir, a pre-vacuumizing system, an ejector pipeline, an ejector system, etc.

(4) Test parameter measurement system

The test parameter measuring system is used to measure and record the parameters of the test system during the test, which include the vacuum capsule pressure, vacuum capsule temperature, propellant pressure, propellant flow rate, propellant temperature, propellant density, temperature of the combustion chamber body, engine thrust, etc.

(5) Test control system

The test control system consists of a test monitoring module, a control scheduling module, a network transmission and timing system module and a test data processing module. The test monitoring module is used for monitoring and recording the operation and test process of the whole test system. The control scheduling module is used for scheduling each operation post of the whole test system and can send remote operation instructions to the test system. The network transmission and timing system module is responsible for the data transmission and unified time calibration of the whole test system. The test data processing module is used for classifying and storing various test data and information generated in the test and comprehensively analyzing various data generated at the same time.

(6) Test bench pipeline

The test bench pipeline is used for the supply of propellant and the supply of driving gas to the cutoff valve. Filters shall be arranged on the propellant supply pipeline and the gas pipeline to avoid the remainder accumulation.

14.4.2.6 Buffering Performance Test of Airbag

1) Test purpose

(1) To verify the structural design and impact resistance of the landing module.
(2) To verify the connection design, material performance and structural process design of the airbag module.
(3) To verify the buffering performance of the airbag system.
(4) To verify the dynamic modeling and simulation results of the airbag, and correct the dynamic model through the test data.
(5) To verify the working process of the air bag and landing module from falling, buffering, bouncing to resting in the Earth environment and accordingly map it to the working process in the environment of the target celestial body.

2) Test constraints and test methods

When designing the test system of airbag buffering performance, the following constraints should be considered:

(1) The working pressure of the airbag and the pressure difference between its inside and outside.
(2) Free boundary conditions.
(3) Mechanical characteristics and surface topography simulation of the landing surface.
(4) Landing structure and landing load state of airbag internal protection.

Generally, the initial dropping speed and attitude are realized by simple-pendulum dropping device, as shown in Fig. 14.24. This test can be carried out in the atmospheric environment of the Earth or in a closed container simulating the atmospheric environment of the target celestial body.

3) Design of test system

The test system is mainly composed of dropping device, airbag system (including internal protection structure and payload simulator), environmental simulation system, simulated terrain, etc.

Dropping device: it is used to fix the airbag system and release the airbag at the required velocity and attitude to establish initial test conditions. Generally, a simple-pendulum-type dropping system is used.

Airbag system: the airbag system and its internal connection structure and payload simulator are in real design state.

Environmental simulation system: according to the test requirements, a sealed chamber that can simulate the test pressure environment should be selected, or the equivalent buffering performance method can be used to carry out the test in the Earth environment.

Simulated terrain: Typical and extreme landing surface terrains need to be simulated to achieve the capability of assessing the airbag buffering, the collision between internal products and the ultimate deformation.

14.4 Test Verification Technology

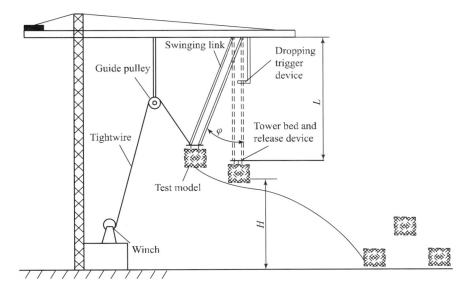

Fig. 14.24 Simple-pendulum dropping device

4) Application examples (Fig. 14.25).

Fig. 14.25 Airbag buffering performance test

14.4.2.7 Test for the Plume Effect on Thermal Control

1) Test purpose

The purpose of this test is to assess the capability of thermal control materials of landing gear, bottom plate and its protrusions to resist the ablation and erosion of engine plume during soft landing by simulating the plume conditions, thus providing an important reference for thermal control design of landing gear and lander bottom plate.

2) Test constraints and test methods

When designing a system to test the plume effect on thermal control, the following constraints need to be considered:

(1) It can simulate the vacuum environment when the engine works on orbit.
(2) It can simulate the heat flux when the engine works on orbit.
(3) The test pieces can simulate the real thermal control state.
(4) The effect of engine plume on thermal control can be simulated.

Due to the restriction of vacuum test conditions, it is difficult to simulate the plume of high-thrust engine on the ground, so a low thrust engine can be adopted in the test for equivalent heat flux simulation. The test run of a small thrust engine in high vacuum simulation system is to place a test piece covered with thermal control material at an appropriate position in the plume of a small engine in order to simulate the effect of the plume of a high-thrust engine on thermal control materials.

3) Design of test system

This test system is mainly composed of vacuum simulation system, three-dimensional moving system, engine, thermal control material specimen and test parameter measurement system, as shown in Fig. 14.26.

(1) Vacuum simulation system

The vacuum simulation system is used to simulate the vacuum environment when the engine is ignited on orbit. The heat flux meter, the temperature measuring channel and the pipeline for propellant supply shall be reserved on the wall of the vacuum reservoir.

(2) Three-Dimensional moving system

The three-dimensional moving system is used to fix the test piece and locate the test piece to the spatial location where the heat flux meets the test requirements in the plume field.

(3) Engine

The engine should be a real product. When installing the engine, the vector line of the engine shall be perpendicular to the tested surface of the test piece (the installation surface of test tooling), and the axis of the engine nozzle shall be coaxial with the axis of the installation tooling of heat flux meter.

(4) Test piece of thermal control material

The test piece is the thermal control coating material for landing gear, bottom plate and its protrusions.

14.4 Test Verification Technology

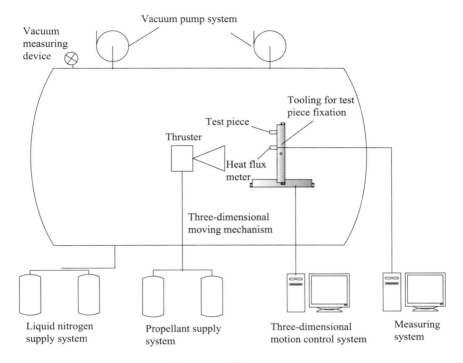

Fig. 14.26 Composition diagram of test system [6]

(5) Test parameter measurement system

The test parameter measuring system is responsible for measuring and collecting the heat flux and temperature during the test. It mainly consists of heat flux meters, pressure sensors, thermocouples, vacuum gauges, resistance gauges, etc.

4) Application examples

During the development of Chang'e 3 lander, the experiment for assessing the plume effect on thermal control was completed. The photograph of test process is shown in Fig. 14.27.

14.4.2.8 Takeoff Plume Effect Test

1) Test purpose

The purpose of the takeoff plume effect test is mainly to examine the plume-guiding scheme, verify the comprehensive force and thermal effect of the takeoff engine plume on the ascender and lander under typical working conditions, demonstrate the

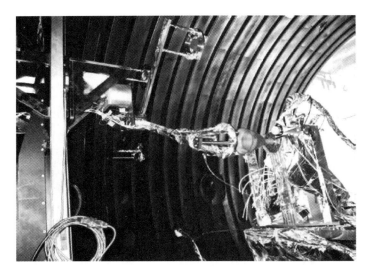

Fig. 14.27 Test for assessing the plume effect on thermal control for Chang'e 3 [6]

compatibility between the ascender engine and the plume-guiding device and assess the performance and plume protection effect of the thermal protection material.

2) Test constraints and test methods

When designing the test system for takeoff plume effect, the following constraints need to be considered:

(1) It can simulate the atmospheric environment when the engine works on orbit.
(2) It can simulate the heat flux when the engine works on orbit.
(3) It can simulate the performance of real thermal protection materials and plume-guiding devices.
(4) The measurement system needs to have a response speed that is fast enough.

There are generally three schemes for takeoff plume effect tests:

(1) Separate simulation test scheme for the engine. By simulating the working environment of the engine, installing a flat plume-guiding plate outside the nozzle of the engine and adjusting the distance and relative angle between the guiding device and the outlet plane of the engine nozzle, the parameters such as the thrust of the engine, the pressure of different areas inside the nozzle, the pressure of the combustor, the heat flux on the guiding device and the like can be measured.
(2) System-level test scheme. Through real simulation of the on-orbit takeoff state and the ignition of the ascending engine, the parameters such as plume interference force and heat flux in the separation process can be verified.

14.4 Test Verification Technology

(3) Flight verification scheme. By launching a test device, the plume effect produced in the takeoff process can be verified, as done in the US "Apollo 5" mission.

Considering the effectiveness, sufficiency and economy of the test, the system-level test scheme is generally selected for the test verification of the takeoff plume effect.

3) Design of test system

The test system for takeoff plume effect mainly consists of a tester, a vacuum simulation system, a pose adjustment system and a test measurement system.

(1) Tester

The tester generally includes an ascender and a lander. The equipment surface state and the state of plume-guiding device that affect the plume diffusion channel of the ascender are real design states.

(2) Vacuum simulation system

The vacuum simulation system is used to provide the low-pressure environment and the installation and operation space of test pieces needed to carry out the test. It has the ability to treat toxic waste gases and waste liquids generated by the ignition of the engine and the ability to measure and alarm the concentration of harmful gases such as propellants. It shall also provide an observation window to observe the plume flux field morphology and shock wave position from the outside, and the necessary thermal protection measures for the test equipment.

(3) Pose adjustment system

The pose adjustment system is used to adjust the position and attitude angle of the tester to realize the adjustment of the relative distance and angle between the lander and the ascender and the simulation of the relative motion characteristics.

(4) Test measurement system

The test measurement system is used to measure plume flux field parameters, plume interference force/moment, vacuum system pressure and temperature, temperature parameters of the tester's points of interest, etc.

4) Application example (Fig. 14.28).

Fig. 14.28 Comprehensive verification test of full-scale plume-guiding device

14.5 Outlook

The adequacy and effectiveness of ground test verification is one of the keys to ensure the success of the mission. How to create a realistic simulated landing mission environment on the Earth is an important task for the overall design of deep space probes. At present, through the implementation of various missions in the lunar exploration project, a series of aerospace industry standards such as the requirements and methods of lunar probe ground test and lunar surface environment simulation have been developed to standardize the items, methods and contents of ground verification tests, effectively ensure the integrity and effectiveness of those tests, and lay a technical foundation for the ground verification of follow-up lunar exploration missions and planetary exploration missions.

With the increasing diversification of deep space exploration approaches, the requirements for test methods and systems become more and more complex. For example, the simulation of the weak gravity environment for the small celestial bodies and the simulation and verification of the strong magnetic field and nuclear power supply in the Jupiter exploration mission need to fully evaluate all kinds of risks in the new mission process, apply advanced technologies such as computers and information, intelligence and software as well as mathematics and physical simulation, develop and configure special ground test equipments and facilities, work out special test schemes and design the verification processes. With these efforts, the accuracy, convenience and economy of the simulation processes can be continuously optimized.

References and Related Reading

1. Taeger Y, Witkowski A (2003) A summary of dynamic testing of the Mars exploration rover parachute decelerator system. In: Monterey: 17th AIAA aerodynamic decelerator systems technology conference and seminar
2. Thomas CO, Donald EH (1967) Operational features of the langley lunar landing research facility. NASA, Houston

3. Donald RB, Gene JM (1965) Design and operational characteristics of a lunar-landing research vehicle. NASA, Houston
4. Ulysse JB (1968) Evaluation of a full-scale lunar-gravity simulator by comparison of landing-impact tests of a full-scale and a 1/6-scale model. NASA, Houston
5. Stanley PW (1973) Apollo experience report Lunar module structural subsystem. NASA, Houston
6. Yu D (2015) Lunar soft landing detector technology. National Defence Industry Press, Beijing (Chinese vision)
7. Ulysse JB (1969) Full-scale dynamic landing-impact investigation of a prototype lunar module landing gear. NASA, Houston